Hydraulic Modelling – an Introduction

CU00405112

Modelling forms a vital part of all engineering design, yet many hydraulic engineers are not fully aware of the assumptions they make. These assumptions can have important consequences when choosing the best model to inform design decisions.

Considering the advantages and limitations of both physical and mathematical methods, this book will help you identify the most appropriate form of analysis for the hydraulic engineering application in question. All models require the knowledge of their background, good data and careful interpretation and so this book also provides guidance on the range of accuracy to be expected of the model simulations and how they should be related to the prototype.

Applications for models include:

- Open channel systems;
- Closed conduit flows;
- Storm drainage systems;
- Estuaries;
- Coastal and nearshore structures;
- Hydraulic structures.

An invaluable guide for students and professionals.

Pavel Novak is Emeritus Professor of Civil and Hydraulic Engineering at the University of Newcastle upon Tyne, UK.

Vincent Guinot is Professor at the University of Montpellier, France.

Alan Jeffrey is Emeritus Professor of Engineering Mathematics at the University of Newcastle upon Tyne, UK.

Dominic E. Reeve is Professor of Coastal Dynamics at the University of Plymouth, UK.

Hydraulic Modelling – an Introduction

Principles, methods and applications

P. Novak, V. Guinot, A. Jeffrey
and D. E. Reeve

Spon Press
an imprint of Taylor & Francis

LONDON AND NEW YORK

First published 2010
by Spon Press
2 Park Square, Milton Park, Abingdon, Oxon OX14 4RN

Simultaneously published in the USA and Canada
by Spon Press
270 Madison Avenue, New York, NY 10016, USA

Spon Press is an imprint of the Taylor & Francis Group, an informa business

© 2010 P. Novak, V. Guinot, A. Jeffrey and D. E. Reeve

Typeset in Sabon by
Integra Software Services Pvt. Ltd, Pondicherry, India
Printed and bound in Great Britain by CPI Antony Rowe, Chippenham,
Wiltshire

British Library Cataloguing in Publication Data
A catalogue record for this book is available from the British Library

Library of Congress Cataloging in Publication Data
Hydraulic modelling – an introduction : principles, methods and
applications / P. Novak . . . [et al.].
p. cm.
Includes bibliographical references and index.
1. Hydraulic engineering--Data processing. 2. Hydrodynamics--
Mathematics. 3. Hydraulic structures--Mathematical models.
I. Novák, Pavel.
TC157.8.I58 2010
627.0285--dc22   2009027719

ISBN10: 0-419-25010-7 (hbk)
ISBN10: 0-419-25020-4 (pbk)
ISBN10: 0-203-86162-0 (ebk)

ISBN13: 978-0-419-25010-4 (hbk)
ISBN13: 978-0-419-25020-3 (pbk)
ISBN13: 978-0-203-86162-2 (ebk)

Contents

Preface

Two related questions had to be considered in the preparation of this text: the need for it and the required level.

There are many good books and research papers dealing with some aspects of hydraulic modelling, but – as far as we are aware – there is no single text available combining the various approaches to the subject. Furthermore, our experience from teaching and consulting is that many students and even practitioners tend to use the various computer packages and results from hydraulic models without being sufficiently aware of their background and limitations; hence the decision to prepare this volume.

As various aspects of hydraulic modelling are the subject of ongoing research and publications, it would be presumptuous to attempt a book of this type at anything but an introductory level. This text is thus not a research monograph, but a textbook aimed at final-year undergraduate and postgraduate students; at the same time, we hope that practitioners in the field will find it a useful source of reference, and that for all of them it can serve as a basis for further study and development.

One notable omission in the text is groundwater modelling, although it is superficially alluded to in Chapters 2, 4 and 5. The reason for this is twofold: first, groundwater modelling, particularly of flow through complicated ground conditions and fissured rocks, is beyond the scope of this text; and, second, its inclusion would have made the book too long.

Professor V. Guinot of Université Montpellier is the author of Chapters 7, 8, 9 and 10; Professor A. Jeffrey of the University of Newcastle upon Tyne wrote Chapters 2 and 3 and Section 6.4; Professor D. E. Reeve of the University of Plymouth is the author of Chapters 11 and 12; Professor P. Novak of the University of Newcastle upon Tyne wrote Chapters 1, 4, 5, 6 and 13, the sections on physical modelling in Chapters 7, 8, 9, 11 and 12, and also edited the whole text.

P. Novak, V. Guinot, A. Jeffrey, D. E. Reeve
April 2009

Acknowledgements

We are grateful to the following individuals who have given valuable advice in the preparation of this text:

R. Bettess, O. Cazaillet, J. Cunge, P. Gabriel, B. Li, M. Pfister, P. G. Samuels, E. M. Valentine.

The following organizations have kindly given permission for the reproduction of copyright material (figure or section number in parentheses):

Bray Town Council, Éire, the Office of Public Works, Dublin and O'Connor Sutton Cronin, Dublin (7.6.2); Halcrow, Swindon (11.1, 11.6, 11.25–29, 12.12, 12.18, 12.20); HR Wallingford (7.28, 12.15, 12.16); Montgomery, Watson, Harza, Newcastle upon Tyne (13.5.4); School of Civil Engineering and Geosciences, University of Newcastle upon Tyne (13.13); Sogreah, Grenoble (8.8, 8.9, 11.19, 13.12, 13.14); T. G. Masaryk, Water Research Institute (VÚV-TGM, Prague) (7.29, 7.30, 8.10, 13.8, 13.9, 13.10, 13.11); VAW-ETH Zurich (13.6, 13.7).

Main symbols

a	acceleration, wave amplitude
A	area
b	width
B	water surface width (in open-channel flow)
c	coefficient, wave celerity
C	constant (with suffix), Chezy coefficient
C'	concentration (suspended sediment, air)
C_d	coefficient of discharge
Ca	Cauchy number
d	sediment diameter
d_s	equivalent sediment diameter
d_{90}	diameter at 90% grain distribution curve
D	diffusion coefficient, pipe diameter
D_{gr}	dimensionless grain diameter
e	work
E	specific energy, Young's modulus
Eu	Euler number
f	function, frequency
F	function, force
Fr	Froude number
Fr_d	grain Froude number
g	acceleration due to gravity
g_s	specific (unit) sediment discharge (N/m/s)
G_s	sediment discharge (N/s)
h	head, stage, depth
H	head, total energy head, wave height
J	Jacobean
k	roughness size, permeability coefficient, wave number
K	bulk modulus
K'	coefficient in Strickler equation

Ka	Karman number
Ke	Keulegan–Carpenter number
l	length
L	dimension of length, length, wave length
La	Lagrange number
m	index denoting model, mass
M	dimension of mass
Ma	Mach number
M_x	scale of x (x prototype/x model)
n	Manning coefficient
N	power, hydraulic exponent, RPM
Ne	Newton number
p	pressure intensity, index denoting prototype, porosity
P	force, wetted perimeter
q	specific (unit) discharge (m^3/m/s)
q_s	specific (unit) discharge of sediment (m^3/m/s)
Q	discharge (m^3/s)
Q_s	sediment discharge (m^3/s)
r	pipe radius, exponent
R	hydraulic radius
Re	Reynolds number
Ri	Richardson number
S	slope
S_o	bed slope
S_e	slope of the energy line
Sh	Strouhal number
t	time
T	dimension of time, wave period
u	(instantaneous) local velocity in x direction
$\bar{u}$	time-averaged local velocity
u'	$(\bar{u} - u)$ or $(u - \bar{u})$
U	depth-averaged velocity
U_*	shear velocity
v	(instantaneous) velocity in y direction, velocity in general
V	mean (cross-sectional) velocity, volume
Ve	Vedernikov number
w	(instantaneous) velocity in z direction, settling velocity
We	Weber number
x	coordinate (longitudinal direction)
y	coordinate (lateral direction), depth
Y	depth
z	coordinate (vertical direction)
α	Coriolis number
β	Boussinesq coefficient

Γ	circulation
γ	specific weight $(= \rho g)$ (N/m^3)
γ	specific weight of sediment
δ	boundary-layer thickness
δ'	laminar sublayer thickness
Δ	difference, relative density of sediment $((\rho_s - \rho)/\rho)$
ε	relative roughness, coefficient of (dynamic) eddy viscosity
ζ	vorticity
η	efficiency, coefficient of (kinematic) eddy viscosity, distance of water level from a reference datum
κ	Karman universal constant
λ	coefficient of friction-head loss $(= 2gDS/V^2)$
λ_R	coefficient of friction-head loss $(= 2gRS/V^2 = \lambda/4)$
μ	coefficient of (dynamic) viscosity
ν	coefficient of (kinematic) viscosity
ρ	specific mass (density) (kg/m^3)
ρ_s	specific mass of sediment
σ	coefficient of surface tension, cavitation number
σ_c	critical cavitation number
τ	shear stress
τ_o	wall shear stress
ϕ	function, sediment transport parameter
φ	function
Φ	function
ψ	function, flow parameter $(= 1/\mathrm{Fr}_d)$
ω	angular velocity

Chapter 1

Introduction

The hydraulic engineer's concerns are liquids, their motion and their interaction with conveyances and structures. Usually, but not exclusively, the liquid in question is water – a viscous, slightly compressible fluid. The science of hydraulics thus works with the real liquids of engineering interest, although it owes much to the laws derived in theoretical hydromechanics for ideal (homogeneous, incompressible, non-viscous) liquids.

It is almost impossible in hydraulic research to draw a clear dividing line between basic and applied research, as both intermingle in the solution of hydraulic problems connected with engineering design. An extraordinary development in experimental methods and the application of computational techniques have also been of great importance.

There are three ways to approach the solution of a problem in hydraulics and hydraulic engineering design: by theory and reasoning; by experience (e.g. derived from similar structures); or by investigating the problem and testing the design on a model. However, our past **experience** may be inadequate due to the uniqueness of the design and circumstance; the complexity of many cases of liquid flow and our still limited analytical abilities permit the strict application of **theory** and basic flow equations only in certain, often schematized, situations and thus methods using **models** are needed to achieve a solution or to test the effect of simplifications. It must be emphasized, however, that a purely experimental approach to the solution of a problem without any theoretical analysis, even if restricted only to a dimensional analysis, is likely to be a waste of effort. Systematic experiments require theoretical guidelines, and in the absence of such they show, at best, only a certain relationship of observed hydraulic parameters within the range of the experiments undertaken. If the physical principles depicted by an empirical function are not elucidated, then the function can neither be safely extrapolated nor generalized for other similar cases of flow.

The term *model* is used in hydraulics to describe a *physical* or *mathematical* simulation of a 'prototype', or field-size situation. The hydraulic engineer's models are tools for predicting the effect of a proposed design and to producing technically and economically optimal solutions to engineering problems. In other words, a model is a system that will convert a given input (geometry, boundary conditions, force, etc.) into an output (flow rates, levels, pressures, etc.) to be used in civil engineering design and operation.

Simulation may be **direct** by the use of *hydraulic models*, **semi-direct** using *analogues* or **indirect** by making use of theoretical and computer-based analysis, including *mathematical, computational* and *numerical models*. The basic distinction is between *physical* and *mathematical* models. Physical models then comprise hydraulic and analogue models; analogue models include also *aerodynamic* models (which really form a transition between hydraulic and other analogue models), and both hydraulic and aerodynamic models can be grouped as *scale* models. Analogue models had their main application in groundwater flow simulation, but have now mainly been replaced by mathematical models. As the application of aerodynamic models also is being confined to special cases, the terms *physical, scale and hydraulic* models have gradually almost become synonymous. (The term 'hydraulic model' is also sometimes used loosely to denote all models – including mathematical ones – instead of the correct overall term 'hydraulic modelling' or 'models in hydraulic engineering'.)

As we are primarily concerned with the reproduction of present or future full-size behaviour, obtaining relevant *field data* is an important and integral part of the modelling process.

It is obviously the basic requirement of any **scale model** to reproduce correctly the behaviour of the situation to be modelled. The success of the solution depends on the accurate formulation of the problem and on the correct identification of the main parameters influencing the phenomena under investigation. This may lead to an intentional suppression of forces and influences, the role of which in the prototype is, in the light of experience, only of secondary importance. It is a possible pitfall that the magnitude of forces neglected in the analysis may assume a disproportionately large significance in the model, a discrepancy that is usually referred to as *scale effect*. The appreciation of similarity laws and of the limits of their validity is, therefore, particularly important if this is to be avoided. All these considerations influence the selection of appropriate methods and techniques of simulation (Novak and Čábelka (1981)).

One of the first to use hydraulic models was Osborne Reynolds, who in 1885 designed and operated a tidal model of the Upper Mersey at Manchester University. In 1898, Hubert Engels established the first River Hydraulics Laboratory at Dresden. Then followed a gradual and, after 1920, an accelerating expansion of laboratories for the study of hydraulic engineering problems using scale models.

The widespread use and role of hydraulic models may have changed somewhat in recent years, mainly due to the advances in computational modelling, but they remain an important modelling tool, especially in the design of hydraulic structures, river and coastal engineering applications, environmental protection and in providing the physical input to mathematical modelling.

An **analogue model** is a system reproducing a prototype situation in a physically different medium. This technique depends on the equations representing the prototype and model being mathematically identical. Thus torsional vibrations of a bar may represent the water-level oscillations of a simple surge tank, and both can be simulated by the voltage changes in an electric circuit, i.e. by an electrical analogue.

Although engineers use the terms *mathematical model, numerical model* and *computational model* as synonyms, there is a clear distinction between them (Samuels (1993)). A **mathematical model** is a set of algebraic and differential equations that represents the interaction between the flow and process variables in space and time. It is based on a certain set of assumptions about the physics of the prototype flow and associated environmental processes. These assumptions will set clear limits to the domain of applicability of the mathematical model and any numerical and computational model that may be derived from it. A prerequisite for the development of a mathematical model is an understanding of the key physical processes involved, leading either to fundamental principles such as Newton's laws of motion or to well-attested empirical relationships such as the Chezy and Manning resistance laws.

It is extremely rare for a mathematical model to be amenable to an exact closed-form solution except for the simplest geometries. Hence, the power of mathematical models was only realized with the availability of cheap, reliable computing from about 1960 onwards. Mathematical models of most physical phenomena are non-linear, necessitating the use of numerical methods when developing approximate solutions with the aid of a digital computer. This leads to the definition of a numerical model.

A **numerical model** is an approximation of a mathematical model of some prototype situation, giving a computable set of parameters that describes the flow at a set of discrete points. Many numerical models can be formulated from the same underlying mathematical model by employing alternative numerical methods and mathematical manipulations. The performance of the numerical models will be determined by the properties of the numerical methods employed, and for the same geometric and boundary data may give significantly different results. These differences are often masked, in part, by the calibration procedure.

A numerical model, like a mathematical model, is not specific to any particular site, and the strength of both these types of model lies in their generality. A specific application will require data from the prototype site and a computer with a program to organize the data and execute the calculations.

A **computational model** is an implementation of a numerical model on a computer system with the relevant data from a specific site. The results of the computational model depend on a variety of factors, including the quality of the prototype data, the details of data processing, possibly the internal organization of the calculations and the type of computer used.

Many computational modelling systems and packages are available for a variety of hydraulic engineering problems. The end user may have to choose which model to use, and certainly will have to be able to interpret the model results critically and responsibly. It is hoped that this book will provide at least some guidance on how to distinguish between models that are appropriate for a particular application and those that are not. It is important that the results of a computational model should not be accepted as definitive just because the numbers were produced by a computer – the results must also make physical sense. Past (field) results should be used to gain a better understanding of what is happening physically and why a given model does not reproduce observations accurately, and to assess imprecisions and/or uncertainty intervals in the results, and always to calibrate the model.

From data handling the discipline of computational hydraulics has grown to *hydroinformatics*, which uses simulation modelling and information and communication technologies (ICT) to help to solve problems in hydraulics, hydrology and environmental engineering for better management of water-based systems. In a further development, *artificial neural networks* attempt to simulate – in a crude way – the working of a human brain by passing on information from one 'neuron' to all other 'neurons' connected with it. The output of the model is related to the input through a set of functions with constants determined during the 'training' of the network; a large set of wide-ranging data is required to train a network to achieve good results.

In conclusion, it may be helpful to identify the principal differences between the types of modelling discussed in this chapter. Physical (scale) models (hydraulic and aerodynamic) are based on full fluid physics but at a reduced geometric scale, whereas a computational model is at full prototype scale but embodies only approximate physics. A physical model provides a continuous representation of the prototype but a computational model offers only a finite dimensional approximation; if a model does not reproduce observations accurately, it is necessary to assess the uncertainty in the results. Physical and computational modelling should not be viewed as conflicting methods of investigation; rather, they have complementary strengths and weaknesses. Often a hydraulic engineering problem will require a combination of these methods, i.e. *hybrid modelling*, to achieve a cost-effective solution.

References

Novak, P. and Čábelka, J. (1981), *Models in Hydraulic Engineering – Physical Principles and Design Applications*, Pitman, London.

Samuels, P. G. (1993), *What's in a Model?* Paper presented to the River Engineering Section IWEM, January 1–12.

Chapter 2

Theoretical background – mathematics

2.1 Ordinary differential equations

2.1.1 Definitions

Physical situations described by quantities that vary continuously with respect to their position in space and possibly with time can usually be described in terms of systems of partial differential equations (PDEs). These are equations that relate the quantities involved to some of their derivatives with respect to space variables and time. In the simplest case, when only a single quantity $u(t)$ depending on a variable t is involved, the variation of $u(t)$ with respect to t is described by an ordinary differential equation (ODE) that relates $u(t)$ to some of its derivatives. If the highest order derivative involved in an ODE is $d^n u/dt^n$, the ODE is said to be of the nth order. The variable t is called the *independent variable*, and in physical situations t is often the time, while the quantity $u(t)$ is called the *dependent variable* because its value depends on t. A general nth-order ODE can be written symbolically as

$$F(t, u, u', u'', \ldots u^{(n)}) = 0, \tag{2.1}$$

where $u' = \frac{du}{dt}$, $u'' = \frac{d^2 u}{dt^2}, \ldots, u^{(n)} = \frac{d^n u}{dt^n}$, and F is an arbitrary function of its $n + 1$ arguments $t, u, u', u'', \ldots, u^{(n)}$. The form of equation (2.1) is too general to be of use when discussing ODEs, so in practice it is necessary to restrict study to some of the most frequently occurring types of ODE that arise in applications. This involves considering special forms that may be taken by the function F, although only the most important of these will be mentioned here.

The simplest type of ODE is of the form $dy/dt = g(y)h(t)$, where $g(y)$ and $h(t)$ are functions of their respective arguments. An ODE of this type is said to have *separable variables*, because it can be written as $\int (\frac{1}{g(y)}) dy = \int h(t) dt$, in which the variables y and t have been separated, after which the general solution follows by integration. Here, the *solution* of an ODE is a relationship between y and t that does not contain derivatives which, when substituted into the ODE, satisfies it identically. For ways of solving other special types of ODE, such as solution by substitution, solution by elimination and the use of integrating factors, we refer readers to any standard text on ODEs such as those by Birkhoff and Rota (1989), Boyce and DiPrima (2005), Edwards and Penney (2001) and Krusemeyer (1999).

An important type of differential equation is that where the function F contains only a sum of terms of the form $u, u', u'', \cdots, u^{(n)}$, each of which occurs linearly (raised to the power one), although u and each of its derivatives may be multiplied by a function of t, and the sum of such terms may be equal to a given function $f(t)$. An equation of this type is said to be a *linear variable coefficient* nth-*order ODE*, and its general form is

$$a_0(t)u^{(n)}(t) + a_1(t)u^{(n-1)}(t) + \cdots + a_{n-1}(t)u'(t) + a_n(t)u(t) = f(t), \qquad (2.2)$$

where coefficients $a_0(t), a_1(t), \ldots, a_n(t)$ are known functions of t. The function $f(t)$ is called the *forcing function* because, after the start of the solution of equation (2.2), its subsequent behaviour is determined (forced) by the nature of the function $f(t)$ that represents some external influence. In the case of equation (2.2), the equation $F(t, u, u', u'', \ldots, u^{(n)}) = 0$ in equation (2.1) takes on the simple form

$$a_0(t)u^{(n)}(t) + a_1(t)u^{(n-1)}(t) + \cdots + a_{n-1}(t)u'(t) + a_n(t)u(t) - f(t) = 0.$$

The simplest ODE of this type is the nth-*order constant-coefficient ODE* where the coefficients a_i, for $i = 0, 1, \ldots, n$, are all constants. When the forcing function $f(t)$ in ODE equation (2.2) is equal to zero the equation is said to be *homogeneous*, otherwise the ODE is said to be *non-homogeneous*.

To understand the meaning and importance of the term *linear* when used to describe ODE equation (2.2) it is necessary to introduce the concept of the linear independence of functions. A set of n functions $u_1(t), u_2(t), \ldots, u_n(t)$ defined for t in some interval I, say $a \leq t \leq b$, that may be finite, semi-infinite or infinite, are said to be *linearly independent* if the linear combination of terms

$$c_1 u_1(t) + c_2 u_2(t) + \cdots + c_n u_n(t) = 0, \qquad (2.3)$$

where the constants c_i are arbitrary, is only true for all t in I when $c_1 = c_2 = \ldots = c_n = 0$. When the n functions are not linearly independent, they are said to be *linearly dependent*, and in that case not all the constants c_i are zero. In the simplest case, when only two functions are involved, linear independence means that the functions are *not* proportional, whereas linear dependence implies their proportionality. So, for example, the functions $u_1(t) = e^t$ and $u_2(t) = e^{2t}$ are linearly independent for all $-\infty < t < \infty$ because they are not proportional, but the functions $u_1(t) = \ln t$ and $u_2(t) = \ln t^2$

for $0 < t < \infty$ are linearly dependent because $\ln t^2 = 2 \ln t$, so $u_2(t) = 2u_1(t)$, showing that the functions are in fact proportional.

It is a standard result in the study of ODEs that the homogeneous form of equation (2.2) (i.e. when $f(t) \equiv 0$) always has n linearly independent solutions. The significance of this result can be understood by considering the fact that if $u_1(t)$, $u_2(t)$, ..., $u_n(t)$ is any set of n suitably differentiable functions, and b_1, b_2, ..., b_n is any set of n arbitrary constants, it follows from the linearity of the operation of differentiation that

$$\frac{d^r}{dt^r} \left(b_1 u_1(t) + b_2 u_2(t) + \cdots + b_n u_n(t) \right) = b_1 \frac{d^r u_1(t)}{dt^r}$$
$$+ b_2 \frac{d^r u_2(t)}{dt^r} + \cdots + b_n \frac{d^r u_n(t)}{dt^r},$$

for $r = 1, 2, \ldots, n$. This has the effect that if the n functions $u_1(t)$, $u_2(t)$, ..., $u_n(t)$ are the n linearly independent solutions of equation (2.2), the general solution of the homogeneous form of equation (2.2) can always be expressed as a sum of its n linearly independent solutions, each of which can be multiplied by an arbitrary constant. The property that a solution of a linear homogeneous equation can always be expressed as a sum of multiples of its n linearly independent solutions $u_1(t)$, $u_2(t)$, ..., $u_n(t)$ is described by saying that the solutions of the equation possess *linear superposition property*.

A test for the linear independence of n solutions u_1, u_2, ..., u_n of a homogeneous linear nth-order ODE defined over an interval $a \le t \le b$ is provided by the *Wronskian test*. This test requires that for the linear independence of the n solutions the determinant $W(u_1, u_2, \ldots, u_n) \neq 0$ over the interval $a \le t \le b$, where

$$W(u_1, u_2, \cdots, u_n) = \begin{vmatrix} u_1 & u_2 & \cdots & u_n \\ u_1' & u_2' & \cdots & u_n' \\ \vdots & \vdots & \vdots & \vdots \\ u_1^{(n)} & u_2^{(n)} & \cdots & u_n^{(n)} \end{vmatrix}.$$

For example, it is easily checked by substitution that the ODE $u''' + 4u'' + 5u' + 2u = 0$ has the three solutions: $u_1 = e^{-t}$, $u_2 = te^{-t}$ and $u_3 = e^{-2t}$. The linear independence of these three solutions can be proved by the Wronskian test, because

$$W(u_1, u_2, u_3) = \begin{vmatrix} e^{-t} & te^{-t} & e^{-2t} \\ -e^{-t} & (1-t)e^{-t} & -2e^{-2t} \\ e^{-t} & (t-2)e^{-t} & 4e^{-t} \end{vmatrix} = e^{-4t},$$

and as $e^{-4t} \neq 0$ for $-\infty < t < \infty$, it follows that the solutions are linearly independent for all finite t. Thus, the general solution of this homogeneous ODE can be written as $u(t) = a_1 e^{-t} + a_2 t e^{-t} + a_3 e^{-2t}$, where a_1, a_2 and a_3 are arbitrary constants.

Clearly, when the functions $u_i(t)$, $i = 1, 2, \ldots, n$ are solutions of the homogeneous form of ODE (2.2), an expression such as

$$u_c(t) = a_1 u_1(t) + a_2 u_2(t) + \cdots + a_n u_n(t) \tag{2.4}$$

cannot represent the solution of the non-homogeneous equation (2.2), because the result of substituting $u(t) = u_c(t)$ into equation (2.2) leads to the contradictory result $0 = f(t)$. Consequently, when $f(t) \neq 0$, because of the linearity property of the ODE the general solution must be of the form

$$u(t) = u_c(t) + u_p(t) \tag{2.5}$$

in which case the function $u_p(t)$ must be such that

$$a_0(t) u_p^{(n)} + a_1(t) u_p^{(n-1)} + \cdots + a_n(t) u_p = f(t). \tag{2.6}$$

The function $u_c(t)$ is called the *complementary function* of ODE equation (2.2) and contains all the arbitrary constants, while the function $u_p(t)$ is called a *particular integral* of the equation. In practical terms, a particular integral $u_p(t)$ is a function that when substituted into the ODE generates the non-homogeneous term $f(t)$. A particular integral is not necessarily unique, because if terms from the complementary function are added to it then equation (2.6) will still be satisfied. However, $u_p(t)$ will become unique once any terms belonging to the complementary function have been deleted. Ways of finding the complementary function $u_c(t)$ and the particular integral $u_p(t)$ for any specific equation are discussed at length in the standard texts on ODEs already mentioned. Further useful references are Farlow *et al.* (2002) and Peterson and Sochacki (2002).

2.1.2 Initial conditions

In specific applications it is necessary to specify how a particular solution $u(t)$ of equation (2.2) may be constructed, assuming that the complementary function $u_c(t)$ and the particular integral $u_p(t)$ are known. This involves recognizing that the general solution $u(t) = u_c(t) + u_p(t)$ of the ODE contains the n arbitrary constants a_1, a_2, $\ldots$, a_n, so if the solution is to start at a time $t = t_0$, then in order to determine these constants it is necessary to specify n conditions that are to be satisfied by the solution when $t = t_0$. This is accomplished by saying how a particular solution must start at the time t_0, and this involves specifying the n starting values k_0, k_1, k_2, ..., k_{n-1} of the function $u(t)$ and its first $n-1$ derivatives at time $t = 0$ by setting

$$u(t_0) = k_0, u'(t_0) = k_1, u''(t_0) = k_2, \ldots, u^{(n-1)}(t_0) = k_{n-1}. \tag{2.7}$$

This leads to the following system of n linear simultaneous equations involving the constants a_1, a_2, ..., a_n,

$$u_c(t_0) + u_p(t_0) = k_0, u'_c(t_0) + u'_p(t_0) = k_1, \ldots, u_c^{(n-1)}(t_0) + u_p^{(n-1)}(t_0) = k_{n-1},$$

the solution of which will yield the n constants a_1, a_2, ..., a_n in $u_c(t)$, after which the required particular solution follows from equation (2.5).

To illustrate this, consider the non-homogeneous ODE $u''' + 4u'' + 5u' + 2u = 1$. It has already been shown that the complementary function is $u_c(t) = a_1 e^{-t} + a_2 t e^{-t} + a_3 e^{-2t}$, and it is easily checked by substitution that the particular integral $u_p(t) = 1/2$, so the general solution is $u(t) = u_c(t) + u_p(t) = 1/2 + a_1 e^{-t} + a_2 t e^{-t} + a_3 e^{-2t}$. If, for example, this solution is to start when $t = 0$ with $u(0) = 1$, $u'(0) = 0$ and $u''(0) = 0$, then setting $t = 0$ and substituting these values into the expression for $u(t)$ gives the following algebraic equations for a_1, a_2 and a_3

$$\frac{1}{2} + a_1 + a_3 = 1, -a_1 + a_2 - 2a_3 = 0 \text{ and } a_1 - 2a_2 + 4a_3 = 0.$$

The solution of these algebraic equations is $a_1 = 0$, $a_2 = 1$ and $a_3 = 1/2$, and thus the required solution of the ODE becomes $u(t) = \frac{1}{2} + te^{-t} + \frac{1}{2}e^{-2t}$.

The quantity t_0 in equation (2.7) is called the *initial time* (the time when the solution starts), and the n constants k_0, k_1, ..., k_{n-1} are called the *initial values* (*conditions* or *data*) to be satisfied by the solution at the initial time. Note that the value of $u^{(n)}(t_0)$ cannot be specified as part of the initial data, because once the n initial values have been specified the ODE itself will determine the value of $u^{(n)}(t_0)$.

To illustrate the need for initial conditions, consider the steady radial flow of water in a discharging well, with r the radial distance from the borehole, $q(r)$ the total radial discharge as a function of r, w the steady water production rate per unit volume, and b the thickness of a confined aquifer. If h_0 is the pre-pumping static water height underground, and $h(r)$ is the hydraulic head at radius r during pumping, both measured from the bottom of the aquifer, then $h_0 - h(r) > 0$ is the *drawdown* at radius r due to pumping. Assuming that $q(r) = -Tdh/dr$, with T a transmission constant depending on the aquifer, then $h(r)$ satisfies the second-order variable coefficient linear ODE

$$\frac{d^2h}{dr^2} + \frac{1}{r}\frac{dh}{dr} = -\frac{wb}{K}, \tag{2.8}$$

where the negative sign is necessary because the larger the drawdown, the less the pressure, with the hydraulic conductivity constant K depending on the nature of the aquifer. The general solution of this ODE is

$$h(r) = a_1 + a_2 \ln r - \frac{1}{4}\frac{wbr^2}{4K}, \quad (r > 0), \tag{2.9}$$

with two arbitrary constants a_1 and a_2. Solution (2.9) is not valid when $r = 0$, because the term $\ln r$ becomes infinite when $r = 0$. To find a particular solution, it is necessary to choose some value $r = r_0 > 0$, and then to specify the two values k_0 and k_1 so that $h(r_0) = k_0$ and $h'(r_0) = k_1$. Even though this solution is independent of the time t, these starting conditions are still called the *initial conditions* for $h(r)$, and once $h(r)$ is known the total radial rate of discharge $q(r)$ follows by using the result $q(r) = -Tdh/dr$.

At this point it is appropriate to draw attention to what is probably the most important first-order ODE, called the *first-order linear ODE*, and to give an example of its application. The most general first-order linear equation is of the form

$$\frac{dy}{dt} + p(t)y = q(t), \tag{2.10}$$

where $p(t)$ and $q(t)$ are arbitrary functions of t. The equation has an *integrating factor*

$$\mu(t) = \exp\left[\int p(t)dt\right], \tag{2.11}$$

in terms of which the solution can be written

$$y(t) = \frac{A}{\mu(t)} + \frac{1}{\mu(t)}\int \mu(t)q(t)dt, \tag{2.12}$$

where A is an arbitrary constant.

Let us apply this result to equation (2.8), after first replacing t by r, and then reducing it to a linear first-order equation for u by setting $u = dh/dr$, when it becomes

$$\frac{du}{dr} + \frac{1}{r}u = -\frac{wb}{K}.$$

This equation is of the form of equation (2.10) with $p(r) = 1/r$ and $q(r) = -wb/K$. The integrating factor is $\mu(r) = \exp\int\left(\frac{1}{r}\right)dr = \ln r$, so from equation (2.12) $u(r) = C_1/r - wbr/(2K)$. As $dh/dr = u$, integration of $u(r)$ shows that $h(r) = C_2 + C_1\ln r - \frac{wbr^2}{(4K)}$, where, apart from the symbols representing the arbitrary constants C_1 and C_2, this result is the same as equation (2.9).

2.1.3 Structure of solutions

The solution of an ODE may vary in many different ways, but in general terms it is said to be *stable* if, although it changes with time and may never settle down, the solution remains bounded for all time. The solution is said to be *unstable* if the solution becomes unbounded as time increases.

It may happen, independently of the initial conditions, that after a suitably long time all solutions of ODE equation (2.2) approach arbitrarily close to a function, say $u = \phi(t)$. When this happens the solutions of equation (2.2) are said to *converge* to the solution $\phi(t)$, which is then called the *steady-state solution*. This name is somewhat misleading, because it does not necessarily mean that the steady-state solution is independent of the time t, and so is an absolute constant. What it really means is that the steady-state solution (possibly time dependent) is the solution to which all solutions of initial-value problems converge after sufficient time has elapsed for the complementary function to decay to zero. Clearly, the complementary function $u_c(t)$ will only decay to zero as $t \to \infty$ if $\lambda < 0$ in each of its exponential terms $e^{\lambda t}$, so the steady-state solution is determined by the particular integral.

An ODE or a system of simultaneous ODEs in which the dependent variables do not appear linearly is said to be *non-linear*. As a rule, analytical solutions of non-linear equations are difficult to find, one reason for which is that solutions of non-linear equations do *not* possess the valuable linear superposition property. Non-linearity often occurs in differential equations describing physical problems, and it can arise in many different ways. For example, the presence of non-linear terms such as $u^{1/2}$, u^2, $(du/dt)^2$ and udu/dt in a differential equation will cause it to become non-linear. As solutions of non-linear equations do not possess the linear superposition property, the terms homogeneous and non-homogeneous have no meaning when studying non-linear ODEs.

A simple example of a non-linear ODE is the horizontal-beam equation that occurs in structural problems, and takes the form

$$\frac{d^2}{dx^2}\left\{ \frac{EI d^2 y / dx^2}{\left[1 + (dy/dx)^2\right]^{3/2}} \right\} = w(x), \tag{2.13}$$

where x is the distance along the beam measured from one end, $y(x)$ is the downward vertical deflection of the beam at a distance x due to a load, $w(x)$ is the line density of the distributed load along the beam, E is Young's modulus of elasticity, and I is the moment of inertia of a cross-section of the beam about its central axis. The type of beam to be described is determined by the conditions imposed at its ends. For example, a beam fastened rigidly at one end but left free at the other is a *cantilevered beam*.

This ODE equation (2.13) can be simplified if the expression $[1 + (dy/dx)^2]^{3/2}$ in the denominator that takes account of the curvature of the beam can be approximated by 1, in which case equation (2.13) becomes the very simple fourth-order linear equation

$$EI\frac{d^4y}{dx^4} = w(x), \tag{2.14}$$

which can be solved by straightforward integration. This type of approximation replacing a non-linear ODE by an approximate linear ODE is called *linearization*, and when it can be justified it allows approximate solutions of non-linear equations to be obtained.

The process of linearization must be used with care because, even when it is permissible, the interval over which the approximation is valid is usually very restricted. In the case of the beam equation, linearization is only valid when the downward displacement $y(x)$ is very small. However, some non-linear equations *cannot* be linearized, because linearization produces an equation that no longer describes the fundamental physical phenomenon that was modelled by the full non-linear equation. A case in point is the non-linear equation

$$\frac{d^3y}{dx^3} + \frac{1}{2}y\frac{d^2y}{dx^2} = 0,$$

which arises in the study of the boundary layer formed when a viscous fluid flows past a horizontal plate. In this case, to study boundary layer flow, it is necessary to use special techniques when working with the full non-linear equation. In general, apart from the use of special analytical methods when examining certain important types of non-linear ODEs, such equations must be solved using numerical methods that will be described in Chapter 3.

Another example of a non-linear equation that cannot be linearized is the flow of water from an orifice of area a in the bottom of a water tank of area A, when at time t water flows into the tank at a rate $f(t)$. If the height of water above the orifice at time t is $h(t)$ and the exit velocity of the water is $v = c_v\sqrt{2gh}$, where c_v is the velocity coefficient, the ODE for $h(t)$ based on *Torricelli's law of flow* is the non-linear first-order equation

$$\frac{dh}{dt} + \frac{ac_d}{A}\sqrt{2gh}^{1/2} = \frac{f(t)}{A}.$$

In this equation the discharge coefficient $c_d(= c_c c_v)$ takes account of the fact that, after the water has passed through the orifice in the bottom of the tank, the cross-sectional area of the emerging jet contracts from its initial area a to a smaller one $c_a a$ (for water $c_a \approx 0.6$ for a sharp-edged orifice).

Various physical situations, such as those modelling mixing phenomena, are too complicated to be described by a single ODE, so they must be described by a *simultaneous system* of ODEs. A typical first-order system in which the dependent variables are $y_1, y_2, \ldots, y_n$, takes the form

$$
\begin{aligned}
dy_1 / dt &= f_1(y_1, y_2, \ldots, y_n, t), \\
dy_2 / dt &= f_2(y_1, y_2, \ldots, y_n, t), \\
&\cdots \\
dy_n / dt &= f_n(y_1, y_2, \ldots, y_n, t),
\end{aligned}
\tag{2.15}
$$

where, in general, the functions $f_1, f_2, \ldots, f_n$ are non-linear functions of the dependent variables. In such cases, solutions of initial-value problems must be obtained by numerical methods. Even when the functions $f_1, f_2, \ldots, f_n$ are linear combinations of the dependent variables and the time t, an analytical solution of equation (2.15) is usually only possible when it simplifies to a simultaneous system of constant-coefficient equations

$$
\begin{aligned}
dy_1 / dt &= a_{11} y_1 + a_{12} y_2 + \cdots + a_{1n} y_n + h_1(t), \\
dy_2 / dt &= a_{21} y_1 + a_{22} y_2 + \cdots + a_{2n} y_n + h_2(t), \\
&\cdots \\
dy_n / dt &= a_{n1} y_1 + a_{n2} y_2 + \cdots + a_{nn} y_n + h_n(t),
\end{aligned}
\tag{2.16}
$$

where the coefficients a_{ij} are constants, and the terms $h_i(t)$ are given functions of t.

This system can be written in the matrix form

$$
\frac{dy}{dt} = \mathbf{A}y + \mathbf{h}(t),
$$

with $\mathbf{y}$ the column vector with elements $y_1, y_2, \ldots, y_n$, dy/dt the column vector with elements $dy_1/dt, dy_2/dt, \ldots, dy_n/dt$, $\mathbf{h}(t)$ the column vector with elements $h_1(t), h_2(t), \ldots, h_n(t)$, and where $\mathbf{A} = [a_{ij}]$ is an $n \times n$ constant matrix with elements a_{ij}. The system can be solved analytically by *diagonalizing* the matrix $\mathbf{A}$. An outline of how a solution is obtained is given in the Appendix to this chapter. However, even in this case, when an analytical solution can be found, if more than three equations are involved it is usually simpler to solve an initial-value problem for a system by numerical methods.

One final class of problems involving ODEs that needs to be mentioned is what are called *two-point boundary-value problems*. These are time-independent problems for which initial conditions are inappropriate; instead an ODE must be solved on a fixed interval $a \leq x \leq b$ with suitable conditions being imposed on the solution at each end (on the boundaries) of the interval. The conditions at the two ends of the interval are called *boundary conditions*, and such problems can be difficult to solve. They usually require numerical solutions. A brief account of how some problems of this type can be solved numerically is given in Chapter 3. A simple two-point boundary-value problem that can be solved by ordinary integration is the linearized beam equation (2.14), although when it was introduced no mention was made of the boundary conditions to be imposed on the solution at the ends $x = a$ and $x = b$ of the beam. A cantilevered beam, clamped rigidly at $x = a$ but free at the end $x = b$, must satisfy the boundary conditions $y(a) = y'(a) = 0$ at $x = a$, and the boundary conditions $y''(b) = y'''(b) = 0$ at $x = b$.

2.2 Partial differential equations and their classification

In hydraulics, usually of more concern than ODEs are the cases where one or more continuously differentiable functions, say u, v and w, depend on both position in space and also on time t. When these functions and their partial derivatives can be connected by a system of equations, the equations become *partial differential equations* (PDEs). The *order* of a PDE or system is the order of the highest derivative that occurs in the PDE or system. When only *one* space variable is involved, say x, and the other independent variable is the time t, the equation is said to be *one-dimensional and time-dependent*, simply called a *1D time-dependent* PDE. Correspondingly, if two or three space variables and also the time t are involved, the equations are said to be *2D* or *3D time-dependent equations*. It may happen that time does not enter into a PDE as an independent variable. In such cases the equations are called *1D, 2D* or *3D time-independent* PDEs or, more simply still, just *1D, 2D* or *3D* PDEs.

Unlike the situation with ODEs, where in the linear case a general solution can be found and then used to solve any given problem, this is not possible with PDEs, as general solutions are seldom available. Instead, it becomes necessary to find ways of solving specific problems. It is to be expected that, as the number of space dimensions increases, so also does the complexity of finding a solution of a PDE, and this is indeed the case. In this case, a *solution* is a relationship free from partial derivatives that relates all the variables involved, satisfies any auxiliary conditions that are imposed (such as initial and boundary conditions, to be described later), and is such that when substituted into the PDE it satisfies it identically.

However, for the moment, let us confine our attention to PDEs involving two independent variables that may be either two space variables, or one space variable and the time. As with ODEs, a *linear* PDE is one in which the dependent variable and its partial derivatives only occur linearly. When this is not the case, a PDE is called *non-linear*, and there is no general theory that can be used when seeking a solution of a non-linear PDE. The situation is somewhat better when in a non-linear PDE or in a system of PDEs the highest-order partial derivatives of the dependent variables occur linearly. This can simplify the task of finding solutions, and equations of this type are called *quasilinear* PDEs. A simple example of a first-order quasilinear PDE is

$$\frac{\partial u}{\partial t} + u\frac{\partial u}{\partial x} + f(u) = 0.$$

This equation is *non-linear* because of the product term $u\partial u/\partial x$, and possibly also because of the term $f(u)$, in which u may occur non-linearly. However, the PDE is *quasilinear* because its highest-order partial derivatives, namely $\partial u/\partial x$ and $\partial u/\partial y$, occur linearly.

The simplest of the linear second-order PDEs that occur most frequently in engineering and physics are of the general type

$$A(x,y)\frac{\partial^2 u}{\partial x^2} + 2B(x,y)\frac{\partial^2 u}{\partial x \partial y} + C(x,y)\frac{\partial^2 u}{\partial y^2} + a(x,y)\frac{\partial u}{\partial x} + b(x,y)\frac{\partial u}{\partial y}$$
$$+ c(x,y)u = f(x,y),$$

$$(2.17)$$

where the coefficients $A(x,y), B(x,y), \ldots, C(x,y)$ are given functions of x and y, and where the independent variables x and y may either both be space variables, or one space variable and the time variable t. As with ODEs, if $f(x,y) \equiv 0$ the linear PDE equation (2.17) is said to be *homogeneous*, otherwise it is *non-homogeneous*. If the equation is homogeneous, and the terms on the left of equation (2.17) are denoted by $L[u]$, the homogeneous equation takes the simple form

$$L[u] = 0,$$

$$(2.18)$$

where $L[.]$ is called a *linear differential operator*. This means that $L[.]$ is, in effect, an instruction to perform certain differentiation operations on whatever function appears in place of the dot that lies between the brackets [ ]. The differential operator only becomes a function when it acts on a suitably differentiable function u. The linearity of the PDE means that, if c is a constant and u is a solution of the homogeneous PDE equation (2.18), then

$$L[cu] = cL[u],$$

$$(2.19)$$

while if u_1 and u_2 are any two solutions of equation (2.18), it follows that

$$L[u_1 + u_2] = L[u_1] + L[u_2].$$

$$(2.20)$$

In terms of the homogeneous equation this means that any solution u may be multiplied by a constant c to form cu and still remain a solution, while if u_1 and u_2 are any two solutions of the homogeneous equation, then so also is their sum $c_1 u_1 + c_2 u_2$ for any arbitrary constants c_1 and c_2. This last result extends to the linear combination of any finite number of solutions and, as in the case of an ODE, this is said to represent the *linear superposition property* of the solutions of the linear homogeneous PDE in equation (2.18). It is this property that forms the basis of the *method of separation of variables* that in especially simple cases can be used to construct analytical solutions of linear PDEs. We will not discuss this method here as it is seldom of use when practical problems in hydraulics need to be solved, so instead we refer the interested reader to the books on advanced engineering mathematics by Jeffrey (2002), Kreyszig (2005) and O'Neil (1999) and to the more advanced books by Garabedian (1999), Keener (1994), Logan (2006) O'Neil (2006 a, b) and Zauderer (2006).

The second-order equation (2.17) with two independent variables, either (x, y) or (x, t), belongs to one of three quite different types of PDE. The three types describe very different physical phenomena, where each PDE has its own quite separate mathematical properties. The classification of PDE equation (2.17) is determined algebraically by considering the expression $\Delta = B^2 - AC$, which is called the *discriminant* of the PDE. The name discriminant is given to Δ because it discriminates (distinguishes) between the three types of PDE that equation (2.17) can be. Equation (2.17) is said to be *hyperbolic* when $\Delta > 0$, *parabolic* when $\Delta = 0$ and *elliptical* when $\Delta < 0$. Note that in equation (2.17) the coefficients A, B and C depend on x and y, so the classification of such an equation as hyperbolic, parabolic or elliptical may vary from point to point in the (x, y) plane, depending on the nature of the coefficients A, B and C, although when the coefficients are constants the classification will, of course, remain unchanged throughout the entire (x, y) plane.

The names hyperbolic, parabolic and elliptical used in this classification arise as a result of the introduction of two new independent variables $\xi = \xi(x, y)$ and $\eta = \eta(x, y)$ in equation (2.17), chosen in such a way that at a given point of the (x, y) plane the coefficients of second-order terms simplify. It is because this simplification involves algebra similar to that of the equations describing a hyperbola, a parabola or an ellipse that the PDEs are given these names. It should be clearly understood that the names hyperbolic, parabolic and elliptical are only convenient names used when classifying PDEs, and that the names have no geometrical implications for the solutions of the associated PDEs.

The result of applying such changes of independent variable to PDE equation (2.17) at a given point in the (x, y) plane is that the simplified forms of the different types of equation are produced. These are called the *canonical forms* or *standard forms* of the PDEs.

In the *hyperbolic case* the canonical form is found to be

$$\frac{\partial^2 u}{\partial \xi^2} - \frac{\partial^2 u}{\partial \eta^2} = F_1(\xi, \eta, u, \partial u/\partial \xi, \partial u/\partial \eta), \tag{2.21}$$

or, equivalently,

$$\frac{\partial^2 u}{\partial \xi \partial \eta} = F_2(\xi, \eta, u, \partial u/\partial \xi, \partial u/\partial \eta). \tag{2.22}$$

For the *parabolic case* the canonical form is

$$\frac{\partial^2 u}{\partial \eta^2} = F_3(\xi, \eta, u, \partial u/\partial \xi, \partial u/\partial \eta), \tag{2.23}$$

and for the *elliptical case* the canonical form is

$$\frac{\partial^2 u}{\partial \xi^2} + \frac{\partial^2 u}{\partial \eta^2} = F_4(\xi, \eta, u, \partial u/\partial \xi, \partial u/\partial \eta), \tag{2.24}$$

where F_1 to F_4 represent functions whose arguments may contain terms in ξ, η, u, u_ξ and u_η.

The way the variables ξ and η are introduced to bring about such a simplification will not be described here, as the process is lengthy and can be found in standard texts such as those mentioned previously.

In applications it is often useful to write ODEs and PDEs in what is called a *non-dimensional form* (see also Chapter 5). This is accomplished by introducing convenient length, mass and time reference units L_0, M_0 and T_0 appropriate to an application, and then, if the equivalent physical quantities involved are x, m and t, the equations are rewritten in terms of the new dimensionless variables $x' = x/L_0$, $m' = m/M_0$ and $t' = t/T_0$. Thereafter, for convenience, the prime is often dropped, it being understood that dimensionless variables are involved. This approach allows the easy interpretation of solutions when they are applied to similar situations, but with different length, mass and time scales.

Familiar examples of constant-coefficient equations that are already in their canonical form when expressed in terms of the Cartesian coordinates x and y are as follows.

The *hyperbolic equation*

$$\frac{\partial^2 u}{\partial x^2} - \frac{\partial^2 u}{\partial t^2} = 0 \tag{2.25}$$

also called the *wave equation*, that with the change of variable $\xi = x + t$, $\eta = x - t$ can be transformed into the equivalent form

$$\frac{\partial^2 u}{\partial \xi \partial \eta} = 0; \tag{2.26}$$

the *parabolic equation*

$$\frac{\partial u}{\partial t} = \frac{\partial^2 u}{\partial x^2}, \tag{2.27}$$

called the *heat equation*, also known as the *diffusion equation*, because it describes both the temperature distribution in a solid and also diffusion phenomena, each of which behave in a similar fashion;

the *elliptical equation*

$$\frac{\partial^2 u}{\partial x^2} + \frac{\partial^2 u}{\partial y^2} = 0, \tag{2.28}$$

called the two-dimensional *Laplace equation*.

2.3 Dispersion and dissipation in hyperbolic linear equations

The hyperbolic equation (2.25) describes the propagation of disturbances (*waves*) in the positive and negative x-directions with respect to the time t. All linear hyperbolic equations describe some form of wave propagation, although, unlike the equation in (2.25), in general they describe waves that distort and may decay as they propagate due to effects called *dispersion* and *dissipation*. To understand these effects it is necessary to generalize the hyperbolic equation (2.25) to

$$\frac{\partial^2 u}{\partial x^2} = \frac{1}{c^2} \left\{ \frac{\partial^2 u}{\partial t^2} + p \frac{\partial u}{\partial t} + qu \right\}, \tag{2.29}$$

where p and q are constants. The linearity of this equation allows us to consider the way in which this equation propagates a sinusoidal wave, because in the linear case any wave can be constructed by the linear superposition of suitable multiples of such waves, as with Fourier series. It will simplify the analysis if the sinusoidal wave is represented in terms of complex variables, because a physical wave can always be considered to be the *real part* of such a representation. Accordingly, we will consider the sinusoidal wave (see also Section 4.5)

$$u(x,t) = a \, \exp\left\{i\left[(kx - \omega t) + \varepsilon\right]\right\},$$

where a is a complex number, i is the square root of -1, and the real number $|a|$ is called the *amplitude* of the wave. The *wavelength* L of the wave is the smallest length by which x can be increased while leaving u unchanged. Thus, because the complex exponential function is periodic with period 2π, it follows that $kL = 2\pi$, where the number $k = 2\pi/L$ is called the *wavenumber* and L is called the *wavelength* of the wave. Similarly, if T is the smallest value by which the time t may be increased while leaving u unchanged, it follows that $\omega T = 2\pi$, where T is called the *period* of the wave, while the number $\omega = 2\pi/T$ is called the *frequency* of the wave. The constant quantity ε in the expression for $u(x,\ t)$ is arbitrary and is called a *phase* shift. It will be convenient to rewrite $u(x,\ t)$ as

$$u(x,t) = A \exp\left\{i(kx - \omega t)\right\}, \quad \text{with} \quad A = \exp(-i\varepsilon). \tag{2.30}$$

Substituting this result into equation (2.29) and factoring out the non-zero complex exponential function leads to the equation

$$\omega^2 + ip\omega - c^2 k^2 - q = 0. \tag{2.31}$$

This equation provides useful physical information about the wave propagation process, because it shows that the frequency ω, the wavenumber k and n are not independent. In physical wave propagation k must be real, so it follows directly that ω may be complex, in which case

$$\omega = -\frac{ip}{2} \pm \frac{1}{2}\left\{4c^2 k^2 + 4q - p^2\right\}^{1/2}. \tag{2.32}$$

Substitution of equation (2.32) into equation (2.30) gives the result

$$u(x,t) = A \exp\left(-\tfrac{1}{2}pt\right) \exp\left\{i\left[kx \pm \tfrac{1}{2}\sqrt{4c^2 k^2 + 4q - p^2}\right]\right\}. \tag{2.33}$$

Inspection of equation (2.33) shows that if $p > 0$ the wave will *decay* as it propagates, and this process is called *dissipation*. However, the wave frequency depends on the wavelength through the wavenumber k, so the *speed* of the wave will also depend on the frequency, and this process is called *dispersion*. When dispersion is present different frequencies propagate with different speeds; as the initial wave is the superposition of waves with different frequencies, as it propagates the dependence of wave speed on frequency will cause the wave to change shape. Note that the generalized wave equation (2.29) reduces to the wave equation (2.25) if the constants $p = q = 0$, and from equation (2.33) it then follows that waves propagated by the ordinary wave equation do so without the effects of dissipation and dispersion.

For an example of this situation involving ocean gravity waves see Holly (1985), and for the effects of dispersion in rivers see Sauvaget (1985).

As any linear wave can be constructed by linear superposition of waves like that in equation (2.30), it follows that, in general, waves satisfying the PDE in equation (2.29) will decay and change shape as they propagate. The name *dispersion* is used with wave propagation because the different speeds with which waves of different frequencies propagate cause waves to 'spread out' and so to 'disperse'. It is for this reason that equation (2.31) is called the *dispersion relation* for PDE equation (2.29).

It will be shown later that hyperbolic equations possess special curves called *characteristic curves* in the plane of their independent variables, and that these have the property that each transmits a disturbance (a point on a wave) at a finite speed. Apart from describing general linear wave propagation, one way in which a hyperbolic equation arises in hydraulics is when the Saint Venant equations are included in the equations describing open-channel flow (see also Section 4.4.3).

2.4 Parabolic and elliptical equations, diffusion, quasilinearity and systems of equations

The *heat equation* (equation 2.27), also called the *diffusion equation*, describes the propagation of heat in a heat-conducting body. It also describes the diffusion of a physical quantity, such as an impurity in water into still water or into the ground, in the x-direction (downward) with the passage of time t. In general, parabolic equations arise when modelling viscous or other diffusive processes. A typical parabolic equation occurs when studying unsteady groundwater flow. There, for an aquifer of constant thickness in which the flow is uniform, the groundwater flow is modelled by the parabolic equation

$$\frac{\partial h}{\partial t} = D\frac{\partial^2 h}{\partial x^2} + S,$$

where h is the head of the aquifer (the height of the water above the lower boundary of a confined aquifer), D is a diffusion constant, and the source term S is the *inflow* of water. See, for example, Chadwick *et al.* (2004), Sen (1995) and Walton (1991).

In the Laplace equation, which is elliptical, time does not enter and the variables x and y are space variables. This important equation arises in a variety of different ways, one of which is in the study of fluid flow governed by a *velocity potential* ϕ that is a solution of the Laplace equation. The velocity potential has the property that the component of the fluid velocity u in the x-direction is given by $\partial\phi/\partial x$, and the component v in the y-direction is given by $\partial\phi/\partial y$. This result will be encountered later in terms of the *stream function* ψ of a fluid flow that is related to the velocity and the velocity potential by $u = \partial\psi/\partial y = \partial\phi/\partial x$ and $v = -\partial\psi/\partial x = \partial\phi/\partial y$. The stream function is so named because in incompressible inviscid (non-viscous) fluids the curves ψ = constant are the fluid-flow lines (see also Section 4.2.5).

We mention here that the *complex potential* for a fluid flow is the complex analytical function $w = \phi + i\psi$, so in the context of complex analysis the functions ϕ and ψ are *conjugate harmonic functions*, where a *harmonic function* is one that satisfies Laplace's equation. See, for example, Jeffrey (2002), Kreyszig (2005) and O'Neil (1995).

Linear second-order equations such as equation (2.17) are important for the following reason. Typically, in the derivation of the PDEs governing many physical situations, two coupled (simultaneous) first-order PDEs arise, each of which describes a fundamental physical property represented by the dependent variables u and v, say. In many cases the structure of these equations is such that one of the dependent variables v, say, can be eliminated by differentiation, leading to a single second-order equation such as equation (2.17), which is satisfied by the other dependent variable u. Once the second-order equation for u has been solved, its solution can be used with the original first-order equations to determine the other dependent variable v, thereby leading to the solution of the original system.

The classification of PDEs can be extended to second-order equations in n independent variables, but a discussion of how this may be achieved will not be appropriate here. This matter is discussed in, for example, Garabedian (1999). We mention in passing that when the multi-independent variable situation arises, although some types of equation can be classified as being of purely hyperbolic, parabolic or elliptical type, still more classifications become possible, so that, for example, equations can arise that are of mixed hyperbolic and elliptical type. This can also happen with coupled first-order systems of PDEs in more than two dependent variables.

Let us now return to the fact that a special type of non-linear PDE or system is said to be *quasilinear* if, although it is linear in the highest-order derivatives, it contains non-linear terms of lower order. An example of a quasilinear second-order PDE for $u(x, y)$ is the equation

$$A(x, y, u, u_x, u_y)\frac{\partial^2 u}{\partial x^2} + 2B(x, y, u, u_x, u_y)\frac{\partial^2 u}{\partial x \partial y} + C(x, y, u, u_x, u_y)\frac{\partial^2 u}{\partial y^2}$$
$$+ F(x, y, u, u_x, u_y) = 0, \tag{2.34}$$

where $u_x = \partial u/\partial x, u_y = \partial u/\partial y$, and A, B, C and F are functions that may contain terms such as x, y, u, u^2, uu_x and $u_x u_y$. As with linear second-order PDEs, the classification of quasilinear PDEs like equation (2.34) is determined by the discriminant $\Delta = B^2 - AC$.

More general than this single quasilinear second-order equation are systems of coupled quasilinear first-order PDEs in two or more dependent variables. Only in special cases can a system of quasilinear equations be solved for one dependent variable in terms of a single higher-order PDE, so when this is not possible the dependent variables must be found by solving the complete coupled system of equations.

The typical example of a first-order quasilinear equation for the variable $u(x, t)$ given earlier was

$$u_t + f(u)u_x = h(x, t, u), \tag{2.35}$$

where $u_t = \partial u / \partial t$ and $u_x = \partial u / \partial x$. The quasilinear equation (2.35) is sometimes called the *advection equation*. This is the simplest example of an equation that describes how a quantity of interest, such as the *vorticity* in fluid mechanics, is transported through a medium. The functions $f(u)$ and $h(x, t, u)$ are usually continuous functions of their arguments. This equation simplifies to a linear equation when $f(u) = c =$ constant, and $h(x, t, u)$ depends linearly on u. As a special case of equation (2.35) we mention the situation when the function f depends only on x and t, while the function h, which may or may not depend on x and t, depends non-linearly on u. An equation of this form, where the non-linearity in u occurs only in the undifferentiated function $h(x, t, u)$ on the right of equation (2.35), is said to be *semilinear*. A typical first-order semilinear equation is $u_t + cu_x = \sin u$. More will be said later about the quasilinear equation (2.35), as it will be used to introduce what is called the *method of characteristics* that leads to both analytical and numerical solutions of hyperbolic equations. This equation will also be used to illustrate how non-linearity in a quasilinear hyperbolic equation can, as time increases, cause a solution that starts in a smooth manner but evolves into a discontinuous solution. In hydraulics this effect leads to the occurrence of *hydraulic jumps*.

An important example of a coupled quasilinear first-order system of PDEs is provided by the one-dimensional form of the so-called *shallow water equations* (see also Section 4.6.2):

$$u_t + uu_x + g\eta_x = 0 \tag{2.36}$$

$$\eta_t + \left[u(\eta + h) \right]_x = 0. \tag{2.37}$$

In these equations the x-axis lies in the surface of the equilibrium level of the water, the y-axis is vertically upward, $u(x, t)$ is the x-component of the water velocity, $\eta(x, t)$ is the elevation of the surface of the water above the equilibrium level, and $y + h(x) = 0$ is the equation of the river bed or seabed, while g is the acceleration due to gravity. The general geometrical configuration of these equations is shown in Figure 2.1.

Here, as usual, suffixes have been used to denote partial derivatives, so that $u_t = \partial u/\partial t$, $u_x = \partial u/\partial x$, $\eta_t = \partial \eta/\partial t$ and $\eta_x = \partial \eta/\partial x$, although

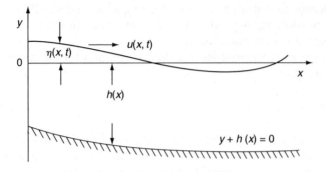

Figure 2.1 The geometry of the shallow-water model

$h_x = \mathrm{d}h/\mathrm{d}x$, because $h(x)$ describes the shape of the river bed or seabed and so only depends on x. It will be seen later that the *speed of propagation* of this surface wave is $c = \sqrt{g(h+\eta)}$, although for historical reasons in fluid mechanics the speed c is often called the wave *celerity* (derived from the Latin *celeritat* meaning swiftness).

The quasilinearity in this system is caused by the product terms uu_x, $u_x\eta$ and $u\eta_x$, and the equations form a *first-order* system of PDEs because the highest-order derivatives u_t, u_x, η_t and η_x that occur only appear linearly. Systems of equations can be classified as hyperbolic, parabolic or elliptical, although the method of classification is more complicated than the one used for a single second-order equation. However, when a second-order PDE such as equation (2.17) is expressed in the form of a system, the classification can be shown to be compatible with the results obtained using the discriminant Δ. It will be shown later that the system of shallow-water equations (2.36) and (2.37) is *unconditionally hyperbolic* (the hyperbolic nature does not change throughout the (x, t) plane), and that the system describes the propagation of surface waves on water with speed $c = \sqrt{g(h+\eta)}$. For more about the shallow-water equations see, for example, Abbott (1979), Abbott and Cunge (1982), Abbott and Minns (1998), Cunge *et al.* (1980), Mader (2004), Verboom *et al.* (1982) and Verwey (1983).

2.5 Initial and boundary conditions for partial differential equations: existence and uniqueness

2.5.1 General

In the study of PDEs, general solutions are hardly ever known, so instead it is necessary to seek the solution of a PDE in the context of a specific application. When doing this, certain auxiliary conditions must be specified that serve to identify the precise nature of the problem. If the solution depends on the time (hyperbolic and parabolic equations), these auxiliary conditions amount to saying how the solution is to start (*initial conditions*), although there may also be conditions that must be satisfied on some fixed boundaries (*boundary conditions*). However, when only space variables are involved so that time is absent, as in the Laplace equation, the auxiliary conditions must describe how the solution is to behave on the boundary of some region D of interest in the (x, y) plane, which then involves the specification of boundary conditions. It is through the properties of solutions, and their dependence on the auxiliary conditions, that equations or systems of hyperbolic, parabolic and elliptical type exhibit fundamental differences.

The types of initial conditions and the boundary-value data that are appropriate for the different types of second-order equations (either linear or quasilinear), in one dependent variable and two independent variables, are shown in the table on page 27. First, though, the most important and frequently occurring types of auxiliary condition that arise in applications must be named.

In each of the following cases (Sections 2.5.2 to 2.5.4) the boundary of a region is specified parametrically. The advantage of a parametric representation is that it enables a simple representation of curves to be given, because were Cartesian coordinates to be used, they often involve many-valued functions. For example, if a circular boundary occurs with radius a, it has the algebraic equation $x^2 + y^2 = a^2$, and to express y in terms of x it is necessary to introduce the two-valued square-root function and to write $y = \pm(a^2 - x^2)^{1/2}$. However, in terms of plane polar coordinates (r, θ), the equation of the circle has the very convenient parametric representation in terms of θ such that $x = a\cos\theta$, $y = a\sin\theta$, with $0 \leq \theta < 2\pi$, so now each point on the perimeter of the circle has a unique representation.

2.5.2 Cauchy conditions

Cauchy conditions are used to specify what is called a *Cauchy problem* for a PDE in some open region (area) D in the (x, t) plane, part of which is bounded by a given curve Γ defined parametrically by $x = g(s)$, $t = h(s)$. Often the parameter s is taken to be the distance measured along Γ from some reference point on Γ. Using the arc length s along Γ as a parameter, the *Cauchy conditions*, or *initial conditions*, for a PDE involve requiring the solution $u(x, t)$ to be equal to a given function $f(s)$ at each point of Γ, and in addition that the directional derivative of u normal to Γ is equal to a given function $n(s)$ at each point of Γ. To explain the terminology used here, a region (area) D is said to be *open* if all, or part of it, has no boundary. A typical open region in the (x, t) plane is the area $t > 0$ that lies above, but excludes, the x-axis. A Cauchy problem is often called a *pure initial-value problem* when the only conditions to be imposed on a PDE are Cauchy conditions on the initial line $t = 0$.

2.5.3 Dirichlet conditions

Dirichlet conditions require the solution u to be equal to a given function $f(s)$ on part or all of a boundary Γ defined parametrically by $x = g(s)$, $y = h(s)$, where s can be taken to be the distance measured along Γ from some convenient reference point on Γ. A region (area) D is said to be *closed* if it is enclosed by a boundary curve, and each point of the boundary curve belongs to D. If area D is closed, then the specification of Dirichlet conditions on all of its boundary defines a *pure boundary-value problem* for the PDE. A typical closed region is the interior and boundary points of a rectangle in the (x, y) plane.

2.5.4 Neumann conditions

Neumann conditions involve first specifying part or all of a boundary curve Γ enclosing a region D in which a PDE for a function $u(x, y)$ is given. As before, it will be assumed that Γ is defined parametrically in terms of s by $x = g(s)$, $y = h(s)$. Then Neumann conditions require that on Γ the directional derivative of u normal to Γ is equal to a given function $n(s)$. If region D is closed then, as with Dirichlet conditions, the specification of Neumann conditions on all the boundary defines a *pure boundary-value problem* for the PDE. In many boundary-value problems Dirichlet and Neumann conditions are prescribed on different parts of the boundary, and sometimes in the combined form $\left(\alpha u + \beta \frac{\partial u}{\partial n}\right) = 0$ on another part of the boundary Γ, where α and β are constants.

2.6 Well-posed problems

A PDE in a region D is said to be *well-posed* if the imposition of initial and boundary values on the boundary of region D leads to a *unique stable solution*. Here, a *unique solution* means there is only one solution that satisfies the PDE and its associated auxiliary conditions. A *stable solution* is a solution that does not depend critically on the choice of the auxiliary conditions, in the sense that a very small change in the auxiliary conditions produces a disproportionately large change in the solution. It is possible to formulate a very precise definition of stability, but the intuitive definition given here will suffice for our purposes.

To illustrate how unique and non-unique solutions can occur it is only necessary to consider the Laplace equation (2.28) in a rectangle. It is not difficult to see that if the Dirichlet condition $u = k = $ constant is imposed on the entire boundary, then $u = k$ satisfies both the Laplace equation and the Dirichlet condition on the boundary, and so $u = k$ is, indeed, the unique solution. If, however, a homogeneous Neumann condition is imposed on the boundary, then $u = k$ is still a solution of the Laplace equation, because it satisfies the homogeneous Neumann condition on the boundary, but the solution is not unique because the constant k may assume any value.

The commonly occurring types of second-order equation and the type of auxiliary conditions that are appropriate, together with the nature of the region (area) D to which the conditions apply, are listed below.

Type of PDE	Nature of the auxiliary conditions	Type of region D
Hyperbolic	Cauchy	Open
Parabolic	Dirichlet, Neumann or a mixture	Open
Elliptical	Dirichlet, Neumann or a mixture	Closed

A typical Cauchy problem for the wave equation (2.25) takes region D to be the upper-half plane $y > 0$, where y is a time-like independent variable, the x-axis is the boundary Γ of D, and u is required to satisfy the conditions $u(x,\ 0) = f(x)$ and $u_y(x,\ 0) = n(x)$ for $-\infty < x < \infty$, where $f(x)$ and $n(x)$ are given functions. In the case of the heat equation (2.27), a typical problem takes D to be the open rectangular (strip) $0 < x < L$, $y > 0$, and its boundary comprising the two sides of the strip and part of the x-axis to be Γ. The solution u is then required to satisfy a condition $u(x,\ 0) = f(x)$ for $0 < x < L$ on the base of the semi-infinite strip, while on the sides of the strip the conditions $u_x(0,\ y) = n_1(y)$ and $u_x(L,\ y) = n_2(y)$ must be satisfied for $y > 0$, where $f(x)$, $n_1(y)$ and $n_2(y)$ are given functions. In this example, Dirichlet conditions are specified on the base of the boundary $0 < x < L$, $y = 0$, and Neumann conditions on the sides $x = 0$, $y > 0$ and $x = L$, $y > 0$ of the semi-infinite strip. Finally, a typical problem for the Laplace equation could take region D to be the rectangle $0 < x < a$, $0 < y < b$, and its finite boundary Γ to be the boundary of D (the sides of the rectangle). Then Dirichlet conditions could be prescribed on three sides of D, and a Neumann condition on the fourth side.

In certain cases analytical solutions can be found for all three types of PDE by the *method of separation of variables*, which is described in detail in standard texts on partial differential equations (see Garabedian (1999), Jeffrey (2003), Lamb (1995), O'Neil (1995), (1999), Zauderer (2006)). However, in practical applications the equations and the shape of the regions involved are usually too complicated for an analytical solution to be found, so a numerical solution becomes necessary.

2.7 The influence of initial and boundary conditions on a solution: characteristics, domains of dependence and determinacy, and the d'Alembert solution

The auxiliary conditions applied to the boundary Γ of the region D through-out which a parabolic or elliptical equation is valid can be shown to influence the solution u at *every* point of region D. The situation is, however, very different when hyperbolic equations are involved, and to understand how initial conditions on the *initial line* Γ influence the solution u in D it will suffice if we consider the following standard problem for the wave equation.

Identify y with the time t, let x be a space variable, and consider the wave equation in the slightly more general form

$$\frac{\partial^2 u}{\partial t^2} = c^2 \frac{\partial^2 u}{\partial x^2}, \tag{2.38}$$

where $c > 0$ is a constant. It is easily checked by direct substitution that the general solution of equation (2.38) may be written as

$$u(x, t) = f(x - ct) + g(x + ct), \tag{2.39}$$

where f and g are any two arbitrary twice differentiable functions of their respective arguments $x - ct$ and $x + ct$. Inspection of equation (2.39) shows that the function f is constant along any straight line $x - ct = \xi = \text{constant}$, while the function g is constant along any straight line $x + ct = \eta = \text{constant}$. Now let $t > 0$, and consider a region D in which the initial line Γ is taken to be the x-axis on which, of course, $t = 0$. Recalling that x is a distance and t is a time, and the expressions x and ct occurring in equation (2.39) must both have the same dimensions, it follows that the dimensions of c must be those of a speed – the speed with which a disturbance (a wave) is propagated.

If the function f is specified, setting $t = 0$ is equivalent to determining the initial profile (shape) of this disturbance or wave as a function of x. As $f(\xi) = \text{constant}$ along any line $x - ct = \xi = \text{constant}$, it follows that the initial profile of f will be transported to the *right* with speed c without change of shape or attenuation, because the slope of each of the parallel lines $x - ct = \xi = \text{constant}$ is simply c. In similar fashion, if the function g is specified, it follows that the initial profile of g is another wave, which this time will be transported to the *left* with speed c, again without change of shape or attenuation. This result is confirmed by the fact that the wave equation has neither dissipation nor dispersion. Thus, the interpretation of the solution, equation (2.39), is that, once the initial profiles of f and g have been specified, the solution at position x and any time $t_1 > 0$ is given by the linear superposition

$$u(x, t_1) = f(x - ct_1) + g(x + ct_1),\tag{2.40}$$

of the two translated profiles, or waves. It is because of this property that equation (2.38) is called the *wave equation*. The straight lines $\xi = x - ct$ and $\eta = x + ct$ are called the *characteristics curves* of the wave equation (equation 2.38), despite the fact that, in this case, the characteristic 'curves' are *parallel* straight lines. The additive property of solutions comprising waves propagated in opposite directions that is exhibited in equation (2.40) is a direct consequence of the linearity of the wave equation.

To discover another fundamental property of the wave equation, let us find the solution corresponding to the Cauchy conditions (pure initial conditions)

$$u(x, 0) = p(x) \quad \text{and} \quad \frac{\partial u}{\partial t}(x, 0) = q(x),\tag{2.41}$$

where $p(x)$ and $q(x)$ are given functions, and the solution is required for all x and $t > 0$, so that in this case region D in which wave propagation takes place is the half-plane $t > 0$.

Setting $t = 0$ in equation (2.39) gives

$$f(x) + g(x) = p(x),\tag{2.42}$$

and, after partial differentiation of equation (2.39) with respect to t, followed by setting $t = 0$, we have

$$-cf'(x) + cg'(x) = q(x).\tag{2.43}$$

Integration of equation (2.43) from an arbitrary point a to x gives

$$-f(x) + g(x) = \frac{1}{c} \int_a^x q(s)\mathrm{d}s + g(a) - f(a),\tag{2.44}$$

where, to avoid confusion with the upper limit x, the symbol s has been used as a dummy variable of integration.

Combining equations (2.42) and (2.44), and using equation (2.39), leads to the two results

$$f(x) = \frac{1}{2}p(x) - \frac{1}{2c} \int_a^x q(s)\mathrm{d}s - \frac{1}{2}(g(a) - f(a)),\tag{2.45}$$

and

$$g(x) = \frac{1}{2}p(x) + \frac{1}{2c}\int_a^x q(s)\mathrm{d}s + \frac{1}{2}\left(g(a) - f(a)\right). \tag{2.46}$$

Replacing x by $x - ct$ in equation (2.45), and by $x + ct$ in equation (2.46), addition of the results leads to a solution in the form

$$u(x, t) = \frac{1}{2}\left[p(x - ct) + p(x + ct) - \frac{1}{c}\int_a^{x-ct} q(s)\mathrm{d}s + \frac{1}{c}\int_a^{x+ct} q(s)\mathrm{d}s \right].$$

This result can be simplified by using the negative sign in the third term on the right to reverse the order of the limits in that integral, and compensating by replacing the negative sign by a positive sign, after which the two integrals can be added to yield the fundamental result

$$u(x, t) = \frac{p(x - ct) + p(x + ct)}{2} + \frac{1}{2c}\int_{x-ct}^{x+ct} q(s)\mathrm{d}s. \tag{2.47}$$

This is called the *d'Alembert formula* for the solution of the Cauchy problem (equation 2.41) for the wave equation (equation 2.38). This result has been derived here because of the insight it gives into the nature of the solution of the wave equation, which in many respects is a typical hyperbolic equation. The result is illustrated in Figure 2.2(a), where AP and BP are the two characteristics through P, one with slope c and the other with slope $-c$. Equation (2.47) shows how the initial conditions influence the solution at a typical point (x_0, t_0) in region D. If P is the point (x_0, y_0), the d'Alembert solution (equation 2.47) shows that the solution at P is influenced only by the Dirichlet condition at point A located at $(x_0 - ct_0, 0)$ and at point B located at $(x_0 + ct_0, 0)$ at opposite ends of the interval $x_0 - ct_0 \leq x \leq x_0 + ct_0$ on the *initial line*. However, the solution at P is influenced by the integral of q over the entire interval AB.

For obvious reasons, in Figure 2.2(a) the interval AB on the initial line is called the *domain of dependence* of the solution at P, while the area APB is called the *domain of determinacy* of the solution, because it represents all points in the (x, t) plane at which the solution is determined by the initial data on AB. The value of the initial condition at point Q in Figure 2.2(b) will influence (although not completely determine) the value of the solution in the region $t > 0$ that lies between the two straight-line characteristics through Q, so this is called the *region of influence* of Q. When x is a space variable and t is the time, this result also shows why the wave equation (equation 2.38) describes the propagation of a wave in space with a *finite* speed c. This is because, at any given time t_0, a point on a disturbance located at x_0 on the initial line has only travelled as far as $x_0 - ct_0$ to the

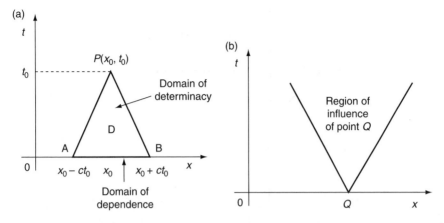

Figure 2.2 (a) The domains of dependence and determinacy of point P. (b) The region of influence of point Q

left and as far as $x_0 + ct_0$ to the right on that line. Later, an understanding of the domain of dependence will turn out to be of fundamental importance when a numerical solution of a hyperbolic or parabolic equation is required.

It is helpful to illustrate the geometrical significance of the superposition of solutions in the d'Alembert formula (equation 2.47). Accordingly, we will consider the wave equation (equation 2.38) subject to artificial and rather simple initial conditions. Specifically, we will set $u(x, 0) = F(x)$ and $u_t(x, 0) = 0$ and, to make the resulting development of the initial waveform easily identifiable geometrically as t increases, the initial waveform is localized by defining $F(x)$ as

$$F(x) = \begin{cases} 0, & x < -1 \\ -1 - x, & -1 \leq x < 0 \\ 1 - x, & 0 \leq x < 1 \\ 0, & x \geq 1. \end{cases}$$

For a time scale it is convenient to use multiples of the wave speed c. The result at $t = 0$, illustrated in Figure 2.3(a), shows as dashed lines the wave f that will move to the right with speed c, and the wave g that will move to the left with speed c, each in their initial position. The initial waveform $F(x)$, shown as the solid line, is then the average of the two dashed waveforms. The result at time $t = 1/(2c)$, illustrated in Figure 2.3(b), shows the translated waves represented by dashed lines, and the actual waveform given by their sum as a solid line, with the interaction between the two waves

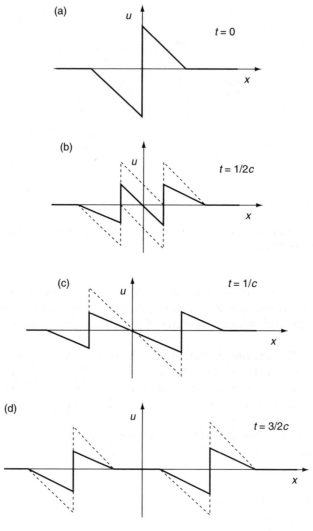

Figure 2.3 (a) $t=0$, (b) $t=1/(2c)$, (c) $t=1/c$ and (d) $t=3/2c$

restricted to the interval $-1/2 < x < 1/2$. The corresponding situation at time $t=1/c$ is shown in Figure 2.3(c), where the two translated waveforms, now shown as solid lines, are seen to have just separated, while at time $t=3/(2c)$ Figure 2.3(d) shows the situation when the two translated waves are well separated, each travelling in opposite directions, and having ceased to interact.

2.8 The method of characteristics and a non-linear first-order equation

To discover how an analytical solution can be obtained for the quasilinear equation

$$u_t + f(u)ux = h(x, t, u), \tag{2.48}$$

(a particular case of equation (2.35)) subject to the continuous initial condition

$$u(x, 0) = g(x) \tag{2.49}$$

we proceed as follows: the rate of change of a function $u(x, t)$ with respect to t is

$$\frac{du}{dt} = \frac{\partial u}{\partial t} + \frac{\partial u}{\partial x}\frac{dx}{dt}, \tag{2.50}$$

so if u is constrained to lie on a curve C in the (x, t) plane, then at a point P on C it follows directly that dx/dt is the gradient of curve C at P. Comparison of equations (2.48) and (2.50) shows that the quasilinear equation (2.48) can be interpreted as the solution of the ODE

$$\frac{du}{dt} = h(x, t, u) \tag{2.51}$$

along any member of the family of curves C defined by the solution of the ODE

$$\frac{dx}{dt} = f(u), \tag{2.52}$$

where the solution u must satisfy the initial condition in equation (2.49) on the initial line $t = 0$. This representation of PDE equation (2.48) as two coupled ODEs is said to be its *characteristic form*, and the family of curves C defined by equation (2.52) determines what are then called the *characteristic curves* of equation (2.48), usually abbreviated to the *characteristics* of the PDE. The fact that the curves C are real, and dx/dt has the dimensions of velocity, justifies the classification of equation (2.48) as *hyperbolic*. In general, ODEs (2.47) and (2.48), subject to the initial condition (2.49), are sufficiently complicated that they have to be solved using a numerical method of solution, which when generalized to hyperbolic systems of equations is called the *method of characteristics*.

2.9 Discontinuous solutions and conservation laws

Examination of a simple case of equation (2.48) will be useful because, although it is a special case of a hyperbolic equation, the fact that it propagates a wave can be used to demonstrate a fundamental property possessed by all quasilinear hyperbolic equations and systems. This property is that, unlike linear systems, such equations *may* allow a solution that starts as a smooth and differentiable solution to evolve to the point where it becomes a *discontinuous solution*. So, even though a solution may start in a completely smooth manner, it is possible for it to develop to the point where an abrupt jump occurs in the solution u itself.

Consider the case where the function f depends only on u, while the function $h \equiv 0$, and apply the method of characteristics to the equation

$$ut + f(u)ux = 0, \tag{2.53}$$

subject to the initial condition (equation 2.49) where $u(x,\ 0) = g(x)$, where $g(x)$ is continuous and smooth.

Equations (2.51) and (2.52) now reduce to

$$\frac{du}{dt} = 0 \text{ along the characteristic curves } C \text{ defined by} \frac{dx}{dt} = f(u), \tag{2.54}$$

with u satisfying the initial condition $u(x,\ 0) = g(x)$ on the initial line $t = 0$. Inspection of the first of these equations shows that $u = \text{constant}$ on a characteristic C, while the second equation shows that because $u = \text{constant}$ the corresponding characteristic curve C must be a straight line. As the slope dx/dt of C depends on u, the straight-line characteristics originating from points on the initial line will, in general, each have a different slope. On the straight-line characteristic C_ξ through a point $(\xi,\ 0)$ on the initial line the constant solution is equal to $g(\xi)$, so on C_ξ we have $dx/dt = f(g(\xi))$, and so the equation of this straight-line characteristic C_ξ is given by

$$x = \xi + f(g(\xi))t \quad \text{for} \quad t > 0. \tag{2.55}$$

It is because equation (2.53) describes wave propagation that it is classified as being hyperbolic, although there are other good reasons why this is so. As the result will be needed in Chapter 3 when considering the numerical solution of hyperbolic equations, we draw attention to the fact that equation (2.48) has a domain of dependence, and in this case it is the single point $(\xi, 0)$ on the initial line from which the characteristic C_ξ originates, while the domain of determinacy is just the characteristic C_ξ itself.

Two completely different situations now arise, according to whether the characteristics emanating from all points on a segment AB of the initial line *diverge* in the (x, t)-plane, or whether they *converge*. The characteristic equations show that this behaviour of the characteristics depends only on the specification of the continuous initial condition $g(x)$ over interval AB, independently of whether or not $g(x)$ is continuously differentiable.

These two situations are illustrated in Figure 2.4(a,b), where examination of Figure 2.4(a) shows that, because $g(x)$ is such that the characteristics diverge, the solution is defined uniquely throughout the region D bounded by the characteristics originating from its end points A and B.

The situation illustrated in Figure 2.4(b) is different, because there the characteristics *converge* and so will intersect after a finite time. Where characteristics intersect the solution cannot be unique, because a different value of u is transported along each characteristic, showing that at some stage a continuous initial condition must evolve to the point where it becomes discontinuous. Figure 2.4(b) illustrates a typical situation where the characteristics form an envelope with its cusp at the point (x_c, t_c) in the (x, t) plane, corresponding to the place and time at which a discontinuous

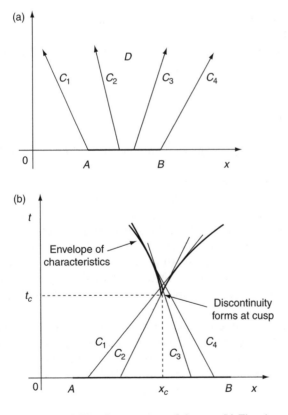

Figure 2.4 (a) The characteristics C diverge. (b) The characteristics C converge

solution first forms. Thus, for the time interval $0 \leq t < t_c$, because the characteristics have not yet intersected, the solution of the initial-value problem that gives rise to this envelope of characteristics is well defined and unique, but for times $t > t_c$ the intersection of characteristics means that a discontinuous solution has been formed, so the unique differentiable solution that exists prior to $t = t_c$ cannot be extended beyond the time $t = t_c$.

An analytical way of demonstrating this last result is by eliminating the parameter ξ between the initial condition $u = g(\xi)$ and equation (2.52), leading to the implicit solution for u given by

$$u = g(x - tf(u)). \tag{2.56}$$

It is a well-known result from the calculus that, when an implicit equation has a solution, the solution is not necessarily unique. So, in the case of Figure 2.4(a) the implicit equation (2.51) has a unique solution, whereas in the case of Figure 2.4(b), at some time $t = T$ the continuous initial solution must evolve to the point where it becomes non-unique due to the formation of a discontinuous solution. In the case of the shallow water equations where this phenomenon can also occur, this non-uniqueness can be shown to correspond to the development of an *hydraulic jump*.

Further evidence that a jump solution can occur can be seen by differentiation of equation (2.56) partially with respect to x to obtain

$$\frac{\partial u}{\partial x} = \frac{g'(x - tf(u))}{1 + tg'(x - tf(u))f'(u)}, \tag{2.57}$$

from which it follows that u_x becomes infinite whenever t is such that the denominator $1 + tg'(x - tf(u))f'(u) = 0$.

Unlike elliptical and parabolic equations, hyperbolic equations can describe the propagation of jump discontinuities and, as already remarked, in the case of the shallow-water equations these discontinuous solutions represent idealized hydraulic jumps. In reality, dissipative effects convert a discontinuous jump into a continuous but very localized solution that changes rapidly from a constant value u_- to the immediate left of a narrow transition region to a constant value u_+ to the immediate right. To illustrate the basic idea of a discontinuous-jump solution it is necessary to introduce the idea of a *conservation law*.

Let $u(\mathbf{r}, t)$ be the density of a physical material Q of interest at a point with position vector $\mathbf{r}$ in space at time t. Typically, such a physical material Q could be a compressible fluid, in which case $u(\mathbf{r}, t)$ could be its density. Then, if M is the amount of material Q present in a volume V at time t

$$M = \int_V u(\mathbf{r}, t) \mathrm{d}V,$$

so the rate of change of the amount of material Q with respect to time is

$$\frac{dM}{dt} = \frac{d}{dt} \int_V u(\mathbf{r}, t) dV. \tag{2.58}$$

Now let $\mathbf{F}$ be a vector function depending on $u(\mathbf{r}, t)$ that is defined at each point of V. Then, from elementary vector calculus, the flow dq of $\mathbf{F}$ through an element of area dA of the surface S that bounds V at position $\mathbf{r}$ and time t is $dq = \mathbf{F}.\mathbf{n} dA$, where $\mathbf{n}$ is the outward-drawn normal to S at position $\mathbf{r}$, and the dot signifies a vector scalar product. Defining the vector element of area on the surface S as $d\mathbf{S} = \mathbf{n} dA$, the amount dq of material Q *leaving* volume V through the vector element of area $d\mathbf{S}$ is $dq = \mathbf{F} \cdot d\mathbf{S}$, and so after integration over the surface S this becomes

$$q = \int_S \mathbf{F} \cdot d\mathbf{S}. \tag{2.59}$$

Combining equations (2.58) and (2.59), and assuming there is no mechanism by which material Q can be added to or removed from V, we arrive at the result

$$\frac{d}{dt} \int_V u(\mathbf{r}, t) dV = -\int_S \mathbf{F} \cdot d\mathbf{S}, \tag{2.60}$$

where the negative sign is necessary because material Q is *leaving* the surface S.

Taking the differentiation with respect to t under the integral sign on the left of equation (2.60) and applying the Gauss divergence theorem to the term on the right (see Jeffrey (2002), Kreyszig (2005), O'Neil (1995)) permits this conservation law to be written as the single integral

$$\int_V \left(u_t + \text{div}\,[\mathbf{F}(u)] \right) dV = 0. \tag{2.61}$$

This balance law, which relates the rate of change in the amount of material Q inside V to its loss due to outflow from V, is called an *integral conservation law* for the material Q.

If the quantity under the integral sign is continuous and differentiable then, because V is arbitrary, the result can only be true if the integrand vanishes, in which case

$$u_t + \text{div}\,[\mathbf{F}(u)] = 0, \tag{2.62}$$

and this is the *differential equation form* of the integral conservation law in equation (2.61).

Let us now consider the one-dimensional case in the (x, t) plane, when $\operatorname{div}[F(u)] = (F(u))_x = f(u)u_x$, with $f(u) = dF/du$. This causes equation (2.62) to reduce to

$$u_t + f(u)u_x = 0, \tag{2.63}$$

which is simply equation (2.53), showing that it is, in fact, a differential form of a conservation law.

If $u(x, t)$ is not continuous, the result of equation (2.63) is no longer true, and to examine the consequence of this we must work with the integral form of the conservation law. This becomes necessary because, although the derivative of a discontinuous function is not defined, the definite integral of a discontinuous function is always well defined. So we must now work with the one-dimensional form of equation (2.60), which becomes

$$\frac{d}{dt} \int_a^b u(x, t)dx + \int_a^b [F(u)]_x dx = 0, \tag{2.64}$$

where the interval $a \le x \le b$ is arbitrary. When $F(u)$ is continuous and differentiable, it follows from the fundamental theorem of calculus that

$$\frac{d}{dt} \int_a^b u(x, t)dx + F[u(b, t)] - F[u(a, t)] = 0.$$

Now let us suppose the interval contains a moving point $x = s(t)$ across which u and hence $F(u)$ is discontinuous. The result of equation (2.64) then becomes

$$\frac{d}{dt} \int_a^{s(t)-} udx + \frac{d}{dt} \int_{s(t)+}^b udx + F[u(b, t)] - F[u(a, t)] = 0,$$

where $s(t)_-$ and $s(t)_+$ represent the left and right sides of the discontinuity at $s(t)$. We now apply the Leibniz theorem from the calculus for differentiation under an integral sign to this last result, where the theorem takes the form

$$\frac{d}{dt} \int_a^{s(t)-} u_t dx + \frac{ds}{dt} u(s(t)_-, t) + \int_{s(t)+}^b u_t dx - \frac{ds}{dt} u(s(t)_+, t)) + F(u(s(t)_-)$$
$$-F(u(s(t)_+).$$

After letting $a \to s(t)_-$ and $b \to s(t)_+$, and using the result from the previous integral, the following result is found to be valid across the discontinuity in u:

$$\frac{ds}{dt}(u_- - u_+) = F(u)_- - F(u)_+,$$

where $ds/dt = U$ is the speed of propagation of the discontinuity in u. This last result can be written more concisely as

$$U[[u]] = [[F(u)]], \tag{2.65}$$

where the double bracket $[[.]]$ signifies the jump in its argument across the discontinuity.

When this result is generalized and applied to the shallow-water equations, the discontinuity in u corresponds to a hydraulic jump, and U corresponds to the speed of propagation of the discontinuity. A corresponding result when applied to supercritical flow describes pressure surges (*waterhammer*).

2.10 The classification of quasilinear and semilinear systems, hyperbolic systems and characteristics

We now show how the classification of a quasilinear first-order system of equations can be defined, and why a system like the shallow-water equations (equations 2.36 and 2.37) that describe wave propagation is classified as hyperbolic. To classify a linear or quasilinear first-order system of PDEs it will be necessary to work with matrices. We will consider the quasilinear first-order system

$$\frac{\partial \mathbf{U}}{\partial t} + \mathbf{A}(\mathbf{U})\frac{\partial \mathbf{U}}{\partial x} + \mathbf{B}(\mathbf{U}) = 0, \tag{2.66}$$

where $\mathbf{U}$ is an n-element column vector with the continuous components $u_1(x, t)$, $u_2(x, t)$, $\ldots, u_n(x, t)$, $\mathbf{A}(\mathbf{U}) = \mathbf{A}[a_{ij}(\mathbf{U})]$ with $i, j = 1, 2, \ldots, n$ is an $n \times n$ matrix with continuous real elements a_{ij} that may depend on $\mathbf{U}$, and $\mathbf{B}(\mathbf{U})$ is an n-element column vector with the continuous elements $b_1(\mathbf{U}), b_2(\mathbf{U}), \ldots, b_n(\mathbf{U})$ that may or may not depend on $\mathbf{U}$. Here we use the notation $\partial \mathbf{U}/\partial t$ to denote the column matrix obtained by partial differentiation of each of the elements of $\mathbf{U}$ with respect to t, and $\partial \mathbf{U}/\partial x$ to denote the column matrix obtained by partial differentiation of each of the elements of $\mathbf{U}$ with respect to x. When $\mathbf{A}(\mathbf{U})$ is only a function of x and t, and $\mathbf{B}(\mathbf{U})$ depends linearly on $\mathbf{U}$, and possibly also on x and t, system (2.66) becomes *linear*, while if $\mathbf{A}(\mathbf{U})$ depends only on x and t, but $\mathbf{B}(\mathbf{U})$ depends non-linearly on $\mathbf{U}$, system (2.66) becomes *semilinear*.

Although the space variable x and the time t are natural variables to choose when deriving a system of equations that describe some physical phenomenon, they are not always the most convenient ones to use when seeking to understand the mathematical properties of such a system. Accordingly, as we wish to study the way in which a solution vector $\mathbf{U}$ evolves with time, we will leave the variable t essentially unchanged, but examine the effect of replacing the space variable x by a new *curvilinear coordinate* $\xi = \xi(x, t)$, the nature of which is to be determined. Thus, the starting point will be to change from the coordinates (x, t) to the new coordinates (ξ, t') where

$$\xi = \xi(x, t), \quad t' = t. \tag{2.67}$$

Later we will see that there is an important difference between the old and new coordinate systems that will be significant in what is to follow. Provided the transformation between points in the (x, t) plane and the (ξ, t')-plane is unique, an application of the chain rule shows that the differential operators $\partial/\partial t$ and $\partial/\partial x$ become

$$\frac{\partial}{\partial t} \equiv \frac{\partial \xi}{\partial t}\frac{\partial}{\partial \xi} + \frac{\partial t'}{\partial t}\frac{\partial}{\partial t'} \equiv \frac{\partial \xi}{\partial t}\frac{\partial}{\partial \xi} + \frac{\partial}{\partial t'}, \quad \text{and} \quad \frac{\partial}{\partial x} \equiv \frac{\partial \xi}{\partial x}\frac{\partial}{\partial \xi} + \frac{\partial t'}{\partial x}\frac{\partial}{\partial t'} \equiv \frac{\partial \xi}{\partial x}\frac{\partial}{\partial \xi}$$

where, of course, $\partial \xi/\partial t$ and $\partial \xi/\partial x$ are scalar quantities. It is here that the difference between the old and new coordinate systems becomes important. To appreciate this, note that $\mathbf{U}_x$ is the partial derivative of $\mathbf{U}$ with respect to x when $t =$ constant, so it is the partial derivative of $\mathbf{U}$ normal to the straight line $t =$ constant in the (x, t) plane, and $\mathbf{U}_t$ is the partial derivative of $\mathbf{U}$ along a line $x =$ constant. However, the partial derivative $\mathbf{U}_{t'}$ is the partial derivative of $\mathbf{U}$ along the *curved line* $\xi =$ constant, while the derivative $\mathbf{U}_\xi$ is the partial derivative of $\mathbf{U}$ normal to this curved line. When system (2.66) is transformed in this way it becomes

$$\frac{\partial \mathbf{U}}{\partial t'} + \left(\frac{\partial \xi}{\partial t}\mathbf{I} + \frac{\partial \xi}{\partial x}\mathbf{A}\right)\frac{\partial \mathbf{U}}{\partial \xi} + \mathbf{B} = 0 \, , \tag{2.68}$$

where $\mathbf{I}$ is the $n \times n$ unit matrix.

To proceed further, and to arrive at definitions of hyperbolicity, parabolicity and ellipticity that can be applied to quasilinear first-order systems of PDEs, it will be necessary to make use of the concept of the eigenvalues and eigenvectors of an $n \times n$ matrix $\mathbf{A}$. A reader who needs to refresh their ideas about these matters is referred to the Appendix to this chapter.

If we now consider equation (2.68) as an algebraic system for which the vector $\partial \mathbf{U}/\partial t'$ is specified on an initial line (an initial condition), it is clear that this information can only be used to determine $\partial \mathbf{U}/\partial \xi$ if the inverse of the matrix premultiplying $\partial \mathbf{U}/\partial \xi$ exists, because only then can $\partial \mathbf{U}/\partial \xi$ be found from $\partial \mathbf{U}/\partial t'$. For this to be true it is necessary that the determinant

$$\det \left| \frac{\partial \xi}{\partial t}\mathbf{I} + \frac{\partial \xi}{\partial x}\mathbf{A} \right|$$

must be non-singular (it must not vanish) in order to ensure that the matrix multiplier of $\partial \mathbf{U}/\partial \xi$ has an inverse. So far the choice of ξ has been arbitrary, but now suppose ξ is identified with a variable φ with the property that the determinant really does vanish, so that

$$\det \left| \frac{\partial \varphi}{\partial t} \mathbf{I} + \frac{\partial \varphi}{\partial x} \mathbf{A} \right| = 0. \tag{2.69}$$

This means that the matrix vector $\partial \mathbf{U} / \partial \varphi$ in equation (2.68) will become indeterminate across the curves $\varphi(x, t) = $ constant, while along these curves $\partial \varphi / \partial x \, dx + \partial \varphi / \partial t \, dt = 0$. Combining this result with equation (2.69) gives

$$\det \left| \mathbf{A} - \frac{dx}{dt} \mathbf{I} \right| = 0. \tag{2.70}$$

Setting $dx/dt = \lambda$, equation (2.70) becomes

$$\det |\mathbf{A} - \lambda \mathbf{I}| = 0. \tag{2.71}$$

However, the dimensions of λ are those of a speed, and equation (2.71) is the condition that determines the *eigenvalues* $\lambda_1, \lambda_2, \ldots, \lambda_n$ of matrix $\mathbf{A}(\mathbf{U})$ (see the Appendix to this chapter). So, if $\mathbf{A}$ is such that these eigenvalues are all real, this equation will determine n speeds, called the *characteristic speeds* of the system. When this occurs, provided $\mathbf{A}$ has a corresponding *full set* of *right eigenvectors* $\mathbf{r}_1, \mathbf{r}_2, \ldots, \mathbf{r}_n$ satisfying the defining matrix equation

$$[\mathbf{A} - \lambda_i \mathbf{I}] \mathbf{r}_i = 0, \quad i = 1, 2, \ldots, n, \tag{2.72}$$

equation (2.66) will be classified as *hyperbolic*. In this case there will be n families of real *characteristic curves* $C_i, i = 1, 2, \ldots, n$ determined by integration of the n equations

$$C_i : \frac{dx}{dt} = \lambda_i, \quad i = 1, 2, \ldots, n, \tag{2.73}$$

where it will be remembered that $\lambda = \lambda(\mathbf{U})$, so in general these characteristic curves can only be determined when $\mathbf{U}$ is known and $\lambda(\mathbf{U})$ is integrable. When $\mathbf{U}$ is not known, which is generally the case, the solution $\mathbf{U}$ and the characteristic curves must be found simultaneously by means of numerical methods.

We have seen that across each characteristic curve $\varphi = $ constant the vector $\partial \mathbf{U} / \partial \varphi$ *may* be discontinuous. So each family of characteristic curves is capable of transporting a discontinuity in the slope of $\mathbf{U}$ (but *not* a discontinuity in $\mathbf{U}$ itself), and we will call this type of disturbance a propagating *wavefront*. This situation is shown in Figure 2.5 for the solution surface S for the element u_j of $\mathbf{U}$, which at $t = 0$ satisfies an initial condition

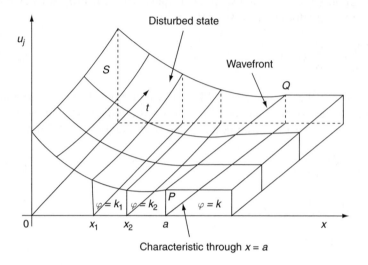

Figure 2.5 The propagation of an initial discontinuity in $\partial u_j / \partial \varphi$ with time

$u_j(x, 0) = \psi_j(x)$, while the characteristic curve $\varphi = k$ transports an initial discontinuity in the derivative $\partial u_j / \partial x$ that exists across the solution surface S along PQ as time t increases, forming a propagating wavefront. It is convenient to describe the characteristic through the point $(a, 0)$ in Figure 2.5 as the *wavefront trace*, because it is the projection of the wavefront PQ onto the (x, t) plane.

Let us now show that the shallow-water equations (equations 2.36 and 2.37) are hyperbolic, and to do this we need to write the equations in the form

$$\frac{\partial \mathbf{U}}{\partial t} + \mathbf{A}(\mathbf{U}) \frac{\partial \mathbf{U}}{\partial x} + \mathbf{B}(\mathbf{U}) = 0, \tag{2.74}$$

with

$$\mathbf{U} = \begin{bmatrix} u \\ \eta \end{bmatrix}, \quad \mathbf{A}(\mathbf{U}) = \begin{bmatrix} u & g \\ \eta + b & u \end{bmatrix}, \quad \text{and} \quad \mathbf{B}(\mathbf{U}) = \begin{bmatrix} 0 \\ ub_x \end{bmatrix}. \tag{2.75}$$

A routine calculation shows that the eigenvalues of $\mathbf{A}$ and its right eigenvectors that satisfy the equation $[\mathbf{A} - \lambda_i \mathbf{I}] \mathbf{r}_i = 0$ are

$$\lambda_1 = u - \sqrt{g(\eta + b)}, \quad \lambda_2 = u + \sqrt{g(\eta + b)}, \tag{2.76}$$

and

$$\mathbf{r}_1 = \left[\begin{array}{c} 1 \\ -\sqrt{(\eta+h)/g} \end{array} \right], \quad \mathbf{r}_2 = \left[\begin{array}{c} 1 \\ \sqrt{(\eta+h)/g} \end{array} \right]. \tag{2.77}$$

The eigenvalues are always real, and because they have the dimensions of speed they represent *wave speeds*. The right eigenvectors are linearly independent so, by our definition of hyperbolicity, this requires the system to have n real eigenvalues and a full set of n linearly independent eigenvectors, and the system is thus seen to be *unconditionally hyperbolic*. Thus, in the shallow-water equations the *speed* of a point on the surface of a surface wave is $\sqrt{g(\eta+h)}$. This result provides a simple explanation of why waves 'break' – because, depending on the change of depth, the crest of a smooth wave may advance faster than a trough, the waves steepen and finally break when the crest overtakes the trough.

The quasilinear system (equation 2.74) will be *elliptical* when all the eigenvectors of $\mathbf{A}$ are complex, so in this case, as the characteristics are not real curves, the method of characteristics does not apply. The *parabolic* case corresponds to the situation where the eigenvalues and eigenvectors are all real, but the structure of $\mathbf{A(U)}$ is such that one of its eigenvectors may be assigned arbitrarily. So, in the case of a parabolic system, there is a degeneracy in the families of characteristic curves. More complicated situations can also arise as, for example, in the case when $\mathbf{A(U)}$ has both real and complex eigenvalues, which corresponds to a *mixed hyperbolic–elliptical system*. The considerable difficulties that arise when seeking a solution to a system of this type will not be discussed here.

Any second-order PDE can always be written as a first-order system by introducing its first-order partial derivatives as new dependent variables. We now use this result to show that the classification described above is compatible with the classification of the wave equation based on the discriminant $\Delta = B^2 - AC$. Let us show the wave equation

$$\frac{\partial^2 u}{\partial t^2} = c^2 \frac{\partial^2 u}{\partial x^2} \tag{2.78}$$

is hyperbolic, while the Laplace equation

$$\frac{\partial^2 u}{\partial x^2} + \frac{\partial^2 u}{\partial y^2} = 0 \tag{2.79}$$

is elliptical.

Introducing the new variables $v = \partial u/\partial t$ and $w = \partial u/\partial x$, the wave equation (2.78) takes the form

$$\frac{\partial v}{\partial t} = c^2 \frac{\partial w}{\partial x}, \tag{2.80}$$

and to connect v and w we use the fact that, as the partial derivatives are continuous, there must be equality of the mixed partial derivatives, so that

$$\frac{\partial v}{\partial x} = \frac{\partial w}{\partial t}. \tag{2.81}$$

When this first-order system is written in matrix form it becomes

$$\frac{\partial \mathbf{U}}{\partial t} + \mathbf{A}\frac{\partial \mathbf{U}}{\partial x} = 0, \tag{2.82}$$

where

$$\mathbf{U} = \begin{bmatrix} v \\ w \end{bmatrix} \quad \text{and} \quad \mathbf{A} = \begin{bmatrix} 0 & -c^2 \\ -1 & 0 \end{bmatrix}, \tag{2.83}$$

from which the eigenvalues λ_1 and λ_2 and the corresponding right eigenvectors $\mathbf{r}_1$ and $\mathbf{r}_2$ of $\mathbf{A}$ are found to be

$$\lambda_1 = -c, \quad \lambda_2 = c, \quad \mathbf{r}_1 = \begin{bmatrix} c \\ 1 \end{bmatrix} \quad \text{and} \quad \mathbf{r}_2 = \begin{bmatrix} -c \\ 1 \end{bmatrix}. \tag{2.84}$$

The eigenvalues representing the wave speeds are $\pm c$, in agreement with the general solution $u(x, t) = f(x - ct) + g(x + ct)$, and the right eigenvectors are linearly independent, so by the criterion introduced for the classification of systems of first-order PDEs, the wave equation (2.78) is *unconditionally hyperbolic*.

To show that the Laplace equation is *elliptical*, we again introduce the new variables $v = \partial u/\partial x$ and $w = \partial u/\partial y$, when equation (2.75) becomes

$$\frac{\partial v}{\partial x} + \frac{\partial w}{\partial y} = 0, \tag{2.85}$$

and once again we connect u and v by the equality of mixed derivatives

$$\frac{\partial v}{\partial y} = \frac{\partial w}{\partial x}. \tag{2.86}$$

When written in matrix form, equations (2.85) and (2.86) become

$$\frac{\partial U}{\partial x} + A\frac{\partial U}{\partial y} = 0 \ , \tag{2.87}$$

where

$$U = \begin{bmatrix} v \\ w \end{bmatrix} \quad \text{and} \quad A = \begin{bmatrix} 0 & 1 \\ -1 & 0 \end{bmatrix}. \tag{2.88}$$

The eigenvalues of A are found to be the *complex conjugates* $\lambda_1 = -i$ and $\lambda_2 = i$, confirming that the Laplace equation is elliptical, as already determined by the discriminant test, because $\Delta = -1$ is negative.

2.11 A fundamental difference between elliptical and hyperbolic equations

This is a suitable place to demonstrate the fundamental difference between the solutions of hyperbolic and elliptical equations, and to do so it will be necessary to make use of two results that will only be quoted. The first is the *Poisson integral formula* for the solution of the Laplace equation in a circular disc of radius r_0 centred on the origin. This result asserts that, in terms of the plane polar coordinates (r, θ), the solution $u(r, \theta)$ of the two-dimensional Laplace equation inside a circle of radius r_0, on the boundary of which the continuous Dirichlet condition $u(r_0, \theta) = f(\theta)$ is imposed, where $f(\theta)$ is periodic with period 2π, is given by

$$u(r, \theta) = \frac{1}{2\pi} \int_0^{2\pi} \frac{(r_0^2 - r^2)f(\psi)\mathrm{d}\psi}{r_0^2 - 2rr_0\cos(\psi - \theta) + r^2}, \tag{2.89}$$

where ψ is a dummy variable of integration (introduced to avoid confusion with θ).

The second result is taken from complex analysis where it is shown that a special type of transformation, called a *conformal transformation*, can always be found. This transforms any simple region D, in which u is the solution of the Laplace equation subject to Dirichlet conditions on its boundary, onto the interior of a circle of radius r_0 centred on the origin. The transformed solution thus obtained is again a solution of the Laplace equation subject to the same Dirichlet conditions. Note that the transformation is such that the boundary conditions for the solution in region D are the same as those for the solution inside the circle. (See, for example, Brown and Churchill (2007).) This means that the fundamental properties of the solutions of the Laplace equation are mirrored by those of the solution described by equation (2.89). The property to be stressed here can be seen immediately from equation (2.89), because the Dirichlet boundary condition enters into the numerator of the integral in equation (2.89), so the boundary condition over any arc of the perimeter of the circle, or indeed over the complete perimeter, will influence the solution of the Laplace equation at *every* point inside the circle. Contrast this with the wave equation, where the solution at any point P only depends on the initial conditions over a finite segment of the initial line that forms the domain of dependence for that point. So, in the hyperbolic case, changes in the initial conditions outside the domain of dependence of point P will not affect the solution at P, or at points in the *range of influence* of P.

A qualitative property of elliptical equations that is often useful is that, if u is a continuous solution of the Laplace equation in some finite region D with boundary Γ, then provided u is not constant (when the result is trivial) the maximum and minimum values of u must occur on the boundary Γ. This result can be found in Garabedian (1999), Zauderer (2006) and elsewhere.

2.12 The derivation of a mathematical model involving partial differential equations – the shallow-water equations

To give an example of how a mathematical model can be constructed, we now derive the shallow-water equations in one space dimension and time. This quasilinear hyperbolic system of equations provides a simple description of the way surface waves on water behave in rivers and when approaching beaches. First, we introduce some definitions (see also Chapter 1). A *mathematical model* of a physical situation is a set of equations representing physical laws that can be considered to describe a real situation. In the construction of a mathematical model it is usually necessary to make certain approximations, and in the case of surface waves on water the approximations may involve ignoring viscosity and assuming the length of a surface wave is large relative to the depth of the water. A *numerical model* is an approximation of a mathematical model based on a grid of points in the (x, t) plane, at each point of which derivatives and partial derivatives in the mathematical model are replaced by finite-difference equations. The numerical model obtained in this way is then a set of algebraic equations that connect the solution at each of the discrete points of the grid to the initial and boundary conditions. A *computational model* is an application of a numerical model to a specific situation in which numerical values for the unknown functions are obtained at each of the grid points. We now illustrate one of the ways in which a mathematical model can be derived from the fundamental equations that govern a physical situation.

For our illustration we choose the derivation of the shallow-water equations (equations 2.36 and 2.37), starting from the basic fluid-mechanics equations and using reasoning similar to that used by Lamb (1993) in the reprint of his classical account of hydrodynamics, and repeated in the reprint of the book by Stoker (1992). Here $u(x, t)$ and $v(x, t)$ are the respective horizontal and vertical components of the water velocity, $y = -h(x)$ is the equation of the variable depth of the river or sea bed relative to the horizontal x-axis and the vertical axis y, with the x-axis taken to lie in the equilibrium level of the water, while $y = \eta(x, t)$ is the vertical displacement of the free surface. For convenience, this situation shown first in Figure 2.1 is repeated in Figure 2.6.

The fundamental equations involved are:

the *equation of continuity*

$$u_x + v_y = 0, \tag{2.90}$$

and

the *free surface kinematical condition*

$$(\eta_t + u\eta_x - v)\big|_{y=\eta} = 0. \tag{2.91}$$

The *free surface dynamical condition* on the pressure p is

$$p\big|_{y=\eta} = 0, \tag{2.92}$$

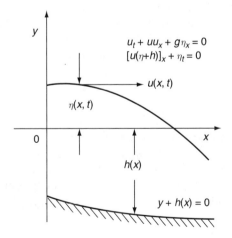

Figure 2.6 The geometry of the shallow-water model

while the *condition at the river or sea bed* is

$$(u h_x + v)\big|_{y=-h} = 0. \tag{2.93}$$

Integration of the continuity condition from the bed to the free surface with respect to y gives

$$\int_{-h}^{\eta} u_x dy + v\big|_{-h}^{\eta} = 0. \tag{2.94}$$

Using the conditions at the free surface $y = \eta$, and at the bed $y = -h$, this becomes

$$\int_{-h}^{\mu} u_x dy + \eta_t + u\big|_\eta \, \eta_x + u\big|_{-h} \, h_x = 0. \tag{2.95}$$

After using the Leibniz theorem for differentiation under the integral sign

$$\frac{\partial}{\partial x} \int_{-h(x)}^{\eta(x)} u dy = u\big|_{y=\eta} \, \eta_x + u\big|_{y=-h} \, h_x + \int_{-h}^{\eta} u_x dy,$$

equation (2.95) takes the simple form

$$\frac{\partial}{\partial x} \int_{-h(x)}^{\eta(x)} u dy = -\eta_t. \tag{2.96}$$

So far the only approximation to have been made, apart from ignoring viscosity, has been to restrict the analysis to one-dimensional flow.

Because the water is shallow, the shallow-water equations assume that the vertical component of acceleration of the water can be neglected, and so it will have a negligible effect on the pressure p, which from hydrostatic reasoning is given by

$$p = g\rho(\eta - y), \tag{2.97}$$

where g is the acceleration due to gravity and ρ is the density of water. Thus, $p_x = g\rho\eta_x$, showing that p_x is *independent* of y. Consequently, the x-component of the acceleration of the water velocity must also be independent of y, with the result that the x-component of the water velocity is also independent of y. Under these conditions the equations governing the flow simplify to

$$u_t + u u_x + g\eta_x = 0, \tag{2.98}$$

which is just the usual Euler form of the equation of motion, and to the equation

$$[u(\eta + h)]_x + \eta_t = 0, \tag{2.99}$$

already quoted in equation (2.37), where the result $\int_{-h}^{\eta} u\,dy = u \int_{-h}^{\eta} dy$ has been used, which is permissible because in the shallow-water approximation u is independent of y.

If the further approximation is made that u and η are small, so that their products and squares can be neglected relative to the linear terms, elimination of η leads to the result

$$[uh]_{xx} = \frac{1}{g} u_{tt}.$$

If now the depth h of the water is constant, we arrive at the result $u_{tt} = ghu_{xx}$, and by setting $c^2 = gh$, where c is the *surface wave speed*, this becomes the familiar wave equation

$$u_{tt} = c^2 u_{xx}. \tag{2.100}$$

A numerical model for the shallow-water equations follows when equations (2.98) and (2.99) are replaced by some form of discrete numerical approximation, such as, for example, when the finite-difference method (described later) is used. A variety of applications of this model to different physical situations can be found in Mader (2004) and Abbott and Minns (1998).

The conditions under which the shallow-water equations have been derived are such that they can be used in hydraulics to study open-channel flows, roll waves, and surges in channels due to a sudden influx of water. In general, the validation of a mathematical model requires a careful examination of the approximations that have been made, followed by a comparison of the numerical results obtained from the equations with experimental data obtained under conditions for which the above approximations are valid. For applications of hydraulics to flood routing, see Price (1973), (1975), (1978), (1985); see also Anderson (1995) and Bürgisson (1999).

References

Abbott, M. B. (1979), *Computational Hydraulics, Elements of the Theory of Free Surface Flows*, Pitman, London.

Abbott, M. B. and Cunge, J. A. (eds) (1982), *Engineering Applications of Computational Hydraulics, Vol. 1: Homage to Alexandre Preissmann*, Pitman, London.

Abbott, M. B. and Minns, A. W. (1998), *Computational Hydraulics*, 2nd Edn, Ashgate, Aldershot.

Anderson, J. D. (1995), *Computational Fluid Dynamics*, McGraw-Hill, New York.

Birkhoff, G. and Rota, G. C. (1989), *Ordinary Differential Equations*, 4th Edn, John Wiley, New York.

Boyce, W. E. and DiPrima, R. C. (2005), *Elements of Differential Equations and Boundary Value Problems*, 5th Edn, John Wiley, New York.

Brown, J. W. and Churchill, R. V. (2007), *Fourier Series and Boundary Value Problems*, 5th Edn, McGraw-Hill, New York.

Bürgisser, M. (1999), *Numerische Simulation der freien Wasseroberfläche bei Ingenieurbauten Mitteilungen* No. 162, Versachsanstalt für Wasserbau, ETH, Zürich.

Chadwick, A., Morfett, J. and Borthwick, M. (2004), Hydraulics in civil environmental engineering. In *Computational Hydraulics*, 4th Edn, Chapter 14, Spon Press, London.

Cunge, J. A., Holly, F. M. Jr and Verwey, A. (1980), *Practical Aspects of Computational River Hydraulics*, Pitman, London.

Edwards, C. H. and Penney, D. E. (2001), *Differential Equations and Linear Algebra*, Prentice Hall, New Jersey.

Farlow, J., Hall, J. E., McDill, J. M. and West, B. H. (2002), *Differential Equations and Linear Algebra*, Prentice Hall, New Jersey.

Garabedian, P. R. (1999 reprint), *Partial Differential Equations*, American Mathematical Society Reprint, Rhode Island.

Holly, F. M. Jr (1985), Dispersion in rivers and coastal waters – 1, Physical principles and dispersion equations. In *Developments in Hydraulic Engineering*, Vol. 3 (ed. P. Novak), Elsevier Applied Science, London.

Jeffrey, A. (2002), *Advanced Engineering Mathematics*, Academic Press, New York.

Jeffrey, A. (2003), *Applied Partial Differential Equations: An Introduction*, Academic Press, New York.

Kreyszig, E. (2005), *Advanced Engineering Mathematics*, 8th Edn, John Wiley, New York.

Krusemeyer, M. (1999), *Differential Equations*, Macmillan, New York.

Lamb, G. L. Jr (1995), *Introductory Applications of Partial Differential Equations*, John Wiley, New York.

Lamb, H. (1993 reprint), *Hydrodynamics*, Cambridge University Press, London.

Mader, C. L. (2004), *Numerical Modeling of Water Waves*, 2nd Edn, CRC Press, Boca Raton, FL.

MAPLE® The product of Waterloo Maple Inc., 57 Erb Street West, Waterloo, ON N2L 6C2, Canada.

MathCAD® The product of MathSoft Inc., One Kendall Square, Cambridge, MA 02139, USA.

MATLAB® The product of The Mathworks Inc., 24 Prime Park Way, Natrick, MA 01760-1500, USA.

O'Neil, P. V. (2006a), *Advanced Engineering Mathematics*, 4th Edn, PWS-Kent, Boston, MA.

O'Neil, P. V. (2006b), *Beginning Partial Differential Equations*, Springer-Verlag, New York.

Peterson, G. L. and Sochacki, J. S. (2002), *Linear Algebra and Differential Equations*, Addison Wesley, New York.

Price, R. K. (1973), Flood routing methods for British rivers. In *Proceedings, Institute of Civil Engineers (London)* Part 2, Paper 7674, pp. 913–930.

Price, R. K. (1975), *Flood Routing*, Vol. III of the *Flood Studies Report*, Natural Environment Research Council, London.

Price, R. K. (1978), A river catchment flood model. In *Proceedings, Institute of Civil Engineers (London)* Part 2, Paper 8141, pp. 655–668.

Price, R. K. (1985), Flood routing. In *Developments in Hydraulic Engineering*, Vol. 3 (ed. P. Novak), Elsevier Applied Science, London.

Sauvaget, P. (1985), Dispersion in rivers and coastal waters – 2, Numerical computation of dispersion. In *Developments in Hydraulic Engineering*, Vol. 3 (ed. P. Novak), Elsevier Applied Science, London.

Sen, Z. (1995), *Applied Hydrology for Scientists and Engineers*, Lewis, Boca Raton, FL.

Stoker, J. J. (1992 reprint), *Water Waves*, John Wiley Reprint, New York.

Verboom, G. K., Stelling, G. S. and Officer, M. J. (1982), Boundary conditions for the shallow water equations. In *Engineering Applications of Computational Hydraulics*, Vol. 1 (eds M. B. Abbott and J. A. Cunge), Chapter 11, Pitman, London.

Verwey, A. (1983), The role of computational hydraulics in the hydraulic design of structures. In *Developments in Hydraulic Engineering*, Vol. 1 (ed. P. Novak), Elsevier Applied Science, London.

Walton, W. C. (1991), *Principles of Groundwater Engineering*, Lewis, Boca Raton, FL.

Zauderer, E. (2006), *Partial Differential Equations of Applied Mathematics*, 2nd Edn, John Wiley, New York.

Zienkiewitz, O. C. and Taylor, R. L. (2000), *The Finite Element Method*, 5th Edn, Butterworth-Heinemann, London.

Appendix

Eigenvalues, eigenvectors, and an application of matrix diagonalization

This appendix reviews the related concepts of the eigenvalues and eigenvectors of $n \times n$ matrices $\mathbf{A}$, and describes their use when diagonalizing a matrix. The use of diagonalization mentioned in Section 2.1.1 in connection with solving linear systems of ODEs is also discussed.

The eigenvalues and eigenvectors of an $n \times n$ matrix $\mathbf{A}$ arise when finding the solution of the matrix equation

$$\mathbf{A}\mathbf{x} = \mathbf{b}, \tag{A2.1}$$

where $\mathbf{A}$ has real elements a_{ij}, $\mathbf{x}$ is a column vector with the n elements $x_1, x_2, \ldots, x_n$, and $\mathbf{b}$ is an n-column element vector that is *proportional* to the vector $\mathbf{x}$.

Denoting the scalar constant of proportionality between the vectors $\mathbf{x}$ and $\mathbf{b}$ by λ reduces this problem to finding a column vector $\mathbf{x}$ such that $\mathbf{b} = \lambda \mathbf{x}$, causing system (A2.1) to simplify to the matrix equation $\mathbf{A}\mathbf{x} = \lambda \mathbf{x}$.

When written out in full, the system $\mathbf{A}\mathbf{x} = \lambda \mathbf{x}$ becomes

$$\begin{aligned}
a_{11}x_1 + a_{12}x_2 + \cdots + a_{1n}x_n &= \lambda x_1 \\
a_{21}x_1 + a_{22}x_2 + \cdots + a_{2n}x_n &= \lambda x_2 \\
&\cdots \\
a_{n1}x_1 + a_{n2}x_2 + \cdots + a_{nn}x_n &= \lambda x_n.
\end{aligned} \tag{A2.2}$$

At first sight this appears to be a non-homogeneous system of algebraic equations, but in each equation the term on the right can be combined with a corresponding term in the expression on the left, leading to the following homogeneous system of algebraic equations, in which λ appears as a parameter

$$(a_{11} - \lambda)x_1 + a_{12}x_2 + \cdots + a_{1n}x_n = 0$$
$$a_{21}x_1 + (a_{22} - \lambda)x_2 + \cdots + a_{2n}x_n = 0$$
$$\cdots \tag{A2.3}$$
$$a_{n1}x_1 + a_{n2}x_2 + \cdots + (a_{nn} - \lambda)x_n = 0.$$

In matrix notation, after introducing the $n \times n$ unit identity matrix $\mathbf{I}$, equation (2.99) becomes

$$[\mathbf{A} - \lambda\mathbf{I}]\mathbf{x} = 0. \tag{A2.4}$$

An example of a problem that leads to a matrix equation of this type occurs when attempting to solve a linear first-order matrix differential equation of the form $d\mathbf{x}/dt = \mathbf{A}\mathbf{x}$, where $\mathbf{A}$ is an $n \times n$ constant matrix and $\mathbf{x}$ is an n-element column vector. Copying the type of solution expected for the scalar ODE $dx/dt = ax$, by choosing for a trial solution the expression $\mathbf{x} = \mathbf{x}_0 e^{\lambda t}$, where $\mathbf{x}_0$ is a constant column vector, substitution into the differential equation gives $\lambda\mathbf{x}_0 e^{\lambda t} = \mathbf{A}\mathbf{x}_0 e^{\lambda t}$. Cancelling the factor $e^{\lambda t}$ produces the matrix equation $[\mathbf{A} - \lambda\mathbf{I}]\mathbf{x}_0 = 0$, which is precisely of the form of equation (A2.4).

As the algebraic system (A2.4) is homogeneous, it is known from the study of matrices that it has two possible types of solution. The obvious solution is $\mathbf{x} = 0$, in which case $x_1 = x_2 = \ldots = x_n = 0$, and this is called the *trivial solution* because it is of no interest. The second solution is non-trivial, and it follows when a row of elements in matrix $\mathbf{A}$ is a sum of multiples of the elements in the other rows, so there is *linear dependence* between the rows of $\mathbf{A}$. When this occurs, not all the elements of vector $\mathbf{x}$ can vanish; this situation can only happen if the determinant of the matrix $\mathbf{A} - \lambda\mathbf{I}$ vanishes, because the vanishing of the determinant is the condition that there is linear dependence between the rows of $\mathbf{A}$. Thus, the condition for a non-trivial solution to become possible is that

$$\det[\mathbf{A} - \lambda\mathbf{I}] = 0. \tag{A2.5}$$

When the determinant in equation (A2.5) is expanded it will give rise to a polynomial in λ of degree n, called the *characteristic polynomial* associated with matrix $\mathbf{A}$, and this polynomial will vanish only when λ is any one of its n zeros. The polynomial equation in λ defined by equation (A2.5) is called the *characteristic equation* associated with matrix $\mathbf{A}$. The n roots $\lambda_1, \lambda_2, \ldots, \lambda_n$ of the characteristic polynomial are called the *eigenvalues* of $\mathbf{A}$. From equation (A2.4) it follows that to each eigenvalue λ_i of $\mathbf{A}$ there corresponds a column vector $\mathbf{x}^{(i)}$ such that

$$[\mathbf{A} - \lambda_i\mathbf{I}]\mathbf{x}^{(i)} = 0. \tag{A2.6}$$

The n vectors $\mathbf{x}^{(i)}$, with $i = 1, 2, \ldots, n$, are called the *eigenvectors* of $\mathbf{A}$ corresponding to the *eigenvalues* λ_i. In general, an $n \times n$ matrix $\mathbf{A}$ will have n different eigenvectors $\mathbf{x}^{(1)}, \mathbf{x}^{(2)}, \ldots, \mathbf{x}^{(n)}$, in the sense that no one eigenvector is proportional to any of the others.

It can happen that a matrix has an eigenvalue λ_j that is repeated r times, in which case $(\lambda - \lambda_j)^r$ is a factor of the characteristic equation. Such a repeated root of the characteristic equation is said to be an eigenvalue with *algebraic multiplicity* r. Our concern will be with the case when $\mathbf{A}$ has n distinct eigenvectors, even though some of the eigenvalues may be repeated. The more complicated situation that arises when $\mathbf{A}$ has fewer than n distinct eigenvectors will not be considered here, although it is of importance in the study of parabolic systems of PDEs, and elsewhere.

Expanding $\det[\mathbf{A} - \lambda\mathbf{I}]$, the eigenvalues λ_i are seen to be the roots of the polynomial of degree n in λ given by

$$\det[\mathbf{A} - \lambda\mathbf{I}] = \begin{vmatrix} a_{11} - \lambda & a_{12} & \cdots & a_{1n} \\ a_{21} & a_{22} - \lambda & \cdots & a_{2n} \\ \vdots & \vdots & \vdots & \vdots \\ a_{n1} & a_{n2} & \vdots & a_{nn} - \lambda \end{vmatrix} = 0, \tag{A2.7}$$

so $\det[\mathbf{A} - \lambda\mathbf{I}]$ can be factored and written as

$$\det(\mathbf{A} - \lambda\mathbf{I}) = (\lambda_1 - \lambda)(\lambda_2 - \lambda) \cdots (\lambda_n - \lambda), \tag{A2.8}$$

For conciseness when displaying eigenvectors in text, as here, the column eigenvectors $\mathbf{x}^{(i)}$ in equation (A2.6) will be written as $\mathbf{x}^{(i)} = [x_1^{(i)}, x_2^{(i)}, \cdots, x_n^{(i)}]^{\mathrm{T}}$, for $i = 1, 2, \ldots, n$. Here the superscript T signifies the *matrix transpose operation*, which when applied to a matrix switches its rows into columns and its columns into rows. When displayed in full, equation (A2.6) becomes

$$\begin{bmatrix} a_{11} - \lambda_i & a_{12} & \cdots & a_{1n} \\ a_{21} & a_{22} - \lambda_i & \cdots & a_{2n} \\ \vdots & \vdots & \vdots & \vdots \\ a_{n1} & a_{n2} & \cdots & a_{nn} - \lambda_i \end{bmatrix} \begin{bmatrix} x_1^{(i)} \\ x_2^{(i)} \\ \vdots \\ x_n^{(i)} \end{bmatrix} = \begin{bmatrix} 0 \\ 0 \\ \vdots \\ 0 \end{bmatrix}, \quad i = 1, 2, \ldots, n.$$

$$\tag{A2.9}$$

The algebraic homogeneity of equations (A2.9) (the vector on the right is 0) means the values of the n quantities $x_1^{(i)}, x_2^{(i)}, \ldots, x_n^{(i)}$ cannot be determined uniquely, because one of the equations in (A2.9) must be linearly dependent on the others. This has the result that $n - 1$ of the elements of an eigenvector can only be expressed in terms of multiples of the remaining element, say $x_r^{(i)}$, the value of which may be assigned arbitrarily. So equation (A2.9) only determines the *ratios* of the elements of the eigenvector $\mathbf{x}^{(i)}$ relative to an arbitrary value for $x_r^{(i)}$ as a parameter. This has the important consequence that once an eigenvector has been found it can be multiplied by an arbitrary constant $k \neq 0$ (scaled by k) and still remain an eigenvector.

Finding the characteristic polynomial and its roots (the eigenvalues) of an $n \times n$ matrix when n is large requires the use of a computer and numerical methods (see Press *et al.* (2007) in Chapter 3).

To illustrate matters, in the example that follows, the 3×3 matrix

$$\mathbf{A} = \begin{bmatrix} 1 & 0 & -1 \\ -2 & -1 & 2 \\ -1 & 2 & 1 \end{bmatrix}. \tag{A2.10}$$

has been constructed so that its characteristic polynomial is easily obtained and its eigenvalues (the roots of the characteristic equation) can be found by inspection. When the characteristic determinant is expanded it yields the characteristic polynomial

$$\det[\mathbf{A} - \lambda \mathbf{I}] = \begin{vmatrix} 1-\lambda & 0 & -1 \\ -2 & -1-\lambda & 2 \\ -1 & 2 & 1-\lambda \end{vmatrix} = 6\lambda + \lambda^2 - \lambda^3 = \lambda(\lambda+2)(3-\lambda),$$

so the eigenvalues of $\mathbf{A}$, that is, the roots of $\det[\mathbf{A} - \lambda \mathbf{I}] = 0$, are seen to be

$$\lambda_1 = -2, \quad \lambda_2 = 0 \quad \text{and} \quad \lambda_3 = 3, \tag{A2.11}$$

and no eigenvalue is repeated, so each has multiplicity of 1.

To find the eigenvector $\mathbf{x}^{(1)}$ corresponding to λ_1 we must solve the matrix equation $[\mathbf{A} - \lambda \mathbf{I}]\mathbf{x} = 0$, with $\lambda = \lambda_1 = -2$. This matrix equation becomes

$$\begin{bmatrix} 1-(-2) & 0 & -1 \\ -2 & -1-(-2) & 2 \\ -1 & 2 & 1-(-2) \end{bmatrix} \begin{bmatrix} x_1^{(1)} \\ x_2^{(1)} \\ x_3^{(1)} \end{bmatrix} =$$

$$\begin{bmatrix} 3 & 0 & -1 \\ -2 & 1 & 2 \\ -1 & 2 & 3 \end{bmatrix} \begin{bmatrix} x_1^{(1)} \\ x_2^{(1)} \\ x_3^{(1)} \end{bmatrix} = \begin{bmatrix} 0 \\ 0 \\ 0 \end{bmatrix},$$

and when this is written out in full it leads to the three simultaneous equations

$$3x_1^{(1)} - x_3^{(1)} = 0, \quad -2x_1^{(1)} + x_2^{(1)} + 2x_3^{(1)} = 0, \quad -x_1^{(1)} + 2x_2^{(1)} + 3x_3^{(1)} = 0. \quad \text{(A2.12)}$$

As the system is homogeneous, any one of these three equations must be linearly dependent on the other two. So, when using any two of these equations, it is only possible for two of the three unknowns to be found in terms of the third unknown, which may be assigned an arbitrary value (it is a parameter). If we take the third of the above equations to be the redundant equation (an arbitrary choice), we are left with the first two equations from which to determine $x_1^{(1)}, x_2^{(1)}$ and $x_3^{(1)}$. This can be done, because these two equations are linearly independent, and so are not proportional. To proceed further we will find $x_2^{(1)}$ and $x_3^{(1)}$ in terms of $x_1^{(1)} = k_1$, where $k_1 \neq 0$ is arbitrary (it is a parameter). The first equation gives $x_3^{(1)} = 3k_1$, and using $x_1^{(1)} = k_1$ and $x_3^{(1)} = 3k_1$ in the second equation we find that $x_2^{(1)} = -4k_1$. Of course, using $x_1^{(1)} = k_1$, and $x_3^{(1)} = 3k_1$ in the third equation again yields $x_2^{(1)} = -4k_1$, confirming the redundancy of this equation, since it is a linear combination of the first two equations.

We have shown that the eigenvector $\mathbf{x}^{(1)}$ can be taken to be $\mathbf{x}^{(1)} = [k_1, -4k_1, 3k_1]^T$, where $k_1 \neq 0$ is arbitrary. As the scaling of an eigenvector is arbitrary, it is usual to set the scale factor k_1 equal to a convenient numerical value, such as $k_1 = 1$, and with this choice the above eigenvector corresponding to $\lambda = \lambda_1 = -2$ becomes $\mathbf{x}^{(1)} = [1, -4, 3]^T$. The remaining eigenvectors follow in similar fashion, and the set of three eigenvalues and their corresponding eigenvectors turn out to be

$$\lambda_1 = -2, \ \mathbf{x}^{(1)} = \begin{bmatrix} 1 \\ -4 \\ 3 \end{bmatrix}, \lambda_2 = 0, \mathbf{x}^{(2)} = \begin{bmatrix} 1 \\ 0 \\ 1 \end{bmatrix}, \lambda_3 = 3, \mathbf{x}^{(3)} = \begin{bmatrix} 1 \\ -\frac{3}{2} \\ -2 \end{bmatrix}.$$

$$\text{(A2.13)}$$

When necessary, the fact that eigenvectors can be scaled and still remain eigenvectors is used to *normalize* them. Typically, an eigenvector $\mathbf{x} = [a, b, c]^T$ is normalized by multiplication by the normalizing factor $k = 1/\sqrt{a^2 + b^2 + c^2}$, when it becomes the normalized eigenvector $\tilde{\mathbf{x}}^{(1)} = [ka, kb, kc]^T$. Other normalizations are also used, although when this particular normalization is applied to the eigenvector $\mathbf{x}^{(1)}$ above, it produces the normalized eigenvector $\tilde{\mathbf{x}}^{(1)} = \left[\frac{1}{\sqrt{26}}, -\frac{4}{\sqrt{26}}, \frac{9}{\sqrt{26}} \right]^T$. The need for normalization arises from the fact that certain procedures involve repeated scaling of eigenvectors, and the effect of normalization is to ensure that after repeated scaling the elements in an eigenvector do not become either arbitrarily large or vanishingly small, thereby preserving accuracy in the calculations.

Consideration of the effort involved when finding the eigenvalues and eigenvectors of the 3×3 matrix used above will have convinced the reader of the need for software when finding the eigenvalues and eigenvectors of the much larger $n \times n$ matrices that often arise in practice. The way in which numerical procedures work when finding eigenvalues and eigenvectors is quite different from the formal method outlined above, and the details of these computationally efficient procedures will not be given here. Usually eigenvalues and eigenvectors are found using professional software packages such as MAPLE®(2007), MATLAB®(2007) and MathCAD® (2007), where the software is designed to be self-adaptive, so it chooses the most appropriate method of calculation to be used at each stage of the procedure.

A *diagonal matrix* is an $n \times n$ matrix in which the only non-zero elements occur on the *leading diagonal*, that is, the diagonal that runs from top left to bottom right of the matrix. The simplest diagonal matrix is, of course, the unit matrix **I**. A convenient notation for a diagonal matrix **D** with the elements $\lambda_1, \lambda_2, \ldots, \lambda_n$ on its leading diagonal is

$$\mathbf{D} = \mathrm{diag}\{\lambda_1, \lambda_2, \ldots, \lambda_n\}.$$

An $n \times n$ matrix **A** can always be diagonalized if it has n linearly independent eigenvectors; that is, if no one of the n eigenvectors is proportional to any other eigenvector. Let the n linearly independent eigenvectors be $\mathbf{v}_1, \mathbf{v}_2, \ldots, \mathbf{v}_n$, and their corresponding n eigenvalues be $\lambda_1, \lambda_2, \ldots, \lambda_n$, some of which may be repeated. Let **P** be a matrix with its columns being the eigenvectors $\mathbf{v}_1, \mathbf{v}_2, \ldots, \mathbf{v}_n$, then **P** is called a *diagonalizing matrix* for **A**. Matrix **A** is diagonalized by forming the matrix product $\mathbf{P}^{-1}\mathbf{AP}$, where $\mathbf{P}^{-1}$ is the inverse matrix of **P**. The result of the matrix product $\mathbf{P}^{-1}\mathbf{AP}$ is to yield the matrix $\mathbf{D} = \mathrm{diag}\{\lambda_1, \lambda_2, \ldots, \lambda_n\}$, where the order in which the eigenvalues appear on the leading diagonal of **D** is the order in which their corresponding eigenvectors appear as the columns of **A**, so we can write

$$\mathbf{P}^{-1}\mathbf{AP} = \mathbf{D}, \tag{A2.14}$$

and, equivalently,

$$\mathbf{A} = \mathbf{PDP}^{-1}. \tag{A2.15}$$

To diagonalize the matrix $\mathbf{A}$ in the above example, we define the diagonalizing matrix

$$\mathbf{P} = \begin{bmatrix} 1 & 1 & 1 \\ -4 & 0 & -\dfrac{3}{2} \\ 3 & 1 & -2 \end{bmatrix}, \quad \text{from which it follows that}$$

$$\mathbf{P}^{-1} = \begin{bmatrix} -\dfrac{1}{10} & -\dfrac{1}{5} & \dfrac{1}{10} \\ \dfrac{5}{6} & \dfrac{1}{3} & \dfrac{1}{6} \\ \dfrac{4}{15} & -\dfrac{2}{15} & -\dfrac{4}{15} \end{bmatrix}.$$

A simple calculation confirms that $\mathbf{P}^{-1}\mathbf{A}\mathbf{P} = \mathbf{D} = \mathrm{diag}\{-2, 0, 3\}$.

We now illustrate how diagonalization can be used to solve the non-homogeneous system of first-order equations

$$\frac{d\mathbf{x}}{dt} = \mathbf{A}\mathbf{x} + \mathbf{b}(t) \tag{A2.16}$$

where the column vector $\mathbf{b}(t)$ has the elements $b_1(t), b_2(t), \ldots, b_n(t)$. Substituting for $\mathbf{A}$ from equation (A2.15), system (A2.16) becomes

$$\frac{d\mathbf{x}}{dt} = \mathbf{P}\mathbf{D}\mathbf{P}^{-1}\mathbf{x} + \mathbf{b}. \tag{A2.17}$$

However, $\mathbf{P}^{-1}$ is a constant matrix, so if system (A2.17) is multiplied from the left (*premultiplied*) by $\mathbf{P}^{-1}$, this matrix may be taken under the differentiation symbol, when the result becomes

$$\frac{d(\mathbf{P}^{-1}\mathbf{x})}{dt} = \mathbf{D}\mathbf{P}^{-1}\mathbf{x} + \mathbf{P}^{-1}\mathbf{b}. \tag{A2.18}$$

Setting $\mathbf{u} = \mathbf{P}^{-1}\mathbf{x}$ this simplifies to

$$\frac{d\mathbf{u}}{dt} = \mathbf{D}\mathbf{u} + \mathbf{P}^{-1}\mathbf{b}, \tag{A2.19}$$

which is now a system of uncoupled ODEs, each of which is of the form

$$\frac{du_i}{dt} = \lambda_i u_i + f_i(t), \tag{A2.20}$$

with $u_i(t)$ being the ith element of $\mathbf{u}_i$ for $i = 1, 2, \ldots, n$, while $f_i(t)$ is the ith element of $\mathbf{P}^{-1}\mathbf{b}$.

Equations (A2.20) are simple first-order linear ODEs that are easily solved for $u_i(t)$ by means of an integrating factor (see result (2.12)), after which $\mathbf{u}(t)$ can then be constructed. The required solution $\mathbf{x}(t)$ then follows by using the result $\mathbf{x} = \mathbf{Pu}$.

Let us apply this approach to system (A2.16), with $\mathbf{A}$ being the 3×3 matrix in equation (A2.10) and $\mathbf{b}(t) = [0, 1, 0]^T$, subject to the initial conditions $x_1(0) = x_2(0) = x_3(0) = 1$. Equations (A2.20) become

$$\frac{\mathrm{d}u_1}{\mathrm{d}t} = -2u_1 - \frac{1}{5}, \quad \frac{\mathrm{d}u_2}{\mathrm{d}t} = \frac{1}{3}, \quad \frac{\mathrm{d}u_3}{\mathrm{d}t} = 3u_3 - \frac{2}{15}.$$

When these are integrated using the initial conditions derived from the result that $\mathbf{u}(0) = \mathbf{P}^{-1}\mathbf{x}(0)$, and $\mathbf{u}$ is constructed, it follows from $\mathbf{x} = \mathbf{Pu}$ that the solution is

$$x_1(t) = -\frac{8}{15}e^{3t} - \frac{1}{10}e^{-2t} + \frac{1}{3}t + \frac{23}{18}, \quad x_2(t) = \frac{4}{15}e^{3t} + \frac{2}{5}e^{-2t} + \frac{1}{3},$$

$$x_3(t) = \frac{16}{45}e^{3t} - \frac{3}{10}e^{-2t} + \frac{1}{3}t + \frac{17}{18}.$$

Chapter 3

Numerical techniques used in hydraulic modelling

3.1 Introduction

Practical applications of hydraulics require numerical and graphical results, and if analytical expressions that describe phenomena of interest cannot be derived from a mathematical model it becomes necessary to obtain the results by purely numerical methods. This can involve the solution of various types of mathematical problems, some of the most important of which involve the solution of ordinary and partial differential equations (PDEs).

A numerical approach may involve the use of more than one technique, and these may range from finding the roots of equations and approximating discrete data in analytical form, to numerical integration, the determination of the eigenvalues and eigenvectors of matrices, and the solution of the large sets of linear algebraic equations that arise when using numerical methods to solve PDEs.

The calculations that arise from practical problems can be complex, time-consuming and, in many cases, impossible to perform by hand. Typically, this happens when solving PDEs, because the methods used give rise to many hundreds of linear algebraic equations that need to be solved by iteration. Consequently, in practice numerical results are obtained and plotted with the aid of a computer, using one of the efficient and highly optimized commercial software packages that are available. The specialist hydraulics packages will be mentioned later, but from among the general and very powerful mathematical software packages that are readily available and easy to use we mention MAPLE® (2007), MATLAB® (2007) with its various special-purpose packages called Toolboxes, and MathCAD® (2007). Each of these software packages is updated regularly to ensure it uses the latest and most efficient numerical techniques, which in computing terminology are called *procedures*. These are often self-adaptive to enable them to switch between different procedures during a calculation in order to maintain a predetermined accuracy and to accelerate the rate of convergence.

The purpose of this chapter is to provide an outline of the numerical solution of ordinary and PDEs, without attempting to describe all the possible methods and the refinements that are built into the general-purpose optimized software packages.

3.2 Solving large sets of algebraic equations

3.2.1 Gaussian elimination

Because of the importance of solving systems of linear algebraic equations that can arise in many different ways, and before showing how such systems of equations arise when solving PDEs, we review the following general problem. In applications of mathematics to hydraulic computation and elsewhere, systems of n non-homogeneous linear algebraic equations in the n unknown real variables $x_1, x_2, \ldots, x_n$ arise, often where n is very large. All such systems take the form

$$
\begin{aligned}
a_{11}x_1 + a_{12}x_2 + \cdots + a_{1n}x_n &= b_1 \\
a_{21}x_1 + a_{22}x_2 + \cdots + a_{2n}x_n &= b_2 \\
&\cdots \\
a_{n1}x_1 + a_{n2}x_2 + \cdots + a_{nn}x_m &= b_n,
\end{aligned}
\tag{3.1}
$$

where the coefficients a_{ij} and the terms b_l, called the non-*homogeneous* terms, are given numbers. When all the terms b_i are zero, system (3.1) is said to be *homogeneous*.

If one or more of the equations in (3.1) can be expressed as a sum of multiples of the other equations, say $r < n$ of them, these equations are said to be *linearly dependent* on the r equations. If, however, there is no linear dependence among the n equations in (3.1) the system is said to be *linearly independent*. The condition for the linear independence and dependence of the equations in (3.1), when the terms b_i are all zero, is a special case of the corresponding conditions for functions given in equation (2.3) of Chapter 2, with the functions $u_1(t), u_2(t), \ldots, u_n(t)$ replaced by $x_1, x_2, \ldots, x_n$. For example, in the system of non-homogeneous equations

$$
\begin{aligned}
2x_1 - x_2 + 3x_3 &= 1 \\
x_1 + x_2 + x_3 &= 2 \\
4x_1 + x_2 + 5x_3 &= 5,
\end{aligned}
$$

the third equation is linearly dependent on the first two equations, because it is the sum of the first equation and twice the second equation. So, although the system appears to comprise three different equations involving the three unknowns x_1, x_2 and x_3, there are in fact only two linearly independent equations.

If the third equation is replaced by $4x_1 + x_2 + 5x_3 = 6$ the situation is different. Although the expression on the left of the third equation is linearly dependent on the expressions on the left of the first two equations, this is *not* true of the non-homogeneous sixth term on the right of the third equation. This means that the equations are *inconsistent* (i.e. they contradict each other), because the result of subtracting the sum of the first equation and twice the second equation from the third equation is to give $0 = 1$, which is impossible, showing that the modified system has *no* solution. Thus, for equation (3.1) to have a solution when there is linear dependence of the expressions on the left of equation (3.1), this same linear dependence must exist for the terms b_i on the right, although when such a dependence exists the solution will not be unique, as will be shown later.

Finding the *solution* of system (3.1) involves finding the numbers $x_1, x_2, \ldots, x_n$ that satisfy the equations, collectively called the *solution set*, and the method to be described by which the solution set may be found is called the *Gaussian elimination process*. This method, with important modifications, forms the basis of all numerical software procedures for computers that solve such systems directly without the use of iteration. To develop the Gaussian elimination process it is convenient to use matrices, and to represent system (3.1) in the form

$$\mathbf{A}\mathbf{x} = \mathbf{b}, \tag{3.2}$$

where $\mathbf{A} = [a_{ij}]$ is the $n \times n$ coefficient matrix, $\mathbf{b} = [b_1, b_2, \ldots, b_n]^T$ is the non-homogeneous column vector, and $\mathbf{x} = [x_1, x_2, \ldots, x_n]^T$ is the solution vector. Here, to save space, the superscript T, signifying the matrix transpose operation in which rows and columns are interchanged, has been used to allow the matrix *column vectors* $\mathbf{b}$ and $\mathbf{x}$ to be written more concisely as the transpose of the corresponding matrix *row vectors*.

System (3.1), correspondingly (3.2), can be displayed more concisely by introducing what is called the *augmented matrix* denoted by $\mathbf{A}|\mathbf{b}$, comprising matrix $\mathbf{A}$ adjoined to which on the right is vector $\mathbf{b}$ to give

$$\mathbf{A} \,|\, \mathbf{b} = \begin{bmatrix} a_{11} & a_{12} & a_{13} & \cdots & a_{1n} & b_1 \\ a_{21} & a_{22} & a_{23} & \cdots & a_{2n} & b_2 \\ a_{31} & a_{32} & a_{33} & \cdots & a_{3n} & b_3 \\ \vdots & \vdots & \vdots & \vdots & \vdots & \vdots \\ a_{m1} & a_{m2} & a_{m3} & \cdots & a_{mn} & b_m \end{bmatrix}. \tag{3.3}$$

This matrix contains all the information in system (3.1), because in the ith row of $\mathbf{A}|\mathbf{b}$ the element a_{ij} is associated with the variable x_j, while b_i is the corresponding non-homogeneous term on the right of system (3.1). When $\mathbf{A}|\mathbf{b}$ is interpreted as the system of equations in (3.1), it *implies* the presence of an *equality sign* between the terms on the left represented by the matrix product **Ax**, and the non-homogeneous terms on the right represented by the column vector **b**.

The idea underlying Gaussian elimination is simple, and it depends for its success on the obvious facts that the order of the equations in (3.1) can be changed, any equation can be multiplied throughout by a non-zero constant, and multiples of equations in (3.1) can be added to or subtracted from other equations in (3.1), all without altering the solution set of the original system. When working with the augmented matrix **A|b**, which is equivalent to the original set of equations (3.1), performing such operations on the original system of equations corresponds to performing what are called *elementary row operations* on the augmented matrix to produce a modified, but equivalent, augmented matrix. The elementary row operations that can be performed on an augmented matrix are as follows:

1   Interchanging rows;
2   Multiplying each element in a row by a non-zero constant k;
3   Adding (or subtracting) a multiple of a row to (or from) another row.

The effect of performing these elementary row operations on an augmented matrix **A|b** is to produce a modified augmented matrix that is in all respects equivalent to the original system of equations in (3.1). In general, the process of transforming matrix **A|b** to an equivalent matrix using elementary row operations is called *matrix row reduction*.

Gaussian elimination starts by assuming that in equations (3.1) the coefficient $a_{11} \neq 0$. This is no restriction, because if this is not the case the order of the equations can be changed to bring into the first row of equations (3.1) an equation for which this condition is true. The method then proceeds by subtracting multiples of row 1 of equations (3.1) from each of the $n-1$ rows below it in such a way that the coefficient of the variable x_1 is made to vanish from each of these subsequent rows. Thus, a_{21}/a_{11} times row 1 is subtracted from row 2, a_{31}/a_{11} times row 1 is subtracted from row 3 and so on, until finally a_{n1}/a_{11} times row 1 is subtracted from row n, leading to a modified augmented matrix $\mathbf{A|b}^{(1)}$ of the form

$$\mathbf{A}\,|\,\mathbf{b}^{(1)} = \begin{bmatrix} a_{11} & a_{12} & a_{13} & \cdots & a_{1n} & b_1 \\ 0 & a_{22}^{(1)} & a_{23}^{(1)} & \cdots & a_{2n}^{(1)} & b_2^{(1)} \\ 0 & a_{32}^{(1)} & a_{33}^{(1)} & \cdots & a_{3n}^{(1)} & b_3^{(1)} \\ \vdots & \vdots & \vdots & \vdots & \vdots & \vdots \\ 0 & a_{n2}^{(1)} & a_{n3}^{(1)} & \cdots & a_{nn}^{(1)} & b_n^{(1)} \end{bmatrix}. \tag{3.4}$$

This same process is now repeated, starting with row 2 of $\mathbf{A}|\mathbf{b}^{(1)}$. This time row 2 with its first non-zero element $a_{22}^{(1)}$ is used to reduce to zero all elements in the column below it, leading to a modification of $\mathbf{A}|\mathbf{b}^{(1)}$ denoted by $\mathbf{A}|\mathbf{b}^{(2)}$ of the form

$$\mathbf{A}\,|\,\mathbf{b}^{(2)} = \begin{bmatrix} a_{11} & a_{12} & a_{13} & \cdots & a_{1n} & b_1 \\ 0 & a_{22}^{(1)} & a_{23}^{(1)} & \cdots & a_{2n}^{(1)} & b_2^{(1)} \\ 0 & 0 & a_{33}^{(2)} & \cdots & a_{3n}^{(2)} & b_3^{(2)} \\ \vdots & \vdots & \vdots & \vdots & \vdots & \vdots \\ 0 & 0 & a_{n3}^{(2)} & \cdots & a_{nn}^{(2)} & b_n^{(2)} \end{bmatrix}. \tag{3.5}$$

Continuing this process will lead to a simplification of the original system of equations in which the numbers $a_{11}, a_{22}^{(1)}, a_{33}^{(2)}, a_{44}^{(3)}, \ldots, a_{n-1,n-1}^{(n-1)}$ used to reduce to zero the entries in the columns below them are called the *pivots* for the Gaussian elimination method.

If, as may happen, at some intermediate stage a pivot becomes zero and so cannot be used to reduce to zero all entries in the column below it, the difficulty is usually overcome by interchanging the row with the zero pivot with a row *below* it in which the corresponding entry is non-zero, after which the process continues as before.

The pattern of entries produced by Gaussian elimination after the completion of its row reduction is said to be the *echelon form* of the matrix, the general definition of which is as follows:

1   All rows containing non-zero elements lie above any rows that contain only zeros.
2   The first non-zero entry in a row, called the *leading entry* in the row and also a *pivot*, lies in a column to the right of the leading entry in the row above.

A matrix is said to be in *reduced echelon form* when, in addition to being in echelon form, each row is scaled so that its pivot is 1 (unity).

As a leading entry is a pivot, condition 2 implies all entries in the column below a leading entry are zero.

The pattern of entries in the echelon form of $\mathbf{A}|\mathbf{b}$ generated by row reduction applied to a 7×8 matrix with a *unique* solution (the equations are linearly independent) is shown in equation (3.17), where the symbol ● represents a leading entry that is always non-zero or 1 if the matrix is in reduced echelon form, while the symbol □ represents an entry that may, or may not, be non-zero.

$$\begin{bmatrix} \bullet & \square & \square & \square & \square & \square & \square & \square \\ 0 & \bullet & \square & \square & \square & \square & \square & \square \\ 0 & 0 & \bullet & \square & \square & \square & \square & \square \\ 0 & 0 & 0 & \bullet & \square & \square & \square & \square \\ 0 & 0 & 0 & 0 & \bullet & \square & \square & \square \\ 0 & 0 & 0 & 0 & 0 & \bullet & \square & \square \\ 0 & 0 & 0 & 0 & 0 & 0 & \bullet & \square \end{bmatrix}. \tag{3.6}$$

Recalling that the augmented matrix in this illustration represents a set of seven equations for the unknowns $x_1, x_2, \ldots, x_7$, with an implied equality sign between the seventh column and the non-homogeneous terms in the eighth column, the solution of the system follows directly by the process of *back substitution*. In this process the last equation determines x_7, the last but one equation determines x_6 after substituting for x_7, and so on, until the first equation determines x_1 after substituting for $x_7, x_6, \ldots, x_2$.

The following simple example illustrates the process of Gaussian elimination applied to the system of four linearly independent equations in the unknowns x_1, x_2, x_3 and x_4:

$$\begin{aligned} 2x_1 + x_3 + 2x_4 &= 1 \\ -2x_1 + x_2 - x_3 + 2x_4 &= 1 \\ x_1 + 2x_2 - 2x_3 - x_4 &= 1 \\ x_1 + x_3 &= 2. \end{aligned}$$

When written in its augmented matrix form, the system becomes

$$A \mid b = \begin{bmatrix} 2 & 0 & 1 & 2 & 1 \\ -2 & 1 & -1 & 2 & 1 \\ 1 & 2 & -2 & -1 & 1 \\ 1 & 0 & 1 & 0 & 2 \end{bmatrix}.$$

After performing elementary row operations on this augmented matrix and using the symbol $\sim$ as an abbreviation for 'is equivalent to', the matrix $A \mid b$ is reduced to the echelon form

$$A \mid b \sim \begin{bmatrix} 2 & 0 & 1 & 2 & 1 \\ 0 & 1 & 0 & 4 & 2 \\ 0 & 0 & -\frac{5}{2} & 10 & -\frac{7}{2} \\ 0 & 0 & 0 & -3 & \frac{4}{5} \end{bmatrix}.$$

The last row of this echelon form corresponds to the equation $-3x_4 = 4/5$, and so $x_4 = -4/15$. The last but one row corresponds to the equation $-5/2x_3 + 10x_4 = -7/2$, but $x_4 = -\frac{4}{15}$, so $x_3 = 37/15$. Continuing this process of back substitution, the solution set $\{x_1, x_2, x_3, x_4\}$ is found to have the elements $x_1 = -7/15, x_2 = 46/15, x_3 = 37/15, x_4 = -4/15$.

A typical modification of the Gaussian elimination process used in computer procedures involves changing the order of the equations at each stage of the process, when necessary, to make the absolute value of the pivot about to be used as large as possible. This is to ensure that at no stage is a pivot with an unnecessarily small absolute value used to reduce to zero a coefficient below it with a much greater absolute value, thereby reducing the build-up of round-off errors as the computation proceeds. An augmented matrix like equation (3.17) with n rows and $n + 1$ columns will always be produced when a finite-difference scheme (to be described later) is used to solve a PDE, and in that case the solution will be unique.

In the more general case, if the solution of a system is not unique, there will be rows containing only zeros. This situation arises when seeking the solution of a general system of n linear non-homogeneous equations in n unknowns among which there is linear dependence. In such a case, if the echelon form contains r rows of zeros, it follows from the structure of the echelon form $A|b$ that $n - r$ of the unknowns can only be found in terms of r of the unknowns as parameters that can be assigned arbitrarily.

To illustrate this situation, consider the system of equations

$$x_1 + 2x_2 + x_3 = 1$$
$$2x_1 - x_2 + 3x_3 = 2$$
$$4x_1 + 3x_2 + 5x_3 = 4,$$

for which the augmented matrix is

$$A|b = \begin{bmatrix} 1 & 2 & 1 & 1 \\ 2 & -1 & 3 & 2 \\ 4 & 3 & 5 & 4 \end{bmatrix}.$$

The echelon form of this augmented matrix is

$$A|b \sim \begin{bmatrix} 1 & 2 & 1 & 1 \\ 0 & -5 & 1 & 0 \\ 0 & 0 & 0 & 0 \end{bmatrix}$$

in which there is one row of zeros (because the third equation is the sum of twice the first equation and the second equation). When written out in full, $A|b$ becomes

$$x_1 + 2x_2 + x_3 = 1 \quad \text{and} \quad -5x_2 + x_3 = 0$$

showing there are only two equations from which to find three unknowns. Setting $x_3 = k$, an arbitrary parameter, the solution set becomes

$$\left\{ x_1 = 1 - \frac{7}{5}k, x_2 = \frac{1}{5}k, x_3 = k \right\},$$

so as $n = 3$ and $r = 1$ we see that $n - 1 = 2$ of the unknowns x_1, x_2 and x_3 have been found in terms of $r = 1$ of the unknowns, namely $x_3 = k$, as a parameter.

We could, of course, equally well have solved for x_1 and x_3 in terms of $x_2 = p$, as an arbitrary parameter, when we would have found the solution set

$$\{ x_1 = 7p, x_2 = p, x_3 = 5p \},$$

or in terms of $x_1 = s$ as an arbitrary parameter, when we would have found the solution set

$$\{ x_1 = s, x_2 = (1/7) - (1/7s), x_3 = (5/7) - (5/7s) \}$$

although, of course, the three solution sets are equivalent. For more about the solution of general systems of algebraic equations, and, in particular, systems of m equations in n unknowns, see Jeffrey (2002), Kreyszig (2005) and O'Neil (2006).

3.2.2 Gauss–Seidel iteration

When the number of equations n is very large, a completely different way of solving system (3.1) is used that involves iteration. The method now described is called the *Gauss–Seidel iterative scheme*, and for the process to converge, that is, for each of the rth iterates $x_1^{(r)}, x_2^{(r)}, \ldots, x_n^{(r)}$ to converge to a specific value as r increases, it is necessary that the system of equations (3.1) is written in diagonally dominant form. The condition of *diagonal dominance* means that in the ith row of matrix **A** the element a_{ii} must be such that $|a_{ii}|$ is greater than the sum of the absolute values of all the other elements in the ith row, for $i = 1, 2, \ldots, n$, so the matrix

$$\mathbf{A} = \begin{bmatrix} a_{11} & a_{12} & a_{13} \\ a_{21} & a_{22} & a_{23} \\ a_{31} & a_{32} & a_{33} \end{bmatrix}$$

will be diagonally dominant if

$$|a_{11}| > |a_{12}| + |a_{13}|, |a_{22}| > |a_{21}| + |a_{23}|, |a_{33}| > |a_{31}| + |a_{32}|.$$

Assuming this set of conditions is satisfied by system (3.1), the equations are first solved successively for $x_1, x_2, \ldots, x_n$, leading to the equivalent system

$$x_1 = (b_1 - a_{12}x_2 - a_{13}x_3 - \cdots - a_{1n}x_n)/a_{11}$$
$$x_2 = (b_2 - a_{21}x_1 - a_{23}x_3 - \cdots - a_{2n}x_n)/a_{22}$$
$$\vdots$$
$$x_n = (b_n - a_{n1}x_1 - a_{n2}x_2 - \cdots - a_{nn-1}x_{n-1})/a_{nn}. \tag{3.7}$$

Now let the $(r-1)$th iterated vector $\mathbf{x}^{(r)}$ (approximation) for the solution vector $\mathbf{x}$ have the elements $x_1^{(r-1)}, x_2^{(r-1)}, \ldots, x_n^{(r-1)}$. Then the rth iterates (the next approximation) are found by substituting the $(r-1)$th iterates into the Gauss–Seidel system as follows:

$$x_1^{(r)} = (b_1 - a_{12}x_2^{(r-1)} - a_{12}x_3^{(r-1)} - \cdots - a_{1n}x_n^{(r-1)})/a_{11}$$
$$x_2^{(r)} = (b_2 - a_{21}x_1^{(r)} - a_{23}x_3^{(r-1)} - \cdots - a_{2n}x_n^{(r-1)})/a_{22}$$
$$x_3^{(r)} = (b_3 - a_{31}x_1^{(r)} - a_{32}x_2^{(r)} - \cdots - a_{3n}x_n^{(r-1)})/a_{33}$$
$$\cdots$$
$$x_n^{(r)} = (b_n - a_{n1}x_1^{(r)} - a_{n2}x_2^{(r)} - \cdots - a_{nn-1}x_{n-1}^{(r)})/a_{nn}. \tag{3.8}$$

The iterative process is started with an arbitrary vector $\mathbf{x} = \mathbf{x}^{(0)}$, which is typically taken to be $\mathbf{x}^{(0)} = [1, 1, 1, \ldots, 1]^T$, and terminated with the rth iterate when $|x_i^{(r)} - x_i^{(r-1)}|$ for $i = 1, 2, \ldots, n$ first becomes less than some pre-assigned error $\varepsilon > 0$, where, for example, $\varepsilon = 0.0001$. Notice that the Gauss–Seidel system uses the rth iterate of an unknown as soon as it has been calculated, so by the time $x_n^{(r)}$ is calculated all the rth iterates for the other variables have been used.

If the system is not diagonally dominant the iterations may not converge, and may oscillate in sign while their absolute values increase without bound. A system that is not diagonally dominant because an equation fails the diagonal-dominance condition can usually be converted into a diagonally-dominant equation by changing the order of the equations, or by adding suitable multiples of the other equations to the equation that is not diagonally dominant.

3.2.3 Successive over-relaxation

When n is large, the rate of convergence of the Gauss–Seidel method (3.8) is too slow for many practical purposes, so it must be modified. The modification that is often used, particularly when seeking numerical solutions of PDEs, is called the *successive over-relaxation (SOR) process*. To understand how the SOR process works, let $\bar{x}_i^{(k)}$ be the ith component of the kth Gauss–Seidel iteration, then in the SOR process the kth iterate $x_i^{(k)}$ is found by the following formula in which the constant ω is a parameter:

$$x_i^{(k)} = x_i^{(k-1)} + \omega\left[\bar{x}_i^{(k)} - x_i^{(k-1)}\right]$$

$$= (1 - \omega)x_i^{(k-1)} + \omega\bar{x}_i^{(k)}. \tag{3.9}$$

So, in the SOR process the value of the ith component in the kth iterate is obtained by *extrapolation* from the kth Gauss–Seidel iterate and the previously calculated estimate $x_i^{(k-1)}$. The number ω is called the *over-relaxation parameter*, and for the iterative process to converge this parameter must be in the interval $1 < \omega < 2$, although the determination of its optimum value to achieve the greatest acceleration of convergence can only be found in special cases. Here it is sufficient to remark that, when $\omega = 1$, the SOR process reduces to the Gauss–Seidel process, and the larger the $n \times n$ matrix $\mathbf{A}$ becomes, the closer the optimum value of ω is to 2, although a typical value that is often used is around 1.5. For a discussion of the SOR and related iterative processes, including the choice of the optimum value of the relaxation parameter ω, see, for example, Ferziger (1998), while for SOR and related codes and a more detailed discussion of the choice of ω, see Press *et al.* (2007). Typically, systems of up to 100 equations are best solved by Gaussian elimination, although when more equations are involved it is best to use iterative methods.

In general, the numerical solution of PDEs requires the solution of very large systems of non-homogeneous linear equations. When these equations are written in the form $\mathbf{Ax} = \mathbf{b}$, with $\mathbf{A}$, $\mathbf{x}$ and $\mathbf{b}$ as defined in equation (3.2), most of the elements in the $n \times n$ matrix $\mathbf{A}$ occur as rectangular blocks that contain only zeros, while the non-zero elements tend to occur in the diagonals above and below, but close to, the leading diagonal of $\mathbf{A}$. Matrices of this type are called *sparse matrices*, and particularly efficient procedures exist for the solution of systems of equations with sparse coefficient matrices. One of the techniques used to minimize the time taken for calculations takes account of the fact that when rows of $\mathbf{A}$ are scaled, no arithmetic need be performed on the blocks of zero elements. As computer time spent on additions and subtractions is very small relative to the time taken for multiplications, by avoiding the unnecessary scaling of large blocks of zero elements, such procedures are faster than ordinary Gaussian elimination or standard iterative methods.

3.3 The numerical solution of ordinary differential equations

3.3.1 The Euler method

A numerical solution of the ordinary differential equation (ODE)

$$\frac{dy}{dx} = f(x, y), \quad y(x_0) = y_0, \tag{3.10}$$

where $y(x_0) = y_0$ is the *initial condition*, approximates the continuous solution $y(x)$ for $x > x_0$ by computing a sequence of discrete approximations $\tilde{y}_r$ to $y(x_r)$ at the points $x_r = x_0 + rh, \ldots$, for $r = 0, 1, 2, \ldots$, where h is called the *integration step length* or *step size*. A numerical procedure that computes the approximate solution $\tilde{y}_{r+1}$ using only the approximate solution $\tilde{y}_{r-1}$ computed in the previous step and the value of h, is called a *one-step* method. This is to distinguish the method from others that at each stage of the computation need to know the values of the approximation $\tilde{y}_r$ at two or more previous stages.

The simplest method is the elementary *Euler method*. This uses a tangent approximation to the function $f(x)$ at the point x_r to find the approximation $\tilde{y}_{r+1}$ at x_{r+1} in terms of h, as illustrated in Figure 3.1.

From the interpretation of a derivative, the slope of the tangent to the solution curve at the point $P(x_r, \tilde{y}_r)$ is $f(x_r, \tilde{y}_r)$, from which it follows that $\tan(\theta_r) = f(x_r, \tilde{y}_r)$, where θ_r is the angle between the tangent line at $P(x_r, \tilde{y}_r)$ and the x-axis. Thus, in Figure 3.1 we can write $QR = hf(x_r, \tilde{y}_r)$, where R has the coordinates $(x_{r+1}, \tilde{y}_{r+1}) = (x_r + h, \tilde{y}_r + hf(x_r, \tilde{y}_r))$. Consequently, we have arrived at the result that

$$\tilde{y}_{r+1} = \tilde{y}_r + hf(x_r, \tilde{y}_r), \tag{3.11}$$

which is true for $r = 0, 1, 2, \ldots$.

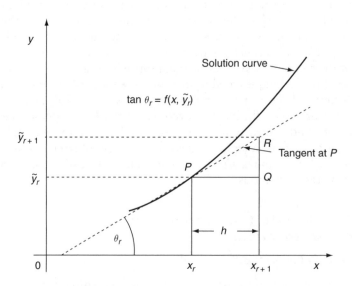

Figure 3.1 The Euler method using $\tilde{y}_r$ at $x = x_r$ and a tangent line, to find the approximate solution $\tilde{y}_{r+1}$ at $x = x_{r+1}$

This result leads to the following algorithm for the integration of a first-order equation by Euler's method, where for convenience $\tilde{y}_r$ has been replaced by y_r.

The Euler integration algorithm

Objective: It is required to integrate numerically the first-order equation $\frac{dy}{dx} = f(x, y)$ over the interval $a \le x \le b$ using n steps, each with length h, subject to the initial condition $y(a) = y_0$.

Method: Let y_r be the numerical approximation to the exact solution $y(x_r)$. As n equal-length steps are to be used, it follows that $h = (b - a)/n$, and so $x_r = a + rh$, with $r = 0, 1, 2, \ldots, n$.

Then the value y_{r+1} is given in terms of r and h by

$y_{r+1} = y_r + hf(x_r, y_r)$ for $r = 0, 1, 2, \ldots, n - 1$.

It can be shown that the Euler method is accurate to the order of h, in the sense that the error will be approximately h, and this is shown by saying that the accuracy is $O(h)$. Although this accuracy can be improved by reducing the value of h, this will result in more steps when advancing a calculation over an interval $a < x < b$, with a corresponding build-up of truncation errors. Consequently, if greater accuracy is required, a different method of integration must be used. In the Euler method the accuracy $O(h)$ is approximated by h raised to the power 1, and because of this the Euler method is called a *first-order method* of integration. The Euler method is an example of a *one-step method*, because the calculation of y_{r+1} is uniquely determined by the value of y_r at the previous step in the calculation. Although the Euler algorithm as given here involves a sequence of steps, all with the same increment h in x, if necessary the step length can be changed at any stage in the calculation.

There are many accurate finite-difference methods of integration based on the representation of derivatives by difference quotients. However, instead of discussing these, the highly versatile fourth-order Runge–Kutta method will be described, and its modification to form the self-adaptive Runge–Kutta–Fehlberg method will be mentioned. For information about finite-difference methods the reader is referred to the books by Atkinson (2007), Burden and Faires (2005), Ferziger (1998), Fröberg (1999), Johnson and Riess (1982), Pearson (1986), Ralston *et al* (2003) and Schwarz (1989).

The poor accuracy of the Euler method can be seen by applying it to the differential equation $dy/dt = y$ over the interval $0 \le x \le 1$ with a step length $h = 0.1$, subject to the initial condition $y(0) = 1$. This will allow a comparison between the numerical solution and the exact solution $y(x) = e^x$. The Euler method gives $y(0.5) = 1.610510$, $y(1) = 2.593742$ and $y(1.5) = 4.177248$, whereas the exact solutions are $y(0.5) = 1.648721$, $y(1) = 2.718282$ and $y(1.5) = 4.481689$.

The accuracy of the Euler method can be improved by reducing the value of h, at the cost of more steps and with a corresponding build-up of truncation errors. Thus, there is a trade-off between the increase in accuracy as h is reduced and the consequent build-up of round-off errors as the number of steps increases. If greater accuracy is required, a different method of numerical integration must be used that is of order $O(h^p)$, with $p > 1$, because the accuracy of a method improves as its order p increases.

3.3.2 Runge–Kutta methods

A family of one-step integration methods exists called *Runge–Kutta methods*, all of which are based on Taylor-series expansions truncated to different orders. The one to be described next is the popular and accurate *fourth-order Runge–Kutta method*, in which the accuracy is approximately $O(h^4)$. This method, usually abbreviated to the *rk4* method, is derived by taking a weighted average of four slopes at specially chosen points between x_r and $x_r + h$ in such a way that the error is optimized at $O(h^4)$. The rk4 method is often the default method found in software packages because of the good results it yields when used with most differential equations.

The fourth-order Runge–Kutta algorithm for a first-order ODE

Objective: To find a numerical approximation for the initial-value problem

$$\frac{dy}{dx} = f(x, y), \quad y(x_0) = y_0$$

for $x > x_0$ using a step-size h at the points $x_{n+1} = x_0 + nh (n = 0, 1, \ldots)$.

Method: Let y_n be a numerical approximation for the exact solution $y(x_n)$. Then

$$y_{n+1} = y_n + \frac{1}{6}(k_1 + 2k_2 + 2k_3 + k_4),$$

where

$$k_1 = hf(x_n, y_n),$$

$$k_2 = hf\left(x_n + \frac{1}{2}h, y_n + \frac{1}{2}k_1\right),$$

$$k_3 = hf\left(x_n + \frac{1}{2}h, y_n + \frac{1}{2}k_2\right),$$

$$k_4 = hf(x_{n+1}, y_n + k_3),$$

with $n = 0, 1, 2, \ldots$.

The rk4 method is illustrated below by applying it to an equation with an analytical solution to permit the exact and the numerical results to be compared. The example chosen is one where the solution changes more rapidly in some intervals than in others, because this subjects the method to a more rigorous test than one where the rate of change of the solution remains fairly constant.

The rk4 method with a step-size $h = 0.1$ is used over the interval $0 \leq x \leq 1.5$ to determine the approximate solution of

$$\frac{dy}{dx} + 4y = 20x \sin 8x, \quad \text{given that } y(0) = 0,$$

which is then compared with the analytical solution

$$y(x) = \frac{1}{20}(3 \sin 8x + 4 \cos 8x) + x(\sin 8x - 2 \cos 8x) - \frac{1}{5}e^{-4x}.$$

Setting $x_n = nh$ and $y_n = y(x_n)$ for $n = 01, \ldots, 15$, and $f(x, y) = 20x \sin 8x - 4y$, the rk4 method gives the following results, where the last column shows the absolute error $|y_n - Y(x_n)|$ between the rk4 solution and the exact solution.

The rk4 calculations, the exact solution and the absolute error

rk4	exact	Absolute error			
x_n	y_n		y_n	$|y_{n+1} - Y(x_n)|$	
0	0	0	0		
0.3	0.5384421		0.5386771	0.0002350	
0.6			−0.8526942	−0.8527660	0.0000718
0.9			−0.1451507	−0.1454756	0.0003249
1.2			1.9286007	1.9293262	0.0007255
1.5			−3.2479248	−3.2486321	0.0007073

◊

The rk4 method is easily adapted to solve systems of first-order differential equations; the steps involved are described in the following rk4 algorithm.

The fourth-order Runge–Kutta algorithm for a system of two first-order ODEs

Objective: To find a numerical approximation for the initial-value problem

$$\frac{dy}{dx}=f(x,y,z), \quad \frac{dz}{dx}=g(x,y,z), \; y(x_0)=y_0, \; z(x_0)=z_0,$$

using a step size h at the points $x_{n+1}=x_0+nh(n=0,1,\dots)$.

Calculate

$$k_{1n}=hf(x_n,y_n,z_n) \quad K_{1n}=hg(x_n,y_n,z_n)$$

$$k_{2n}=hf\left(x_n+\frac{1}{2}h,y_n+\frac{1}{2}k_{1n},z_n+\frac{1}{2}K_{1n}\right)$$

$$K_{2n}=hg\left(x_n+\frac{1}{2}h,y_n+\frac{1}{2}k_{1n},z_n+\frac{1}{2}K_{1n}\right)$$

$$k_{3n}=hf\left(x_n+\frac{1}{2}h,y_n+\frac{1}{2}k_{2n},z_n+\frac{1}{2}K_{2n}\right)$$

$$K_{3n}=hg\left(x_n+\frac{1}{2}h,y_n+\frac{1}{2}k_{2n},z_n+\frac{1}{2}K_{2n}\right)$$

$$k_{4n}=hf(x_n+h,y_n+k_{3n},z_n+K_{3n})$$

$$K_{4n}=hg(x_n+h,y_n+k_{3n},z_n+K_{3n}).$$

Set

$$k_n=\frac{1}{6}(k_{1n}+2k_{2n}+2k_{3n}+k_4), \; K_n=\tfrac{1}{6}(K_{1n}+2K_{2n}+2K_{3n}+K_{4n}).$$

The numerical solutions $y_{n+1}=y(x_{n+1})$, $z_{n+1}=z(x_{n+1})$ are given by $y_{n+1}=y_n+k_n$ and $z_{n+1}=z_n+K_n$.

When this approach is used with a step size $h=0.1$ to solve the initial-value problem

$$\frac{dy}{dx}=z-2y+x, \quad \frac{dz}{dx}=y-2z+1, \quad y(0)=1, \quad z(0)=2$$

over the interval $0\le x\le 2$, the results of the calculations for $y(x)$ and $z(x)$ are shown below for $x=0$, 0.5, 1.0, 1.5 and 2.0. For conciseness, the intermediate calculations leading to the determination of the constants k_i and K_i have been omitted.

x	$y(x)$	$z(x)$
0	1.0	2.0
0.5	0.95891885	1.36067395
1.0	1.98243077	1.12120856
1.5	1.10938644	1.06000495
2.0	1.31342550	1.09258108

The analytical solution is $y(x) = \frac{2}{3}x - \frac{2}{9} + \frac{3}{22}e^{-x}$, $z(x) = \frac{1}{3}x + \frac{2}{9} + \frac{3}{2}e^{-x} + \frac{5}{18}e^{-3x}$ and the agreement between the analytical values

$$y_{\text{exact}}(2) = 1.31342549 \quad \text{and} \quad z_{\text{exact}}(2) = 1.09258036,$$

and the rk4 numerical solutions are seen to be excellent.

It is a simple matter to adapt this algorithm to solve an initial-value problem for the general second-order equation

$$a(x)\frac{d^2y}{dx^2} + b(x)\frac{dy}{dx} + c(x)y = 0 \quad \text{with} \quad y(x_0) = y_0, \quad y'(x_0) = z_0 . \quad (3.12)$$

All that is necessary is to define $dy/dx = z$, when $d^2y/dx^2 = dz/dx$ and then to replace the ODE by the equivalent first-order system

$$\frac{dy}{dx} = z, \quad \frac{dz}{dx} = -\left(\frac{b(x)}{a(x)}z - \frac{c(x)}{a(x)}y\right), \quad \text{with} \quad a(x) \neq 0 \quad (3.13)$$

subject to the initial conditions

$$y(x_0) = y_0 \quad \text{and} \quad z(x_0) = z_0. \quad (3.14)$$

The numerical results will give both $y(x)$ and $z(x) = dy/dx$, although usually only $y(x)$ is required. To illustrate the method, and to show how it may equally well be applied to non-linear ODEs, this same approach is used to solve the second-order non-linear initial-value problem

$$\frac{d^2y}{dx^2} + (1 + x^2)\frac{dy}{dx} + \sin y = 0, \quad y(1) = 2, \quad z(1) = 3$$

for $1 \leq x \leq 1.4$, using a step size $h = 0.1$, where the non-linearity is due to the term $\sin y$.

Converting the second-order non-linear equation to a first-order system leads to the following non-linear initial-value problem for the first-order system

$$\frac{dy}{dx} = z, \quad \frac{dz}{dx} = -\left[(1+x^2)z + \sin y\right], \quad y(1) = 2, \quad z(1) = 3.$$

The result of using the rk4 method over the interval $1 \le x \le 1.4$ with a step size $h = 0.1$ is shown in the table below.

x_n	y_n	z_n
1.0	2.0	3.0
1.1	2.266989	2.355528
1.2	2.474258	1.805809
1.3	2.631244	1.349200
1.4	2.746991	0.979803

In this case there is no known analytical solution with which these results can be compared, but a repetition of the calculation with $h = 0.05$ confirms that they are accurate to four decimal places.

In many differential equations, changes in the solution occur rapidly over some intervals and slowly over others, so if a uniform step size h is maintained it is not possible to ensure that the magnitude of the error remains the same as the computation proceeds. To overcome this difficulty a variant of the rk4 method called the *Runge–Kutta–Fehlberg 45 method* is now used in most software packages, where it is often abbreviated to the *rkf45* method. The rkf45 method uses a *self-adaptive approach* to maintain accuracy by finding two solutions, one using the fourth-order rk4 method and the other a complicated but more accurate fifth-order Runge–Kutta method. Because of its complexity, the rkf45 algorithm will be omitted and in its place only an outline of the method is given.

The method starts by specifying that throughout the integrations the magnitude ε of the error in the rk4 calculations must lie between a minimum error ε_{min} and a maximum error ε_{max} where, typically, $\varepsilon_{min} = 1 \times 10^{-4}$ and $\varepsilon_{max} = 5 \times 10^{-4}$. Then, with a step length of size h, the value of the solution $y(x)$ at the next step $x = x_i + h$ is computed, first by the rk4 method to yield $y_{i+1}^{(4)}$ and then by the fifth-order Runge–Kutta method to yield $y_{i+1}^{(5)}$. The absolute value $\left| y_{i+1}^{(4)} - y_{i+1}^{(5)} \right| = \varepsilon$ is then taken to be an estimate of the magnitude of the error in the rk4 computation.

It is at this stage that the self-adaptive process comes into operation:

1   If the estimated error lies within the accepted limits, so that $\varepsilon_{min} < \varepsilon < \varepsilon_{max}$, the calculation proceeds using the step length h.
2   If $\varepsilon > \varepsilon_{max}$, the acceptable error is exceeded, so the step length is reduced to $h/2$.
3   If $\varepsilon < \varepsilon_{min}$, the accuracy is greater than required, so the step length is increased to $2h$.
4   Thereafter the calculation continues in this self-adaptive manner.

References have already been given to books that describe various other ways of solving ODEs, most of which are based on the use of finite-difference techniques, so here we only mention the useful reference work by Press *et al.* (2007), which lists the details of the computer codes for all these methods, including the rkf45 method.

3.3.3 Numerical instability and improperly posed problems

Before proceeding to the next topic, it is necessary to mention that for some initial-value problems simply increasing the accuracy of a numerical method of integration cannot produce a correct solution. To illustrate this assertion, consider the following simple initial-value problem:

$$y'' - 7y' - 8y = 0,$$

subject to the initial conditions

$$y(0) = 1 \quad \text{and} \quad y'(0) = -1.$$

It is easily shown that this initial-value problem has the solution $y(x) = e^{-x}$, so the solution tends to zero as $x \to \infty$.

Now consider the related initial-value problem in which the initial condition $y(0) = 1$ is replaced by the condition $y(0) = 1 + \varepsilon$, where $\varepsilon > 0$ is an arbitrarily small number. This modified initial-value problem has the solution

$$y_\varepsilon(x) = \left(1 + \frac{8}{8}\varepsilon\right)e^{-x} + \frac{1}{9}\varepsilon e^{8x}.$$

Inspection of this result shows that, however small ε may be, eventually the factor e^{8x} will always cause the solution $y_\varepsilon(x)$ to tend to infinity as $x \to \infty$. This has an immediate implication for a numerical solution $y(x)$ of the first initial-value problem, because however many decimal places are used in the calculations, the build-up of round-off and truncation errors that are unavoidable will eventually cause a numerical solution to tend to infinity instead of to zero, as $x \to \infty$.

By way of illustration, an application of the rkf45 method to the initial-value problem $y'' - 7y' - 8y = 0$, with $y(0) = 1$, and $y'(0) = -1$, yields a good approximation to the exact solution $y_{\text{exact}}(x) = e^{-x}$ in the interval $0 \le x \le 4$, with $y_{\text{exact}}(4) = 0.0183$ and $y_{\text{rkf}}(4) = 0.0183$. However, for larger x the numerical instability begins to manifest itself, because $y_{\text{exact}}(5) = 0.0067$ and $y_{\text{rkf}}(5) = 0.0412$, after which $y_{\text{exact}}(6) = 0.0025$ and $y_{\text{rkf}}(6) = 102.7$. The solution $y_{\text{rkf}}(x)$ can be forced to approximate the true solution $y_{\text{exact}}(x)$ for $x > 6$ by using double-precision arithmetic, but eventually the numerical instability will again manifest itself.

This is an example of what is called the numerical *instability* of a solution. A differential equation such as this, the solution of which exhibits extreme sensitivity to change in an initial condition, is said to be *ill-conditioned* or *improperly posed*.

3.4 Two-point boundary-value problems

3.4.1 The shooting method

Not all problems involving ODEs are initial-value problems and a different type of problem for a second-order equation is a *two-point boundary-value problem* for an ODE over an interval $a \leq x \leq b$. In such problems a solution of the ODE is required to satisfy a single boundary condition at $x = a$ and another boundary condition at $x = b$. So, unlike initial-value problems, instead of the functional value and its derivative being imposed at $x = a$, one functional value is prescribed at $x = a$ and another at $x = b$. Two-point boundary-value problems may either involve linear or non-linear ODEs, but whereas solving linear problems is straightforward, solving non-linear problems is more difficult and requires iteration.

If the general solution of the second-order linear equation

$$\frac{d^2y}{dx^2} + P(x)\frac{dy}{dx} + Q(x)y = R(x), \quad a \leq x \leq b, \tag{3.15}$$

subject to the two-point boundary conditions

$$y(a) = y_a \quad \text{and} \quad y(b) = y_b \tag{3.16}$$

cannot be found analytically, one way to solve the problem numerically is by using the rk4 or the rkf45 method. At first sight this leads to a difficulty, because a Runge–Kutta method solves initial-value problems and not boundary-value problems. A way of overcoming this difficulty, while still finding a numerical solution by means of a Runge–Kutta method, or indeed any other numerical method for solving the initial-value problem, starts by assigning an arbitrary value d_1 to the unknown derivative of $y(x)$ at $x = a$, so that $y'(a) = d_1$. A Runge–Kutta method can then be used to obtain a numerical solution $y_1(x)$, subject to the initial conditions $y_1(a) = y_a$ and $y_1'(a) = d_1$.

If the interval $a \leq x \leq b$ is divided into N equal parts, a Runge–Kutta method will determine a solution $y_1(x)$ for $x > a$, at intervals with a step size $h = (b - a)/N$, and as a result at the end $x = b$ of the interval the solution will be $y_1(b) = \beta_1$, say, where β_1 is the last value to be computed. This is not likely to satisfy the second boundary condition $y(b) = y_b$, so this process is repeated using the same step size h, but with a *different* arbitrary value d_2

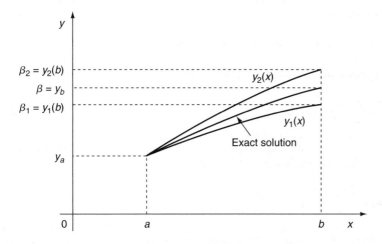

Figure 3.2 An illustration of the shooting method

assigned to the derivative $y'(x)$ at $x = a$, so that this time $y'(a) = d_2$. Now, subject to the initial conditions $y_2(a) = y_a$, $y_2'(a) = d_2$, a Runge–Kutta method will again determine a solution $y_2(x)$ for $x > a$ at intervals with a step size h, and this time at the end $x = b$ of the interval the solution will be $y_2(b) = \beta_2$, where β_2 is the last value to be computed. Once again this is unlikely to satisfy the boundary condition at $x = b$. The relationship between $y_1(x)$ and $y_2(x)$ is illustrated in Figure 3.2, where for the purpose of illustration the required boundary condition $y(b) = y_b$ is seen to lie between β_1 and β_2. In fact the position of y_b relative to the values of β_1 and β_2 is unimportant, because it will be seen from what follows that the method works whatever the value of y_b.

A linear combination of the solutions $y_1(x)$ and $y_2(x)$ is now formed by setting

$$y(x) = k_1 y_1(x) + k_2 y_2(x), \tag{3.17}$$

where it is convenient, although not necessary, to choose the constants k_1 and k_2 (to be found later) so that

$$k_1 + k_2 = 1. \tag{3.18}$$

As ODE equation (3.15) is linear, the substitution of equation (3.17) into equation (3.18), followed by grouping the terms, shows that equation (3.17) is also a solution of the boundary-value problem.

It is required to find a solution so that $y(b) = y_b$, so substituting this condition into equation (3.17) gives

$$y_b = k_1 y_1(b) + k_2 y_2(b) \quad \text{or, equivalently,} \quad y_b = k_1 \beta_1 + k_2 \beta_2. \qquad (3.19)$$

Solving equations (3.18) and (3.19) for k_1 and k_2 gives

$$k_1 = \frac{y_b - \beta_2}{\beta_1 - \beta_2} \quad \text{and} \quad k_2 = \frac{\beta_1 - y_b}{\beta_1 - \beta_2}. \qquad (3.20)$$

The solution of the boundary-value problem is now known, because using equation (3.20) in equation (3.17) gives

$$y(x) = \left[\frac{y_b - \beta_2}{\beta_1 - \beta_2} \right] y_1(x) + \left[\frac{\beta_1 - y_b}{\beta_1 - \beta_2} \right] y_2(x), \quad a \leq x \leq b. \qquad (3.21)$$

Thus, when a Runge–Kutta method of solution is applied to equation (3.15), the boundary-value problem subject to equation (3.16) is solved over the interval $a \leq x \leq b$ at intervals with a step size h in terms of the two numerical solutions $y_1(x)$ and $y_2(x)$.

This method of solving a boundary-value problem is called the *shooting method*, because of the analogy between the way the numerical problem is approached and shooting at a target from a fixed point using a different angle of elevation for each shot in order to hit the target at the required point.

As already mentioned, the numerical solution of boundary-value problems becomes harder when the ODE is non-linear, because the linear superposition condition (equation 3.17) no longer applies. To make the method work it is necessary to modify the shooting method and again to start by using two different (guessed) values X_1 and X_2 for the derivatives $y_1'(a)$ and $y_2'(a)$. Thus, the calculation starts with $y_1(a) = y_a$, $y_1'(a) = X_1$, and $y_2(a) = y_a$, $y_2'(a) = X_2$, each satisfying the same boundary value at the left, but with different (guessed) slopes at $x = a$. Solutions $y_1(x)$ and $y_2(x)$ are then calculated, usually by using either the rk4 or rkf45 method, to determine $y_1(b)$ and $y_2(b)$, both of which will differ from the required boundary value $y(b) = y_b$ at the right. If the approximations are iterated, the subsequent iterations can be produced by using the sequence of values X_n given by linear interpolation:

$$X_n = X_{n-1} - \frac{\left(y_{n-1}(b) - y_b\right)(X_{n-1} - X_{n-2})}{y_{n-1}(b) - y_{n-2}(b)}, \quad \text{for} \quad k = 2, 3, \dots, n \qquad (3.22)$$

The iteration is terminated when for some $n = N$ the absolute value of the difference between X_N and X_{N-1} is such that $|X_N - X_{N-1}| < \varepsilon$, for some pre-assigned error $\varepsilon > 0$, say $\varepsilon = 10^{-4}$. The required solution is then taken to be the solution corresponding to the initial conditions $y_1(a) = y_a$, $y_1'(a) = X_N$. Worked examples of this approach and of the use of Newton's method instead of using the linear interpolation method given above can be found in Burden and Faires (2005).

3.4.2 Finite-difference methods

A completely different approach involves the use of finite differences with the linear equation (3.15). To understand this method we must first define the meaning of the different types of finite-difference approximations that arise. Let a differentiable function $y(x)$ be defined at the points $x = x_0 - h$, $x = x_0$ and $x = x_0 + h$. Then $y'(x_0)$ can be approximated in the three different ways

$$y'(x_0) \approx \frac{1}{2h} \left[y(x_0 + h) - y(x_0 - h) \right], \tag{3.23}$$

$$y'(x_0) \approx \frac{1}{h} \left[y(x_0 + h) - y(x_0) \right], \tag{3.24}$$

and

$$y'(x_0) \approx \frac{1}{h} \left[y(x_0) - y(x_0 - h) \right]. \tag{3.25}$$

Approximation (3.23) represents $y'(x_0)$ as the quotient of the difference between $y(x)$ at equally spaced values h of x to the left and right of x_0, and the interval $2h$ between them, and is called the *centred-difference* approximation for $y'(x_0)$. Approximation (3.24) represents $y'(x_0)$ as the quotient of the difference between the values if $y(x)$ at x_0 and at $x_0 + h$ ahead of x_0, and the interval h between them, and is called the *forward-difference* approximation, while for obvious reasons approximation (3.25) is called the *backward-difference* approximation.

The formal derivation of these results follows from Taylor's theorem with a remainder. So, for example, to derive equation (3.23) we have

$$y(x_0 + h) = y(x_0) + y'(x_0) + \frac{1}{2} h^2 y''(\xi),$$

where ξ is unknown, but such that $a < \xi < b$; after rearrangement

$$y'(x_0) = \frac{1}{h} \left[y(x_0 + h) - y(x_0) \right] - \frac{h}{2} y''(\xi), \tag{3.26}$$

which is equation (3.23) with an error term $-1/2hy''(\xi)$ of magnitude $O(h)$. Similarly,

$$y(x_0 - h) = y(x_0) - hy'(x_0) + \frac{1}{2}h^2y''(\xi),$$

and after rearrangement

$$y'(x_0) = \frac{1}{h}\left[y(x_0) - y(x_0 - h)\right] - \frac{h}{2}y''(\xi_2), \tag{3.27}$$

which is equation (3.24), again with an error term of magnitude $O(h)$. Adding equations (3.24) and (3.25) gives

$$y'(x_0) = \frac{1}{2h}\left[y(x_0 + h) - y(x_0 - h)\right] - \frac{h}{2}y''(\xi), \tag{3.28}$$

which is equation (3.25) with an error term of magnitude $O(h)$. A corresponding argument shows the centred-difference approximation for $y''(x_0)$ is given by

$$y''(x_0) = \frac{1}{h^2}\left[y(x_0 - h) - 2y(x_0) + y(x_0 + h)\right] - \frac{h^2}{12}y^{(4)}(\xi), \tag{3.29}$$

although this time the order of magnitude of the error term is smaller, because it is $O(h^2)$.

In a finite-difference approach to the solution of the boundary-value problem (3.15) and (3.16), the interval $a \leq x \leq b$ is divided into N equal subintervals with divisions at $x_i = a + kh$, for $k = 0, 1, \ldots, N + 1$, with $h = (b - a)/N$, at each point of which the solution $y(x_k) = y_k$. The first- and second-order derivatives $y'(x)$ and $y''(x)$ in equation (3.15) are then replaced by centred-difference approximations (3.27) and (3.29), and after rearrangement the ODE is replaced by the *finite-difference approximation*

$$\left(\frac{1}{2}hP(x_k) - 1\right)\tilde{y}_{k-1} + \left(2 - h^2Q(x_k)\right)\tilde{y}_k - \left(1 + \frac{1}{2}P(x_k)\right)\tilde{y}_{k+1} = h^2R(x_k),$$
$$\tag{3.30}$$

where $\tilde{y}_{k-1}, \tilde{y}_k$ and $\tilde{y}_{k+1}$ are the finite-difference approximations for y_{k-1}, y_k and y_{k+1}.

The two-point boundary-value problem for equations (3.15) and (3.16) can then be written in the matrix form

$$\mathbf{Ax = b}, \tag{3.31}$$

where, after using the boundary conditions $\tilde{y}_0 = y_a$ and $\tilde{y}_{N+1} = y_b$, we have

$$A = \begin{bmatrix} 2 - h^2 Q(x_1) & -1 - \frac{h}{2}P(x_1) & 0 & \cdots & & 0 \\ -1 + \frac{h}{2}P(x_2) & 2 - h^2 Q(x_2) & -1 - \frac{h}{2}P(x_2) & \cdots & & 0 \\ 0 & 0 & \vdots & \vdots & & 0 \\ \vdots & \vdots & \vdots & \vdots & & -1 - \frac{h}{2}P(x_{N-1}) \\ 0 & 0 & 0 & & -1 - \frac{h}{2}P(x_N) & 2 - h^2 Q(x_N) \end{bmatrix},$$

and

$$\mathbf{x} = \begin{bmatrix} \tilde{y}_1 \\ \tilde{y}_2 \\ \tilde{y}_3 \\ \vdots \\ \tilde{y}_N \end{bmatrix}, \quad \mathbf{b} = \begin{bmatrix} \left(1 - \frac{h}{2}P(x_1)\right)y_a - h^2 R(x_1) \\ -h^2 R(x_2) \\ \vdots \\ -h^2 R(x_{N-1}) \\ \left(1 + \frac{h}{2}P(x_N)\right)y_b - h^2 R(x_N) \end{bmatrix}. \tag{3.32}$$

In the linear algebraic system (3.31), $\mathbf{A}$ is a matrix in which non-zero entries only occur on the leading diagonal and on the diagonals immediately above and below it. This is called a *tri-diagonal matrix* and the equation $\mathbf{Ax = b}$ can be solved for the solution vector $\mathbf{x}$ using Gaussian elimination or by iteration, and possibly by using the SOR method if matrix $\mathbf{A}$ is very large.

As the error in this finite-difference approach is of the order h^2, the accuracy of the result will depend on the number N of subdivisions in the interval $a \leq x \leq b$. Note that the number of calculations involved in the finite-difference method will increase very rapidly as N increases. Contrast this with the effort involved and the accuracy attained when the shooting method is applied to a linear ODE, because there the solution is easily constructed as a linear combination of two solutions obtained either by using the rk4 or the rkf45 method. This method can be adapted for use with non-linear two-point boundary-value problems, but iteration again becomes necessary and can lead to far more computational effort than the generalization of the shooting method.

For more information about the solution of two-point non-linear boundary-value problems, see Atkinson (2007), Johnson and Riess (1982) and Schwarz (1989), while for a detailed explanation and application of the method to a non-linear problem, see Burden and Faires (2005).

3.5 The numerical solution of partial differential equations

3.5.1 Two-dimensional incompressible flow

The numerical solution of PDEs is an important and complicated subject, and detailed descriptions of numerical procedures are to be found in many textbooks, some of which will be mentioned later. These books also describe the different ways by which the large systems of equations produced by the finite-difference and finite-element methods of solution can be solved. Consequently, the limited space available here is only sufficient to outline the basic ideas involved.

The study of two-dimensional incompressible fluid flow in Cartesian coordinates (see also Sections 4.2 and 4.3) is based on the two fundamental *Navier–Stokes equations* (the momentum equations)

$$\frac{\partial u}{\partial t} + u\frac{\partial u}{\partial x} + v\frac{\partial u}{\partial y} = -\frac{1}{\rho}\frac{\partial p}{\partial x} + v\left(\frac{\partial^2 u}{\partial x^2} + \frac{\partial^2 u}{\partial y^2}\right),\tag{3.33}$$

$$\frac{\partial v}{\partial t} + u\frac{\partial v}{\partial x} + v\frac{\partial v}{\partial y} = -\frac{1}{\rho}\frac{\partial p}{\partial y} + v\left(\frac{\partial^2 v}{\partial x^2} + \frac{\partial^2 v}{\partial y^2}\right),\tag{3.34}$$

and the *continuity equation*

$$\frac{\partial u}{\partial x} + \frac{\partial v}{\partial y} = 0.\tag{3.35}$$

In these equations u and v are the velocity components in the x- and y-directions, t is the time, p is the pressure, ρ is the fluid density and v is the kinematic viscosity.

We mention in passing that, in computational work, as elsewhere, it is usual to express equations (3.33–3.35) in *non-dimensional form*. This is accomplished by introducing a characteristic length $\bar{L}$, and a characteristic velocity $\bar{U}$, when the non-dimensional unit of time becomes $\bar{L}/\bar{U}$. Other dimensionless quantities can then be defined as (see Chapter 5):

$$\bar{x} = \frac{x}{\bar{L}}, \quad \bar{y} = \frac{y}{\bar{L}}, \quad \bar{u} = \frac{u}{\bar{U}}, \quad \bar{v} = \frac{v}{\bar{U}}, \quad \bar{\zeta} = \frac{\zeta}{(\bar{U}/\bar{L})} \quad \text{and} \quad \bar{t} = \frac{t}{(\bar{L}/\bar{U})},\tag{3.36}$$

and, where appropriate, the non-dimensional *Reynolds number* $\mathbf{Re} = \bar{U}\bar{L}/\bar{v}$ can be introduced, where $\bar{v}$ is the dimensionless viscosity.

Various simplifications can be made in the governing equations, such as neglecting viscosity by setting $v = 0$ and by confining attention to steady-state problems by setting $\partial u/\partial t = \partial v/\partial t = 0$. However, instead of attempting to study the specific types of PDE that arise from various approximations, the problem will be simplified by examining instead some model equations that are typical of the PDEs that occur.

The model PDE equations to be considered are:

The *Laplace equation*

$$\frac{\partial^2 u}{\partial x^2} + \frac{\partial^2 u}{\partial y^2} = 0,$$ (3.37)

the solutions of which are boundary-value problems.

The *Poisson equation*

$$\frac{\partial^2 u}{\partial x^2} + \frac{\partial^2 u}{\partial y^2} = f(x, y),$$ (3.38)

the solutions of which are boundary-value problems. Note that the Poisson equation reduces to the Laplace equation when $f(x, y) \equiv 0$.

A *transport equation*

$$\frac{\partial u}{\partial t} = -u\frac{\partial u}{\partial x} + \kappa\frac{\partial^2 u}{\partial x^2}, \quad (\kappa \geq 0).$$ (3.39)

This equation is also called the one-dimensional second-order *advection* equation and it involves both initial conditions and boundary conditions.

The *diffusion equation* (also called the *heat equation*)

$$\frac{\partial u}{\partial t} = \kappa\frac{\partial^2 u}{\partial x^2}. \quad (\kappa > 0 \text{ is the diffusivity constant})$$ (3.40)

This equation involves both initial conditions and boundary conditions.

The *wave equation*

$$\frac{\partial^2 u}{\partial t^2} = c^2\frac{\partial^2 u}{\partial x^2}.$$ (3.41)

This equation can involve only initial conditions, when it becomes a *pure initial-value problem*, or initial and boundary conditions, when it becomes an *initial boundary-value problem*.

3.5.2 Finite-difference approximations for partial derivatives

All finite-difference methods are based on the approximation of partial derivatives obtained by the truncation of Taylor series. The approach for two-dimensional problems starts by considering a two-dimensional region D of the (x, y)-plane with boundary B in which a solution of a PDE is required. For the purpose of illustration let us suppose the region D is a rectangle $a \leq x \leq b$ and $c \leq y \leq d$. Now divide the interval $a \leq x \leq b$ into m subintervals each of length $h = (b - a)/m$, and the interval $c \leq y \leq d$ into n subintervals each of length $k = (d - c)/n$. Region D is then covered by a grid of lines $x = a + ih$, and by another grid of lines $y = c + jh$, with $i = 0, 2, \ldots, m$ and $j = 0, 1, 2, \ldots, n$. It is then required to approximate the solution of the PDE by discrete values at each of the grid points formed by the intersection of the grid lines.

To obtain approximations for partial derivatives, let $u(x, y)$ be a suitably differentiable function of x and y, and denote by u_{ij} the functional value of $u(x_i, y_j)$, where $x_i = a + ih$ and $y_j = c + jk$. Then the Taylor-series expansion of $u(x_i + h, y_j)$ about the point (x_i, y_j) in terms of x can be written as

$$u_{i+1,j} = u_{ij} + h \left(\frac{\partial u}{\partial x} \right)_{ij} + \frac{1}{2} h^2 \left(\frac{\partial^2 u}{\partial x^2} \right)_{ij} + \text{remainder.} \tag{3.42}$$

When this is solved for $\partial u / \partial x$, and only terms of order h are retained, we arrive at the *forward-difference approximation*

$$\left(\frac{\partial u}{\partial x} \right)_{ij} = \frac{u_{i+1,j} - u_{ij}}{h} + O(h), \tag{3.43}$$

where $O(h)$ represents a remainder with an order of magnitude h. Here the expression $O(h)$ includes not only terms of order h, but also all smaller terms of still higher orders such as $h^2, h^3, \ldots$.

A similar argument applied to the function $u(x_i - h, y_j)$ yields the corresponding *backward-difference approximation*

$$\left(\frac{\partial f}{\partial x} \right)_{ij} = \frac{f_{ij} - f_{i-1,j}}{h} + O(h). \tag{3.44}$$

The *centred-difference approximation* for $(\partial u / \partial x)_{ij}$, obtained by differencing the forward- and backward-difference approximations over an interval of length $2h$ while taking proper account of the behaviour of the higher-order terms, is found to be

$$\left(\frac{\partial u}{\partial x} \right)_{ij} = \frac{u_{i+1,j} - u_{i-1,j}}{2h} + O(h^2), \tag{3.45}$$

showing that the truncation error is now much smaller, because it is $O(h^2)$.

Further manipulation of Taylor series involving the expansion of $u_{i-1,j}$ and $u_{i+1,j}$ to higher terms followed by addition of the results shows that

$$\left(\frac{\partial^2 u}{\partial x^2}\right)_{ij} = \frac{u_{i+1,j} - 2u_{ij} + u_{i-1,j}}{h^2} + O(h^2). \tag{3.46}$$

where once again the truncation error is $O(h^2)$. In similar fashion it follows that

$$\left(\frac{\partial^2 u}{\partial y^2}\right)_{ij} = \frac{u_{i,j+1} - 2u_{ij} + u_{i,j-1}}{k^2} + O(k^2), \tag{3.47}$$

where this time the truncation error is $O(k^2)$.

By equating the mixed derivatives $\partial^2 u/\partial x \partial y$ and $\partial^2 u/\partial y \partial x$ it is found that

$$\left(\frac{\partial^2 u}{\partial x \partial y}\right)_{ij} = \frac{u_{i+1,j+1} - u_{i+1,j-1} - u_{i-1,j+1} + u_{i-1,j-1}}{4hk} + O(h^2, k^2). \tag{3.48}$$

These results can be combined to give finite-difference representations of more general operators. For example, the addition of equations (3.46) and (3.47) yields the following centred-difference approximation for the Laplacian $\Delta u \approx \partial^2 u/\partial x^2 + \partial^2 u/\partial y^2$:

$$\Delta u \approx \frac{u_{i+1,j} - 2u_{ij} + u_{i-1,j}}{h^2} + \frac{u_{i,j+1} - 2u_{ij} + u_{i,j-1}}{k^2} + O(h^2, k^2). \tag{3.49}$$

The same approach applies to time-dependent problems where $u = u(x, y, t)$, when the time t is advanced in discrete steps of magnitude τ. So, for example, at the nth time step, the centred-difference approximation $(\partial u/\partial t)_{ij}^n$ for $(\partial u(x, y, t)/\partial t)_{(x_i, y_j, t=n\tau)}$ is given by

$$\left(\frac{\partial u}{\partial t}\right)_{ij}^n = \frac{u_{ij}^{n+1} - u_{ij}^{n-1}}{2\tau} \tag{3.50}$$

with i and j fixed. The calculation of this derivative illustrates something of the difficulties associated with time-dependent problems. This is because, to find an approximation for $\left(\partial f/\partial t\right)_{ij}^n$ at the time $t = n\tau$, it is necessary to know by estimation the functional value at the *next* time step, although this difficulty can be overcome by using a different approach.

The above results can be combined to yield finite-difference approximations for PDEs, although with time-dependent problems the straightforward substitution of finite differences in place of partial derivatives does not always yield a useful numerical approximation. This is because some numerical schemes are **unstable** in the sense that their forward predictions become *chaotic*. As a result, the stability of a numerical scheme must always be investigated to obtain stability conditions, which when satisfied ensure the scheme yields stable results.

3.5.3 The finite-difference method for elliptical equations

We now give a simple example of how an elliptical PDE can be solved in a bounded region D using a finite-difference approach. Although the method is illustrated by applying it to the Laplace equation, the approach is quite general and extends immediately to other elliptical equations.

Example

The problem to be solved requires the solution of the Laplace equation in a unit square D, on the boundary of which Dirichlet conditions are imposed.

We will make use of result (3.49), where, because region D is a square, we may set $h = k$. Using the centred-difference approximation (3.49), and setting $\Delta f = f_{xx} + f_{yy} = 0$ to form the two-dimensional Laplace equation simplifies it to the finite-difference equation

$$u_{i+1,j} + u_{i-1,j} + u_{i,j+1} + u_{i,j-1} - 4u_{ij} = 0, \tag{3.51}$$

and solving this for u_{ij} gives the following simple result:

$$u_{ij} = \frac{1}{4}\left(u_{i+1,j} + u_{i-1,j} + u_{i,j+1} + u_{i,j-1}\right). \tag{3.52}$$

This result is illustrated in Figure 3.3(a), which shows the distribution of points around the grid point (i, j), while Figure 3.3(b) shows the weight to be attached to each of the points relative to the central point. Thus, when $h = k$, it is seen that the functional value at the grid point (i, j) is the average of the functional values at the four surrounding points.

For convenience of reference the pattern of points shown in Figure 3.3(a) and (b) is sometimes called a *computational molecule*.

The boundary-value problem to be considered involves solving the Laplace equation in the unit square D with the boundaries $0 \leq x \leq$, $0 \leq y \leq 1$ shown in Figure 3.4, subject to the Dirichlet boundary conditions

$$u(x, y) = \begin{cases} 2x & \text{on} \quad y = 0, 0 \leq x \leq 1 \\ 2 & \text{on} \quad x = 1, 0 \leq y \leq 1 \\ 2\sin\frac{1}{2}\pi x & \text{on} \quad y = 1, 0 \leq x \leq 1 \\ 0 & \text{on} \quad x = 1, 0 \leq y \leq 1. \end{cases}$$

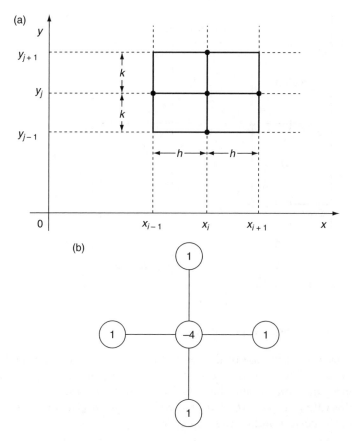

Figure 3.3 (a) The distribution of points around the grid point (i, j). (b) The weights attached to the grid points in (a)

As the example only serves to illustrate the application of the finite-difference method for an elliptical equation, we will not seek high accuracy and set $h = k = \frac{1}{3}$; the corresponding grid lines are shown in Figure 3.4.

To formulate the finite-difference scheme we will apply (3.52) to each of the points P_1 to P_4 in Figure 3.4; the boundary conditions at the points Q_i for $i = 1, 2, \ldots, 12$ follow from the Dirichlet boundary conditions. This leads to the following four equations:

Point P_1: $4u_1 - u(Q_{12}) - u_2 - u(Q_2) - u_4 = 0$
Point P_2: $4u_2 - u_1 - u(Q_5) - u(Q_3) - u_3 = 0$
Point P_3: $4u_3 - u_4 - u(Q_6) - u_2 - u(Q_8) = 0$
Point P_4: $4u_4 - u(Q_{11}) - u_3 - u_1 - u(Q_9) = 0,$

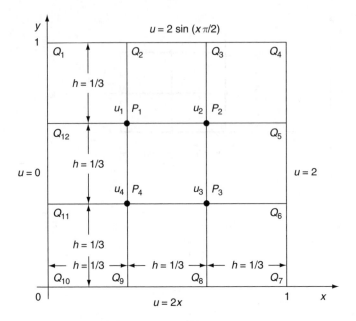

Figure 3.4 The region $0 \le x \le, 0 \le y \le 1$

where, because of the use of the centred-difference approximation, the values of u at the four corners Q_1, Q_4, Q_7 and Q_{10} do not appear in these difference equations. Using the Dirichlet conditions with $h = 1/3$ to determine the values $u(Q_2), u(Q_3), u(Q_5), u(Q_8), u(Q_9), u(Q_{11})$ and $u(Q_{12})$ then leads to the corresponding four equations

Point $P_1: 4u_1 - u_2 - u_4 = 2 \sin\left(\dfrac{\pi}{6}\right)$

Point $P_2: 4u_2 - u_1 - u_3 = 2 + 2$

Point $P_3: 4u_3 - u_4 - u_2 = 2 + \left(\dfrac{4}{3}\right)$

Point $P_4: 4u_4 - u_3 - u_1 = \dfrac{2}{3}$.

The solution of these equations is easily found to be $u_1 = 0.7971$, $u_2 = 1.4774$, $u_3 = 1.3804$ and $u_4 = 0.7111$, so an approximate solution of the boundary-value problem has been obtained. Naturally, to obtain a better approximation a finer grid must be used, but the crude grid used here is adequate to illustrate the general numerical approach.

Had a boundary B of region D not coincided with a grid line, as happened in the above example, the problem would have become more complicated. In such a case the finite-difference approximations at grid points close to the boundary will require modification. This is to allow for the fact that close to the boundary at least one of the points in the computational molecules will lie outside boundary B. We will not discuss how this is done, and refer instead to the books by Ames (1978), Schwarz (1989) and Smith (1985).

The imposition of a Neumann boundary condition on a boundary also requires consideration. Suppose, for example, that the condition $\partial u/\partial x = 0$ is to be imposed on a boundary $x = $ constant. Then, depending on whether the boundary is to the left or right of the region D, either the forward-difference quotient in equation (3.43) equated to zero can be used to approximate the Neumann boundary condition on the left, or the backward-difference quotient in equation (3.44) can be equated to zero to approximate the boundary condition on the right. This causes a slight loss of accuracy, because the centred-difference quotients used to represent the Laplacian operator have the second-order accuracy $O(h^2 + k^2)$, while results (3.43) and (3.44) only have the first-order accuracy $O(h)$.

On occasion, symmetry can be used to represent a Neumann condition, albeit at the cost of doubling the computational effort required to obtain a numerical solution. The symmetry approach works as follows. Suppose, for example, that the Neumann condition $\partial u/\partial x = 0$ is $\partial x = 0$ is to be imposed on the boundary $x = 1$ of a region like D in $x = 1$, together with its boundary conditions on its other three sides to form a region $\widetilde{D}$ double the size of D, while leaving unspecified the values of u along the line $x = 1$ that bisects region $\widetilde{D}$. Then, from symmetry considerations, the solution in D will also be reflected into the region to the right of $x = 1$. As the solution of the Laplace equation must be continuous with continuous first-order partial derivatives, the symmetry of the solution across $x = 1$ implies that $\partial u/\partial x = 0$ on $x = 1$, although of course it is not necessary that $u = 0$ on $x = 1$. So, if the solution is found in $\overline{D}$ as in the example, the solution in the part of $\overline{D}$ that forms D will be the required solution. This approach causes no loss of accuracy when allowing for the Neumann condition $\partial u/\partial x = 0$ on $x = 1$, although the more accurate result is only obtained at the cost of doubling the size of the set of linear equations that must be solved.

The finite-difference method extends in an obvious way to boundary-value problems for the Poisson equation

$$\frac{\partial^2 u}{\partial x^2} + \frac{\partial^2 u}{\partial y^2} = f(x, y). \tag{3.53}$$

In this case, it is not difficult to see that the finite-difference result corresponding to equation (3.49) must be changed to

$$\frac{1}{h^2}\left[u(x_0 - h, y_0) - 2u(x_0, y_0) + u(x_0 + h, y_0)\right]$$
$$+ \frac{1}{k^2}\left[u(x_0, y_0 - k) - 2u(x_0, y_0) + u(x_0, y_0 + k)\right] = f(x_0, y_0), \tag{3.54}$$

after which the computation proceeds as before.

Finite-difference schemes can also be derived for boundary-value problems with more than two independent variables, although the numerical calculations become extremely lengthy, particularly when curved boundaries are involved.

A clear and straightforward account of the solution of linear elliptical equations in general is to be found in Ames (1978), Ferziger (1998), Pearson (1986), Smith (1985) and see also Ortega and Poole (1983).

3.5.4 The first-order transport equation

It is now necessary to comment on the solution of problems that depend on both space variables and time, a typical example being the non-linear transport equation

$$\frac{\partial z}{\partial t} = -z\frac{\partial z}{\partial x} + \kappa\frac{\partial^2 z}{\partial x^2}, \quad (\kappa \geq 0) \tag{3.55}$$

subject to the initial condition

$$z(x, 0) = F(x). \tag{3.56}$$

An equation like this can arise in various ways, one of which occurs when deriving the PDE that describes how vorticity is transported through a fluid.

To avoid difficulties due to non-linearity, the equation will be linearized by setting $z = U + u$, where $U = $ constant. The result obtained after linearization is

$$\frac{\partial u}{\partial t} = -U\frac{\partial u}{\partial x} + \kappa\frac{\partial^2 u}{\partial x^2}, \quad (\kappa > 0) \tag{3.57}$$

with the initial condition

$$u(x, 0) = f(x), \quad \text{where} \quad f(x) = F(x) - U. \tag{3.58}$$

To examine the constraints imposed on the space step size h and the time step size τ if a finite-difference equation representing (3.57) is to yield a satisfactory approximation, it will suffice to simplify equation (3.57) still further by setting $\kappa = 0$ to arrive at the hyperbolic equation

$$\frac{\partial u}{\partial t} + U\frac{\partial u}{\partial x} = 0 \tag{3.59}$$

subject to the initial condition $u(x, 0) = f(x)$.

It follows from the discussion in Chapter 2 that the characteristics of this equation in the (x, t)-plane determined by $dx/dt = U$ are the parallel straight lines $x = x_0 + Ut$, where x_0 is the intercept of the characteristic on the initial line $t = 0$, while the constant value

$$u(x, t) = f(x_0) \tag{3.60}$$

is transported along the characteristic through the point $(x_0, 0)$ on the x-axis. This shows that the *domain of determinacy* of any point P on the characteristic through $(x_0, 0)$ is the characteristic itself that lies between $(x_0, 0)$ and P, while the *domain of dependence* of point P is the point $(x_0, 0)$. Let us now examine how these concepts apply to a finite-difference approximation of equation (3.59).

Let the functional values at $x = ih$ at the times $t = n\tau$ and $t = (n + 1)\tau$ be u_i^n and u_i^{n+1}, and the functional values at $x = (i - 1)h$ and $x = ih$ at the time $t = n\tau$ be u_{i-1}^n and u_i^n.

Then a simple finite-difference approximation for equation (3.59) is

$$\frac{u_i^{n+1} - u_i^n}{\tau} + U \frac{u_i^n - u_{i-1}^n}{h} = 0. \tag{3.61}$$

When solved for u_i^{n+1} we find that

$$u_i^{n+1} = (1 - \lambda) u_i^n + \lambda u_{i-1}^n, \quad \text{where} \quad \lambda = \left(\frac{U\tau}{h} \right). \tag{3.62}$$

Now consider the network of points in the (x, t)-plane that mark the intersection of grid lines. Let the discrete values of u at a given time $t = (m + 1)\tau$ be represented in the (x, t)-plane by dots on the horizontal line $t = (m + 1)\tau$, with the spacing between the dots equal to h, and consider the structure of the approximation u_i^{m+1} at the dot P located at the point $(x_0 + ih, (m + 1)\tau)$. Then, on the previous line where the time $t = m\tau$, the structure of equation (3.61) is such that the solution at points influencing the solution at P will lie at the two points $(x_0 + (i - 1)h, m\tau)$ and $(x_0 + ih, m\tau)$. Thus, on the line where $t = m\tau$, one dot will lie directly below P, with another one on the same line to its immediate left, as shown in Figure 3.5, which illustrates a particular case.

Considering the successive grid lines that lie below $t = m\tau$, and continuing in this manner, it can be seen that the dots corresponding to solutions that influence the discrete solution at P will lie in a right-angled triangle, the right side of which corresponds to the line $x = x_0 + ih$, while all the other dots will lie to the left of $x = x_0 + ih$. By way of example, the pattern of dots through the representative point $(x_0 + 4h, 4\tau)$ in the (x, t)-plane is illustrated in Figure 3.5. The discrete approximate solutions in the triangular region with its vertex at P in Figure 3.5 form the discrete analogue of the *domain of*

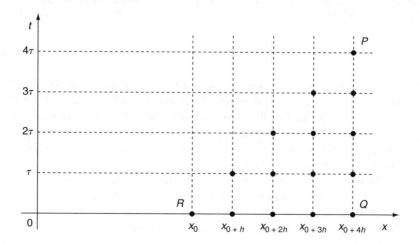

Figure 3.5 The solutions at discrete points in the (x, t)-plane that influence the solution at point P located at $(x_0 + 4h, 4\tau)$

determinacy in the continuous case, while the points from Q to R represent the discrete *domain of dependence* of the discrete solution at P.

Now consider the two different situations shown in Figure 3.6(a) and (b), where the straight-line characteristic through P is shown as the solid line that has been traced back to the x-axis at point S. The difference between the two cases is that in Figure 3.6(a) the sizes of the spatial step length h and the time step length τ are not the same as those in Figure 3.6(b). In Figure 3.6(a) the characteristic intersects the x-axis at the point S that lies between Q and R, so that S lies in the discrete domain of dependence. The result of (3.60) is the solution along the straight-line characteristic through the point $(x_0, 0)$.

In Figure 3.6(a) the solution at P will be determined correctly by the finite-difference scheme. However, in Figure 3.6(b) point S lies outside the discrete domain of dependence, so in this case the solution at P *cannot* be properly determined by the discrete form of the initial condition for the equation.

Consideration of Figure 3.6(a) shows that for the solution at P to be properly determined by the initial condition, the slope of the line QP must be less than or equal to the slope of the characteristic through the point S. In Figure 3.6(a) the x-axis is horizontal and the t-axis is vertical; as we already know that $dx/dt = U$ on the characteristic in the (x, t)-plane, because of the orientation of the axes it follows that the slope of the characteristic through S in Figure 3.6(a) must be $dt/dx = 1/U$. For the solution at P to be determined correctly it is necessary that $\tau/h \leq 1/U$, which leads to the condition

$$U\tau/h \leq 1. \tag{3.63}$$

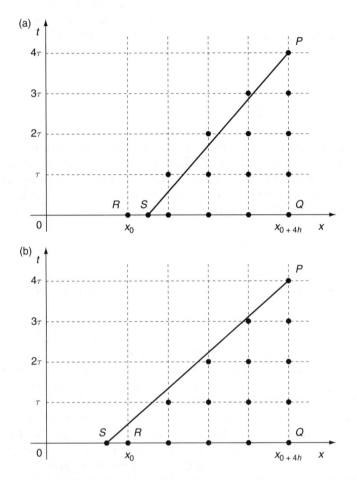

Figure 3.6 (a) Point S lies inside the discrete domain of dependence of point P. (b) Point S lies outside the discrete domain of dependence of point P

This important result is known as the *Courant, Friedrichs and Lewy condition*, abbreviated to the CFL condition, that must be satisfied by the condition on h and τ in equation (3.61) in order that the solution at P is properly determined. The CFL condition is a *necessary*, but not a *sufficient*, condition for a finite-difference scheme involving the space variable x and the time variable t to take proper account of the initial conditions.

Just as the solution of an ODE can be unstable, so also can the solution of a finite-difference approximation in space and time. Conditions can be derived that determine if and when a finite-difference scheme will be stable and converge to the solution of the PDE as the sizes of both h and τ

are reduced. For a scheme to produce correct results it is necessary that both its stability condition and the corresponding CFL condition are satisfied.

An example of a finite-difference scheme for equation (3.59), that even when the CFL condition is satisfied can be shown always to produce chaotic results, is

$$\frac{u_i^{n+1} - u_i^n}{\tau} + U\frac{u_i^n - u_{i-1}^n}{h} = 0. \tag{3.64}$$

So it is appropriate to say this scheme is *unconditionally unstable*, because no choice of τ and h can be made that will avoid this chaotic behaviour. A scheme for equation (3.59) that uses a centred-difference expression for the space derivative, instead of the forward difference used in equation (3.64), and which does not lead to chaotic behaviour, is

$$\frac{u_i^{n+1} - u_i^n}{\tau} + U\frac{u_{i+1}^n - u_{i-1}^n}{h} = 0. \tag{3.65}$$

This scheme can be said to be *stable*, and we now explain the exact meaning of this term.

For any finite-difference scheme to be useful it is necessary that it is both stable and convergent, so let us now make clear what the requirements of stability and convergence really mean. A finite-difference scheme is said to be *stable* if when it is applied to two closely related problems, say to two Dirichlet problems with slightly different boundary conditions or to two slightly different initial-value problems, the difference between the two sets of boundary or initial conditions and between the two solutions is small. A finite-difference scheme is said to be *convergent* if, as the size of the integration steps is reduced, the scheme generates solutions that approach a limiting form. It can be seen that if these ideas are to be applied successfully it is necessary that in the case of stability the meaning of the 'closeness' of boundary or initial conditions, and also of solutions, must be made precise; while for convergence the exact meaning of 'tending to a limiting solution' must be expressed in mathematical terms.

In both cases of stability and of convergence the mathematical meanings of 'closeness' and of 'tending to a limiting solution' can be made precise by introducing a concept called a *norm*, which can be related to both the closeness of boundary and the initial conditions, and also to the closeness of solutions. There are many different types of norm, but one that will be familiar is the Euclidean norm where the 'closeness' between two space vectors $\mathbf{u} = u_1\mathbf{i} + u_2\mathbf{j} + u_3\mathbf{k}$ and $\mathbf{v} = v_1\mathbf{i} + v_2\mathbf{j} + v_3\mathbf{k}$ is measured by the distance between their tips, given by $d = \sqrt{(u_1 - v_1)^2 + (u_2 - v_2)^2 + (u_3 - v_3)^2}$. The mathematical notation for this Euclidean norm involves writing $d = \|\mathbf{u} - \mathbf{v}\|$, where the symbol $\|\ \|$ is to be read 'the norm of'. This idea easily extends to the n-component vectors $\mathbf{u}$ and $\mathbf{v}$, when the norm of $\mathbf{u} - \mathbf{v}$ becomes $d = \|\mathbf{u} - \mathbf{v}\| = \sqrt{(u_1 - v_1)^2 + (u_2 - v_2)^2 + \ldots + (u_n - v_n)^2}$.

To develop these ideas a little further, we can write a general linear difference scheme as $\mathbf{A}U^{n+1} = \mathbf{B}U^n + \mathbf{C}^n$, where $\mathbf{A}$, $\mathbf{B}$ and $\mathbf{C}$ are matrices that describe the finite-difference operations, while U^n and U^{n+1} are matrix vectors containing the finite-difference solutions at grid points at the nth and $(n+1)$th steps. Provided $\mathbf{A}^{-1}$ exists, which it does if $\det \mathbf{A} \neq 0$, the finite-difference scheme can be written as $U^{n+1} = \mathbf{A}^{-1}[\mathbf{B}U^n + \mathbf{C}^n]$. Then the norm of the difference of the solution vectors $\|U^{n+1} - U^n\|$ provides a measure of the closeness of the two vectors U^{n+1} and U^n. This can be compared with the corresponding norm of the difference between the two boundary or initial conditions. So the ability of the norm to measure the closeness of the boundary or initial conditions also shows how the closeness of solutions can be measured. A condition based on the norm can then be derived which, when satisfied, ensures the stability of the scheme. We will not attempt to explain how a stability condition can be derived, but suffice it to say that one way of finding such a condition makes use of a Fourier-type approach.

At first sight, a process similar to the one used to determine a stability condition for a finite-difference scheme needs to be repeated in order to establish the convergence of the scheme. Fortunately, a theorem proved by Lax in 1953 has shown that the stability of a scheme implies (is equivalent to) the convergence of the scheme, so only a stability analysis becomes necessary.

This important result, called the *Lax equivalence theorem*, has wide-ranging implications. In straightforward terms the theorem says that for a finite-difference scheme for a PDE whose solution is not unreasonably sensitive to changes in the boundary or initial data, or to round-off error, the stability of the scheme is a sufficient condition to ensure its convergence.

Before proceeding to the next section, we mention in passing that a stable finite-difference scheme for the linearized parabolic equation (3.57) is

$$\frac{u_i^{n+1} - u_i^n}{\tau} + U \frac{u_{i+1}^n - u_{i-1}^n}{2h} = \kappa \frac{u_{i+1}^n - 2u_i^n + u_{i-1}^n}{h^2}. \tag{3.66}$$

3.5.5 The finite-difference method for parabolic equations

We will take as a typical example of a parabolic equation the diffusion equation

$$\frac{\partial u}{\partial t} = \kappa \frac{\partial^2 u}{\partial x^2}, \quad (\kappa > 0 \text{ is the } \textit{diffusivity constant}) \tag{3.67}$$

in the strip $0 \leq x \leq 1$ subject to the initial condition $u(x, 0) = f(x)$ and the boundary conditions $u(0, t) = F(t)$ and $u(1, t) = G(t)$. Following an approach due to Fröberg (1999), a way of solving this problem that illustrates an important property of parabolic equations is as follows: divide the interval $0 \leq x \leq 1$ into N equal subintervals of length $h = 1/N$ with their end points at $x_n = nh$ with $n = 0, 1, \ldots, N$, and replace u_{xx} by its centred-difference approximation. Leaving the time derivative u_t unmodified, we arrive at the coupled system of first-order ODEs

$$\frac{\mathrm{d}u_n}{\mathrm{d}t} = \frac{\kappa}{h^2} (u_{n-1} - 2u_n + u_{n+1}), \quad \text{for} \quad n = 1, 2, \ldots, N-1, \tag{3.68}$$

where now the x variation has become discrete, and $\partial u_n / \partial t$ has been replaced by $\mathrm{d}u_n / \mathrm{d}t$. This system of ODEs can be solved subject to the conditions that $u_0 = f(t)$ and $u_N = g(t)$, where at $t = 0$ we have $u_1 = f(h)$, $u_2 = f(2h), \ldots, u_{n-1} = f([N-1]h)$. In terms of matrices, the system of equations becomes

$$\frac{h^2}{\kappa} \frac{\mathrm{d}U}{\mathrm{d}t} = \mathbf{A}U + \mathbf{B}, \tag{3.69}$$

where

$$U = \begin{bmatrix} u_1 \\ u_2 \\ \vdots \\ u_{N-1} \end{bmatrix}, \quad B = \begin{bmatrix} u_0 \\ 0 \\ \vdots \\ 0 \\ u_N \end{bmatrix}, \quad A = \begin{bmatrix} -2 & 1 & 0 & \cdots & 0 & 0 \\ 1 & -2 & 1 & \cdots & 0 & 0 \\ \vdots & \vdots & \vdots & \vdots & \vdots & \vdots \\ 0 & 0 & 0 & \cdots & 1 & -2 \end{bmatrix}. \tag{3.70}$$

The solution of this system of equations will involve a sum of time-dependent exponential functions, each of the form $e^{\lambda t}$, where the numbers λ are the eigenvalues of matrix $\mathbf{A}$ which can be shown to be given by

$$\lambda_n = 2 (\cos (\pi n/N) - 1).$$

It can be seen from this that when N is large the quotient of the absolute values of the largest and smallest eigenvalues of $\mathbf{A}$ is approximately $4N^2/\pi^2$. Thus, the smaller the space step h becomes, the larger becomes the quotient of the absolute values of the largest and smallest eigenvalues. This demonstrates the fact that, as the size of the space step h is reduced, in order to preserve the correct magnitudes of the effects due to all the time-dependent exponential terms, it is necessary that the size of the time step must be reduced much *faster* than the space step h. A more careful analysis of the problem shows that if the time step is τ, then for the method to converge and give correct results, it is necessary that the following *stability condition* is satisfied

$$\frac{\tau\kappa}{h^2} < \frac{1}{2}. \tag{3.71}$$

This approach can be converted into a forward finite-difference scheme by replacing the term $\partial u/\partial t$ by its forward finite-difference approximation. In what follows, if h is the step in the space variable x, and τ is the step in the time variable t, the discrete values $u(mh, n\tau)$ of the solution will be represented by u_m^n. Relative to the point $(mh, n\tau)$, the forward finite-difference approximation for u_t is given by $\left(u_m^{n+1} - u_m^n\right)/\tau$. Substituting this result into the left of equations (3.68) and modifying the notation we find that

$$\frac{u_m^{n+1} - u_m^n}{\tau} = \frac{\kappa}{h^2}\left(u_{m-1}^n - 2u_m^n + u_{m+1}^n\right), \tag{3.72}$$

which should be compared with (3.67).

So the value of u_m^{n+1} at $x = mh$ and at the forward time $t = (n+1)\tau$ is given in terms of u_{m-1}^n, u_m^n and u_{m+1}^n by

$$u_m^{n+1} = u_m^n + \frac{\tau\kappa}{h^2}\left(u_{m-1}^n - 2u_m^n + u_{m+1}^n\right), \tag{3.73}$$

and from equation (3.71) it follows that this simple scheme will only converge if $\tau\kappa/h^2 < 1/2$. This approach is called an *explicit method*, because the solution at the time stage $t = (n+1)\tau$ is completely determined by the solution at points occurring during the previous time stage. The distribution of points in this scheme is shown in Figure 3.7, where the point u_m^{n+1} is shown as a solid dot, while the three points on which it depends are shown as circles.

To solve an initial-value and boundary-value problem for equation (3.67) using the forward-difference explicit scheme, the space–time region $0 \le x \le 1$, $t > 0$ is first covered by a grid in which the space step length is h and the time step is τ, chosen to satisfy the stability condition (3.71). The values of the solution at the grid points on the boundary of the space–time region

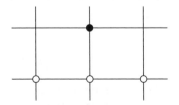

Figure 3.7 An explicit finite-difference scheme

follow from the initial condition $u(x, 0) = f(x)$, and the boundary conditions $u(0, t) = F(t)$ and $u(1, t) = G(t)$. Taking a specific number of space steps M_0, and N_0 time steps, and applying the finite-difference scheme (3.73) to each internal grid point produces a system of linear equations for the approximate solution at each internal grid point in terms of the known values at the grid points on the boundary of the space–time region. Solving this system by Gaussian elimination, by an iterative scheme, or if necessary using SOR, will then give the required solution.

However, this explicit approach suffers from the disadvantage that in order to represent the x variation accurately by taking h sufficiently small, it becomes necessary that the time step τ must be taken to be extremely small if the method is to remain stable. Thus, to obtain an approximate solution that is accurate over a large interval of time will lead to a very large system of equations. It can be shown that the error in this explicit method of solution is $O(\tau + h^2)$, so the method is described by saying it is only *first order in time*, but *second order in space*.

To illustrate an application of this finite-difference method we will model a simple hypothetical physical problem for the diffusion equation. Let us suppose that a liquid pollutant is distributed throughout a lake of uniform depth Y in which the water is at rest, and that when the x-coordinate is measured vertically downward from the surface of the water, at a time $t = 0$ the concentration of the pollutant at a depth x is $f(x)$. Then a natural diffusion process will occur throughout the water, across the surface of the lake into the air, or into the impervious material at the bottom of the lake. The question we now ask is: what is the distribution of the pollutant as a function of x and t? Physical intuition suggests that regions of high concentration will diffuse into regions of lower concentration. As no diffusion occurs at the top or bottom of the lake, the total amount of the pollutant will remain constant, and eventually the concentration must become uniform throughout the water.

When formulating the problem it will be convenient to replace the distance x, the time t and the concentration $u(x, t)$ of the pollutant at depth x and time t by the corresponding non-dimensional quantities X, T and $U(X, T)$. In what follows we will choose these to be $X = x/Y$, $T = t/T_0$, where T_0 is some convenient reference time, say 1 day, and $U = u/u_0$, where u_0 is a representative value of the concentration of the pollutant. Then the equation governing diffusion is

$$\frac{\partial U}{\partial T} = \kappa \frac{\partial^2 U}{\partial X^2},$$

where $\kappa = $ constant is the diffusion coefficient. The initial condition to be imposed that specifies the distribution of concentration at $t = 0$ is

$$U(X, 0) = f(X), 0 \leq X \leq 1.$$

As $\partial U/\partial X$ represents the change of U with respect to X, and no diffusion can occur at the top or bottom of the lake, the boundary conditions are

$$U_X(0, T) = U_X(1, T) = 0, T > 0.$$

To formulate a finite-difference approximation it is necessary to specify the function $f(X)$ and to assign a value to κ, so we will set $f(X) = X(1 - X)$ for $0 \leq X \leq 1$ and the diffusivity constant $\kappa = 0.2$. Before proceeding to examine the numerical results, let us first find the steady-state concentration that is to be expected. The amount M of pollutant in a column of water with a unit cross-sectional area is

$$M = \int_0^1 X(1 - X)dX = \frac{1}{6}.$$

As the non-dimensional height of the column of water is 1 and the amount of pollutant is conserved, the steady-state concentration m must be such that $m \times 1 = M$, showing that $m = 1/6$. Although it is a simple matter to find the concentration when the distribution of pollutant is uniform (in equilibrium), this simple analysis gives no information about the time (non-dimensional) for this condition to be reached. To answer this question it is necessary to solve this initial-value and boundary-value problem numerically.

The three-dimensional surface plot obtained from the finite-difference solution is shown in Figure 3.8, where the solution $U(X, T)$ is shown over the region $0 \leq X \leq 1$ and $0 \leq T \leq 1$ in the (X, T)-plane. In this computation the space step was $h = 0.05$ and the time step was $\tau = 0.005$. Note that, as required, this choice of parameters satisfies the general stability condition $\tau \kappa / h^2 < 1/2$, because in this case $\tau \kappa / h^2 = 0.4$. The number N of grid points used in the calculation was $N = 1/(hk) = 4,000$.

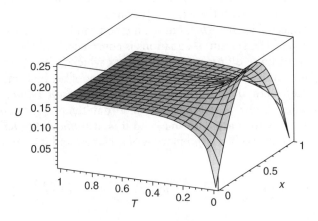

Figure 3.8 The distribution of U as a function of X and T

Figure 3.8 shows that the diffusion process redistributes the initial concentration very rapidly, and by $T = 0.6$ it has almost reached the uniform steady-state value $U = 1/6$, confirming the result found by the previous elementary conservation argument.

In the more general case where the diffusivity is a slowly varying function of X, so that $\kappa = \kappa(X)$, the explicit formula to be used becomes more complicated than the one in equation (3.73), and it can be approximated by the result

$$u_m^{n+1} = \left(1 - \frac{2\tau\kappa(mh)}{h^2}\right) u_m^n + \frac{\tau\kappa(mh)}{h^2} \left(u_{m-1}^n + u_{m+1}^n\right). \tag{3.74}$$

A different approach is needed if, as in the case of Figure 3.8, the introduction of a step length h that is small enough to produce a good approximation in space necessitates the introduction of the very large number of small time steps of length τ if the essential stability condition in equation (3.71) is to be satisfied. The methods to be outlined next that overcome this difficulty are called *implicit backward finite-difference methods*, because they relate three approximate solutions at the $(n+1)$th time step to a solution or solutions at the previous nth time step. In the first of these methods u_t is represented by a backward finite-difference approximation, with the result that it yields the *implicit backward finite-difference approximation*

$$u_m^n = -\frac{\tau\kappa}{h^2} u_{m-1}^{n+1} + \left(1 + \frac{2\tau\kappa}{h^2}\right) u_m^{n+1} - \frac{\tau\kappa}{h^2} u_{m+1}^{n+1}. \tag{3.75}$$

The points used in this approximation are shown in Figure 3.9(a), where the three forward points are shown as solid dots, while the other point is shown as a circle. As with the explicit forward finite-difference scheme the error to be expected is $O(k + h^2)$, but the method can be shown to be unconditionally stable. The system of equations is constructed in the same way as for the forward finite-difference scheme, after which it is solved either by Gaussian elimination, by an iterative scheme, or possibly by using SOR.

A far better method is the famous *Crank–Nicholson implicit scheme*, which has an error $O(\tau^2 + h^2)$ that is smaller than the previous ones as it is second order in both space and time, and it is also *unconditionally stable*. This scheme is essentially a combination of schemes, and it takes the form

$$-\frac{\kappa\tau}{2h^2} u_{m-1}^{n+1} + \left(1 + \frac{\kappa\tau}{h^2}\right) u_m^{n+1} - \frac{\kappa\tau}{2h^2} u_{m+1}^{n+1} = \frac{\kappa\tau}{2h^2} u_{m-1}^n + \left(1 - \frac{\kappa\tau}{h^2}\right) u_m^n + \frac{\kappa\tau}{2h^2} u_{m+1}^n. \tag{3.76}$$

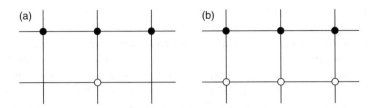

Figure 3.9 (a) The implicit backward finite-difference approximation. (b) The Crank–Nicholson implicit scheme

In Figure 3.9(b), the three forward points used in this scheme are shown as solid dots, while the three backward points are shown as circles.

The construction and solution of the system of linear algebraic equations from which the approximate solution is to be found proceeds as before. A straightforward account of the solution of parabolic equations, together with examples, can be found in Smith (1985), who also shows how to deal with different types of boundary conditions.

The numerical schemes used in professional software packages are usually more complicated than the Crank–Nicholson scheme, although they use the same basic approach and are also implicit. Although implicit schemes involve more computation than do explicit ones, they are usually used because of their unconditional stability.

3.5.6 Finite-difference methods and hyperbolic equations

A stable finite-difference scheme for the elementary linear hyperbolic equation

$$u_t + Uu_x = 0,$$

subject to the CFL condition being satisfied, has already been given in equation (3.65). If the more complicated linear equation $u_t + f(x, t)u_x = g(x, t)$ is encountered, it is a simple matter to modify result (3.65) by replacing the constant U by the discrete form of $f(x, t)$ and adding the discrete form of $g(x, t)$ to the scheme. The scheme can also be modified in a straightforward manner to deal with two linear first-order simultaneous PDEs, again subject to the CFL condition being satisfied, so this situation will not be considered here.

Finding a numerical solution for a non-linear first-order ODE such as $u_t + f(u, x, t)u_x = g(u, x, t)$ is difficult, because of the result established in Chapter 2 that discontinuous solutions may arise from smooth initial conditions. Special techniques have been developed to deal with the occurrence of discontinuities in solutions (tidal bores in rivers and shock waves in gas dynamics), because when these occur a finite-difference scheme is no longer valid. One approach involves introducing a *pseudo-viscous* term into the hyperbolic equations, the objective being to smooth out the discontinuity so that a finite-difference scheme can be used. This method is not always successful, because when the pseudo-viscous term is large the discontinuity is smoothed out over a distance that is physically unrealistic, whereas when it is small physically unrealistic oscillations may occur in the solution. In this short chapter it is not possible to elaborate on the various techniques that have been developed to deal with the evolution of discontinuous solutions. For more information about this topic, we refer the reader to the work of LeVeque (1992), wherein other references can be found, and Chung (2003), and to the early work by Richtmyer and Morton (1994).

In what follows we will take the linear wave equation as our model hyperbolic equation

$$\frac{\partial^2 u}{\partial t^2} = c^2 \frac{\partial^2 u}{\partial x^2}, \quad (c = \text{constant}) \tag{3.77}$$

subject to suitable initial and boundary conditions. The wave equation (3.77) can be solved by replacing the second-order partial derivatives $\partial^2 u/\partial x^2$ and $\partial^2 u/\partial t^2$ by their centred finite-difference approximations, which leads to the *explicit scheme*

$$u_m^{n+1} = 2\left(1 - \frac{c^2\tau^2}{h^2}\right)u_m^n + \frac{c^2\tau^2}{h^2}\left(u_{m+1}^n + u_{m-1}^n\right) - u_m^{n-1} \tag{3.78}$$

with an error $O(\tau^2 + h^2)$, where h is the space step length and τ is the time step. This scheme can be shown to be stable provided the condition $c^2\tau^2/h^2 \leq 1$ is satisfied, which contains the CFL condition.

An *implicit scheme* that can be derived, the details of which will be omitted, is

$$-u_{m+1}^{n+1} + \left(1 + 2\frac{c^2\tau^2}{h^2}\right)u_m^{n-1} - \frac{c^2\tau^2}{h^2}u_{m-1}^{n+1} = 2u_m^n - u_m^{n-1}. \tag{3.79}$$

This scheme has an error $O(\tau^2 + h^2)$ and, like other implicit schemes, it is unconditionally stable but, for it to be applied, the CFL condition must be satisfied.

A difficulty arises when dealing with the pure initial-value problem for equation 3.77 subject to the initial conditions

$$u(x, 0) = f(x) \quad \text{for} \quad -\infty < x < \infty \tag{3.80}$$

and

$$u_t(x, 0) = g(x). \tag{3.81}$$

This is because in the finite-difference approximation it is necessary to know the values of $u_t(x, \tau)$ at the first time step $t = \tau$, although this information is not part of the initial conditions. The difficulty is overcome by using the following approximation, which can be derived directly from the initial conditions

$$u_m^1 = u_i^0 + \tau g(mh) + \frac{c^2 \tau^3}{2h^2} \left(f((m-1)h) - 2f(mh) + f((m+1)h) \right), \tag{3.82}$$

for which the error is

$$O(\tau^2 + \tau h^2). \tag{3.83}$$

This scheme requires modification if the wave speed c is a function of x, although the details of how this is to be done will be omitted. For more information, we refer the reader to Ames (1978), Pearson (1986), Richtmyer and Morton (1994) and Smith (1985). For an account of the application of numerical methods to conservation laws and to shock formation, we again refer to LeVeque (1992), and to the extensive list of references therein. Applications to hydraulics can be found in Abbott (1979), Abbott and Cunge (1982), Abbott and Minns (1998), Anderson (1995), Benqué *et al.* (1982), Chadwick *et al.* (2004), Chung (2003), Cunge *et al.* (1980), Guinot (2008), Mader (2004) and Verwey (1983), in the following chapters.

3.5.7 The finite-element method for elliptical equations

We return to the solution of elliptical equations to describe a very effective method of solution called the *finite-element method*, which is accurate and deals very effectively with arbitrarily shaped boundaries. Although originally intended for the solution of elliptical problems, the finite-element method can be adapted to solve parabolic and hyperbolic problems, and also non-linear problems. The approach is computationally highly intensive as it leads to the solution of extremely large systems of equations, and is always performed by computer. Professionally developed software is usually used, which may be either of a very general and flexible nature, like MATLAB® (2007), or developed for specific purposes that take account of special features of a class of problems, thereby reducing the amount of computation involved. For a detailed description of the finite-element method, specialist texts such as the one by Schwarz (1988) should be consulted, while Burden and Faires (2005) provide an introductory account that includes some very simple applications. See also Brebbia (1983) and Zienkiewitz and Taylor (2000).

Because the finite-element method can be implemented in many ways, only a brief outline of the general approach will be given here. By way of illustration we will consider a boundary-value problem for the (elliptical) Poisson equation

$$\frac{\partial^2 u}{\partial x^2} + \frac{\partial^2 u}{\partial y^2} = f(x, y), \tag{3.84}$$

in the interior of a plane region D of arbitrary shape with boundary Γ, subject to the Dirichlet condition $u = 0$ on Γ. The approach starts by multiplying the PDE (3.84) by a smooth function $v(x, y)$, called a *test function*, with the special property that it vanishes on the boundary Γ, and then integrating the result over D to get

$$\int_D \left(\frac{\partial^2 u}{\partial x^2} + \frac{\partial^2 u}{\partial y^2} \right) v \, dx dy = \int_D f v \, dx dy. \tag{3.85}$$

To proceed further, the two-dimensional form of Green's theorem is needed (see Jeffrey (2002), Kreyszig (2005) and O'Neil (2006)) that takes the form

$$\int_D \left(\frac{\partial Q(x, y)}{\partial x} - \frac{\partial P(x, y)}{\partial y} \right) dx dy = \int_\Gamma \left(P(x, y) dx + Q(x, y) dy \right). \tag{3.86}$$

This is applied to the integral on the *left* of equation (3.86), with $P(x, y) = -v u_y$ and $Q(x, y) = v u_x$, where the integral on the right of equation (3.86) is a line integral around Γ, so this integral is to be evaluated *on* and *around* Γ. This leads to the result

$$\int_D \left(\frac{\partial v}{\partial x} \frac{\partial u}{\partial x} + \frac{\partial v}{\partial y} \frac{\partial u}{\partial y} \right) dx dy + \int_D \left(v \frac{\partial^2 u}{\partial x^2} + v \frac{\partial^2 u}{\partial y^2} \right) dx dy$$

$$= \int_\Gamma \left(-v \frac{\partial u}{\partial y} dx + v \frac{\partial u}{\partial x} dy \right),$$

but the integral on the right vanishes because $v = 0$ on Γ, so using this result in equation (3.85) allows it to be replaced by

$$-\int_D \left(\frac{\partial u}{\partial x} \frac{\partial v}{\partial x} + \frac{\partial u}{\partial y} \frac{\partial v}{\partial y} \right) dx dy = \int_D f v \, dx dy. \tag{3.87}$$

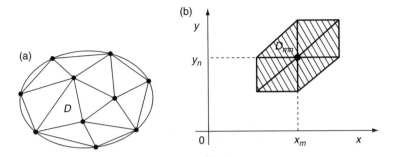

Figure 3.10 (a) Triangularization of an arbitrary region D. (b) Triangularization of the hexagonal region D_{mn} about a node at (x_m, y_n)

This expression is called the *weak* or *variational form* of the Poisson equation (3.84), and is to be interpreted to mean that u must be such that equation (3.87) is true for *all* test functions v.

The next step is to subdivide region D into smaller regions, usually of a triangular nature, as shown in Figure 3.10(a), although all without obtuse internal angles. This allows the curved boundary Γ to be approximated arbitrarily closely by a polygonal line. Each of the small interior triangular regions is called a *finite element*; the vertices of these elements are called *nodes*. A typical triangularization of a unit square is shown in Figure 3.10(b). Note that neighbouring triangles around the point (x_m, y_n) either share a common side or meet at a common node. If for some positive integer N the grid lines in Figure 3.10(b) are $h = 1/N$ apart, then the grid points occur at $x_m = mh$, $y_n = nh$, so there are $(N-1)^2$ grid points inside the unit square. The six finite elements with the common node at (x_m, y_n) representing an area D_{mn} are shaded in the diagram.

In the simplest case, the approximation $\hat{u}(x, y)$ for the true solution $u(x, y)$ above each of the shaded regions in Figure 3.10(b) is represented by a triangular planar surface of the form

$$\hat{u}(x, y) = a_1 + b_1 x + c_1 y, \tag{3.88}$$

and it is required that the approximate solution is continuous across the six edges of the pyramid function shown in Figure 3.11. For convenience, because of subsequent scaling, the height of the vertex of the pyramid function above the central node is taken to be unity.

The six-faced pyramid function $\varphi_{mn}(x, y)$ shown in Figure 3.11 that approximates the solution above the six shaded triangles forming the region D_{mn} is called a *basis function*, and the finite-element approximation above

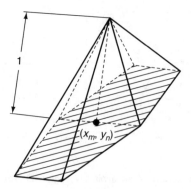

Figure 3.11 The approximate solution in region D_{mn} represented by the surface of the pyramid above the point (x_m, y_n)

the entire region D is taken to be the following linear combination of the basis functions

$$\hat{u}(x, y) = \sum_{m,n=1}^{N-1} c_{mn}\varphi_{mn}(x, y), \tag{3.89}$$

where the coefficients c_{mn} have still to be determined.

The finite-element method was first introduced to deal with the elastic deformation of solids, so for historical reasons the square matrix $\mathbf{C} = [c_{mn}]$ is usually called the *stiffness matrix*, although this name has no physical significance when working with hydraulic problems.

The conditions that must be satisfied by the basis functions $\varphi_{mn}(x, y)$ if the approximate solution is to be continuous over D are:

(a) $\varphi_{mn}(x_r, y_s) = \begin{cases} 1, & \text{when } r = m, \ s = n \\ 0 & \text{at every other grid point of } D_{mn}, \end{cases}$

(b) $\varphi_{mn}(x, y) = 0$ when the point (x, y) does not belong to D_{mn}.

The coefficients c_{mn} are found by requiring the approximate solution in equations (3.89) to satisfy the variational form of the problem in equation (3.87) for each function $\varphi_{mn}(x, y)$. Consequently, when $\hat{u}(x, y)$ is substituted for $u(x, y)$ and $\varphi_{mn}(x, y)$ is substituted for $v(x, y)$ in equation (3.87), a linear system of equations for the coefficients c_{mn} is obtained of the form

$$\sum_{m,n=1}^{N-1} c_{mn}A(\varphi_{mn}, u_{rs}) = B(u_{rs}), \quad \text{for} \quad r, s = 1, 2, \ldots, N-1, \tag{3.90}$$

where

$$A(\varphi_{mn}, \varphi_{rs}) = \int\limits_{A_{rs}} \left(\frac{\partial \varphi_{mn}}{\partial x} \frac{\partial \varphi_{rs}}{\partial x} + \frac{\partial \varphi_{mn}}{\partial y} \frac{\partial \varphi_{rs}}{\partial y} \right) \mathrm{d}x\mathrm{d}y \quad \text{and}$$

$$B(u_{rs}) = \int_{A_{rs}} f\varphi_{rs}\mathrm{d}x\mathrm{d}y. \tag{3.91}$$

When the analytical expressions for the basis functions are used in equations (3.90) and the necessary integrations are performed, a large system of linear equations is generated for the elements c_{mn}. Once this system has been solved, substituting the constants c_{mn} and the basis functions $\varphi_{mn}(x, y)$ in equation (3.89) gives the required finite-element approximation.

Because of the way the approximate solution has been constructed, it will generate a continuous solution over the triangularization of region D formed by a large number of small planar surfaces. A smoother approximate solution can be found if the linear expression in x and y in equation (3.88) is replaced by a quadratic expression, although at the cost of a significant increase in the amount of computation that is required.

Professional finite-element software packages provide the facility for automatic triangularization of regions, the ability to allow for more general boundary conditions, the efficient solution of the system of equations for the coefficients c_{mn}, and the plotting of the approximate solution surface, together with the ability to apply the method to different types of PDE. It will be clear from this brief outline of the method that the amount of computation involved is such that the finite-element method can only be implemented by computer.

References

Abbott, M. B. (1979), *Computational Hydraulics, Elements of the Theory of Free Surface Flows*, Pitman, London.

Abbott, M. B. and Cunge, J. A. (eds) (1982), *Engineering Applications of Computational Hydraulics, Vol. 1: Homage to Alexandre Preissmann*, Pitman, London.

Abbott, M. B. and Minns, A. W. (1998), *Computational Hydraulics*, 2nd Edn, Ashgate, Aldershot.

Ames, W. F. (1978), *Numerical Methods for Partial Differential Equations*, 2nd Edn, Nelson, London.

Anderson, J. D. (1995), *Computational Fluid Dynamics*, McGraw-Hill, New York.

Atkinson, K. E. (2007), *An Introduction to Numerical Analysis*, 2nd Edn, Wiley, New York.

Benqué, J. P., Hauguel, A. and Viollet, P. L. (1982), *Engineering Applications of Computational Hydraulics*, Vol. 2, Pitman, London.

Brebbia, C. A. (1983), The finite element method in the design of hydraulic structures. In *Developments in Hydraulic Engineering*, Vol. 1 (ed. P. Novak), Elsevier Applied Science, London.

Burden, R. L. and Faires, J. D. (2005), *Numerical Analysis*, 5th Edn, PWS Publishing, Boston, MA.

Chadwick, A., Morfett, J. and Borthwick, M. (2004), *Hydraulics in Civil and Environmental Engineering*, 4th Edn, Chapter 14: *Computational Hydraulics*, Spon, London.

Chung, T. J. (2003), *Computational Fluid Dynamics*, Cambridge University Press, London.

Cunge, J. A., Holly, F. M. (Jr) and Verwey, A. (1980), *Practical Aspects of Computational River Hydraulics*, Pitman, London.

Ferziger, J. H. (1998), *Numerical Methods for Engineering Application*, Wiley, New York.

Fröberg, C.-E. (1999), *Numerical Mathematics: Theory and Computer Applications*, Benjamin/Cummings, California.

Guinot, V. (2008), *Wave Propagation in Fluids: Methods and Numerical Techniques*, ISTE, London and Wiley, Hoboken, NJ.

Jeffrey, A. (2002), *Advanced Engineering Mathematics*, Academic Press, New York.

Jeffrey, A. (2003), *Applied Partial Differential Equations*, Academic Press, New York.

Johnson, L. W. and Riess, R. D. (1982), *Numerical Analysis*, 2nd Edn, Addison Wesley, Reading, MA.

Kreyszig, E. (2005), *Advanced Engineering Mathematics*, 8th Edn, Wiley, New York.

LeVeque, R. L. (1992), *Numerical Methods for Conservation Laws*, Birkhauser, Boston.

Mader, C. L. (2004), *Numerical Modeling of Water Waves*, 2nd Edn, CRC Press, Boca Raton, FL.

MAPLE® The product of Waterloo Maple Inc., 57 Erb Street West, Waterloo, Ontario, Canada ON N2L.

MathCAD® The product of MathSoft Inc., One Kendall Square, Cambridge, MA 02139, USA.

MATLAB® The product of Mathworks Inc., 24 Prime Park Way, Nantick, MA 01760-1500, USA.

O'Neil, P. V. (2006), *Advanced Engineering Mathematics*, 4th Edn, PWS-Kent, Boston, MA.

Ortega, J. M. and Poole, W. G. (Jr) (1983), *Numerical Methods for Partial Differential Equations*, Pitman, London.

Pearson, C. E. (1986), *Numerical Methods in Engineering and Science*, Van Nostrand Reinhold, New York.

Press, W. H., Flannery, B. P., Teukolsky, S. A. and Vetterling, W. T. (2007), *Numerical Recipes: The Art of Scientific Computing*, 3rd Edn, Cambridge University Press, London.

Ralston, A. and Rabinowitz, P. (2003), *A First Course in Numerical Analysis*, McGraw-Hill, New York.

Richtmyer, R. D. and Morton, K. (1994), *Difference Methods for Initial-value Problems*, Interscience-Wiley, New York.

Schwarz, H. R. (1988), *Finite Element Methods*, Academic Press, London.

Schwarz, H. R. (1989), *Numerical Analysis: A Comprehensive Introduction*, Wiley, New York.

Smith, G. D. (1985), *Numerical Solution of Partial Differential Equations: Finite Difference Methods*, Clarendon Press, Oxford, London.

Verwey, A. (1983), *The Role of Computational Hydraulics in the Hydraulic Design of Structures*. In *Developments in Hydraulic Engineering*, Vol. 1 (ed. P. Novak), Elsevier Applied Science, London.

Zienkiewitz, O. C. and Taylor, R. L. (2000), *The Finite Element Method*, 5th Edn, Butterworth-Heinemann, London.

Chapter 4

Theoretical background – hydraulics

4.1 Introduction

As this text is not intended to be a treatise on fluid mechanics or hydraulics, only the essentials of these disciplines necessary as a background to further discussion of the main theme, i.e. modelling in hydraulics, will be dealt with here; furthermore, a basic knowledge and understanding of fluid mechanics and hydraulics and of their principles are assumed throughout. In most cases the equations presented are given without formal proofs in their derivation – for further details and references, see, for example, Batchelor (2000), Chadwick *et al.* (2004), Daily and Harleman (1966), Douglas *et al.* (2005), Fox (1977), Goldstein (1965), Rouse (1961), Tritton (1988) and Vardy (1990). Some parts of (applied) hydraulics developed from statements in this section and used in the formulation and application of scaling laws, in mathematical modelling of various types of flow or engineering applications, are best dealt with immediately before discussing the modelling procedure and problems, and therefore have mainly been placed in the relevant 'applied' Chapters (7–13).

4.2 Some basic concepts and equations in hydrodynamics

4.2.1 The continuity equation

For a cubical element of fluid (control volume) of density ρ and velocities u, v, w along the axes x, y, z, from considerations of mass conservation

$$\frac{\partial \rho}{\partial t} + \frac{\partial}{\partial x}(\rho u) + \frac{\partial}{\partial y}(\rho v) + \frac{\partial}{\partial z}(\rho w) = 0 \tag{4.1}$$

or

$$\frac{\partial \rho}{\partial t} + \nabla.(\rho u) = 0 \tag{4.1a}$$

For steady incompressible fluid flow ρ is constant; thus

$$\frac{\partial u}{\partial x} + \frac{\partial v}{\partial y} + \frac{\partial w}{\partial z} = 0 \tag{4.2}$$

or

$$\nabla.u = 0 \tag{4.2a}$$

4.2.2 The Euler equations

The differential form of Newton's second law of motion for fluid flow in a pressure field without friction results in

$$-\frac{\partial p}{\partial x} + \rho X = \rho \left(u\frac{\partial u}{\partial x} + v\frac{\partial u}{\partial y} + w\frac{\partial u}{\partial z} + \frac{\partial u}{\partial t} \right) \tag{4.3}$$

$$-\frac{\partial p}{\partial y} + \rho Y = \rho \left(u\frac{\partial v}{\partial x} + v\frac{\partial v}{\partial y} + w\frac{\partial v}{\partial z} + \frac{\partial v}{\partial t} \right) \tag{4.4}$$

$$-\frac{\partial p}{\partial z} + \rho Z = \rho \left(u\frac{\partial w}{\partial x} + v\frac{\partial w}{\partial y} + w\frac{\partial w}{\partial z} + \frac{\partial w}{\partial t} \right) \tag{4.5}$$

where $\rho X, \rho Y$ and ρZ are body forces/unit volume in the directions x, y, and z. These equations can be written more concisely as

$$\frac{du}{dt} = X - \frac{\partial p}{\rho \partial x} \tag{4.3a}$$

$$\frac{dv}{dt} = Y - \frac{\partial p}{\rho \partial y} \tag{4.4a}$$

$$\frac{dw}{dt} = Z - \frac{\partial p}{\rho \partial z} \tag{4.5a}$$

In equations (4.3)–(4.5) the groups of derivatives $u(\partial u/\partial x) + v(\partial u/\partial y) + w(\partial u/\partial z)$, etc. are convective accelerations describing velocity changes by movements in space. As the Euler equations describe relationships between forces and accelerations they are differential forms of momentum equations, the integration of which with respect to distance yield energy relationships.

4.2.3 The Navier–Stokes equations

Introducing the definition of dynamic viscosity as the ratio of shear intensity τ in the x–y plane to the rate of angular deformation

$$\tau_{xy} = \mu \left(\frac{\partial u}{\partial y} + \frac{\partial v}{\partial x} \right) \tag{4.6}$$

(with similar expressions for τ_{yz} and τ_{zx}), and by adding the effect of viscosity to account for frictional forces and shear stresses to the Euler equations (4.3)–(4.5) we obtain the Navier–Stokes equations:

$$-\frac{\partial p}{\partial x} + \rho X + \mu \left(\frac{\partial^2 u}{\partial x^2} + \frac{\partial^2 u}{\partial y^2} + \frac{\partial^2 u}{\partial z^2} \right) = \rho \left(u\frac{\partial u}{\partial x} + v\frac{\partial u}{\partial y} + w\frac{\partial u}{\partial z} + \frac{\partial u}{\partial t} \right)$$

$$(4.7)$$

Similar expressions could be written for the y and z axis (see equations (4.4) and (4.5)); or more concisely

$$\frac{du}{dt} = X - \frac{1}{\rho}\frac{\partial p}{\partial x} + v\nabla^2 u \tag{4.7a}$$

4.2.4 Vorticity, irrotational flow, velocity potential, Laplace equation

The rate of rotation (angular velocity) of a fluid particle in the x–y plane is $\omega_{xy} = 1/2((\partial v/\partial x) - (\partial u/\partial y))$

The *vorticity* ζ (intensity of circulation Γ) is equal to 2ω.

Irrotational flow is flow without vorticity; thus, for this type of flow $\omega_{xy} = \omega_{xz} = \omega_{zy} = 0$, and

$$\frac{\partial v}{\partial x} = \frac{\partial u}{\partial y} \tag{4.8}$$

$$\frac{\partial u}{\partial z} = \frac{\partial w}{\partial x} \tag{4.9}$$

$$\frac{\partial w}{\partial y} = \frac{\partial w}{\partial y} = \frac{\partial v}{\partial z} \tag{4.10}$$

Denoting $u = \partial\varphi/\partial x$, $v = \partial\varphi/\partial y$, $w = \partial\varphi/\partial z$ we get from equation (4.2)

$$\frac{\partial^2\varphi}{\partial x^2} + \frac{\partial^2\varphi}{\partial y^2} + \frac{\partial^2\varphi}{\partial z^2} = 0 \tag{4.11}$$

or

$$\nabla^2\varphi = 0 \tag{4.11a}$$

Irrotational flow with a *velocity potential* ($u = \partial\varphi/\partial x$, etc.) is called *potential flow* and equation (4.11) is the *Laplace equation*.

4.2.5 The stream function

The stream function ψ defines streamlines in fluid motion as lines with a constant value of ψ ($\partial\psi = 0$) given by:

$$u = \frac{\partial\psi}{\partial y}\left(=\frac{\partial\varphi}{\partial x}\right) \tag{4.12a}$$

and

$$v = -\frac{\partial\psi}{\partial x}\left(=\frac{\partial\varphi}{\partial y}\right) \tag{4.12b}$$

From this definition it follows that no flux can cross the streamlines, and that between adjacent streamlines

$$\delta Q = \delta\psi \tag{4.13}$$

(ψ and φ for two-dimensional flow are conjugate harmonic functions).

4.3 Hydraulics – basic concepts, boundary layer, turbulence

4.3.1 Some basic concepts and equations in hydraulics

4.3.1.1 Continuity

From equation (4.13) for steady incompressible fluid flow between two streamlines, and applying the concept to a stream tube of cross-sectional area A, we get for two consecutive sections 1 and 2:

$$Q = A_1 v_1 = A_2 v_2 = \text{constant} \tag{4.14}$$

where v is the velocity in section A, etc.

Replacing velocity v with the mean cross-sectional velocity V in a conveyance area A results in

$$Q = AV = \text{constant} \tag{4.14a}$$

4.3.1.2 Energy

Energy considerations result in the Bernoulli equation for the 'total' energy head

$$H = \frac{p}{g} + \frac{v^2}{2g} + z = \text{constant} \tag{4.15}$$

i.e. the sum of pressure, kinetic and potential energy components per unit weight of fluid is constant ('total' energy is an oversimplification, as this should include other forms, e.g. turbulent and thermal energy).

For a real fluid taking into account the velocity distribution in the cross-section with the local (time averaged) velocity differing from the cross-sectional velocity V, the kinetic energy head has to be written as $\alpha V^2/2g$, where α is the *Coriolis coefficient*

$$\alpha = \frac{A^2 \int\limits_A u^3 \mathrm{d}A}{\left(\int\limits_A V \mathrm{d}A\right)^3}$$

4.3.1.3 Force and momentum

From Newton's second law for steady flow and an incompressible fluid

$$\Sigma P = \rho Q(v_2 - v_1) = \rho A v(v_2 - v_1) = \rho Q^2 \left(\frac{1}{A_2} - \frac{1}{A_1}\right) \tag{4.16}$$

where ΣP is the sum of external forces acting on a segment of fluid between sections 1 and 2 of mass $\rho A \delta l$ (with $v = \mathrm{d}l/\mathrm{d}t$).

For a real fluid with non-uniform velocity distribution, equation (4.16) has to be modified by the introduction of the *Boussinesq coefficient*

$$\beta(<\alpha)(\beta(<\alpha)\left(\beta = \frac{A \int\limits_A u^2 \mathrm{d}a}{\left(\int\limits_A V \mathrm{d}A\right)^2}\right).$$

$$\Sigma P = \rho Q(\beta_2 v_2 - \beta_1 v_1) \tag{4.16a}$$

Applications of the mass conservation, energy and momentum equations (4.15) and (4.16) lead to many equations used in hydraulics; e.g. equations for flow through orifices, over notches, through venturimeters and contractions (energy principle), and for forces acting on vanes, in pipe bends, and the hydraulic jump (momentum principle), etc.

4.3.1.4 Vortex

By introducing rotational motion into a flow (e.g. the Earth's rotation) a *free vortex* is formed which, ideally, does not dissipate any energy, and thus for $dH/dr = 0$, where r is the radius of rotation of a fluid particle,

$$\frac{dv}{dr} + \frac{v}{r} = 0 \tag{4.17}$$

and

$$vr = c \tag{4.17a}$$

where c is a constant – the *strength* of the free vortex ($2\pi c$ is the circulation Γ).

Combining a free cylindrical vortex and radial flow (sink) results in a *free spiral vortex*.

If the fluid is forced to rotate (e.g. by a stirring device) with an angular velocity ω a forced vortex results, with

$$\frac{v}{r} = \omega \tag{4.18}$$

At the centre of a free vortex there are high velocities and velocity gradients, and thus large viscous shears are present, causing energy 'losses' and decay of the free vortex near its centre. This results in a forced vortex surrounded by a free vortex, i.e. the *Rankine vortex*. (Energy 'losses' are really energy transfers from bulk-flow kinetic energy to other forms of energy, e.g. turbulent, heat and sound energy.)

4.3.2 Boundary layer and resistance

The flow field may be generally considered to consist of two parts – influenced or not influenced by frictional (viscous) effects caused by the presence of flow boundaries (in reality there is, of course, no sharp dividing line between the two parts). In the former case there is a boundary layer with a velocity gradient, while in the latter there is a velocity potential and the Euler equations may be applied.

When a fluid flow approaches a boundary, it is slowed down (to zero at the contact with the boundary). First, a laminar boundary layer (with parabolic velocity distribution) develops, and after some distance (from the leading edge of the boundary) there is a transitional boundary layer that shortly develops into a turbulent boundary layer with a laminar sublayer (where there is a large velocity gradient and strong viscous shear). The thickness of the boundary layer may be taken as the distance from the boundary to the point where the velocity attains 99% of the undisturbed velocity u_0 (Figure 4.1).

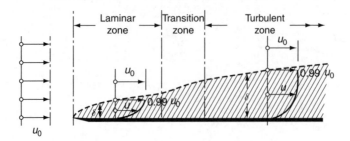

Figure 4.1 Development of the boundary layer along a flat plate

From experimental observation of, for example, flow past a flat plate, the boundary layer thickness δ for a *laminar layer* can be written as

$$\delta = 5\left(\frac{vx}{u_0}\right)^{1/2} = x5\mathrm{Re}_x^{-1/2} \tag{4.19}$$

or

$$\mathrm{Re}_\delta = 5\mathrm{Re}_x^{1/2} \tag{4.19a}$$

where $\mathrm{Re}_\delta = u_0\delta/v$ and x is the distance from the leading edge. The corresponding equation for the *turbulent boundary layer* with $\mathrm{Re}_x > 10^5\text{--}10^6(\mathrm{Re}_\delta > 3,000)$ is

$$\delta = x0.37\mathrm{Re}_x^{-1/5} \tag{4.20}$$

The velocity distribution in a turbulent boundary layer (in the x–y plane) can for $\mathrm{Re} < 20 \times 10^6$ be closely approximated by

$$\frac{u}{u_0} = \left(\frac{y}{\delta}\right)^{1/7} \tag{4.21}$$

and the *displacement thickness* δ^* (the distance by which a surface would have to be moved in order to reduce the discharge of an ideal fluid at velocity u by the same amount as the reduction caused by the velocity reduction in the boundary layer) (from equation (4.21)):

$$\delta^* = \frac{\delta}{8} \tag{4.22}$$

The *boundary resistance* may be expressed in terms of the kinetic energy of the undisturbed flow, using a 'friction coefficient' c_f, as

$$\tau_0 = \frac{c_f \rho u_0^2}{2} \tag{4.23}$$

where c_f will be a function of Reynolds number. Generally, the *drag resistance* of immersed bodies can then be formulated as

$$P = \frac{CA\rho u_0^2}{2} \tag{4.24}$$

where A is the projected area of the body on a plane perpendicular to the flow vector u_0, and C is a function of Reynolds number and the shape of the body.

Using the Prandtl *mixing-length* concept and writing $u = \bar{u} + u'$ (where u' is the instantaneous velocity deviation from the time average $\bar{u}$; the same applies for v and w – see also Section 4.3.3), the velocity fluctuation between adjacent streamlines in the *x-y* plane becomes

$$u' = l\frac{\partial u}{\partial y} \tag{4.25}$$

Assuming homogeneity with $u' = v'$, the shear stress

$$\tau = \rho u' v' \tag{4.26}$$

becomes

$$\tau = \rho l^2 \left(\frac{\partial u}{\partial y}\right)^2 \tag{4.27}$$

where l is the mixing length, or

$$U_* = l\frac{\partial u}{\partial y} \tag{4.28}$$

where $U_* = (\tau/\rho)^{1/2}$ is the *shear velocity*.

Taking $l = \kappa y$ with $\kappa = 0.4$ (von Karman's 'universal constant') and $\tau = \tau_0$ (shear stress at the boundary) results in

$$u = 2.5\left(\frac{\tau_0}{\rho}\right)^{1/2} \ln y + C = 5.75 U_* \log\left(\frac{y}{c}\right) \tag{4.29}$$

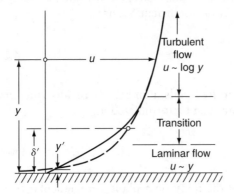

Figure 4.2 Definition sketch for evaluating the thickness of the laminar sublayer

where c is a 'constant', which in fact is a function of the surface roughness and shape of the conveyance. We can thus conclude that the growth of the turbulent boundary layer (and the boundary resistance) are a function of the Reynolds number, relative surface roughness and shape of the conveyance.

Equation (4.29) can also be written as

$$\frac{u}{U_*} = \frac{1}{\kappa} \ln \frac{y}{y'} \tag{4.30}$$

where y' is the distance from the boundary at which the velocity u (as given by equation (4.30)) reduces to zero (Figure 4.2).

Equating the velocity from the velocity distribution given by equation (4.29) with that given by a laminar velocity distribution (derived from $\tau = \mu\, du/dy$) (and using Nikuradse's work – see Section 4.4.1) results in the equation for the *laminar sublayer* thickness

$$\delta' = 11.6 \frac{\nu}{U_*} \tag{4.31}$$

Further development and applications of equations (4.30) and (4.31) will be given in Section 4.3.1.

If the flow over a boundary is in a situation of decreasing pressure in the direction of flow, the fluid will accelerate, the boundary layer will become thinner and the flow will be stable (e.g. flow between convergent boundaries). In the opposite case of a positive pressure gradient (e.g. in divergent flow) boundary layer separation will occur after a stagnation point (a point where the transverse velocity profile begins to exhibit an inflexion) has been reached. Flow separation is characterized by increased energy losses, with the flow being inherently unstable.

For a flow velocity u_0 around a cylinder of diameter D the flow behaviour depends on the Reynolds number $u_0 D / v$ changing from no separation at and behind the cylinder (Re < 0.5) through symmetrical attached vortices (2 < Re < 30) to an alternative shedding of vortices (Re > 70–120) – the *Karman vortex street* – with a frequency f given by the Taylor equation:

$$f = \frac{0.198 u_0}{D(1 - 19.7/\text{Re})} = \frac{0.2 u_0}{D(1 - 20/\text{Re})}$$

(4.32)

At Re > 10^5, the wake behind a cylinder narrows (Francis 1975).

4.3.3 Turbulence

Reynolds' classical experiment demonstrated that originally stable laminar flow, when the velocity is increased, passes through an ill-defined transitional stage to turbulent flow with chaotic violent eddying and mixing (of originally adjacent fluid layers).

Each time a flow changes as the result of an instability, the ability to predict the details of motion – using the Navier–Stokes equations describing the flow – is reduced. When successive instabilities have reduced the level of predictability so much that it is appropriate to describe the flow statistically and globally rather than in detail, it is considered to be turbulent with a predominance of random features. Most of the knowledge of turbulent flow comes from the interaction between theory and experimental results.

For a statistical description of turbulent flow, the velocity (and pressure fields) are divided into mean and fluctuating parts, e.g. $u = \bar{u} + u'$ (see also Section 4.3.2) (the time average of velocity u is over a long period of time relative to turbulence but could be over a short period relative to variations in the primary unsteady flow). For the description of the flow conditions $\bar{u}$ and the root-mean-square (RMS) value of u', the energy spectrum and the probability function of the velocity fluctuations are important.

Returning to the Navier–Stokes equation (4.7) and substituting for each component of the instantaneous velocity vector u the sum of the mean and its fluctuation $(\bar{u} + u')$ results in

$$\frac{d\bar{u}}{dt} = X - \frac{1}{\rho}\frac{\partial \bar{p}}{\partial x} + v\nabla^2 \bar{u} - \frac{\partial \overline{u'^2}}{\partial x} - \frac{\partial \overline{u'v'}}{\partial y} - \frac{\partial \overline{u'w'}}{\partial z}$$

(4.33)

(and similar expressions for the y–z axis).

It is important to appreciate that these modifications first formulated by Reynolds (the *Reynolds equations*) do not apply more accurately to conditions of turbulence than do the original Navier–Stokes equations. They permit, however, the differentiation between primary and secondary characteristics of motion.

It is evident from equation (4.33) that the magnitude of the terms for purely viscous stress are reduced and that the additional term implicitly expresses the viscous action in the secondary motion of eddies, i.e. the actual stress due to viscous action is replaced by the viscous stress given by the mean velocity gradient plus the apparent stress due to the exchange of momentum in the mixing process. The stress intensity τ expressed by

$$\tau = \rho u' v' (\text{etc.}) \tag{4.34}$$

(see also equation (4.26)) is called the *Reynolds stress*. Reynolds stresses are thus the mathematical representation of the transport of momentum across a hypothetical surface due to random turbulent velocity fluctuations.

Analoguous to the laminar shearing flow with stress $\tau = \mu du/dy$, Prandtl introduced the *eddy viscosity* concept for turbulent flow with $\bar{\tau} = \tau + \tau''$, which for $\tau_\mu = 0$ and $\tau' = \rho l^2 (d\bar{u}/dy)^2$ gives

$$\bar{\tau} = 4\rho \left(\frac{\Delta y d\bar{u}}{dy} \right)^2 = \frac{\eta d\bar{u}}{dy} \tag{4.35}$$

where $\eta = \rho(2\Delta y)^2 d\bar{u}/dy$ is a function of the shear rate and eddy size $2\Delta y = l$ (mixing length). Here, η is the coefficient of the dynamic eddy viscosity and $\varepsilon = \eta/\rho$ is the coefficient of kinematic eddy viscosity depending only on flow properties (i.e. the eddy size and eddy velocity). Since ε provides a direct measure of the mixing process, it is also called the *diffusion coefficient*.

The '$k - \varepsilon$'' turbulence model based on the use of the turbulent kinetic energy k and the rate of its dissipation due to viscous damping ε' is frequently used in computational modelling, and will be discussed in later chapters.

In all cases of turbulent motion there is a tendency for the mean square values of the three components of turbulent motion to become equal to one another, leading to *isotropic turbulence*. This might be expected if the time elapsed since the turbulence was formed is so great that there is no correlation between the motion of a particle and its initial motion.

We can conclude that turbulence is a system of eddies – large ones generated by the main flow, with a size of the order of the flow field (e.g. depth of flow), where viscous effects are negligible; and small eddies (generated by the large ones) dissipating energy due to viscosity.

It is beyond the scope of this text to proceed further in the development and application of the turbulence equation and discussion of turbulence models, and the reader is referred to the many publications on the subject, some of which are listed in the references to this chapter (see particularly Rodi 1993). It is, however, pertinent to note here that closure of turbulence equations can only be achieved by using empiricism at some level.

4.4 Flow in conduits

4.4.1 Steady flow and friction coefficients

The pressure head loss due to friction h_f for flow in a pipe of diameter D over a length l is given by the *Darcy–Weisbach* equation:

$$h_f = \frac{\lambda l V^2}{2gD} \qquad (4.36)$$

where the coefficient of friction head loss λ is given by

$$\lambda = \Phi\left(\frac{VD}{\nu}, \frac{k}{D}\right) = \Phi\left(\text{Re}, \frac{k}{D}\right) \qquad (4.37)$$

where k is the 'effective' wall roughness size. Introducing the hydraulic radius $R = A/P$ (for a pipe of circular cross-section $R = D/4$) and substituting into equation (4.36) gives:

$$h_f = \frac{\lambda l V^2}{8gR} = \frac{\lambda_R l V^2}{2gR} \qquad (4.36a)$$

with $\lambda = 4\lambda_R$.

Equating the wall resistance over a length l ($\tau_0 Pl$, where τ_0 is the mean wall shear stress) with the pressure loss Δp over area $A(\Delta pA = \rho g h_f A)$ yields

$$h_f = \frac{\tau_0 l P}{\rho g A} = \frac{4\tau_0 l}{\rho g D} \qquad (4.38)$$

From equations (4.36) and (4.38)

$$\tau_0 = \frac{\lambda \rho V^2}{8} = \rho U_*^2 \qquad (4.39)$$

(see also equation (4.23)).

Integrating the (Hagen–Poiseuille) equation for the distribution of velocity u in laminar flow in a pipe (see Section 5.5.2)

$$u = \frac{\Delta p}{(4\mu l)(D^2/4 - y^2)}$$

results in:

$$h_f = \frac{32lV^2\nu}{gD^2} \tag{4.40}$$

From equations (4.36) and (4.40), for laminar flow λ is given by

$$\lambda = \frac{64\nu}{VD} = \frac{64}{\text{Re}} \tag{4.41}$$

The laminar regime is maintained for flow through a circular cross-section pipe up to the critical value $\text{Re}_{cr} = 2,300 (VR/\nu = 580)$. Below this limit, flow at a sufficient distance from the inlet into the pipe is always laminar. In favourable circumstances (e.g. very quiet water in the inlet reservoir) laminar flow may also be achieved for $\text{Re} > \text{Re}_{cr}$.

For values of the Reynolds number above the critical limit Re_{cr} the flow passes through the transitional zone into the region of turbulent flow, where energy losses are much greater than for laminar flow and where there are three possible regimes: smooth turbulent flow, transition from 'smooth' to 'rough', and rough turbulent flow (see Figure 4.4). The exponent of the mean cross-sectional velocity in the head-loss equation changes from 1 (see equation (4.40)) up to 2 for fully developed rough turbulent flow. In this case, the Reynolds number disappears from equation (4.37) and the resistance coefficient is only a function of the relative roughness k/D. The Reynolds number for which rough turbulent flow begins (i.e. for which the friction loss is proportional to the square of mean velocity) is denoted by Re_{sq}. For $\text{Re} > \text{Re}_{sq}$ the resistance coefficient λ is, therefore, independent of the Reynolds number.

When stating the general equation for the friction-loss coefficient λ (equation 4.37), the effective roughness size k of the pipe wall was used. It is evident that the roughness coefficient will be influenced not only by the magnitude of the wall protuberances, but also by their shape, homogeneity and concentration. For homogeneous roughness formed by regularly spaced sand grains of equal size k glued to the inside wall of smooth bronze pipes, equation (4.37) was solved experimentally by Nikuradse (for roughness values $30 < D/k < 1,014$). The resultant graph illustrating the relationship between the resistance coefficient λ, Reynolds number $\text{Re} = VD/\nu$ and the relative roughness value r/k (where r is the radius of the pipe) is shown in Figure 4.4.

For $\text{Re} < \text{Re}_{cr}$ the Lagrange number $\text{La} = \text{constant}$, for $\text{Re} > \text{Re}_{sq}$ the Euler number $\text{Eu} = \text{constant}$ (see Section 5.5.2) and for the intermediate region $\text{Re}_{cr} < \text{Re} < \text{Re}_{sq}$ equation (4.37) applies. The velocity distribution in turbulent flow for $\text{Re} < 20 \times 10^6$ can be approximated by a power law

$$u = y^{1/m} \tag{4.42}$$

where $m = 7$ (y is the distance from the wall); it is, however, better to use the velocity distribution equation (4.30), which can be written as

$$\frac{\bar{u}}{\sqrt{\tau_0/p}} = 5.75 \log \left(\frac{y}{y\prime} \right) \tag{4.43}$$

On the basis of Nikuradse's experiments, and using equation (4.31) for 'smooth' pipes

$$y\prime = \frac{\delta\prime}{107} = \frac{11.6v}{U_*107} = \frac{0.108v}{\sqrt{\tau_0/\rho}} \tag{4.44}$$

For 'rough' wall pipes experiments have shown that

$$y\prime = \frac{\kappa}{30} \tag{4.45}$$

Substituting for $y\prime$ in equation (4.43) and integrating the results in the following equations for the mean velocity of flow,

$$\frac{V}{U_*} = 5.75 \log \left(\frac{rU_*}{v} \right) + 1.75 \tag{4.46}$$

for *smooth pipes*, and

$$\frac{V}{U_*} = 5.75 \log \left(\frac{r}{k} \right) + 4.75 \tag{4.47}$$

for *rough* pipes.

Introducing the friction coefficient λ from equation (4.39) (and carrying out a small correction in the coefficients according to experimental results) gives the *Karman–Prandtl* equations for the friction coefficient for turbulent flow in *smooth pipes* ($\mathrm{Re} = VD/v$):

$$\frac{1}{\sqrt{\lambda}} = 2 \log \left(\frac{\mathrm{Re}\sqrt{\lambda}}{2.51} \right) \tag{4.48}$$

and in *rough pipes*

$$\frac{1}{\sqrt{\lambda}} = 2 \log \left(\frac{3.71D}{k} \right) \tag{4.49}$$

We also obtain a single equation for the relative velocity distribution in rough and smooth pipes

$$\frac{(\bar{U} - V)}{(V\sqrt{\lambda})} = 2 \log\left(\frac{y}{r}\right) + 1.32 \tag{4.50}$$

Furthermore, from equations (4.31) and (4.39) we obtain the relationship

$$\frac{\delta\prime}{r} = \frac{11.6\nu}{rV\sqrt{\lambda/8}} = \frac{65.6\nu}{VD\sqrt{\lambda}} = \frac{65.6}{\text{Re}\sqrt{\lambda}} \tag{4.51}$$

From the above equations it follows that every pipe of a certain roughness will behave like a smooth pipe for low Reynolds number and like a rough pipe for high values of Reynolds number; between the two cases there is a transitional zone in which loss due to friction is a function both of the Reynolds number and the relative roughness of the pipe. As long as the protuberances on the pipe wall are submerged in the laminar sublayer they do not influence the magnitude of the losses at all; however, as soon as the thickness of the laminar sublayer decreases to such an extent that, because of the protruding parts of the wall, it becomes unstable, the resistance to flow will increase.

The transition from smooth to rough pipe flow begins at about $\delta\prime = 4k$ and finishes at $\delta\prime = k/6$. Using the latter value in equation (4.51) results in

$$\text{Re}_{sq} > \frac{65.6 \times 6r}{k\sqrt{\lambda}} > \frac{400r}{k\sqrt{\lambda}} \tag{4.52}$$

From this the condition for the influence of viscosity to become negligible is

$$\frac{U*k}{\nu}(=\text{Re}_k^*) > 70 \tag{4.53}$$

The above considerations refer to a uniform roughness of pipe walls as used by Nikuradse. However, as the roughness of the wall of every commercial pipe is caused by protuberances of various sizes, shapes and density, using an *equivalent roughness k* (corresponding to a uniform grain size and resulting for a pipe of given diameter with the same value of coefficient λ as the actual heterogeneous roughness) Colebrook and White proposed a single semi-empirical relation that included not only the transitional zone but also closely approximated the experimental results and equations for smooth and rough pipes:

$$\frac{1}{\lambda} = -2 \log\left(\frac{2.51}{\text{Re}\sqrt{\lambda}}\right) + \frac{k}{3.71D} \tag{4.54}$$

On the basis of these studies and further measurements on industrial pipes, Moody drew up a diagram (reproduced in most hydraulics textbooks) that shows the relationship between λ, Re and k/D, i.e. it represents equation (4.37). The values of k (effective roughness size) for various materials and new pipes are quoted in published tables (e.g. for steel k is 0.005–0.05 mm, for cast iron k is 0.12–0.60 mm, and for concrete k is 0.30–3.0 mm). For 1 year's operation of metal and concrete pipes an increase in the value of λ by 0.0005–0.001 should be allowed for.

Combining equation (4.54) with equation (4.36) results in an explicit equation for velocity (and discharge) ($S_f = h_f/l$):

$$V = -2\sqrt{2gDS_f} \log \left(\frac{k}{3.71D} \right) + \frac{2.51v}{D\sqrt{2gDS_f}} \tag{4.55}$$

This equation forms the basis for the charts and tables for the hydraulic design of pipes (e.g. as given in Wallingford and Barr (1994)). Approximating the logarithmic smooth turbulent part of equation (4.54), Barr (1975) proposed the explicit equation for λ:

$$\frac{1}{\lambda} = -2 \log \left(\frac{k}{3.71D} + \frac{5.1286}{\text{Re}^{0.89}} \right) \tag{4.56}$$

For Re > 10^5, equation (4.56) provides a solution for the head-loss coefficient to an accuracy of $\pm 1\%$.

In practice, the method of solution will depend on which two of the three variables (Q, D and $S_f = h_f/l$) are known.

Equations (4.46)–(4.56) apply to pipes of circular cross-section. For *other shapes or circular conduits not flowing full*, the shape of the conveyance will affect the value of the coefficient λ, as the integration of equation (4.43) – even for the same value of k – will produce a different constant of integration. Thus, for example, for flow between two parallel planes at a distance a (where the second dimension of the cross-section is theoretically infinite, or at least too great to influence the velocity distribution in the major part of the cross-section), we obtain for *smooth* surfaces

$$\frac{1}{\sqrt{\lambda}} = 2.03 \log \left(\frac{2aV}{\sqrt{\lambda/v}} \right) - 0.47 \tag{4.57}$$

and for *rough* surfaces

$$\frac{1}{\sqrt{\lambda}} = 2.03 \log \left(\frac{a}{2k} \right) + 2.11 \tag{4.58}$$

Comparing these equations with equations (4.48) and (4.49) (for a large second dimension $R = a/2$), we find that λ for a cross-section other than circular is always larger than for the circular one. This increase in the friction coefficient and the resultant energy losses are caused primarily by secondary currents, the streamlines of which are in a plane perpendicular to the axis of the pipe. The resultant is a spiral flow leading to increased mixing of particles from the zones of higher and lower velocities, and to an increase in turbulence, and thus also to an increase in the friction coefficient of the pipe. The change in the cross-section also influences (to a greater extent) the magnitude of the critical Reynolds number $Re_{cr} = VR/\nu$; for a circular cross-section $Re_{cr} = 580 (= 2320/4)$, for a square conduit $Re_{cr} = 525$ and for a wide open channel $(R = h) Re_{cr} = 500$.

In general, we can express the above equations also in the form

$$V = \frac{U_*}{\kappa} \ln\left(\frac{bR}{\delta'}\right) \tag{4.59}$$

for *smooth* pipes and

$$V = \left(\frac{U_*}{\kappa}\right)\left(\frac{cR}{k}\right) \tag{4.60}$$

for *rough* pipes, where b and c are constants depending on the conduit shape. For a circular pipe $b = 46.6$ and $c = 14.3$, and for a wide rectangular conduit $b = 38.4$ and $c = 11.0$.

Equations (4.59) and (4.60) lead also to the 'Delft' equation, which simplifies the coefficients in a single equation (for the Chezy coefficient C):

$$C = 18 \log\left(\frac{6R}{(k/2 + \delta'/7)}\right) \tag{4.61}$$

Finally, a comparison of the Darcy–Weisbach equation with the empirical Chezy, Manning and Strickler equations

$$V = C\sqrt{RS} \tag{4.62}$$

$$V = \frac{1}{n}(R^{2/3}S^{1/2}) \tag{4.63a}$$

$$V = \frac{c}{k^{1/6}}(R^{2/3}S^{1/2}) \tag{4.63b}$$

results in

$$C = \frac{1}{nR^{1/6}} = \sqrt{\frac{8g}{\lambda}} = \sqrt{\frac{2g}{\lambda_R}} = 26\left(\frac{k}{R}\right)^{1/6} \tag{4.64a–d}$$

(putting $c = 26$). The range within which this simple relationship can well approximate the more complicated logarithmic one is $5 < y/k < 500$.

4.4.2 Local head losses

'Local (head) losses' occur in conduits at changes of direction of flow or cross-section (contractions, expansions, valves, gates, bends, etc.). They are usually expressed in the form

$$h_l = \frac{\xi V^2}{2g} \tag{4.65}$$

where ξ is a 'coefficient' depending on the type of local loss, geometry, wall roughness, Reynolds number and upstream velocity distribution, and V is the mean cross-sectional velocity downstream of the feature causing the local loss.

In the more common cases (bends, changes in pipe diameter, orifices, etc.) the relationship between ξ, Reynolds number and geometric parameters of the change has been well investigated and published in tables and/or graphs (Lencastre 1987, Miller 1994). In cases where this relationship is not known, it has to be investigated experimentally in the laboratory.

Local losses may be substantially influenced (i.e. increased or decreased) by the upstream configuration unless the distance between the items (e.g. fittings or appliances) causing the loss is sufficiently large (usually more than about $40D$).

As in the case of the friction coefficient, ξ will be independent of the Reynolds number only if this exceeds a certain limiting value. Considering that every feature introducing a local loss causes secondary currents and increased turbulence upstream and, particularly, far downstream, it may be safely assumed that this Reynolds number will be smaller or at most equal to $Re_{sq} = 50$–100 with $\xi = 700$. Thus, if equation (4.53) is satisfied for the friction coefficient, the relevant values of ξ will also be given by equation (4.53) (e.g. for a 90° bend, $Re_{sq} = VD/\nu = 5 \times 10^3$, with $\xi = 1$, for an orifice R independent of viscosity).

4.4.3 Basic concepts and equations for non-uniform and unsteady flow in pipes and open channels

In *unsteady flow* the discharge is a function both of position and time $(Q = f(x, t))$. In *open channel flow* (of depth y) $S_f = dH/dx$ results in the *Saint Venant equation*:

$$S_f = \frac{\partial H}{\partial x} = \frac{-\partial z}{\partial x} - \frac{\partial y}{\partial x} - \frac{v}{g}\frac{\partial v}{\partial x} - \frac{v}{g}\frac{\partial v}{\partial t}\frac{dt}{dx} \tag{4.66}$$

The first term on the right of equation (4.66) signifies *uniform flow* and the first three terms signify *non-uniform (gradually varied)* flow.

From continuity (with no lateral discharge in Δx and B the water surface width),

$$\frac{A \partial v}{\partial x} + \frac{v \partial A}{\partial x} + \frac{B \partial y}{\partial t} = 0 \tag{4.67}$$

The first term in equation (4.67) represents the *prism storage* and the second term the *wedge storage*.

From equation (4.66) the equation for *gradually varied non-uniform flow* ($\partial t = 0$) can be written as (where $Fr^2 = Q^2 B/(gA^3)$ and S_0 is the bed slope):

$$\frac{dy}{dx} = \frac{S_0 - S_f}{1 - Fr^2} \tag{4.68}$$

A special case of equations (4.66) and (4.67) is the *shallow water* formulation for surface waves (using y for depth)

$$\frac{\partial u}{\partial t} + \frac{u \partial u}{\partial x} + \frac{g \partial y}{\partial x} = 0 \tag{4.69}$$

$$\frac{\partial y}{\partial t} + \frac{u \partial y}{\partial x} + \frac{y \partial u}{\partial x} = 0 \tag{4.70}$$

For further development and use of equations (4.66)–(4.70), see Chapters 7, 8, 11 and 12.

Considering a *surge* produced by sudden changes in depth and/or a discharge moving with velocity V in a flow with section 1 in front and section 2 behind (Figure 4.3), the surge momentum and continuity result in

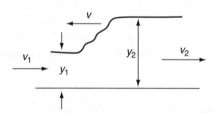

Figure 4.3 Definition sketch of a surge

$$Fr_1^2 = \frac{(V_1 + V)^2}{g y_1} = \frac{1}{2} \frac{y_2}{y_1} \left(\frac{y_2}{y_1} - 1 \right) \tag{4.71}$$

$$(V_1 + V) y_1 = (V_2 + V) y_2 \tag{4.72}$$

From these equations the velocity of the surge is

$$V = -V_1 \pm \sqrt{gy_1 + \frac{3}{2}(y_2 - y_1) + \frac{(y_2 - y_1)^2}{2y_1}} \tag{4.73}$$

For small surges $(y_2 - y_1)$ is small and

$$V = V_1 \pm \sqrt{gy_1} \tag{4.74}$$

For the initial velocity $V_1 = 0$, $V = \sqrt{(gy_1)}$ (the sign convention in equations (4.72)–(4.74) indicates a surge moving against the direction of flow).

The energy loss in the surge (per unit weight of fluid) is

$$\Delta E = \frac{(y_2 - y_1)^3}{4y_2 y_1} \tag{4.75}$$

From the point of view of an observer moving with velocity V the surge becomes a stationary one, and equations (4.71)–(4.73) (with $V = 0$) become the equations for a hydraulic jump denoting the transition from supercritical to subcritical flow.

Fast changes of discharge in *pipelines* caused by, for example, fast changes in flow at a hydroelectric power station, by manipulation of valves or by pump start or shut-down, produce pressure waves in the pipeline system. These transients, with the flow velocity a function of both time and position and influenced by the elastic properties of the fluid and pipeline material, are called *waterhammer*.

Changes in discharge in cases where there is a (very elastic) interface of the liquid (water) with air result in slow transients without the influence of elasticity – *mass oscillation* – where the flow velocity is a function of time only.

The basic differential equations for waterhammer, derived from momentum and continuity principles are:

$$\frac{\partial H}{\partial t} + \frac{1}{g}\left(\frac{v}{c} + 1\right)\frac{\partial v}{\partial t} + S_f = 0 \tag{4.76}$$

$$\frac{\partial V}{\partial x} + \frac{g}{c^2}\frac{\partial H}{\partial t} = 0 \tag{4.77}$$

where H is the pressure head and c is the pressure wave velocity given by

$$c = \sqrt{\frac{K/\rho}{1 + DK(eE)}} = \frac{c_0}{\sqrt{1 + DK/(eE)}} \tag{4.78}$$

where K is the bulk modulus of the fluid, E the Young modulus of the pipeline material and e the wall thickness of the pipeline of diameter D. The velocity of the pressure wave in the liquid (without the effect of the pipe) $c_0 = \sqrt{(K/\rho)}$ is 1,425 m/s for water (this may be greatly reduced even by a small amount of entrained air, e.g. a concentration of air of 0.01 reduces the pressure wave celerity to about 400 m/s). As in most cases $V << c$, the term in the bracket in equation (4.76) may be neglected.

The above system of equations is usually solved by numerical computation using a discrete (finite-difference) formulation (method of characteristics) but can also be solved by mathematical or graphical methods based on their Riemann invariants solution (Schnyder–Bergeron). The main difficulties are in complicated pipe networks with pressure-wave reflections and transmissions, and in the determination of the value of a in the presence of dissolved and/or dispersed air in the liquid.

For *mass oscillation* (slow changes in discharge, e.g. in a system with a surge tank) the dynamic equation is

$$\frac{L}{g}\frac{dV}{dt} + z \pm PV^2 \pm RV_s^2 = 0 \tag{4.79}$$

and continuity results in

$$VA = V_s A_s + Q' \tag{4.80}$$

with

$$V_s = \frac{dz}{dt} \tag{4.81}$$

where z is the difference between the water levels in the surge tank and in the reservoir, $P = \lambda L/2gD$, R is given by the head loss coefficient at the entry into the surge tank, A_s is the surge-tank area, V_s the velocity of flow in the surge tank and Q' is the turbine (pump) discharge. Differentiating equation (4.79) and substituting from equations (4.80) and (4.81) results in a second-order differential equation which, for a general case, cannot be solved. We have to resort, therefore, either to simplifying assumptions (e.g. $Q' = 0$) or to solving the above system of equations by graphical or numerical finite-difference methods.

Further development and applications of equations (4.71)–(4.81) are discussed in Chapter 9.

4.5   Introduction to ocean wave motion

Most ocean waves are examples of periodic progressive gravity waves generated by the action of wind on water. There is little translation involved, with water particles describing elliptical orbits in a vertical plane and only wave crests moving in a horizontal direction.

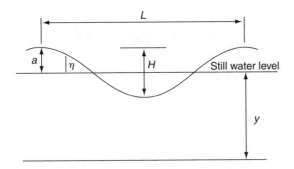

Figure 4.4 Definition sketch for a sinusoidal wave

Consider a wave of amplitude a, height $H(=2a)$, length L (distance between wave crests), period T and celerity (or velocity) c moving on a water surface with a still water depth y (Figure 4.4). Thus

$$c = \frac{L}{T} \tag{4.82}$$

For a sinusoidal wave, excluding the effect of viscosity and surface tension (Airy wave) from two-dimensional ideal fluid flow, the variation η (distance from the still water surface) is

$$\eta = a \cos 2\pi \left(\frac{x}{L} - \frac{t}{T} \right) = a \cos (kx - ft) \tag{4.83}$$

where $k = 2\pi/L$ is the *wave number* and $f = 1/T$ or $\omega = 2\pi f$ (the *wave frequency*).

The Stokesian and cnoidal wave theories give a better approximation of the form of steep waves than the above Airy equation, which is, however, sufficiently accurate for many engineering purposes.

For $H \ll L$ and $H \ll y$, equation (4.83), after substitution into the Laplace equation for two-dimensional irrotational flow, yields the wave celerity c:

$$c = \frac{gT}{2\pi} \tan h\left(\frac{2\pi y}{L}\right) = \sqrt{\frac{gL}{2\pi} \tan h\left(\frac{2\pi y}{L}\right)} \tag{4.84}$$

or

$$c = \sqrt{\frac{gT}{2\pi} \tan h(ky)} \tag{4.84a}$$

Including the effect of surface tension modifies equation (4.84) into

$$c = \sqrt{\left(\left(\frac{gL}{2\pi}\right) + \frac{2\sigma\pi}{\rho L}\right) \tan h\left(\frac{2\pi y}{L}\right)} \tag{4.85}$$

For large values of $2\pi y/L$, $\tan h(2\pi y/L) = 1$, and thus

$$c = \sqrt{\frac{gL}{2\pi} + \frac{2\psi}{\rho L}} = \sqrt{\frac{g}{k} + \frac{k\sigma}{\rho}} \tag{4.86}$$

Evidently, whereas the celerity of gravity waves increases with wave length, the opposite is true for capillary waves. The minimum celerity with which a surface wave may be propagated follows from equation (4.86) as

$$c_{min} = \left(\frac{4g\sigma}{\rho}\right)^{1/4} \tag{4.87}$$

corresponding to a minimum wave length for gravity waves of

$$L_{min} = 2\pi \sqrt{\left(\frac{\sigma}{\rho g}\right)} \tag{4.88}$$

(for $gL/(2\pi) < 2\pi\sigma/(gL)$ the surface tension is dominant). For an air–water interface ($\sigma = 0.075 \, \text{N/m}$), equations (4.87) and (4.88) result in $c_{min} = 0.23 \, \text{m/s}$ and $L_{min} = 0.017 \, \text{m}$. For $L \gg L_{min}$ the effect of surface tension becomes negligible.

Equation (4.84) demonstrates the effect of depth on wave celerity. For $y > 0.5L$, $\tan h(2\pi y/L) = 1$ and thus for a *deep water* (or *short*) wave

$$c = \frac{gT}{2\pi} = \sqrt{\frac{gL}{2\pi}} = \sqrt{\frac{g}{k}} \tag{4.89}$$

For $y < 0.05L$, $\tan h(2\pi y/L) = 2\pi y/L$ and for a *shallow water* (or *long*) wave

$$c = \frac{gTy}{L} = \sqrt{gy} \tag{4.90}$$

(see also equation (4.74)). There is, of course, a transition between these two ratios of y/L. Equation (4.90) is the equation of a *Lagrangian wave*. (Note: waves can move upstream only if $V < c$.)

The *kinetic energy* of one wavelength per unit width of wave crest E_k is given by

$$E_k = \frac{1}{16}(H^2 \rho g L) \tag{4.91}$$

As (for Airy waves) the potential and kinetic energies are equal, the *total energy per unit area E* is

$$E = \frac{1}{8}(\rho g H^2) \tag{4.92}$$

Summing the potential, kinetic and pressure energies of individual particles and multiplying by particle velocity for all particles in a wave leads to the equation for *wave power N*, which for Airy waves is

$$N = \frac{\rho g H^2}{16} = \left(\frac{1 + (4\pi y/L)}{\sin h(4\pi y/L)} \right) = E c_g \tag{4.93}$$

where c_g is the *group wave celerity*:

$$c_g = \frac{c}{2} \left(\frac{1 + (4\pi y/L)}{\sin h(4\pi y/L)} \right) \tag{4.94}$$

$c_g = c/2$ for deep-water waves and $c_g = c$ for shallow-water waves. Thus, for deep-water waves individual wave crests travel twice as fast as the group as a whole (the dispersive property of the waves).

It has been shown that for waves in shallow water $c = \sqrt{(gy)}$; thus, in this case, c is independent of the wave number, and gravity waves (of finite amplitude) in shallow water are non-dispersive.

Regarding a *solitary wave of finite amplitude* as a series of small waves superimposed on each other it is clear that, relative to the undisturbed fluid, each element of the wave must move more rapidly than that directly below it, resulting in progressive steepening of the wave front, which could eventually form a shock wave unless counteracted by the smoothing effects of dispersion.

Further discussion and application of these equations, as well as of wave refraction, shoaling, breaking, diffraction and reflection, are given in Chapter 12.

4.6 Environmental processes – hydrodynamic factors, sediment mechanics, water quality and air–water flows

4.6.1 General

The increase in the social and political awareness of the impact of engineering on the environment and the resulting legislation requiring environmental impact assessments or environmental statements of the effects of works led to further research into the many complex interactions between the physical, chemical and biological factors involved. In the aquatic environment, the simulation modelling of the current and post-project conditions played a major role in the recognition of *environmental hydraulics* as a branch of hydraulics dealing with the interaction of the water movement with the suspended and boundary sediments, marginal vegetation, effluent inputs, and the water chemistry and biological processes within the flow. This section considers in outline the influence of hydrodynamics on other important environmental parameters, often called *transport modelling*, including the motion of sediments and water-quality modelling. However, the general ecological or habitat models are not covered. Further development and applications of the concepts mentioned here are included in Chapters 7–13.

A common feature of transport modelling is that the other processes of interest are strongly affected by the imposed bulk-flow field and, in some cases, can themselves alter the hydrodynamics. Experimental techniques are of great value in developing an understanding of the processes involved, through observations under precisely controlled conditions designed to capture the parameters of interest. Physical models may be used to simulate and investigate the effects of some environmental processes (e.g. saltwater intrusion in estuaries, gas transfer at the air–water interface, buoyant jets, coherent structures in shallow-water flows) and their engineering applications (Jirka 1992). However, the complex interactions between the many processes that occur in, say, determining resultant water quality can usually be best undertaken using a computational model (Falconer 1992). This section will, therefore, also discuss some of the consequences of the approximations inherent in the development of a computational model of these processes.

4.6.2 Hydrodynamic factors

The most important hydrodynamic factor in environmental hydraulics is the influence of turbulence (see Section 4.3.3). Its effect is to introduce a much greater effective diffusion of any transported parameter than would be estimated from molecular viscosity alone. The properties of the turbulent flow field depend strongly on the local geometry and can vary across the domain of interest. In descriptions of the hydrodynamics of environmental and mixing processes, two pairs of similar sounding terms are frequently encountered: *advection* and *convection*, and *diffusion* and *dispersion*.

Convection is the process of transport of a quantity that in itself determines the bulk flow field (e.g. momentum, and also heat if the variation in density with temperature is modelled). The term $u.\nabla u$ in equations (4.3)–(4.5) is often called the 'convection term', as are the corresponding terms in the shallow-water and Saint Venant equations for two- and one-dimensional flow, respectively (see equations (4.66) and (4.69)).

Advection is the transport of any other scalar quantity (e.g. a tracer) by an imposed bulk fluid motion. In a transport equation, the advection term for the concentration of a quantity C takes the conservative form $\nabla.(Cu)$ or a rearrangement of this.

Diffusion is the transport of a quantity from an area of high concentration to an area of lower concentration through random mixing processes within the fluid. Random molecular motion is weakly diffusive, while turbulent motion is strongly diffusive. Molecular diffusion can occur in fluid that has no mean motion, whereas turbulent diffusion requires flow at a high Reynolds number. The diffusive term in a process equation will often appear as the second derivative of the concentration field, thus

$$\varepsilon \nabla^2 C$$

where ε is the diffusivity coefficient.

A related notion is that of the turbulent eddy viscosity, which appears in the turbulent closures of the Reynolds momentum equations (see Section 4.3.3). Often the assumption is made that the eddy viscosity (for momentum) and the eddy diffusivity (coefficient of turbulent diffusion) of, say, a passive tracer are equal. This need not be the case. In the simplest representations of the flow, these coefficients may be assumed to be constants over the flow field or related in a simple manner to the characteristics of the primary flow. In higher-order turbulence the closure properties of the turbulence field are formulated in terms of differential equations that are coupled to the hydrodynamic equations of the primary flow. This is discussed further in Chapters 8 and 11.

Dispersion is the result of a combination of advection and diffusion; it is the differential transport of the distribution of a quantity by spatial variations in the bulk flow velocity. Shear flows are particularly dispersive, and dispersion distorts the shape of the initial concentration profile. This is the key distinction from diffusion, which usually entails only a gradual change in the overall shape of the concentration profile.

In many circumstances, a *computational* model with reduced spatial dimensionality will be used in engineering practice. This reduction in dimensionality has a number of important consequences. All features of the flow physics in the directions that have been excluded from the model will not appear in the model simulations, nor will processes that are on a smaller scale than those resolvable by the model grid. Examples of such processes that may be explicitly excluded from the modelling are the helical secondary currents in river bends, which are removed by depth and area integration, and zones of circulation in the lee of an obstruction, which are removed by area averaging. Whereas these may not be of great significance in determining the principal features of the primary flow, they can have a profound effect on the simulation of other environmental processes, as, in prototype, they can introduce dispersion, and this has to be incorporated in the model.

It is common to analyse flow in a river using a one-dimensional model based on the Saint Venant equations of open channel flow (equation (4.66)) to determine the hydrodynamics (see Chapter 7). The transport of pollutants in the flow is then simulated using an advection–diffusion equation. The coefficient of diffusion needed to simulate the propagation of observed releases of tracers is often found to be much larger than the coefficient determined from turbulent and molecular diffusion alone. Clearly, there are other sources of dispersion in the one-dimensional model. One is the area integration of the deviation from their averages of the contaminant concentration and flow velocity profiles across the section. Another process is the existence of *dead zones*, which act as temporary sinks for the pollutant. The dead zones are areas of recirculation driven by separated flow past irregularities in the channel shape; they take in the pollutant as the main pollutograph passes and then release the trapped pollutant on the tail of the concentration profile. The *aggregated dead zone* (ADZ) dispersion models of river flows are based on this concept.

The *shallow-water equations* (SWEs) (see Chapter 2) represent the hydrodynamics of unstratified flow in shallow seas, estuaries and lakes. These equations are obtained from the Navier–Stokes or Reynolds equations by integrating through the depth of the flow. It is well known that the physical flow may contain an area of strongly sheared fluid near the bed; this introduces dispersion.

In the formulation of the SWEs, the x-component of the convection term of the Reynolds equation (equation (4.33)) $\bar{u}\frac{\partial \bar{u}}{\partial x} + \bar{v}\frac{\partial \bar{u}}{\partial y} + \bar{w}\frac{\partial \bar{u}}{\partial z}$ is integrated through the depth of the flow (i.e. with respect to z), and the order of differentiation and integration is reversed using the continuity equation to simplify the results. Using u, etc. (instead of $\bar{u}$), to represent the mean velocities over the timescale of turbulence, with h being the water surface level and z the bed level, results in

$$\int_{z_b}^{h}\left(u\frac{\partial u}{\partial x}+v\frac{\partial u}{\partial y}+w\frac{\partial u}{\partial z}\right)\mathrm{d}z=\frac{\partial}{\partial z}\int_{z_b}^{h}u^2\mathrm{d}z+\frac{\partial}{\partial y}\int_{z_b}^{h}uv\mathrm{d}z+uw\updownarrow_h-u^2\frac{\partial h}{\partial x}$$

$$-uv\frac{\partial h}{\partial y}-uw\updownarrow_{z_b}+u^2\frac{\partial z_b}{\partial x}+uv\frac{\partial z_b}{\partial y}$$

The contributions to the integrations at the free surface and the bed cancel identically from the kinematic boundary condition, which expresses zero normal flow at these surfaces. Expanding the integrals in terms of the depth ($y=h-z_b$) and average velocity (U, V),

$$U(x,y)=\frac{1}{y}\int_{z_b}^{h}u(x,y,z)\mathrm{d}z;\quad V(x,y)=\int_{z_b}^{h}v(x,y,z)\mathrm{d}z$$

and using $u^2=(U-(U-u))^2$ and $\int_{z_b}^{h}U(U-u)\mathrm{d}z=0$ and similar relationships for the other products uv and v gives

$$\frac{\partial}{\partial x}\int_{z_b}^{h}u^2\mathrm{d}z+\frac{\partial}{\partial y}\int_{z_b}^{h}uv\mathrm{d}z$$

$$=\frac{\partial}{\partial x}\int_{z_b}^{h}(U-(U-u))^2\mathrm{d}z+\frac{\partial}{\partial y}\int_{z_b}^{h}(U-(U-u))\,(V-(V-v))\,\mathrm{d}z$$

$$=\frac{\partial}{\partial x}(U^2y)+\frac{\partial}{\partial y}(UVy)+\frac{\partial}{\partial x}\int_{z_b}^{h}(U-u)^2\mathrm{d}z+\frac{\partial}{\partial y}\int_{z_b}^{h}(u-u)(V-v)\mathrm{d}z$$

$$(4.95)$$

The convection term of the SWEs is thus seen to have two components. The first (first two terms in equation (4.95)) is the analogue of the convection term in strictly two-dimensional flow, and the second (third and fourth terms in equation (4.95)) is dispersive and non-zero if the three-dimensional velocity profile in the vertical is not uniform, as physically must be the case. A common means of accounting for this dispersion is to aggregate it with the turbulent diffusion by using an effective diffusion coefficient. The traditional reasoning is as follows: the terms look like the Reynolds stress terms in the formulation of turbulent flow (i.e. they involve integrals of deviations from the depth mean velocity), and hence they might be modelled using a

Boussinesq-type model relating the integrated values of these deviations to derivatives of the primary flow velocity. This, of course, lacks any rigorous justification, as we are dealing with spatial integrals and not time averages, and the velocity deficit $(U - u)$ etc. will have a systematic structure rather than the chaotic random nature of turbulence.

Dispersive processes dominate the overall effective diffusion coefficient, and their representation by the formula for turbulent diffusion is questionable. Nevertheless, because of its simplicity, this formulation of the dispersion term in the SWEs has been used widely in modelling practice.

In many cases, the flow *density* is not homogeneous throughout the depth (e.g. at outflows from cooling systems into a recipient, or in estuaries with salt-water penetration into the river flow). In the former example, a *buoyant jet* is formed with diffusion across the jet boundary and further dilution due to the entrainment of the surrounding fluid. The degree of *stratification*, or mixing, in the second example is a function of the estuary geometry and the tidal range. The strength of turbulent diffusion varies with the degree of stratification in the flow, with vertical diffusion being depressed across the interface between layers of different density.

A measure of the importance of stratification in the flow is given by the local *densimetric Froude number* Fr_ρ ($1/Fr_\rho^2$ is sometimes called the Richardson number) and the *Richardson number* Ri (sometimes called the *gradient Richardson number* Ri_g), both derived from the continuity equation and the equations of motion of stratified flow with a density difference $\Delta\rho$. These numbers quantify the relative importance of inertia and buoyancy forces (see also Chapter 5) and are defined by (where L is a length term, typically the depth of flow, u is the turbulent mean velocity and z is the vertical coordinate):

$$Fr_\rho = \frac{u}{\sqrt{gL\Delta\rho/\rho}} \tag{4.96}$$

$$Ri = \frac{g\Delta\rho/\Delta z}{\rho(\Delta u/\Delta z)^2} \tag{4.97}$$

The stability of the stratification increases as the gradient Richardson number increases (i.e. for larger density gradients or a smaller vertical gradient of horizontal velocity). The rate of entrainment of the upper layer of flow into the lower depends on the Richardson number defined using the appropriate length and velocity scales of the turbulence. The role of the density gradient on the generation of the turbulence depends on its sign. If the density increases upwards, the buoyancy forces provide an additional source of energy for turbulence. Thus, negative values of the Richardson number indicate an unstable density gradient, with both shear and buoyancy forces generating turbulence (only opposed by viscous dissipation).

For further treatment of and references on the above principles, see, for example, Hino (1994), Raudkivi and Callander (1975), Rutherford (1994) and Singh and Hager (1996); further development and applications are discussed in Chapters 8 and 11.

4.6.3 Basic sediment mechanics

A full discussion of sediment mechanics is clearly outside the scope of this text. Thus, only a few basic concepts necessary for further development of the modelling methods and their application in engineering design (see Chapters 8–13) are included here.

From the point of view of *source*, sediment transported by flow can be divided into *washload*, comprising very fine material derived mainly from overland flow and moving in rivers and canals in suspension, and *bed-material load*, moving as *bedload* and/or *suspended load* depending on sediment size and flow velocity. Thus, washload is limited only by supply, whereas bed-material load depends also on hydraulic factors.

Catchment processes supply a wide range of particle sizes to channels and bedload transport leads to size sorting in rivers both downstream (*fining* of sediments) and vertically within the bed (*armouring*).

Sediment concentration in rivers varies enormously between continents, countries and even catchments (e.g. from 15,000 ppm at the mouth of the Hwang Ho River to 10 ppm in the Rhine delta (de Vries 1985)).

Bedload is the important element of sediment transport for river engineering, navigation canals and sewers, as it determines the morphology of rivers and bed erosion and sedimentation aspects; suspended load is important in river engineering only in reservoir sedimentation and, exceptionally, in sedimentation at canal intakes; it is also important in sewer outfall design.

The important *properties of sediment* and the parameters used in sediment computations are the sediment size d, the shape and grading, the density ρ_s (usually $2,650\,\text{kg/m}^3$), the fall velocity w, the bulk density and porosity, and the sediment concentration C (volumetric, or ppm, or mg/L). According to size, we usually distinguish between clay particles $(0.5 < d(\mu\text{m}) < 5)$, silt $(5 < d(\mu\text{m}) < 60)$, sand $(0.06 < d(\text{mm}) < 2)$ and gravel $(2 < d(\text{mm}) < 60)$.

The *fall velocity* can be approximately expressed by the equation

$$w = \left(\frac{4gd\Delta}{3C_D}\right)^{1/2} \tag{4.98}$$

where $\Delta = (\rho_s - \rho)/\rho$ and C_D is a drag coefficient dependent on the Reynolds number $\text{Re} = wd/\nu$. For $\text{Re} < 1$ (very fine sediment), $C_D = 24/\text{Re}$, which leads to Stokes' law $w = g\Delta d^2/(18\nu)$; for large sizes with $\text{Re} > 10^3$,

C_D becomes constant and is a function of grain shape only (usually $C_D = 1.3$ for sand particles). The fall velocity varies, therefore, with $d^{1/2}$ to d^2.

The *threshold of sediment motion (incipient motion)* is given by a critical value of the shear stress, which for a plane sediment bed is given by the Shields criterion:

$$\tau_c = c(\rho_s - \rho)gd \qquad (4.99)$$

where c, according to various authors, varies between 0.04 and 0.06. The condition of validity of equation (4.99) is that $\text{Re}(= wd/\nu) > 10^3$.

As $\tau_0/\rho = U_*^2$, equation (4.99) can also be written as

$$\text{Fr}_d^2 = \frac{U_*^2}{gd\Delta} = c \qquad (4.99a)$$

where Fr_d is the grain or densimetric Froude number.

For a sediment particle on a slope (e.g. the side slope of a canal) inclined at an angle β to the horizontal, the critical shear stress is reduced by a factor $\{1 - (\sin^2 \beta / \sin^2 \varphi)\}$, where φ is the natural angle of stability of the non-cohesive material. (For stability, naturally $\beta < \varphi$.) The average value of φ is about 35°.

On the other hand, the maximum shear stress induced by the flow on a side slope of the canal is usually only about $0.75\rho gyS$ (instead of ρgRS as applicable for the bed). In a channel that is not straight, the critical shear stresses are reduced further by a factor of 0.6 to 0.9 (0.6 applies to very sinuous channels).

Investigations into *bedload and bed material transport* have been going on for decades, without producing a really satisfactory all-embracing equation connecting the fluid and sediment properties. This is due mainly to the complexity of the problem, including the effect of different *bed forms*, on the mode and the magnitude of bedload transport, the stochastic nature of the problem, and the difficulty of field data collection to verify laboratory investigations. Nevertheless, substantial advances have been made.

Sediment transport equations deal either separately with one component of the transported sediment phase (i.e. bedload or suspended load) or with the total sediment load. There are two approaches to the total sediment load determination: by addition of the two components (bedload and suspended load) – the 'microscopic' methods; or by using a single equation for both types of transport based on a shear parameter and representative size for the entire sediment mixture (although methods based on fractionwise computations have also been developed) – the 'macroscopic' methods. The situation is further complicated by the fact that the suspended sediment load in rivers always includes washload and the suspended part of bed material load, whereas the suspended part of the total sediment load in laboratory flume measurements excludes washload. As the total load equations were developed from results obtained in a controlled laboratory environment, it is more correct in these cases to speak of 'bed material load' rather than 'total load'.

Most of the used approaches can be reduced to a correlation between the *sediment transport parameter* $\phi = q_s/(d^{3/2}\sqrt{(g\Delta)})$ (where q_s is the sediment transport in m^3/s/m) and $Fr_d^2 = 1/\psi = U_*^2/\Delta gd$ (where ψ is called the *flow parameter*; ψ can also contain an additional parameter – the *ripple factor* – to account for the effect of bed form (Graf (1984)). The power of Fr_d^2 in many correlations varies between 2 and 3, i.e. q_s varies as V^n with $4 < n < 6$, demonstrating the importance of a good knowledge of the velocity field in the modelling and computation of bedload transport, particularly when using two- or three-dimensional models.

Examples of simplified ϕ and ψ correlations are the Meyer–Peter and Muller equation (bedload only):

$$\phi = \left(\frac{4}{\psi} - 0.188\right)^{3/2} \tag{4.100}$$

the Einstein–Brown equation (bedload):

$$\phi = 40\psi^{-3} \tag{4.101}$$

and the Engelund–Hansen equation ('total load'):

$$\phi = \frac{0.4\psi^{-5/2}}{\lambda} \tag{4.102}$$

It must be emphasized that the full application of the above and other more sophisticated equations requires further reading (e.g. Garde and Ranga Raju (1985), Graf (1984), (1998), Simons and Senturk (1992) and Yalin (1992)) and the equations are quoted here only to demonstrate the correlation and trend.

The Ackers–White equation correlates the modified transport and flow parameters by the equation ('total load')

$$G_{gr} = C\left(\frac{F_{gr}}{A-1}\right)^m \tag{4.103}$$

where F_{gr} (modified Fr), called the *mobility number*, is given by $F_{gr} = U_*^n V^{1-n} / \left((\Delta gd)^{1/2}(\sqrt{32}\log(10y/d))^{1-n}\right)$ and $G_{gr} = Xy/\left((\Delta+1)d\right)(U_*/V)^n$, where X is the concentration expressed as the mass flux of the solid (sediment) phase divided by the mass flux of the fluid phase $(X = q_s\rho_s/(Vy\rho))$. The coefficients C, A, m and n are functions of a dimensionless grain size $D_{gr} = d(\Delta g/v^2)^{1/3}$; for water $D_{gr} = 25d$ (with d in mm).

Although the relationship between the transport of bed sediment and the flow, and even the relationship between the hydraulic resistance and channel sedimentary features, particularly the bed configuration (which, in turn, is a function of the sediment characteristics and discharge), are broadly known, the third equation required for computations on alluvial channels, relating the flow parameters and the erosive resistance of the banks, still by and large eludes a physically based formulation. Nevertheless, the minimum stream power concept or other optimization methods show much promise in this area, where we otherwise fall back on regime equations that synthesize the physical functions into groups of formulae describing the channel geometry (Ackers 1983). This, as well as river morphology, are briefly discussed in Chapter 8.

The computation of the *suspended sediment load* alone is best related to the discussion of the turbulent eddy viscosity and the coefficient of turbulent diffusion presented in the preceding paragraphs. The equation for a concentration C in a (two-dimensional) flow with diffusivity coefficients ε_x and ε_z (see equation (4.35)) is

$$\frac{\partial C}{\partial t} + \frac{u \partial C}{\partial x} + \frac{v \partial C}{\partial y} - \frac{w \partial C}{\partial z} = \frac{\partial}{\partial x}\left(\frac{\varepsilon_x \partial C}{\partial x}\right) + \frac{\partial}{\partial \varepsilon_z}\left(\frac{\partial C}{\partial z}\right) \tag{4.104}$$

For steady conditions $\partial C/\partial t = \partial C/\partial x = 0$; furthermore, if $v << w$, equation (4.104) reduces to

$$\frac{w \partial C}{\partial z} = \frac{-\varepsilon_z \partial^2 C}{\partial z^2} \tag{4.105}$$

or

$$wC = \frac{\varepsilon_z \partial C}{\partial z} \tag{4.105a}$$

stating that the upward rate of sediment movement due to turbulent diffusion is balanced by the downward volumetric rate of sediment transfer due to gravity. As from equations (4.28) and (4.35) $\varepsilon_z = l^2 du/dz$ and $du/dz = U_*/(\kappa z)$, for $l = \kappa z \sqrt{(1 - z/y)}$

$$\varepsilon_z = \kappa z U_* \left(1 - \frac{z}{y}\right)$$

Integration of equation (4.105) then results in

$$\frac{C}{C_a} = \left(\frac{a(y - z)}{(z(y - a)}\right)^{w/(\kappa U_*)} \tag{4.106}$$

giving the concentration at a height z above the bed as a function of the concentration at a distance a from the bed. As equation (4.106) gives a zero concentration at the water surface $(z = y)$ and $C = \infty$ *at* $z = O$, which evidently is not correct, it is better to use in equation (4.105) a constant average value of $\bar{\varepsilon}_z$, which from above and for $\kappa = 0.4$ is $\bar{\varepsilon}_z = yU_*/15$. Equation (4.105) then leads to

$$\frac{C}{Ca} = e^{-15w/U_*(z-a)/y} \tag{4.107}$$

Combining equation (4.107) with the logarithmic velocity distribution equation (equation (4.29)) and integrating over the depth y results in the equation for suspended sediment discharge:

$$q_s = q C_a \xi e^{15wa/(U_*y)} \tag{4.108}$$

where ξ is a function of w/U_* and relative channel-bed roughness (typically, for $w/U_* = 0.01$, $\xi = 1$; for $w/U_* = 0.1$, $\xi = 0.45$; and for $w/U_* = 1$, $\xi = 0.035$).

To use equation (4.108) it is necessary to measure or estimate C_a. For $a = 2d$ Einstein suggests C_a equal to the computed bedload divided by $2d$ and the velocity in the layer of thickness $2d$.

Equation (4.107) is valid only for particles of equal w, and therefore each part of a suspension with varying grain size has to be computed separately. It also contains among other assumptions the notion that the eddy viscosity and eddy diffusivity of the sediment are equal, which need not be the case (see also Section 4.6.2), as it has been shown that for the same depth, velocity (slope) and boundary roughness the flow resistance of sediment-laden streams is often smaller than that of clear water.

Many empirical equations linking the suspended load and water discharge have been proposed. Engelund suggested a simple form:

$$Q_s = 0.5q\left(\frac{U_*}{w}\right)^4 \tag{4.109}$$

The *washload* phase of sediment transport will depend primarily on the erosion of soil particles by raindrop and leaf-drip impact and overland flow, with the sediment particles reaching the river channels through this mechanism. The final outcome will thus be largely dependent on the soil properties, hydrological and land morphological parameters, as well as on vegetation cover and land use, and is best modelled by computational methods using either empirical equations (e.g. the universal soil equation) or more sophisticated methods (e.g. the SHE system with a physically based distributed sediment yield model (Wicks and Bathurst (1996))).

The washload is generally supply limited and the flow in the watercourse will transport as much of it as arrives there.

Urbanization also brings about the problems of *sediment transport in storm drainage systems*, which has received increased attention in recent years.

For the *critical mean velocity of flow V_c for the initiation of motion* of sediment particles resting on a fixed rough or smooth bed (e.g. sewer invert) Novak and Nalluri (1984) suggested the equation

$$\frac{V_c}{\sqrt{gd\Delta}} = a\left(\frac{d}{R}\right)^b \tag{4.110}$$

for $0.008 < d/R < 1.0$ and $3.5 < d/k$, where the coefficients a and b are primarily functions of d/k (k is the roughness size) and the degree of isolation of individual sediment particles (single or touching particles resting on the bed) with $0.5 < a < 0.61$ and $-0.27 < b < -0.4$. (Later, May (Ackers *et al.* (1996)) suggested $a = 0.125$ and $b = -0.47$ using depth of flow over pipe invert instead of R.)

For the *critical bedload transport*, defined as the maximum possible rate of transport along the channel without any deposition, Novak and Nalluri (1984) proposed the equation

$$\phi = 11.6\psi^{-2} \tag{4.111}$$

which translates into

$$\frac{V_L}{\sqrt{gd\Delta}} = 1.77C_V^{1/3}\left(\frac{d}{R}\right)^{-1/3}\lambda^{-2/3} \tag{4.112}$$

where C_V is the sediment volumetric concentration, λ the friction coefficient for clear water flow and V_L the limiting flow velocity. In a further development, Mayerle *et al.* (1991) proposed the equation

$$\frac{V_L}{\sqrt{gd\Delta}} = 14.43D_{gr}^{-0.14}\lambda_s^{0.18}\left(\frac{d}{R}\right)^{-0.56}C_V^{0.18} \tag{4.113}$$

where λ_s is the friction coefficient in the presence of sediment transport.

May (Ackers *et al.* (1996)) suggested the equation

$$C_V = 3.03 \times 10^{-2}\left(\frac{D^2}{A}\right)\left(\frac{d}{D}\right)^{0.6}\left(\frac{V_L^2}{gd\Delta}\right)^{1.5}\left(1 - \frac{V_c}{V_L}\right) \tag{4.114}$$

where A is the area of flow with a (centre-line) depth z; for a small deposit of about $0.01D$ a doubling of the numerical coefficient is proposed. Ackers

(1991) suggests using the Ackers–White equation (equation (4.103)) and assuming an effective sediment transport width $0.04D$.

The above equations are based on laboratory experimental data and the whole area of sediment transport in sewers is still an active research topic, particularly for cases of varying deposition and degree of cohesion of deposited sediment. Ackers *et al.* (1996) have published a comprehensive review of the whole subject area.

Sediment transport under wave action follows the principles developed for unidirectional flow. On a microscopic scale the velocity of fluid particles orbiting under the waves and developed from equation (2.84) will be appropriate for estimating the effect on a non-cohesive bed, as from an overall point of view the unsteady motion of the fluid can be treated as a quasi-steady flow over a long period.

The net movement of sediment in coastal zones is generally classified as *longshore movement* under the action of waves and currents parallel to the shoreline, and *onshore/offshore* transport of sediment normal to the coast.

The deep-water and shallow-water waves were defined in Section 4.5 (equations (4.89) and (4.90)). Closer to the shore, the *breaker, surf and swash* zones (see Chapter 12) can be identified, where the wave energy (equation (4.92)) is spent at a rate sufficiently large to cause movement of the loose bed and thus littoral drift. Sediment transport in a coastal environment is an active research area (see EC MAST programme); for further discussion of this topic and its application to modelling, see Muir Wood and Fleming (1981) and Chapter 12.

Sediment data are essential for the study of morphological problems and applications of sediment transport equations; although it is possible to estimate by computation the rates of sediment transport, the results obtained with different methods could differ by several orders of magnitude. Thus, actual *sampling in situ*, whenever possible, is the more reliable method of assessing sediment transport rates.

Suspended sediment (concentration) samples can most simply be collected using spring-loaded, flap-valve traps or samplers consisting of a collecting pipe discharging into a bottle. Continuous or intermittent pumped samplers are also used. Point-integrating or depth-integrating sediment samplers with nozzles oriented against and parallel to the flow and samplers shaped to achieve a true undistorted stream velocity at the intake are used for measuring suspended sediment discharge; the US series of integrating samplers (particularly the US P-61) developed by the US Geological Survey are frequently used.

Sediment from the bed is collected for further analysis of size, shape, etc., using various types of grabs.

Quantitative *measurement of bedload transport* is extremely difficult and there is probably no universally satisfactory method, although some reasonably well-functioning *samplers* (devices placed temporarily on the bed and disturbing the bedload movement as little as possible) have been developed. Their efficiency (i.e. the ratio of actually measured sediment transport to that occurring without the presence of the sampler) has to be tested in the laboratory for the range of field conditions under which they are to be used. The VUV sampler (Novak 1957), developed for a wide range of sediment sizes (1–100 mm) and velocities (up to 3 m/s), has an efficiency of about 70% for the sampler filled up to about one-third (for a fuller sampler the efficiency decreases); similar figures have been quoted for the Helley–Smith sampler (Hubbell *et al.* (1981)). In gravel-bed streams wire-mesh baskets attached to a special frame can be used as bedload samplers.

Other methods of measuring bedload transport are: (i) surveying sediment deposits at river mouths or, in smaller streams, collected in trenches; (ii) differential measurement between normally suspended sediment load and total load, including the bedload brought temporarily into suspension in a river section with naturally or artificially increased turbulence (turbulence flumes); (iii) dune tracking; (iv) remote sensing; (v) use of tracers; and (vi) use of acoustic detectors.

4.6.4 Water quality

Sediment and contaminant transport processes are closely linked, as the contaminant load can be in a dissolved and/or sediment-adsorbed form; sediment behaviour thus influences a whole range of chemical, biological and bacteriological reactions. Furthermore, metal contaminants that enter a fluvial system follow the same environmental pathways as any other ion or sediment-associated element. In rivers, and particularly in coastal and estuarine waters, the water quality is also governed by the interaction of a number of hydrodynamic and other processes (e.g. diffusion and dispersion, gravitational circulation, stratification, wind effects, air entrainment etc.).

Wastewater contains the constituents of the water supply with additional input from waste-producing processes. To understand the nature of a particular water sample a range of *characteristics* may have to be considered. The more frequently used parameters are (Tebbutt 1992):

> *physical* – temperature, density, viscosity, taste and odour, colour, turbidity, suspended solids, dissolved solids, radioactivity and electrical conductivity;
> *chemical* – pH, oxidation–reduction potential, alkalinity, acidity, hardness, dissolved oxygen, oxygen demand (e.g. biological, oxygen demand, BOD), nitrogen, chloride and phosphate;
> *biological* – viruses, bacteria, fungi, algae, protozoa and rotifers.

The quality of water is judged in relation to its use (e.g. irrigation, drinking) and its suitability for aquatic life. According to its ultimate usage the quality is defined by several parameters such as pH, suspended solids, BOD, etc. (see e.g. Chapman (1996), James (1993), Sawyer *et al.* (1994)). For each criterion there can be a variety of determinants which are more or less suitable, and easier or more difficult to measure.

These parameters are usually determined by means of laboratory tests on samples collected from the river water, and are carried out either *in situ* (mobile laboratory) or in central laboratory facilities. Continuous *in situ* monitoring is preferable in order to avoid possible changes in characteristics due to transportation and the time taken between sampling and analysis. The quality standards are formulated on a statistical basis, and are more flexible for the uses of abstracted water as it is possible to improve this by treatment.

One of the *processes* most frequently represented by modelling is the *biological oxidation of organic matter and self-purification* in rivers.

From the equation for the rate of change (with time *t*) in the concentration of organic matter remaining in the water (or ultimate BOD) L

$$\frac{\mathrm{d}L}{\mathrm{d}t} = -K_1 L \tag{4.115}$$

where K_1 is a constant:

$$\frac{L_t}{L} = \mathrm{e}^{-K_1 t} = 10^{-K_1 t} \tag{4.115a}$$

Thus, the oxygen uptake is given by

$$\mathrm{BOD}_t = L_0 - L_t = L_0(1 - 10^{-k_1 t}) \tag{4.115b}$$

where L_0 is the initial load, and the value of k_1 is substance and temperature dependent (e.g. for domestic sewage $k_1 = 0.17/\mathrm{day}$ at temperature $20\,^\circ\mathrm{C}$).

Any body of freshwater can assimilate a certain amount of pollution because the natural biological cycle allows for certain adjustments to changed conditions, and self-purification will eventually stabilize organic matter. Only if the capacity to assimilate organic matter in surface waters is exceeded will serious water quality problems arise.

Waste properties of particular interest in pollution studies (Tebbutt 1992) are: toxic compounds in industrial effluents or toxins released by blue-green algae (which inhibit biological activity); components that affect the oxygen balance of water by consuming dissolved oxygen, by hindering reoxygenation from the atmosphere (oils, detergents), or by increasing the temperature, which in turn decreases the saturation concentration of

dissolved oxygen (heated effluents); or by high concentrations of inert suspended or dissolved solids affecting the bed of streams, thus preventing the
growth of fish food. If the dissolved oxygen falls below saturation, on exposure to the atmosphere the water will dissolve more oxygen to restore the
balance. From Henry's law the rate of solution of oxygen is proportional to
the deficit D

$$\frac{\mathrm{d}D}{\mathrm{d}t} = -K_2 D \tag{4.116}$$

or

$$D_t = D_0 e^{-K_2 t} \tag{4.116a}$$

where D_0 is the initial dissolved oxygen deficit and K_2 the reaeration constant. Using the dissolved oxygen concentration C, equation (4.116a) can
be written as

$$\frac{\ln(C_s - C)}{(C_s - C_0)} = -K_2 t \tag{4.117}$$

or

$$K_2 - \frac{1}{t} \frac{\ln(C_s - C_1)}{C_s - C_2} = \frac{1}{t} \ln r \tag{4.117a}$$

where C_1 and C_2 are concentrations at two downstream stations and r is the
deficit ratio.

Assuming that only BOD removal by biological oxidation and dissolved
oxygen replenishment by reaeration from the atmosphere are involved,
combining equations (4.115) and (4.116) leads to the *Streeter–Phelps*
equation – the '*sag curve equation*':

$$\frac{\mathrm{d}D}{\mathrm{d}t} = K_1 L - K_2 D \tag{4.118}$$

or, after integration and changing to base 10 ($k = 0.4343K$)

$$D_t = \frac{k_1 L_0}{k_2 - k_1} (10^{-k_1 t} - 10^{-k_2 t}) + D_0 10^{-k_2 t} \tag{4.119}$$

giving a critical maximum deficit D_c at time t_c

$$D_c = \frac{k_1}{k_2} L_0 10^{-k_1 t_c} \tag{4.120}$$

$$t_c = \frac{1}{k_2 - k_1} \log \left(\frac{k_2}{k_1} \frac{1 - D(k_2 - k_1)}{L_0 k_1} \right) \tag{4.121}$$

Equations (4.115)–(4.121) represent, of course, only a simplified one-dimensional approach. The dissolved oxygen and BOD and the whole self-purification process in a river will also be influenced by settlement and resuspension of bottom sediments, diffusion of dissolved oxygen into bottom muds, longitudinal dispersion and plant activities (addition of dissolved oxygen by photosynthesis in the daytime and removal at night). To achieve a BOD < 4 mg/L an effluent with 30 mg/L suspended solids and 20 mg/L BOD (30:20 standard) has to be diluted with clean river water in a ratio 1:8.

The oxygen balance of a river may be improved by using artificial aeration by aerators; alternatively, hydraulic structures using overfalls (weirs) or hydraulic jumps (gates) also represent a possible source of local dissolved oxygen improvement. Predictive equations for the deficit ratio r as a function of hydraulic parameters at hydraulic structures are given in Chapter 13.

For further discussion of modelling water-quality processes, see Chapters 8, 10 and 11; for a detailed discussion of estuarine water-quality management and modelling, see O'Kane (1980).

4.6.5 Air–water flow and cavitation

4.6.5.1 Air–water flow

In general, there are several possibilities for air–water flows, both in free surface and in closed conduit systems (Kobus 1991):

i   air flow without mixing – e.g. flow of air into (or out of) an air vessel, or air flow in response to air demand in conduits flowing partially full;

ii   air entrainment with mixing – e.g. at and downstream of a hydraulic jump or at the transition from free surface to conduit flow with vortex action;

iii formation of air–water mixtures by air coming out of solution – e.g. in hydraulic transients with waterhammer and cavitation action;

iv   formation of air–water mixture by injection of air – e.g. in designed aeration systems;

v   air in closed conduits can flow in several forms – e.g. bubbly flow, slug flow or stratified flow.

There is an interrelation between *air entrainment* governed by upstream conditions and *air transport capacity* and, possibly, *detrainment*, governed by downstream conditions.

Air entrainment requires a velocity of flow exceeding a minimum value V_i, i.e. the *inception limit* (e.g. for plunging jets V_i is in the region of 0.8 m/s, although this is also influenced by the turbulence of the approach flow; for air entrainment from a free surface flow $V_i = 6$ m/s).

After exceeding the inception limit the *entrainment limit* is principally a function of the approach Froude number, which must be higher than a critical value (e.g. Fr > 1 for a jump), and the *air supply limit*, which in turn depends on the characteristics of the air-supply system.

The upper limit for air-transport capacity is governed by the maximum possible bubble concentration in the downstream flow, where the ratio of flow velocity to bubble-rise velocity v_b is decisive. The bubble-rise velocity v_b is a function of bubble diameter d_b, v, ρ_a/ρ and $Re = v_b d_b/v$. For small bubbles ($d_b < 0.2$ mm, $Re < 1$) Stokes' law results in $v_b = 0.36 \, d_b^2$. For intermediate sizes, $0.2 < d < 20$, $0.1 < v_b < 0.4$, and for large air bubbles with a spherical cap ($d_b > 20$)$v_b = 0.07d^{1/2}$. All the above values apply for fluid properties of air and water at $10\,°C$ with d_b in millimetres and v_b in metres per second.

Neglecting air properties, the air–water flow ($Q_a/Q = \beta$ = ratio of air and water flow) will be a function of the geometry, the size of the conduit, the velocity of the flow and its turbulence, the characteristics of the air-supply system and the physical properties of the water (density, viscosity, surface tension and, in pressure transients, also compressibility).

The effects of the entrainment of air on the flow are manifold and can be beneficial as well as detrimental, and include: a change in the density and elasticity of the fluid; changes in turbulence, wall shear and flow field due to changes in the discharge and pressure distribution in the pipe systems, the effect on pressure transients and the performance of hydraulic machinery with regard to oxygen and nitrogen transfers. A specific problem for closed conduits is the possibility of blow-outs of accumulated air in the system.

For further discussion of dimensionless numbers involved, see Chapter 3 and of air entrainment and transport capacity in hydraulic structures, see Chapter 13.

4.6.5.2 Cavitation

Cavitation occurs when the pressure in a flow drops to the value of the saturated vapour pressure p_v. This may be the result of separation of flow (usually of a high velocity) or of a large pressure fluctuation in highly turbulent flow with a low mean pressure.

As a consequence of these conditions, cavities are formed that are filled with saturated vapour and gases excluded from the fluid due to a severe pressure drop. These bubbles are carried downstream by the flow until they reach areas of higher pressure ($p >> p_v$), where the vapour condenses and the bubbles suddenly implode. The result is not only noise and instability in the flow pattern, resulting possibly in vibrations of structures, but also, and more importantly, damage and pitting (cavitation corrosion), where the implosion, accompanied by violent impact at high pressure of water particles filling the imploded cavitation bubble, occurs against a part of a structure (conduit, spillway) or machine (turbine, pump, screw impeller).

Cavitation, if sustained over a period of time, can cause substantial damage, which may lead to a complete failure of the structure or machine. The cavitation strength (pressure from implosion) is linked to the concentration and size distribution of the gas bubbles as well as to the dissolved gas content of the liquid.

Cavitation may conveniently be characterized by the *cavitation number σ* (a form of Euler number), which combines the two parameters (pressure p and velocity u) that influence the onset of cavitation:

$$\sigma = \frac{(p - p_v)}{1/2\rho u^2} \qquad\qquad (4.122)$$

In free surface flow, p is usually the atmospheric pressure increased by the hydrostatic pressure.

The saturated vapour pressure p_v depends on the temperature and atmospheric pressure p_0. (At normal atmospheric pressure, for $100\,°\mathrm{C}$ $p_v = 10\,\mathrm{m}\,H_2O = p_0$, for $60\,°\mathrm{C}$ $p_v = 2.0\,\mathrm{m}\,H_2O$, and for normal water temperature $(10\text{–}20\,°\mathrm{C})p_v = 0.2\,\mathrm{m}\,H_2O$.)

Cavitation occurs if the cavitation number σ falls below a critical value σ_c (i.e. due to an increase in velocity u or drop in pressure p). The value of σ_c will be a function of geometry (and other factors – see Chapter 13), and has to be ascertained experimentally, usually under laboratory conditions and using special equipment (see Chapter 6).

It must be borne in mind that p and u refer to undisturbed instantaneous values of pressure and velocity, and that fluctuations in these parameters in highly turbulent flow may cause cavitation, even if the time-averaged values do not (see also Chapter 13).

Cavitation damage may be mitigated by use of special materials, or the danger eliminated by a change in the design and/or the introduction of air to the flow boundary at subatmospheric pressures.

References

Ackers, J. C., Butler, D. and May, R. W. P. (1996), *Design of Sewers to Control Sediment Problems*, Report 141, CIRIA, London.

Ackers, P. (1983), Sediment transport problems in irrigation systems design. In *Developments in Hydraulic Engineering*, Vol. 1 (ed. P. Novak), Elsevier Applied Science, London.

Ackers, P. (1991), Sediment aspects of drainage and outfall design. In *Proceedings of the International Symposium on Environmental Hydraulics, Hong Kong* (eds J. H. W. Lee and Y. K. Cheung), A. A. Balkema, Rotterdam.

Barr, D. I. H. (1975), Two additional methods of direct solution of the Colebrook–White function. *Proceedings – Institution of Civil Engineers Part 2*, TN 128, 3, p. 827.

Batchelor, G. K. (2000), *An Introduction to Fluid Dynamics*, Cambridge University Press, London.

Chadwick, A., Morfett, J. and Borthwick, M. (2004), *Hydraulics in Civil and Environmental Engineering*, 4th Edn, E & FN Spon, London.

Chapman, D. (ed.) (1996), *Water Quality Assessment*, 2nd Edn, E & FN Spon, London.

Daily, J. W. and Harleman, D. R. F. (1966), *Fluid Dynamics*, Addison-Wesley, Reading, MA.

Douglas, J. F., Gasiorek, J. M. and Swaffield, J. A. (2005), *Fluid Mechanics*, 5th Edn, Prentice Hall, Harlow.

Falconer, R. (1992), Flow and water quality modelling in coastal and inland water, *J. Hydraulic Res.*, 30(4), 437–452.

Fox, J. A. (1977), *An Introduction to Engineering Fluid Mechanics*, 2nd Edn, Macmillan, London.

Francis, J. R. D. (1975), *Fluid Mechanics for Engineering Students*, Edward Arnold, London.

Garde, R. J. and Ranga Raju, K. G. (1985), *Mechanics of Sediment Transportation and Stream Problems*, Wiley Eastern, New Delhi.

Goldstein, S. (ed.) (1965), *Modern Developments in Fluid Dynamics*, Dover, New York.

Graf, W. H. (1984), *Hydraulics of Sediment Transport*, Water Resources Publications, Littleton, CO.

Graf, W. H. (1998), *Fluvial Hydraulics*, John Wiley & Sons, Chichester.

Hino, M. (ed.) (1994), *Water Quality and its Control*, Hydraulic Structures Design Manual No. 5, IAHR, A. A. Balkema, Rotterdam.

HR Wallingford and Barr, D. I. H. (1994), *Tables for the Hydraulic Design of Pipes Sewers and Channels*, Vols I and II, 6th Edn, Thomas Telford, London.

Hubbell, D. W., Stevens, H. H., Skinner, J. V. and Beverage, J. P. (1981), Recent refinements in calibrating bedload samplers. *Proceedings of Special Conference, Water Forum 81*, San Francisco, CA, ASCE, New York.

James, A. (1993), *An Introduction to Water Quality Modelling*, 2nd Edn, John Wiley & Sons, Chichester.

Jirka, G. H. (1992), In support of experimental hydraulics: three examples from environmental fluid mechanics, *J. Hydraulic Res.*, 30(3), 293–302.

Kobus, H. (1991), Introduction to air–water flows. In *Air Entrainment in Free Surface Flows*, pp. 1–28, (ed. I. R. Wood), Hydraulic Structures Design Manual No. 4, IAHR, A. A. Balkema, Rotterdam.

Lencastre, A. (1987), *Handbook of Hydraulic Engineering*, Ellis Horwood, Chichester.

Mayerle, R., Nalluri, C. and Novak, P. (1991), Sediment transport in rigid bed conveyances, *J. Hydraulic Res.*, 29(4), 475–495.

Miller, D. S. (ed.) (1994), *Discharge Characteristics*, IHR Hydraulic Structures Design Manual, Vol. 8, Balkema, Rotterdam.

Muir Wood, A. M. and Fleming, C. A. (1981), *Coastal Hydraulics*, 2nd Edn, Macmillan, London.

Novak, P. (1957), Bedload meters-development of a new type and determination of their efficiency with the aid of scale models, *Proceedings of 7th Congress IAHR, Lisbon*, A9.

Novak, P. and Nalluri, C. (1984), Incipient motion of sediment particles over fixed beds, *J. Hydraulic Res.*, 27(3), 181–197.

O'Kane, J. P. (1980), *Estuarine Water-Quality Management*, Pitman, London.

Raudkivi, A. J. and Callander, R. A. (1975), *Advanced Fluid Mechanics – An Introduction*, Edward Arnold, London.

Rodi, W. (1993), *Turbulence Models and their Application in Hydraulics*, 3rd Edn, A. A. Balkema, Rotterdam.

Rouse, H. (1961), *Fluid Mechanics for Hydraulic Engineers*, Dover, New York.

Rutherford, J. C. (1994), *River Mixing*, J. Wiley & Sons, Chichester.

Sawyer, C. N., McCarthy, P. L. and Parker, G. F. (1994), *Chemistry for Environmental Engineering*, 4th Edn, McGraw Hill, New York.

Simons, D. B. and Senturk, F. (1992), *Sediment Transport Technology: Water and Sediment Dynamics*, Water Resources Publications, Littleton, CO.

Singh, V. P. and Hager, W. H. (eds) (1996), *Environmental Hydraulics*, Kluwer Academic, Dordrecht.

Tebbutt, T. H. Y. (1992), *Principles of Water Quality Control*, 4th Edn, Pergamon Press, Oxford.

Tritton, D. J. (1988), *Physical Fluid Dynamics*, 2nd Edn, Clarendon Press, Oxford.

Vardy, A. (1990), *Fluid Principles*, McGraw Hill, Maidenhead.

Vries, M. de (1985), *Engineering Potamology*, IHE, Delft.

Wicks, J. M. and Bathurst, J. C. (1992), SHESE D: a physically based distributed erosion and sediment yield component for the SHE hydrological modelling system, *J. Hydrol.*, 175(1–4), 213–238.

Yalin, M. S. (1992), *River Mechanics*, Pergamon Press, Oxford.

Chapter 5

Development of physical models

5.1 Introduction

Research on physical (scale) models is based on the theory of similarity between the model and prototype. This theory provides guidance on the preparation of experiments, computation of model parameters, processing of results, limits of their validity and likely scale effects. In technical disciplines, including hydraulics, the theory of similarity is generally based on one of three approaches (or their combination):

1 The first approach determines the criteria of similarity from a system of homogeneous (differential) equations, which express the investigated phenomenon mathematically (see e.g. Section 5.5).
2 If no equations are available, we have to resort to the second path – dimensional analysis – which, on its own or together with sound empirical equations, may form the basis for determining the conditions of similarity. The use of dimensional analysis requires a careful preliminary appraisal of the physical basis of each investigated phenomenon and of the parameters influencing it; these may have to be determined by separate experiments. A combination of physical and dimensional analyses may have to be used to achieve the required results. For further details, see Section 5.2.
3 The third route could be denoted as the method of synthesis – see Section 5.3.

Sections 5.4–5.9 cover in some detail the theory of similarity, dimensionless numbers and the modelling (scaling) laws most relevant to physical models in hydraulic engineering.

As *analogue* models reproduce a prototype situation in a physically different medium (i.e. in this sense they are *physical* models), a brief treatment of analogue models has been added to this chapter as Section 5.10.

If we denote the dimensions of mass as M, length as L and time as T, we can express the physical dimensions of almost all parameters used in hydraulics with these symbols; for example, velocity has the dimension LT^{-1}, specific mass (mass/volume, density) ML^{-3} and the coefficient of dynamic viscosity $ML^{-1}T^{-1}$. These dimensions are easily derived from the definitions of the above quantities and from the physical laws that characterize them.

For example, the coefficient of viscosity μ is characterized by the equation

$$\tau = \frac{\mu \, du}{dz}$$

where τ is the tangential stress. Stress has the dimension of force acting on a unit area; therefore, $MLT^{-2}L^{-2} = ML^{-1}T^{-2}$. From the above equation we thus obtain

$$ML^{-1}T^{-2} \sim \mu LT^{-1}L^{-1}$$

and therefore $\mu - ML^{-1}T^{-1}$

Expressing the parameter in physical dimensions also permits safe transformation from one system of units to another.

5.2 Dimensional analysis

5.2.1 General

Dimensional analysis provides some basic information about the investigated phenomenon on the assumption that it can be expressed by a dimensionally correct equation containing the variables influencing it. Obtaining a certain grouping of variables allows, for example, a wider application and interpretation of experimental results.

An equation is dimensionally homogeneous if it is independent of the basic units used. For example, in the well-known form of Bazin's equation for discharge Q over a rectangular notch

$$Q = m\sqrt{2g}bH^{3/2}$$

where H is the head, b the length of the notch crest and g the acceleration due to gravity. The coefficient m is independent of the units used for b, H, g and Q as long as the same units of length and time (cm, m, ft, s, etc.) are used throughout, as in the dimensional form both sides of the equation are identical:

$$Q - L^3T^{-1}; \quad g - LT^{-2}; \quad b - L; \quad H - L$$
$$L^3T^{-1} = L^{1/2}T^{-1}LL = L^3T^{-1}$$

It will be noted that in the above example viscosity μ and surface tension σ have not been introduced. It follows that Bazin's equation will be correct only if the dimensionless coefficient m is either independent of these (and other) parameters or includes their influence in the variation of its value. A dimensionally examined equation is, therefore, correct only if it contains *all the necessary variables*, as it would in an analytical derivation. Thus, in the application of dimensional analysis the basic step is the correct choice of variables that might influence the phenomenon under observation.

In hydraulics, the relationship between variables, any one of which may be dependent on the others, is usually experimentally investigated using a physical model. In most cases, all quantities are easily independently controlled, with the exception of one, which becomes the dependent variable. In the choice of variables we may introduce a parameter that under the conditions investigated is in fact constant, but which in connection with different variables forms a dimensionless number.

To include more than one dependent variable in the investigated relationship is almost as serious a mistake as omitting some of the participating quantities. But no harm is done in introducing quantities that do not influence the investigated problem, because they mostly eliminate themselves in further analysis. To make a correct choice of variables we must, first of all, formulate a certain theory about the phenomenon under consideration and accordingly assess which independent variables must be taken into account. This means that we must at least know, either from experience or by analogy, why the phenomenon might be influenced by a particular variable.

It must be appreciated that, apart from sometimes giving incomplete solutions, the pitfalls of the purely simple dimensional approach are that the analysis can lead to spurious correlations or only obvious conclusions. To avoid drawing wrong conclusions from, for example, statistical analysis of experimental data by means of dimensionless products, it is important that the parameters present in the problem and having a strong stochastic character appear only in one dimensionless product.

Although dimensional analysis is often applied in hydraulic research it does not provide an insight into the physical process, and its use must not be exaggerated and regarded as a replacement for, rather than a primary aid to, the analytical solution. The correct use of dimensional analysis is always linked to experience and previous critical analysis of the investigated phenomenon.

The three kinds of variables used in hydraulic investigations describe the geometry, the flow, and the properties of the fluid used. In scale models with geometric similarity (see Section 3.4) it is enough to introduce a single variable into the experimentally investigated relationship for the determination of the linear scale (model scale) on which the other dimensions depend.

Usually only two or three fluid properties are chosen that may play an important part in the case under investigation (e.g. density, viscosity, surface tension). Nearly every flow characteristic comes within the group of dependent variables, but always one only can be considered, according to the type of acting force.

The choice of variables included in the investigated relationships should be kept as simple as possible. Thus, of the three interdependent variables length, velocity and discharge, it is usually more suitable to include length (or area A) and velocity V rather than velocity and discharge $Q(Q = VA)$.

Two conventional and related methods of dimensional analysis are of the greatest importance in hydraulics – Rayleigh's (indicial) method and Buckingham's method (π theorem) – and both are well documented in the literature and briefly explained below.

5.2.2 Rayleigh's method

Rayleigh's method is based on the fact that in each dimensionally homogeneous equation the exponent of every dimension on the left-hand side of the equation must be equal to the sum of the exponents of the corresponding dimension on the right-hand side. If the equation expressing a dependent variable p contains n independent variables and if the dependent variable contains r basic dimensions, we can write r equations for n unknown exponents; from these it is possible to calculate r exponents for arbitrary values of the remaining $(n - r)$ exponents, and the result can be expressed as a general function containing r independent variables with known exponents and $(n - r)$ dimensionless arguments with unknown exponents. This unknown function is then determined on the basis of experimental results.

Let the general expression for the dependent variable p be given by equation

$$p = a^x b^y c^z d^u e^v \qquad (5.1)$$

where a, b, c, d and e are independent variables and x, y, z, u and v are unknown exponents. On the assumption that the dimensions of quantities p, a, b, c, d and e, are P, A, B, C, D and E, the equation

$$P \sim A^x B^y C^z D^u E^v \qquad (5.2)$$

is valid. If we substitute for P, A, B, C, D and E the corresponding dimensions M, L and T, we obtain three homogeneous linear equations for the exponents of the dimensions M, L and T, from which three unknown exponents may be calculated, for example x, y and z expressed by means of the remaining exponents u and v. The result may then be expressed by the equation

$$p = a^{x_1} b^{y_1} c^{z_1} \pi_1^{u} \pi_2^{v} \tag{5.3}$$

where x_1, y_1 and z_1, are integers or fractions calculated from the equations for the exponents of M, L and T, and π_1 and π_2 are dimensionless arguments. Equation (5.3) is, however, only a special form of the general equation

$$p = a^{x_1} b^{y_1} c^{z_1} \left(\sum_1^{\infty} k_n \pi_1^{n} \right) \left(\sum_1^{\infty} k_m \pi_2^{m} \right) \tag{5.4}$$

where k_n and k_m are dimensionless constants.

Equation (5.4) can also be written as

$$p = a^{x_1} b^{y_1} c^{z_1} f(\pi_1, \pi_2) \tag{5.4a}$$

Rayleigh's method may best be illustrated by the following example. Let us consider the equation for the determination of resistance $P(MLT^{-2})$ of a sphere of diameter $D(L)$ moving with a velocity $v(LT^{-1})$ in an incompressible and unlimited medium of specific mass $\rho(ML^{-3})$ and coefficient of viscosity $\mu(ML^{-1}T^{-1})$. The equation will have the following form:

$$P = f(D, v, \rho, \mu) \tag{5.5}$$

or

$$P = D^a v^b cd \tag{5.5a}$$

where a, b, c and d are constants. If we rewrite equation (5.5a) dimensionally, we obtain

$$MLT^{-2} = L^a, L^b T^{-b}, M^c L^{-3c}, M^d L^{-d} T^{-d}$$

For this equation to be dimensionally homogeneous the exponents for M, L and T on both sides of the equation must be identical. Therefore, for

M: $1 = c + d$

L: $1 = a + b - 3c - dL$

T: $-2 = -b - d$

We have three equations for four unknowns, and can therefore express three of them by means of the fourth:

from the equation for M: $c = 1 - d$, $T = 1 - d$

from the equation for T: $b = 2 - d$

from the equation for L: $a = 1 - b + 3c + d = 1 - 2 + d + 3 - 3d + d = 2 - d$.

Substituting into equation (5.5a) we obtain

$$P = D^{2-d} v^{2-d} \rho^{1-d} \mu^d = \rho D^2 v^2 \left(\frac{\mu}{vD} \right)^d \tag{5.6}$$

or, according to equation (5.4),

$$P = \rho D^2 v^2 f \left(\frac{vD}{v} \right) \tag{5.6a}$$

thus

$$\frac{P}{\rho D^2 v^2} = f \left(\frac{vD}{v} \right) \tag{5.6b}$$

In equation (5.6b) both sides are dimensionless numbers:

$$\frac{P}{\rho D^2 v^2} \sim \frac{MLT^{-2}}{ML^{-3} L^2 L^2 T^{-2}} \quad \text{(Newton Number)}$$

$$\frac{vD}{v} = \frac{vD\rho}{\mu} \sim \frac{(LT^{-1} LML^{-3})}{(ML^{-1} T^{-1})} \quad \text{(Reynolds Number)}.$$

The area of the projection of the sphere is $A = \pi D^2 / 4$; we can therefore rewrite equation (5.6a) as

$$P = \frac{1}{2} C_f \rho v^2 A \tag{5.6c}$$

where the resistance coefficient is $C_f = 8/\pi f(vD/v)$ (see also equation (4.24)).

Equation (5.6c) is generally used to express the resistance of a sphere moving in an incompressible medium, and the relationship between C and the Reynolds number has to be investigated experimentally. To obtain the same result without dimensional analysis we would have to analyse many experimental results to show the mutual influence of the variables v, D, ρ and μ.

The correct result in the derivation of equation (5.6) also depended on the choice of the unknown d to express the other three unknowns a, b and

c in equation (5.5a); use was made here of the fact that a force can best be expressed from the given four variables by the use of ρ, D, v because the inertia force

$$P = ma = \rho l^3 \frac{l}{t^2} = \rho l^2 v^2$$

(where l represents length); the Newton number is thus a ratio of the resistance (pressure) and inertia forces and the Reynolds number is the ratio of inertia and the viscous forces.

5.2.3 Buckingham's method (π theorem)

By applying the π theorem, the relationship describing the investigated problem in terms of variables a, b, c, ...

$$f(a, b, c, \ldots, n) = 0 \tag{5.1a}$$

is transformed into another relatively simpler relationship between a smaller number of variable dimensionless arguments π_1, π_2, ..., which are established from the variables participating in the problem:

$$F(\pi_1, \pi_2, \pi_3 \ldots) = 0 \tag{5.7}$$

If n independent variables a, b, c..., n participate in a problem and their dimensions A, B, C, ..., N can be expressed with the aid of r basic magnitudes, it is usually possible to establish $(n - r)$ dimensionless arguments π_1, π_2, ... When these arguments are regarded as new variables, instead of equation (5.1a) a new dimensionally homogeneous relationship (equation (5.7)) with $(n - r)$ variables can be written. This second function is much simpler to investigate, as the reduction in the number of variables by r also reduces the number of experiments necessary for the solution of the problem. The closer the number of participating quantities n to the number of basic dimensions r, the simpler the solution.

The number of basic magnitudes used to express the dimensions of independent variables participating in the investigated problem is usually $r \leq 3$, for three basic units of length, mass and time. If all three participate in the problem (the most frequent case in practice), then $r = 3$.

The dimensionless arguments π_1, π_2, ..., π_{n-r} are products of various powers of variables participating in the problem

$$\pi_1 = a^{x_1} b^{y_1} c^{z_1} \ldots; \quad \pi_2 = a^{x_2} b^{y_2} c^{z_2} \ldots$$

In every argument π there should be $(r + 1)$ of these variables. In the choice of variables that are to occur in every dimensionless argument two further conditions must be fulfilled: *the recurring (governing) variables must together include all the basic dimensions and they must not in themselves form a dimensionless argument*. In the general case ($r = 3$), these conditions are complied with in such a way that among four variables three (the characteristic length, velocity and specific mass) will be repeated in every argument. The fourth variable will differ in every argument (with the exponent 1), so that in the solution all n variables that influence the problem might be applied.

In general, the dimensionless arguments π are simple numbers. Therefore,

$$[\pi] = [A]^x[B]^y[C]^z \ldots [N]^v = 1$$

If the dimensions A, B, etc., of the variables participating in the problem contained in every dimensionless argument π_1, π_2, π_3, etc., are now expressed using the basic dimensions (e.g. L, M and T), we can add the exponents of each basic dimension for each dimensionless argument; this sum is then, in each case, equal to zero, as the product of the various powers of the dimension from which we set out was equal to one (π is a dimensionless argument). Thus, we obtain r equations for the unknown exponents $x, y, z, \ldots, v$; the expressions for all dimensionless arguments having been determined, we can proceed with the experimental solution of the problem.

In practice, certain dimensionless arguments are usually introduced, so that the investigation of unknown exponents and thus also of dimensionless arguments is guided and simplified to a considerable degree.

When dealing with a greater number of dimensionless arguments and lacking sufficient experience in the choice of suitable exponents, it is possible arbitrarily to combine various dimensionless arguments (raise to a power, multiply or divide) so that we obtain, for example, the relationship:

$$F'\left(\pi_1^2, \frac{\pi_1}{\pi_2}, \pi_1, \pi_2, \pi_3 \ldots\right) = 0 \tag{5.8}$$

All these combinations are again dimensionless numbers. However, *their number must be the same as before, i.e. (n − r), and all the original arguments must be used*. This procedure usually aims at cancelling out some of the variables common to two or more combined arguments.

The use of Buckingham's method of dimensional analysis is again best demonstrated by a simple example. With the aid of a geometrically similar reduced model, the resistance of a body (ship) moving with a steady velocity on an unlimited surface of an ideal (non-viscous) liquid of unlimited depth is to be determined. The effect of viscosity is omitted to simplify the problem, as only a method of dimensional analysis is being demonstrated; for further treatment of the subject, see Section 5.9.

The following variables are involved: resistance $P(MLT^{-2})$, velocity $v(LT^{-1})$, gravitational acceleration $g(LT^{-2})$, specific mass of the liquid $\rho(ML^{-3})$ and a basic dimension (e.g. the length of the vessel) $l(L)$, i.e. altogether five variables ($n = 5$). Their dimensions may be expressed by three ($r = 3$) basic dimensions (M, L and T), so that ($n - r$), i.e. two dimensionless arguments may be established, according to the principles stated above.

$$\pi_1 = l^{x_1} v^{y_1} \rho^{z_1} g; \quad \pi_2 = l^{x_2} v^{y_2} \rho^{z_2} P$$

The variables are now expressed in terms of the basic dimensions L, M and T, and the exponents of the same basic dimension added:

$$\pi_1 \sim L^{x_1} \left(\frac{L}{T}\right)^{y_1} \left(\frac{M}{L^3}\right)^{z_1} \left(\frac{L}{T^2}\right)$$

For

$$L: x_1 + y_1 - 3z_1 + 1 = 0$$
$$T: -y_1 - 2 = 0$$
$$M: z_1 = 0$$

Therefore, $x_1 = 1$, $y_1 = -2$, $z_1 = 0$ and
$\pi_1 = gl/v^2$, which is the reciprocal of the square of Froude number Fr (the ratio of gravity and the inertia forces).

$$\pi_2 \sim L^{x_2} \left(\frac{L}{T}\right)^{y_2} \left(\frac{M}{L^3}\right)^{z_2} \left(\frac{ML}{T^2}\right)$$

For

$$L: x_2 + y_2 - 3z_2 + 1 = 0$$
$$T: -y_2 - 2 = 0$$
$$M: z_2 + 1 = 0$$

Therefore, $x_2 = -2$, $y_2 = -2$, $z_2 = -1$ and $\pi_2 = P/(\rho l^2 v^2)$ (the Newton number Ne). The relation between the two dimensionless arguments

$$F\left(\frac{P}{l^2 v^2 \rho}, \frac{gl}{v^2}\right) = 0 \tag{5.9}$$

may be found experimentally (e.g. by dragging a model on the surface of the (ideal) fluid with a varying velocity v and measuring the resistance P).

For pairs of measured values v and P the corresponding pairs of numbers Ne and Fr are calculated; these represent points on the graph of the investigated correlation $F(\text{Ne, Fr}) = 0$.

If the resistance of a geometrically similar body of length l_p dragged at velocity v_p over the surface of an ideal fluid is to be determined, the Froude number $\text{Fr}_p = v_p / \sqrt{(gl_p)}$ is calculated, and the corresponding value of Newton's number Ne_p is read from the graph of $F(\text{Ne, Fr})$. The resistance against the movement of this body on the surface of an ideal liquid is then represented by the relation

$$P_p = \text{Ne}_p l_p^2 v_p^2 \rho$$

From the above it will be seen that there is no fundamental difference between the two methods of dimensional analysis. The advantage of Buckingham's method is that it avoids the use of the general equation (5.4), the introduction of which is part of the derivation of Rayleigh's method. In practice, however, this step can be omitted without difficulty, and we easily change from an equation of the type (5.1) (with a numerical coefficient) to an equation of type (5.5).

An advance on the two conventional methods of dimensional analysis is presented by the basic echelon matrix procedure (Barr (1979)). This integrated procedure cannot be undertaken without simultaneously encompassing a check on the rank of the matrix of dimensions in formal mathematical terms, and therefore it cannot lead to an incomplete set of non-dimensional products.

Barr (1983) lists five procedures for dimensional analyses where the first four – Rayleigh, Buckingham, basic stepwise and echelon matrix – involve the direct formulation of pi-term non-dimensional functional equations. The fifth procedure, the proportionalities-stepwise procedure, is associated with the third route, the method of synthesis.

5.3 Method of synthesis

The conventional methods of dimensional analysis guide the analysis to a correct but not necessarily convenient solution; although convenient solutions may be obtained by compounding (combining) parameters, the full range of solutions is not easily apparent. The method of synthesis was developed to overcome these disadvantages, bridging the dimensional and similitude analysis.

In presenting the method, Barr (1969) originally introduced an intermediate step by formulating a dimensionally homogeneous equation with a redundancy, which then allowed flexibility in the development of the final dimensionless equation. Instead of using force terms, as is the case in conventional similitude analysis, Barr initially suggested the use of 'dynamic velocities'; at this stage the method was really only a variant of the normal one. As the use of 'velocities' was cumbersome, a change was made to linear measures, 'linear proportionalities' (e.g. v^2/g, $v^{2/3}/g^{1/3}$, $Q^{2/5}/g^{1/5}$, etc.), which proved to be easily handled and appropriate.

Although there are similarities with the normal pi-method (functional dimensional equations are formed by combining variables into terms having a dimension of length and then combining these terms with any relevant length), the advantages of this method are that more combinations can be formed than are necessary. Early in the analysis the resulting redundancies lead to the choice of the most convenient terms to be used and the most useful form of the dimensionless equation. Thus, a solution can be obtained where the dependent variables appear as infrequently as possible, which of course is the solution most appropriate for the study of these variables. For example, in a situation involving viscous forces and gravity, a length and a dependent velocity:

$$v = f(g, \mu, \rho, l) \tag{5.10}$$

A dimensionally homogeneous equation with linear proportionalities can now be written as:

$$f\left(\frac{v^2}{g}, \frac{v}{v}, \frac{v^{2/3}}{g^{1/3}}, l\right) = 0 \tag{5.11}$$

To specify the system only two proportionalities (out of the three) are required with each variable included at least once. Therefore,

$$f\left(\frac{v^2}{g}, \frac{v}{v}, l\right) = 0 \tag{5.12a}$$

or

$$f\left(\frac{v^2}{g}, \frac{v^{2/3}}{g^{1/3}}, l\right) = 0 \tag{5.12b}$$

or

$$f\left(\frac{v}{v}, \frac{v^{2/3}}{g^{1/3}}, l\right) = 0 \tag{5.12c}$$

Conventional dimensional analysis starting with equation (5.10) would lead directly to one of the above equations (most likely to $V^2/gl = \phi(vl/v)$ – see equation (5.12)), whereas the method of synthesis, using equation (5.11) as an interim step, proceeds directly to any of the (nine) possible solutions (dimensionless equations).

A full exposition and development of the method with applications to the resistance to flow in pipes and densimetric phenomena, and with examples of formulation of model laws has been given by Sharp (1981).

5.4 Basic concepts and definitions in the theory of similarity

The ratio of a certain variable in the prototype and the corresponding value in the model (e.g. the ratio of lengths) is called the *module*, or *scale*, or *scale factor* (of length) and will be referred to throughout this book as M. Some authors use the reciprocal value of M as the scale factor.

Scaling laws are conditions that must be satisfied to achieve desired similarity between model and prototype.

Distortion is a conscious departure from a scaling law that is often necessitated by a complex set of prototype and laboratory conditions; the term is most frequently used for geometric distortion in which the vertical and horizontal scales are different.

In the theory of similarity, a *dimensionless number* is usually regarded as a physically meaningful ratio of parameters that is dimensionless (e.g. force ratios and 'ratios which are of particular physical significance') (ASCE (1982)). Although the standard form of the principal dimensionless numbers is well established, confusion can arise by the same number being given two different names or the same name being used for different powers of the same number (e.g. Froude number – see Section 3.5.1). It is, therefore, important to define clearly all dimensionless numbers and terms used, particularly the velocity and length terms.

If not all pertinent dimensionless numbers are the same in the model and the prototype, the result is a *scale effect*. The scale effect is thus a consequence of non-similarity between the model and the prototype, an error arising by using the model according to the main determining law (e.g. for forces) and neglecting others. The cause (as opposed to the consequence) of non-similarity is best described as *scale defect* (as opposed to scale effect – see Novak (1984)).

The consequence of necessary laboratory simplifications or physical constraints on the model are best described by the term *laboratory effect* (ASCE (1982)).

If the shape of a reduced model (m) of any object corresponds exactly to the prototype (p), all dimensions are reduced at the same scale and the corresponding angles are the same, and this is referred to as *geometric similarity* of the model and prototype. In geometry, the corresponding points of two formations (which need not necessarily be geometrically similar) are called 'homologous points'. Homologous parts of a model and prototype are thus parts composed of homologous points.

Let us now consider two formations – the real formation and its model in motion – and let us observe the routes described by the homologous points. If the homologous points of both formations lie on the homologous points of their routes in proportional (i.e. homologous) times, this is referred to as *kinematic similarity*. The proportionality of times means that the ratio of times in the prototype and the model in which the homologous points travel a homologous part of their routes is constant. Kinematic similarity thus assumes the similarity of the corresponding components of velocity and acceleration.

For the ratio of lengths in three coordinates

$$\frac{l_{px}}{l_{mx}} = M_{lx}; \quad \frac{l_{py}}{l_{my}} = M_{ly}; \quad \frac{l_{pz}}{l_{mz}} = M_{lz}$$

we can write for the scale of velocities

$$M_u = \frac{u_p}{u_m} = \frac{M_{lx}}{M_t} \tag{5.13a}$$

$$M_v = \frac{v_p}{v_m} = \frac{M_{ly}}{M_t} \tag{5.13b}$$

$$M_w = \frac{w_p}{w_m} = \frac{M_{lz}}{M_t} \tag{5.13c}$$

and for the scale of acceleration

$$M_{ax} = \frac{a_{px}}{a_{mx}} = \frac{M_{lx}}{M_t^2} \tag{5.14a}$$

$$M_{ay} = \frac{a_{py}}{a_{my}} = \frac{M_{ly}}{M_t^2} \tag{5.14b}$$

$$M_{az} = \frac{a_{pz}}{a_{mz}} = \frac{M_{lz}}{M_t^2} \tag{5.14c}$$

If in equations (5.13a)–(5.13c) and (5.14a)–(5.14c) M_u, M_v, M_w and M_{ax}, M_{ay}, M_{az} are constants, kinematic similarity is obtained. Furthermore, if the formations are geometrically similar, i.e. $M_{lx} = M_{ly} = M_{lz} = M_l$, this results in the equality of the velocity scales $M_u = M_v = M_w = M_l/M_t$ and of the acceleration scales $M_{ax} = M_{ay} = M_{az} = M_l/M_t^2$. In this special case the streamlines passing through homologous points of the two formations are also geometrically similar.

Lastly, when the homologous parts of the model and prototype are exposed to proportional total forces, this is referred to as *dynamic similarity*. As, according to Newton's law, force = mass × acceleration, the following equation may be written for the prototype and the model:

$$P_{px} = m_p a_{px}; \quad P_{py} = m_p a_{py}; \quad P_p = m_p a_{pz}$$

$$P_{mx} = m_m a_{mx}; \quad P_{my} = m_m a_{my}; \quad P_m = m_m a_{mz}$$

The scale of forces can be written as

$$M_{Px} = \frac{M_m M_{lx}}{M_t^2} \tag{5.15a}$$

$$M_{Py} = \frac{M_m M_{ly}}{M_t^2} \tag{5.15b}$$

$$M_{Pz} = \frac{M_m M_{lz}}{M_t^2} \tag{5.15c}$$

For geometric similarity of the two formations

$$(M_{lx} = M_{ly} = M_{lz} = M_l)M_P = \frac{M_m M_l}{M_t^2} \tag{5.16}$$

Substituting for $Mm = M_\rho M_V = M_\rho M_l^3$ results in

$$M_P = \frac{M_\rho M_l^4}{M_t^2} \tag{5.16a}$$

or, after substituting for $M_l/M_t = M_v$,

$$M_P = M_\rho M_l^2 M_v^2 \tag{5.16b}$$

If in equation (5.15) M_{Px}, M_{Py} and M_{Pz} are constants (condition of proportionality of total forces), the expression for dynamic similarity is obtained, which then apart from kinematic similarity also contains proportionality of the distribution of mass *m* in both formations. If both formations are geometrically similar, equations (5.16), (5.16a) or (5.16b) express their *mechanical similarity*, which includes their geometric, kinematic and dynamic similarity.

Mechanical similarity may, therefore, be defined as follows: two formations (prototype p and model m) are mechanically similar if they are geometrically similar and if, for proportional masses of homologous points, their paths described in proportional times are also geometrically similar. The definition based on Newton's law thus includes geometric similarity of the two formations, the proportionality of times and the geometric similarity of the paths travelled (kinematic similarity), as well as the proportionality of masses, and thus also of forces (dynamic similarity). Thus, for geometric similarity M_l is a constant, M_{lx}, M_{ly}, M_{lz} and M_t are constants for kinematic similarity, and M_m as well as M_{Px}, M_{Py} and M_{Pz} are constants for dynamic similarity; and for mechanical similarity $M_{Px} = M_{Py} = M_{Pz}$ and $M_{lx} = M_{ly} = M_{lz}$, and in addition M_t and M_m are naturally constants.

It must be stressed that *mechanical similarity always includes dynamic (and thus also kinematic) and geometric similarity, whereas dynamic similarity always includes kinematic but not necessarily geometric similarity* (as equations (3.14) and (3.15) demonstrate). This is particularly important in hydraulic engineering, where geometric similarity often cannot be adhered to (e.g. in models of rivers) and distorted models have to be used. However, the use of distorted models does not exclude the possibility of attaining dynamic similarity (see above). In the literature this distinction between mechanical and dynamic similarity is often confused or not made at all. Obviously, if the condition is stated that the streamlines passing through homologous points of the two dynamically similar formations must also be geometrically similar, the distinction between mechanical and dynamic similarity disappears.

5.5 General law of mechanical similarity in hydrodynamics

5.5.1 Derivation of the law of mechanical similarity by dimensional analysis

The motion of fluids under various conditions, the motion of solid bodies in fluids, or the motion and interaction of both may be investigated using scale models. The physical parameters that, in a general case, may influence the body and fluid motion are: for the *body*, a characteristic length d (e.g. diameter of sediment) and its specific mass ρ_s; for the *fluid*, the specific mass ρ, the coefficient of viscosity μ, the coefficient of surface tension σ, the bulk modulus K and the velocity of flow v; in addition, acceleration due to gravity must also be taken into consideration. Lastly, the entire phenomenon occurs in a *medium* of length l, width b and depth h (e.g. a river channel).

The force acting on the body in the fluid flow can be expressed as:

$$P = c' \mu^a \rho^c K^e \sigma^f v^i b^k l^n h^p d^x \rho^y g^z \tag{5.17}$$

where c' is a constant and $a, \ldots, z$ are unknowns.

The dimensions of the above quantities in the basic system of physical units L, M and T are:

$$P - MLT^{-2} \sigma - MT^{-2}$$
$$\mu - ML^{-1}T^{-1} v - LT^{-1}$$
$$\rho, \rho_s - ML^{-3} D, b, l, d - L$$
$$K - ML^{-1}T^{-2} g - LT^{-2}$$

Rewriting equation (5.17) in terms of dimensions and following the procedure in Section 5.2.1 we obtain three equations (for the exponents of M, L and T) with 11 unknowns. Solving for c, i and n (i.e. the powers of ρ, l and v) in terms of the remaining eight unknown powers, and substituting into equation (5.17) results in

$$P = c' \rho l^2 v^2 \left(\frac{\mu}{\rho l v}\right)^a \left(\frac{K}{\rho v^2}\right)^e \left(\frac{\sigma}{\rho v^2 l}\right)^f \left(\frac{gl}{v^2}\right)^z \left(\frac{h}{l}\right)^p \left(\frac{b}{l}\right)^k \left(\frac{d}{l}\right)^x \left(\frac{\rho_s}{\rho}\right)^y$$

(5.18)

or

$$P = \rho l^2 v^2 \phi \left(\left(\frac{v^2}{gl}\right), \left(\frac{\rho l v}{\mu}\right), \left(\frac{\rho v^2 l}{\sigma}\right), \left(\frac{\rho v^2}{K}\right), \left(\frac{h}{l}\right), \left(\frac{b}{l}\right), \left(\frac{d}{l}\right), \left(\frac{\rho_s}{\rho}\right)\right)$$

(5.18a)

Equation (5.18) is the general equation for the force acting on a body in a fluid during their relative motion, both in the prototype and in the model. It follows from the manner of deriving the equation that all expressions in brackets are dimensionless numbers (this can easily be verified by substituting the dimensions of the various parameters).

As it has been shown that for mechanical similarity $M_P = M_\rho M_l^2 M_v^2$ (equation 5.16), it follows that for mechanical similarity between the model and the prototype the ratio of all corresponding dimensionless numbers in equation (5.18) for the prototype and the model must be equal to one.

The condition

$$\frac{M_h}{M_l} = \frac{M_b}{M_l} = \frac{M_d}{M_l} = 1$$

expresses the condition of geometric similarity. The condition $M_{\rho s}/M_\rho = 1$ contains the condition of the proportionality of masses.

Further dimensionless numbers occurring in equation (5.18) are:

(a) $P/(\rho l^2 v^2)$ – the *Newton number* (Ne) (see also Section 5.2.2). *The New-ton number is given by the ratio of resistance and inertia forces.* As $P/l^2 = p$ (i.e. specific pressure), Ne can be written in the form $p/(\rho v^2)$, which is the *Euler number* (Eu). Sometimes the square root of half the reciprocal value is also used as the Euler number (i.e. $v/\sqrt{(2p/\rho)}$). As pointed out in Chapter 4, the cavitation number σ (equation 4.122) is a form of the Euler number.

(b) $v/\sqrt{(gl)}$ – the *Froude number* (Fr). The same name and notation (Fr) is sometimes used for the square of this ratio ($v^2/(gl)$), particularly in Russian and east European literature (see also Section 5.2.3). *The Froude number is derived from the ratio of gravity and iner-tia forces*; it can also be interpreted as the ratio of the veloc-ity of flow to the celerity of a gravity wave (see Section 4.5). (The ratio of *velocities* gives the traditional expression for the Froude number (Fr), whereas the ratio of *forces* results in the square (Fr^2).)

(c) $\rho l v/\mu = l v/\nu$ – the *Reynolds number* (Re) (see also Section 5.2.2). *The Reynolds number is the ratio of inertia and viscous forces.*

(d) $\rho v^2 l/\sigma$ – the *Weber number* (We). Again the square root of We is some-times quoted as the Weber number. *The Weber number is given by the ratio of inertia and surface tension forces.* The square root of the Weber number represents the ratio of the velocity of flow and the celerity of a capillary wave (see Section 4.5; see also the above comment on the Froude number).

(e) $\rho v^2/K$ – the *Cauchy number* (Ca). (Sometimes the term Cauchy num-ber is used in reference to the elasticity of a rigid body with elas-ticity modulus E instead of K.) The square root, which is called the *Mach number* (Ma), is more often used in its place (the pres-sure wave velocity is $c = \sqrt{(K/\rho)}$); the Mach number is sometimes also referred to as the *Bairstow–Booth* or *Majevskij* number. *The Mach number is thus given by the ratio of inertia and elasticity forces.*

Equation (5.18) can therefore be written as

$$Ne = \left(Fr, Re, We, Ma, \frac{h}{l}, \frac{B}{l}, \frac{d}{l}, \frac{\rho_s}{\rho} \right) \tag{5.18b}$$

The condition of mechanical similarity may now be expressed from equation (5.18b) as

$$M_{Ne} = 1 \tag{5.19a}$$

$$M_{Fr} = 1 \tag{5.19b}$$

$$M_{Re} = 1 \tag{5.19c}$$

$$M_{We} = 1 \tag{5.19d}$$

$$M_{Ma} = 1 \tag{5.19e}$$

$$\frac{M_b}{M_l} = 1 \tag{5.19f}$$

$$\frac{M_b}{M_l} = 1 \tag{5.19g}$$

$$\frac{M_d}{M_l} = 1 \tag{5.19h}$$

$$\frac{M_{\rho s}}{M_\rho} = 1 \tag{5.19i}$$

Equation $M_{Fr} = 1$ (equation 5.19b) expresses the *Froude law* of mechanical similarity under the exclusive action of *gravity*; equally $M_{Re} = 1$ (equation 5.19c) expresses the *Reynolds law* and applies in the case of exclusive action of *viscosity*, $M_{We} = 1$ (equation 5.19d) is the *Weber law* and applies in case of exclusive action of *surface tension*, and $M_{Ma} = 1$ (equation 5.19e) is the *Mach law* which is valid under the exclusive effect of *compressibility*. Section 5.7 discusses the implications of these laws.

It follows from equation (5.18) that for full mechanical similarity for a real fluid (i.e. under the simultaneous effect of all types of forces) all of equations (5.19a)–(5.19d) must be fulfilled, together with geometric similarity and proportionality of masses. Only under these conditions can equation (5.16) be satisfied, with equality of the Newton and Euler number on the model and in prototype.

5.5.2 Derivation of the law of mechanical similarity in hydrodynamics from the basic Navier–Stokes equations

The Navier–Stokes differential equations were introduced in Section 4.2 as equations (4.7) or (4.7a):

$$\frac{du}{dt} = X - \frac{1}{\rho}\frac{\partial p}{\partial x} + \nu\nabla^2 u \quad \text{etc} \tag{4.7a}$$

Considering the equations as valid both for the prototype and the model, we obtain for the force of gravity ($X = g$) and the x direction, using corresponding scales, the equation

$$\frac{M_v}{M_t}\frac{du}{dt} = M_g X - \frac{M_p}{(M_\rho M_l)\partial p/\partial x} + \frac{M_v M_v}{M_l^2}\nu\nabla^2 u$$

or

$$\left(\frac{M_l}{M_v M_t}\right)\frac{du}{dt} = \left(\frac{M_g M_l}{M_v^2}\right)X - \left(\frac{M_p}{M_\rho M_v^2}\right)\frac{\partial p}{\partial x} + \left(\frac{M_v}{M_v M_l}\right)\nu\nabla^2 u$$

For mechanical similarity between model and prototype all the multipliers must be equal to one. Thus

$$\frac{M_l}{M_v M_t} = \frac{M_g M_l}{M_v^2} = \frac{M_p}{M_\rho M_l} = \frac{M_v}{M_v M_l} = 1$$

Obviously, this procedure leads back to the already derived equations (equations 5.19a–5.19c), but it also yields a new condition:

$$\frac{M_l}{M_v M_t} = 1 \tag{5.20}$$

l/vt is another dimensionless expression analogous to fl/v (f is the frequency), referred to as *Strouhal number* (St); the reciprocal vt/l is sometimes (particularly in Russian literature) called the 'homochronometry number'.

As, for steady flow

$$\frac{\partial u}{\partial t} = \frac{\partial v}{\partial t} = \frac{\partial w}{\partial t}$$

condition (5.20) will in this case be irrelevant. The Strouhal number thus is associated with unsteady flow and processes. It represents also a measure of the *ratio of inertia forces due to the unsteadiness of the flow (local acceleration) and the inertia forces due to changes in velocity* from point to point in the flow field (convective acceleration) (Munson *et al.* (1998)). In the study of wake-train formation (a Karman vortex trail shed from a cylinder placed in the flow), the square of the Strouhal number, with f denoting the frequency of the wake, is sometimes referred to as the *Richardson number of the wake train*.

The Navier–Stokes equations are equally valid for the flow past any body as for the flow of liquid in a pipe or channel. The system of differential equations thus expresses a whole class of phenomena with an infinitely large number of solutions. However, dimensionless numbers obtained in the above manner can only be the criteria of similarity if the initial equations have an unambiguous solution. This can only be attained if the differential equations are limited by certain conditions so that, after solution, they give values of variables characterizing the given phenomenon. The boundary conditions that must be added to the differential equations thus attain the character of conditions of unambiguity of solution. These conditions take into account the geometry and dimensions of the space in which the given phenomenon occurs, the physical properties of the medium and, lastly, the conditions determining the values of the variables at the limits of the system. If the various phenomena within the limited group now differ from one another only in scale, they fulfil the main presuppositions of the theory of similarity.

Most practical cases deal with turbulent flow. It will, therefore, be useful to note the consequences of applying the previously stated modifications of the Navier–Stokes equations in Section 4.3.3 to the similarity of turbulent flow.

Recalling equation (4.33a) we can write:

$$\frac{d\bar{u}}{dt} = X - \left(\frac{1}{\rho}\right)\frac{\partial \bar{p}}{\partial x} + \left(\frac{\mu}{\rho}\right)\nabla^2\bar{u} + \left[\partial\left(\frac{\overline{u'\bar{u}}}{\partial x}\right) + \partial\left(\frac{\overline{u'\bar{v}}}{\partial y} + \partial\frac{\overline{u'\bar{w}}}{\partial z}\right)\right]$$

In the same manner as before we obtain

$$M_{St} = M_{Fr} = M_{Eu} = M_{Re} = 1$$

but the last term in equation (4.33a) results in a further equation:

$$\frac{M_{v'}}{M_{\bar{v}}} = 1 \tag{5.21}$$

In the theory of similarity, the dimensionless number v'/v is referred to as the *Karman number* (Ka).

Equation (5.21) introduces the criterion of constant velocity scale for mechanical similarity in turbulent flow (i.e. the condition that the scale of velocity fluctuations should be equal to the scale of mean velocities).

Let us consider the case that inertia and external forces may be neglected. The Navier–Stokes equation (4.7a) for steady flow and the x-axis would then reduce to:

$$0 = -\left(\frac{1}{\rho}\right)\frac{\partial p}{\partial x} + v\nabla^2 u$$

By the same process as before, this gives

$$\frac{M_p}{M_\rho M_l} = \frac{M_v M_v}{M_l^2} = 1$$

and, therefore,

$$\frac{M_p M_l}{M_\mu M_v} = M_{\mathrm{La}} = 1 \tag{5.22}$$

$pl/(\mu v)$ is the *Lagrange number* (La) and is the product of the Euler $(p/(\rho v^2))$ and Reynolds $(lv\rho/\mu)$ numbers.

Equation (5.22) ($M_{\mathrm{La}} = 1$) is thus the consequence (rather than the condition) of the existence of similarity.

The general equation

$$\mathrm{Eu} = f(\mathrm{Re}) \tag{5.23}$$

has two special cases:

(a) $f(\mathrm{Re}) = c/\mathrm{Re}$, where c is a constant, i.e.

$$\mathrm{Eu} = \frac{c}{\mathrm{Re}} \tag{5.24}$$

(equation (5.24) with $\mathrm{EuRe} = pl/(v\mu) = \mathrm{La} = \text{constant}$ complies with equation (5.22)); and

(b) $f(\mathrm{Re}) = \text{constant}$, resulting in

$$\mathrm{Eu} = \text{constant} \tag{5.25}$$

The above applies to a situation where for steady flow and neglecting the volumetric forces in the Navier–Stokes equation the viscous forces are negligible in comparison with inertia:

$$\frac{u\partial u}{\partial x} + \frac{v\partial u}{\partial y} + \frac{w\partial u}{\partial z} = -\left(\frac{1}{\rho}\right)\frac{\partial p}{\partial x}$$

Thus

$$\frac{M_{v^2}}{M_l}\left(\frac{u\partial u}{\partial x} + \frac{v\partial u}{\partial y} + \frac{w\partial u}{\partial z}\right) = \frac{-M_p}{M_\rho M_l}\left(\frac{1}{\rho}\right)\frac{\partial p}{\partial x}$$

and

$$\frac{M_p}{M_\rho M_v^2} = M_{\mathrm{Eu}} = 1 \tag{5.26}$$

which is equivalent to equation (5.25) applied to the model and the prototype.

Examples of equations (5.23)–(5.26) can be found in Section 4.4.1. The Hagen–Poiseulle law for the distribution of velocity in laminar flow in a pipe:

$$u = \frac{p_1 - p_2}{4\mu l}(r^2 - y^2)$$

or the equation for head loss in laminar flow

$$h_f = \frac{32 l \nu V}{gD^2} \tag{4.40}$$

or the equation for the coefficient of friction in laminar flow

$$\lambda = \frac{64}{Re} \tag{4.41}$$

are all examples of equation (5.24) (La = constant).

Equally, the fully developed rough turbulent flow in conduits with Re > Re$_{sq}$, or equation (4.49), where the friction coefficient λ is independent of the Reynolds number, are examples of equation (5.25) (Eu = constant).

From the modelling point of view, equation (5.25), excluding the effect of viscosity (e.g. in the computation of λ) characterizes an *automodelling* region where $M_{Eu} = 1$.

Let us now consider $\nabla^2 u$ in the last term of the Navier–Stokes equation:

$$\nabla^{2+}u = \frac{\partial^2 u}{\partial x^2} + \frac{\partial^2 u}{\partial y^2} + \frac{\partial^2 u}{\partial z^2}$$

After adding and subtracting

$$\frac{\partial^2 v}{\partial x \partial y} + \frac{\partial^2 w}{\partial x \partial z}$$

this results in

$$\nabla^2 u = \frac{\partial^2 u}{\partial x^2} + \frac{\partial}{\partial y}\left(\frac{\partial u}{\partial y} - \frac{\partial v}{\partial x}\right) - \frac{\partial}{\partial z}\left(\frac{\partial w}{\partial x} - \frac{\partial u}{\partial z}\right) + \frac{\partial}{\partial x}\left(\frac{\partial v}{\partial y} + \frac{\partial w}{\partial z}\right)$$

Introducing from continuity $\partial v/\partial y + \partial w/\partial z = -\partial u/\partial x$, and recalling the definition of vorticity from Section 4.2.4 ($\omega_{xy} = 1/2(\partial v/\partial x - \partial u/\partial y)$), we obtain

$$\nabla^2 u = -2\left(\frac{\partial \omega_{xy}}{\partial y} - \frac{\partial \omega_{xz}}{\partial z}\right)$$

with similar expressions for the y- and z-axes, and $\nabla^2 v$ and $\nabla^2 w$.

As shown in Section 4.2, the potential flow of a viscous fluid is irrotational (equations 4.8–4.11), with $\omega_{xy} = \omega_{zy} = \omega_{zx} = 0$, and therefore also $\nabla^2 u, \nabla^2 v, \nabla^2 w = 0$. This means that the Reynolds number is eliminated from the conditions of similarity and only the Froude number (possibly also the Strouhal number for unsteady flow) remains as a criterion. For steady irrotational flow of a viscous liquid with a free surface (i.e. flow with a velocity potential), similarity is thus governed exclusively by the Froude law, which can also be expressed as

$$\mathrm{Eu} = f(\mathrm{Fr}) \tag{5.27}$$

In the case of viscous fluid flow under pressure, the influence of gravity may be included in the pressure differential and not connected physically with the mass of the fluid. Mathematically this assumption may easily be expressed in the Navier–Stokes equations by rewriting them for the force field of gravity with the z-axis vertical and positive downwards. The pressure term then becomes

$$- (1/\rho)\, \partial p^* / \partial x = - (1/\rho)\, \partial (p - p') / \partial x = - (1/\rho)\, \partial (p - \rho g z) / \partial x$$

In the above statement p is the pressure intensity at an arbitrary point of the cross-section during motion, and $p' = \rho g z$ is the hydraulic pressure at the same point at rest under otherwise equal conditions. If we now apply the above-used procedure with the introduction of scales into the modified Navier–Stokes equation, we find that only the Reynolds number (possibly also the Strouhal number for unsteady flow) remains as a criterion of similarity.

Using the Navier–Stokes equations as a basis for the derivation of the law of mechanical similarity we have not only established the condition of equality in the model and the prototype of the Strouhal, Froude, Reynolds and Euler numbers, but have also identified two cases where the Euler number is a function of only the Reynolds number (or even where the Euler number is a constant) (equations 5.22–5.26) or of only the Froude number (equation 5.27). Obviously, the Weber and the Mach numbers have not been considered in this section as the Navier–Stokes equations do not contain the effects of surface tension or compressibility.

5.6 Approximate mechanical (dynamic) similarity

The condition of complete mechanical similarity requires that, while using real fluids, equations (5.19a)–(5.19e) should be simultaneously satisfied. However, it is practically impossible to obtain for the model fluids possessing the necessary physical properties that comply with the given conditions. If we choose the same liquid for the model as in the prototype, it is impossible to attain complete mechanical similarity since, for $M_g = 1$ (the same gravitational acceleration acts on the model as in prototype) we obtain from equation (5.19b) $M_v = \sqrt{M_l}$, and for $M_v = 1$ (the same liquid in the model and prototype) we obtain from equation (5.19c) $M_v = 1/M_l$. Both these equations for M_v can be satisfied only for $M_l = 1$ (i.e. a model of the same size as the prototype).

Hydraulic models thus work only with *approximate mechanical (or dynamic) similarity based on the ratio of those forces that determine the type of motion (or which are predominant in the given situation).*

Accordingly, the Reech–Froude, Reynolds, Weber or Mach law is then chosen for the investigation. By working on the model according to one main and determining law and neglecting the others, errors occur due to scale effects, and the reduction of the model against reality must be chosen to make these errors as negligible as possible in the given situation. These *limits of model reduction* determine the *limits of similarity* or the *limiting boundary conditions* for scaling procedure (for a more detailed analysis of these limits, see Section 5.8.2 and Chapters 7–13, as appropriate).

There are, of course, cases where two or more forces are equally important in the prototype (or the influence of which cannot be neglected); in this situation special procedures have to be adopted (see e.g. Section 5.9).

Good knowledge of established physical laws, judgement and experience all are important in selecting the appropriate method for a modelling problem. In making the compromises that are often needed in selecting scales and the programme of tests, modelling becomes an art as well as a science. A too rigid attitude based only on theory may lead to a rejection of a scale model in a situation where even an imperfect model may be of substantial benefit (Ackers (1987)). Model tests of different design problems are bound to be associated with different confidence levels in the model–prototype correlation (see Chapters 7–13). However, a lower level of confidence in quantitative results does not necessarily negate the use of a physical model but requires care and engineering judgement in the interpretation of model data (Hay (1988)).

5.7 The main similarity laws

5.7.1 Froude law

The Froude law (first expressed by Reech and later independently by Froude, and therefore sometimes also referred to as the Reech–Froude law) represents the condition of mechanical (and dynamic) similarity for flow in the model and prototype governed exclusively by gravity. Other forces, such as the frictional resistance of a viscous liquid, capillary forces and the forces of volumetric elasticity, either do not affect the flow or their effect may be neglected. With certain limitations on the choice of model scales this is permissible, especially for free-surface flow (e.g. when modelling the discharge over notches and weirs, the movement of long surface waves (see equation (4.90)), hydraulic jumps, flow in short open-channel sections, etc.).

In the flow of a real (viscous) fluid, however, internal friction always acts simultaneously with gravity. If the model is geometrically similar to the prototype and the boundary conditions are also similar (e.g. inflow or outflow conditions, wall roughness), then similarity not only between forces due to gravity but also to a large extent between the resistances due to friction is ensured in many cases (see Sections 5.5.2, 5.8 and Chapters 7–13). The Froude law is thus the most widely used similarity law in physical modelling.

The basic equation for similarity under the exclusive or overwhelming action of gravity was given by equation (5.19b) ($M_{Fr} = 1$); from this condition and the definition of the Froude number as $v/\sqrt{(gl)}$ the scale of *velocities* can be determined for a chosen length scale M_l as

$$M_v = \frac{v_p}{v_m} = M_l^{1/2} M_g^{1/2} \tag{5.28}$$

As practically always the scale of acceleration $M_g = 1$ (g is the same on the model and in prototype),

$$Mv = M_l^{1/2} = \left(\frac{l_p}{l_m}\right)^{1/2} \tag{5.28a}$$

The scales of other parameters follow from continuity, identity of Euler numbers and physical definitions:

for discharge: $MQ = M_v M_A = M_l^{1/2} M_l^2 = M_l^{5/2}$

for time: $M_t = \dfrac{M_l}{M_v} = M_l^{1/2}$

for frequency: $M_f = \dfrac{1}{M_t} = M_l^{-1/2}$

for *force* v: $M_P = M_\rho M_v^2 M_l^2 = M_\rho M_l^3$

(for the same liquid in the model as in the prototype, $M_\rho = 1$); thus:

for *specific (unit) pressure*: $M_p = \dfrac{M_P}{M_A} = M_l$

for *power*: $M_N = M_v M_Q M_H = M_l^{7/2}$

for *work(energy)*: $M_e = M_N M_t = M_l^{7/2} M_l^{1/2} = M_l^4$

From the above it follows that the scale of the Reynolds numbers for M_v 1 will be $M_{Re} = M_v M_l = M_l^{3/2}$.

5.7.2 Reynolds law

Reynolds law expresses the criterion of mechanical (and dynamic) similarity of the motion of two incompressible viscous liquids under the exclusive (or predominant) effect of internal friction. It is valid, for example, in modelling flow around bodies submerged in the liquid without surface waves, laminar flow in conduits (see equation 4.41) or turbulent flow in a smooth conduit (see equation 4.48), etc.

The basic equation has already been given by equation (5.19c) ($M_{Re} = 1$) from which the scale of velocities for Re $= vl/v$ and a chosen length scale is

$$M_v = M_v M_l^{-1} \tag{5.29}$$

For the same fluid in the model and the prototype $M_v = 1$, and thus

$$M_v = M_l^{-1} \tag{5.29a}$$

The scales for the other parameters can be derived from equation (5.29) in the same manner as in the previous paragraph.

It is opportune to note here that the coefficient of kinematic viscosity v depends on temperature and pressure; it decreases with increasing temperature in liquids and increases with increasing temperature in gases. At a pressure of 760 mmHg and temperatures of 0 °C, 10 °C and 20 °C, v for air is 1.33×10^{-5}, 1.40×10^{-5} and 1.49×10^{-5} m²/s and v for water is 1.78×10^{-6}, 1.31×10^{-6} and 1.01×10^{-6} m²/s, respectively. Air compressed to 100 atm at 0 °C has $v = 1.33 \times 10^{-7}$ m²/s and at 100 °C $v = 2.45 \times 10^{-7}$ m²/s. At a temperature of 20 °C v for mercury is 1.77×10^{-7}, petroleum 0.74×10^{-4}, engine oil 3.8×10^{-4} and glycerine 6.8×10^{-4} m²/s.

5.7.3 Weber law

The Weber law represents the condition of similarity for the exclusive or prevailing effect of capillary forces causing surface tension. This manifests itself both on the liquid–gas interface and on the interface between two different liquids by the formation of a (curved) membrane due to the effect of molecular forces.

Surface tension decreases with rising temperature, and influences, for example, the flow and shape of small nappes, and the formation of short free-surface waves (equation 4.87).

Surface tension is given by the capillary constant σ, which acts on the unit of length of the surface. For example, on a water surface above which there is air, it is 7.29×10^{-2} N/m at a temperature of 20 °C.

The basic equation for similarity with the exclusive effect of capillary forces is given by equation (5.19d) ($M_{We} = 1$). Defining the Weber number as $\rho v^2 l / \sigma$ results in the velocity scale

$$M_v = \left(\frac{M_\sigma}{M_\rho M_l} \right)^{1/2} \tag{5.30}$$

For the same fluid in the model and the prototype, $M_\rho = M_\sigma = 1$; thus

$$M_v = M_l^{-1/2} \tag{5.30a}$$

From equation (5.30) the scales for all other parameters can again be derived as in Section 5.7.1.

5.7.4 Mach law

The Mach law expresses the criterion of similarity for the exclusive or prevailing effect of volumetric elasticity (compressibility) of the medium (liquid). The modulus of compressibility (bulk modulus) K of a liquid is given by the ratio of increase in stress to the reduction in volume (which causes an increase in specific mass (density)). Thus

$$K = \frac{\rho \Delta p}{\Delta \rho}$$

Again, as in the preceding cases, the basic equation for this law of similarity has already been stated by equation (5.19e) ($M_{Ma} = 1$). For Mach number $Ma = v \sqrt{(\rho / K)}$ this gives for the velocity scale

$$M_v = \left(\frac{M_K}{M_\rho} \right)^{1/2} \tag{5.31}$$

or for $M_\rho = M_K = 1$

$$M_v = 1 \tag{5.31a}$$

The scales for all other parameters can be derived as in previous paragraphs.

5.7.5 Comparison of the main similarity laws

It is interesting at this stage to compare the scales of some important parameters in *geometrically similar* models for the four similarity laws (Table 5.1). In compiling Table 5.1 it was assumed that $M_g = 1$. The comparison clearly demonstrates why modelling according to the Froude law is considerably

Table 5.1

Scale	Froude	Reynolds	Weber	Mach
M_v	$M_l^{1/2}$	$M_l^{-1} M_v$	$M_l^{-1/2} M_\sigma^{1/2} M_\rho^{-1/2}$	$M_K^{1/2} M_\rho^{-1/2}$
M_t	$M_l^{1/2}$	$M_l^2 M_v^{-1}$	$M_l^{3/2} M_\sigma^{-1/2} M_\rho^{1/2}$	$M_l M_\rho^{1/2} M_K^{-1/2}$
M_Q	$M_l^{5/2}$	$M_l M_v$	$M_l^{3/2} M_\sigma^{1/2} M_\rho^{-1/2}$	$M_l^2 M_K^{1/2} M_\rho^{-1/2}$
M_p	$M_l M_\rho$	$M_l^{-2} M_\mu^2 M_\rho^{-1}$	$M_l^{-1} M_\sigma$	M_K

simpler than that according to the other laws. Assuming the same fluid in the model as in the prototype, with $M_\rho = M_v = M_\sigma = M_K = 1$, the Froude law is the only one where the velocities and the pressure intensities in the model are smaller than in the prototype. Even for a different fluid in the model it is difficult to find a suitable one that would reverse this conclusion. Modelling according to the other three laws can thus often result in considerable technical difficulties.

5.7.6 Distorted models

Any deviation from geometric similarity means that the model is distorted, and because of this we can no longer speak of mechanical similarity but only of full or approximate dynamic similarity (see Section 5.4).

By far the most common case of distortion is adopting a vertical scale different from the horizontal one. This is practically always the case in river and estuarine/coastal engineering when modelling a longish river stretch or a large coastal area. Because of the physical laboratory limitations, the horizontal scales M_b and M_l are rather large (usually $100 < M_l < 1,000$). The scale of width is practically always the same as that of length ($M_b = M_l$), although there can be exceptions to this (see Chapter 7). If the same scale were adopted for the vertical scale (i.e. scale of depth M_h), the resulting Reynolds number of the flow would be too small for viscosity to be neglected in a model operated according to the Froude law, i.e. $\mathrm{Re} < \mathrm{Re}_{sq}$ (see Sections 5.5.2 and 5.8.2). Therefore, a value of $M_h < M_l$ is chosen to satisfy the condition $\mathrm{Eu} = \mathrm{constant}$ (equation 5.25).

Considering now the resulting velocity, discharge, time and pressure intensity scales, with $\mathrm{Fr} = v/\sqrt{(hg)}$, we obtain (for $M_g = M_\rho = 1$ and $M_l = M_b$)

for *velocity of flow*: $M_v = M_h^{1/2}$ $\qquad\qquad$ (5.32a)

for *discharge*: $M_Q = M_v M_A = M_h^{3/2} M_b = M_h^{3/2} M_l$ $\qquad\qquad$ (5.32b)

for *time*: $M_t = \dfrac{M_l}{M_v} = M_l M_h^{-1/2} M$ $\qquad\qquad$ (5.32c)

for *pressure intensity*: $M_p = M_h$ $\qquad\qquad$ (5.32d)

It is important to appreciate that, in the case of modelling parameters varying with time, the Strouhal criterion requires a uniform time scale on the model in all directions; therefore, for the *scale of vertical velocities* M_w (velocity of rise and fall in tidal levels, settling of suspensions, lifting and lowering of gates, etc.) we obtain

$$M_w = \frac{M_h}{M_t} = \frac{M_h M_h^{1/2}}{M_l} = M_h^{3/2} M_l^{-1} \tag{5.33}$$

A comparison of equations (5.32) and (5.33) demonstrates that in a model where the vertical velocity components are important, distortion is not acceptable (for $M_w = M_v$ the two scales M_h and M_l must be equal). Nevertheless, where we can assume with good approximation a hydrostatic pressure distribution in the vertical and where we can accept that some details of the flow conditions will not be modelled correctly (e.g. the modelling of eddies, spreading of jets, the flow around some bodies where separation may occur in the model but not in prototype) (Kobus (1980)), vertically distorted models are acceptable not only for cases of uniform and non-uniform flow but also for unsteady flow conditions with relatively slow vertical motion (e.g. fall and rise in tidal water levels) (see also Chapters 7, 11 and 12).

The above is only a brief general outline of the method of handling one type of distortion in physical models; other problems associated with a departure from geometric similarity (e.g. in roughness, sediment size and shape) will be elaborated further in Chapters 7–13.

5.8 Some further dimensionless numbers and limits of similarity

5.8.1 Some further dimensionless numbers

In the preceding text some of the dimensionless numbers important in hydraulics were defined: the *Newton, Euler, Froude, Reynolds, Weber and Mach numbers* (Section 5.5.1), and the *Strouhal, Lagrange, Karman numbers* (Section 5.5.2). In Chapter 4 the *densimetric Froude number* (equation 4.96) and the (gradient) *Richardson* number (equation 4.97), the *sediment transport parameter* and the *flow parameter* (equations 4.100–4.102), the *mobility number* (equation 4.103) and the *cavitation number* (equation 4.122) were introduced. We have also seen that sometimes the same number has several different names (Mach, Majevskij, Bairstow–Booth – Section 5.5.2), different names may apply to a different power of the same number (Mach–Cauchy, Strouhal–Richardson number of the wake train – Section 5.5.2) or the same name is given to different powers of the same number (Froude, Weber – Section 5.5.1). Although it would be helpful to standardize the notation for scale (model/prototype or prototype/model) and the named dimensionless numbers, probably it is not feasible to achieve this and we have fundamentally two choices (Novak (1984)): either to use rigorously the original definitions in the literature, or to accept the inevitable and whenever using a 'named' dimensionless number to *define it clearly and particularly to define the velocity* (mean, local, shear, fall, etc.) *and the length* (size, depth, diameter, grain size, etc.) *parameters* used. A good example is the frequent use of the 'densimetric Froude number' with, for example, the local velocity and a length term (equation 4.96), or the shear velocity and grain diameter (equations 4.99–4.102), or the mean velocity of flow and the grain diameter, etc.

The *Kolf number* (Kf) or *circulation number* (N_Γ) are expressions of vortex flow: $N_\Gamma = \Gamma d/Q$, where Γ is the circulation $= 2\pi c$ (c is the circulation constant $= vr$ – see Chapter 4) and $Kf = \Gamma/vd = \Pi/4N_\Gamma$.

The *Mosonyi number* ($vv/(gl^2)$) is given by the ratio of viscous and gravity forces; the ratio of the (sediment) fall velocity w and the shear velocity U_* (see e.g. equations 4.106–4.109) is the *Rouse* number. The *Galileo number* (gl^3/v^2), derived from $(Re/Fr)^2$, eliminates the velocity term, and the *liquid parameter* (also called the *Morton* number) ($g\mu^4/(\rho\sigma^3)$), derived from $We^3/(Fr^2Re^4)$, excludes both the velocity and length terms.

The *Keulegan–Carpenter number* is used in modelling the flow of waves past cylindrical bodies, and is linked to the vortex-shedding process; there are several versions of this number (which is essentially linked to the Strouhal number (equation 5.20); see also Section 5.5.2) UT/D, $2\pi H/D$, $\pi H/L/(D/L)\tan h(2\pi D/L)$, where U is the relative velocity between body and fluid, T is the wave period, D is the body parameter (diameter), H is the wave height and L is the wave length.

The *Vedernikov number* (Ved) is the Froude number multiplied by a channel-shape factor $\varphi(= 1 - RdP/dA)$ (for 'wide' channels $\varphi = 1$, and for 'narrow' channels $\varphi = 0$) and the exponent k of the hydraulic radius in the uniform-flow equation (usually $k = 2/3$); it is used to express the condition of stability of uniform supercritical flow ($-1 < Ved < 1$) (see also Section 13.2.3).

Schuring (1977) documents 57 named principal dimensionless numbers used in modelling in various engineering disciplines, and many of these are pertinent to hydraulic modelling; even this list is by no means complete, particularly as Schuring does not include derived numbers. Some of the numbers of importance quoted by Schuring, but not yet mentioned here, are the *Bingham* (slow flow of viscoplastic material), *Hedstrom* (rapid flow of viscoplastic material), *Peclet* (mass transfer by diffusion), *Rossby* (large-scale atmospheric or oceanic motion), *Schmidt* (flow with momentum and mass transfer) and *Sherwood* (mass transfer by convection) *numbers*.

In the treatment of *debris flow*, several dimensionless numbers obtained by the ratio of collision stresses to other stresses are in use: the ratio to the grain friction stresses is the *Savage number*, to the liquid viscous stresses is the *Bagnold number*, to the turbulent stresses is the *mass number* and to liquid grain interaction is the reciprocal of the *Darcy number*.

Some of the above numbers will be discussed further in Chapters 7–13.

5.8.2 Some limits of similarity

The limits of similarity are the limiting boundary conditions for the scaling procedure according to one of the laws of similarity (see Section 5.6). These limits may be in the form of a limiting value of a dimensionless number (see (a)–(c) below) or a limiting value of a hydraulic parameter (see (d), (e) and (g) below), or stated as a limiting dimension (see (f) below):

(a) An obvious condition is that the regime of flow in a scale model and in the prototype should be the same: laminar or turbulent ($Re < Re_{cr}$, $Re > Re_{cr}$ – see Section 4.4.1), uniform or non-uniform, steady or unsteady, irrotational or rotational, subcritical or supercritical ($Fr < 1$, $Fr > 1$).

(b) The most frequent case in modelling is the situation where, when using the Froude law, it is necessary to eliminate (or reduce to an insignificant proportion) the effect of viscosity (Reynolds number). We have seen already that this is automatically the case when the flow is irrotational (Section 5.5.2, equation 5.27) or when (in pipe and open-channel flow) $Re > Re_{sq}$ (equations 4.49 and 5.25). Should we not be able to comply with this condition, we can still obtain the same frictional head-loss coefficient in the model as in the prototype but only by distorting the relative roughness (see Figure 4.4 and Chapters 7 and 9).

(c) For cavitation to occur, the cavitation number (equation 4.122) must fall to a critical value σ (see Section 5.6.5). Cavitation tunnels (see Section 6.1.2) with free-water surface should operate with $Re > 10^6$. Although in experimentation using the Froude law the value of the Euler number is maintained automatically, for the cavitation number special additional conditions will apply in order to model σ_c correctly (see Chapter 13).

(d) Air entrainment in a flow of water requires a minimum value of velocity – the inception limit v_i – which should also be exceeded in the model (e.g. for plunging jets and siphons a value $v_i = 0.8\,\text{m/s}$ is quoted) (Kobus and Koschitzky (1991)) (see also Section 4.6.5 and Chapter 13).

(e) Surface tension is not often important in a prototype but can become significant in models if they are too small. Although the critical values of the Weber number are not as clearly defined as, say, those of the Reynolds number (Peakall and Warburton (1996)), some guidance about the limiting values has been quoted:

   (i) In Section 4.5 it was stated that the minimum celerity with which a surface wave may be propagated is $c = 0.23\,\text{m/s}$ (equation 4.87), corresponding to a minimum wavelength (for gravity waves) of $L = 0.017\,\text{m}$ (equation 4.86); this results in a Weber number of about 12.

(ii) From the propagation speed, one can postulate that a minimum water depth is required in the model to eliminate surface-tension effects. Kobus (1980) quotes a limit of 0.03 m (which can be decreased by the addition of surface-active agents that increase the Weber number); Novak and Čábelka (1981) quote a minimum acceptable depth of 0.015 m.

(iii) For the correct reproduction of flow conditions (vortex formation) at vertical intakes, the condition $We > 120 (We = \rho v^2 l / \sigma)$ has been quoted (Ranga Raju and Garde (1987)).

(f) For modelling flow over notches, under gates or through orifices (see also Chapter 13), Novak and Čábelka (1981) quote:

(i) The head on a sharp-edged notch should be at least 60 mm for the shape of the nappe to be capable of extrapolation (Ghetti and D'Alpaos (1977) quote 40 mm); for a head less than 20 mm the overflow parabola of the free jet due to influence of capillary tension is deformed almost into a straight line.

(ii) For flow under a gate the smallest height of its opening a for which the shape of the outflow jet may be extrapolated is 60 mm and the jet shape is independent of the head h for $h > 3.3a$.

(iii) For discharge from a circular orifice of diameter D the smallest size for which the jet shape can be extrapolated is $D > 70\,\text{mm}$, and the jet is independent of the head for $h > 6D$. (The conditions specified in (ii) and (iii) give a Reynolds number of the order 10^5.)

(g) It is generally accepted that when using aerodynamic models (see Section 5.10.2) the effect of compressibility (Mach number) is negligible for air speeds below 50–60 m/s (Kobus (1980)).

In the above notes only some of the more frequently occurring limits of similarity have been summarized; others – particularly those associated with movable beds – are best dealt with in (the applied) Chapters 7–13.

5.9 Methods of modelling complex phenomena

It is sometimes difficult to decide in advance whether the influence of the forces we want to eliminate as unimportant in the choice of criteria of similarity is indeed negligible. Independent experimental work must then determine whether the model scale was chosen correctly or whether in the model (or indeed the prototype) a complex phenomenon occurs that is influenced simultaneously by several types of force.

The simplest method in this case is to investigate the problem using two or more models of different size (i.e. a 'family' of models). If the results from all the models are then plotted (e.g. the relationship of discharge and head, $Q = f(h)$) after conversion to the prototype, we obtain a number of experimentally determined points. These may lie on one curve – with a scatter due to measurement errors – or they must be joined by two or more curves, always one for each model; the latter case then shows a systematic deviation signifying that similarity has not been obtained in the models. When preparing experiments on two models, we must ensure that (after conversion to the prototype) the experiments on both models include, at least in part, values of variables within the same limits.

If the experiment is carried out on three models of different sizes, and the systematic deviation appears only in the case of the smallest model, whereas for the other two the dispersion is within experimental errors, it may be concluded that in the smallest model the systematic error is caused only by exceeding the limit of model similarity. We then exclude it and continue to work with one of the larger models. However, if systematic deviation appears in all results and does not decrease with an increase in the model size, we are in all likelihood faced with a complex phenomenon in the prototype, and the safety of extrapolating from the model to the prototype will be the smaller, the greater this systematic deviation.

If deviations are considerable, separate experiments must be carried out or attempts be made to determine the influence of one type of force by calculations and the results introduced into the experimental procedure. Froude's method of investigating the resistance of a towed ship is an example of solving a complex phenomenon by model experiments.

In the resistance of a towed ship both gravity and viscosity apply, and neither of the resulting forces may be regarded as subordinate. Gravity influences the shape resistance of the ship, which is proportional to the main rib section and the square of the velocity of the wave formed by the movement of the ship. Viscosity affects the frictional resistance of the water on the wetted surface of the ship. The total resistance P is thus made up of the *shape resistance P_1 (Froude law)* and *frictional resistance P_2 (Reynolds law)*: $P = P_1 + P_2$.

During the towing of the model ship the total resistance of the model P_m may be ascertained for various shapes and draughts and different speeds of towing. Froude proved that the tangential resistance of the surface of the ship is practically the same as the resistance of a vertical towed board of the same surface area and roughness as the ship surface. The board must have a sharp leading edge and be thin enough for its shape resistance to be negligible. From experiments with the board we may obtain the frictional resistance for a unit of area and calculate P_2 for the model ship as well as for the prototype. Therefore,

$$P_{1m} = P_m - P_{2m}$$

As the experiment has been carried out according to the Froude law

$$P_{1p} = M_l^3 P_{1m}$$

and for the total resistance of the prototype ship we obtain

$$P_p = M_l^3(P_m - P_{2m}) + P_{2p} \tag{5.34}$$

In equation (5.34) the values P_{2m} and P_{2p} are obtained by computation, and P_m is obtained experimentally.

Froude's method is only approximate, as in the prototype the frictional forces P_2 depend also on the shape and dimensions of the ship, but it demonstrates the possibilities of separate experimental investigation of the various parameters involved.

5.10 Analogue models

Analogues, or semi-direct models, have already been defined in Chapter 1 as a system reproducing a prototype situation in a physically different medium. The prerequisite for this type of modelling is that equations representing the prototype and analogue model are similar expressions.

Consider the equation of changes in x with time:

$$\frac{d^2x}{dt^2} + \frac{adx}{dt} + bx + c = 0 \tag{5.35}$$

where for $t = O d^2x/dt = dx/dt = 0$ and $x = -x_0 (= -c/b)$.

In equation (5.35) x could denote the oscillating water level (relative to the reservoir level) in a hydroelectric power station surge tank (with automatic turbine regulation) (Figure 5.1a); it could be the angle through which a disc attached to a bar fixed at the other end turns in torsional oscillations (Figure 5.1b); or it could stand for the voltage fluctuations after disconnecting from a DC supply a circuit with resistance, capacitance and inductance (Figure 5.1c).

Thus, the torsional vibrations of the bar become the analogue of the water-level oscillations in the surge tank, and vice versa. We can therefore use one to analyse the other. The surge tank is a *hydromechanical analogue* of the bar, and the electric circuit is an *electrohydrodynamical analogue* of the surge tank.

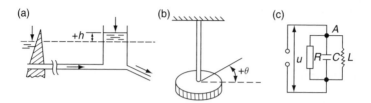

Figure 5.1 Analogues (Novak and Čábelka (1981))

Similarly, an electrical resistance network may represent a groundwater aquifer; here the voltage corresponds to water levels, the current to discharge, the resistance to the reciprocal of transmissivity, and the capacitance to storage capacity.

It is necessary to differentiate between an *analogue model*, which physically represents a flow situation, and an *analogue computer*, which uses an electric circuit for the solution of simultaneous differential equations.

The developments in numerical modelling and in the power of (digital) computers have now, however, rendered analogue computers (and to a certain extent analogue models) obsolete. Nevertheless, in special cases analogue models are still useful, and a combination of analogue and digital techniques may provide good and economical results.

Returning to equation (5.35) and using the procedures established in Section 5.5.2 the following criteria, or 'indicators', of analogy can be derived:

$$M_b M_t^2 = M_a M_t = 1 \tag{5.36}$$

Therefore, also

$$\frac{M_b}{M_a^2} = 1 \tag{5.36a}$$

From the boundary conditions it follows that

$$M_b(= M_\theta) = 1 \tag{5.36b}$$

Equation (5.36) enables us to determine the numerical relationship between analogous physical parameters; it is evident that a single system of units has to be used throughout (Hálek (1965)).

Hydraulic analogues (Novak and Čábelka (1981)) utilizing the similarity of flow patterns in a two-dimensional potential flow and in a thin-layered viscous flow are often used together with flow-visualization methods to study flow past obstacles or streamlines in seepage flow. Typical examples are horizontal or vertical *thin-layer models*, or so-called *Hele–Shaw models*. In these models a viscous fluid flows in a thin slot between two smooth plates with a small, constant velocity, giving laminar flow throughout. Neglecting inertia forces, the velocities in the horizontal (x) and vertical (z) directions (in the case of the vertical models) or in the horizontal (x, y)-plane (horizontal models) are given by the equations

$$u = \frac{(a^2 g \partial p)}{(12 v \gamma \partial x)} = \frac{(a^2 g \partial h)}{(12 v \partial x)} = \frac{k \partial h}{\partial x} \tag{5.37}$$

$$v = \frac{a^2 g \partial p}{12 v \gamma \partial y} = \frac{a^2 g \partial h}{12 v \partial y} = \frac{k \partial h}{\partial y} \tag{5.37a}$$

where a is the distance between the plates and k is a constant.

Equations (5.37) and (5.37a) correspond to, for example, the equation of laminar flow through a porous medium and, when combined with continuity, satisfy the Laplace equation (4.11).

Hele–Shaw models using a fairly viscous fluid (e.g. glycerine) usually have a slot thickness of about 2 mm and a length of 1–2 m, with overfalls at the inlet and outlet simulating the boundary conditions, which may also be time-dependent. Models using water have appreciably smaller thickness, of about 0.1–0.3 mm. Figure 5.2 shows a vertical Hele–Shaw model used to

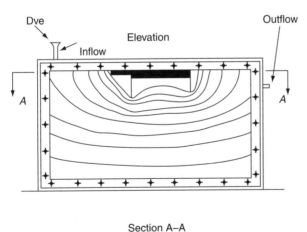

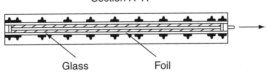

Figure 5.2 Hele–Shaw model (Novak and Čábelka (1981))

study seepage flow under a foundation. A change in the space between the plates could be used to simulate conditions of different permeabilities in stratified soils. The flow between the plates must, of course, remain laminar (i.e. $\text{Re} = ua/v < 500$). Slot models are simple to construct and operate, although care and accuracy in the positioning of the plates are required.

The use of Hele–Shaw models is limited to the solution of two-dimensional flow problems. No such limitations apply to *sand seepage models*, which, however, have disadvantages resulting from large capillary elevation and difficulties in the determination of the free-water surface, and problems of entrapped air. Uniform glass balls or pellets of up to 4 mm diameter are sometimes used to overcome these difficulties.

The *membrane analogy* (Karplus and Soroka (1959)) can be used to study the flow field around obstacles. A rubber membrane is fixed under uniform tension in a metal ring. If placed in an inclined position, the lines of equal elevation (broken lines in Figure 5.3) show the streamlines of parallel flow. If a rod of similar cross-section as the shape of the body, whose effect on the flow field is under investigation, is then pressed from below into the inclined membrane, its deflections at various points give a very good indication of the resultant streamlines (the full lines in Figure 5.3 indicate the resulting flow net of streamlines and equipotential lines). By changing the orientation of the body to the flow, an optimum position with no separation of flow can be determined.

As mentioned earlier, *electrical analogues* may be best used for the solution of groundwater flow problems, particularly those with linear equations.

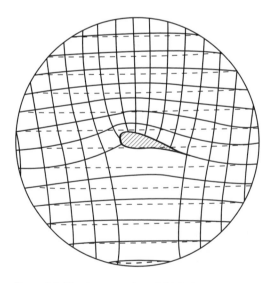

Figure 5.3 Membrane analogy (Hálek (1965))

In principle, a two-dimensional model uses a conductor in the shape of the considered porous medium, with the contours simulating the boundary between the permeable and impermeable ground. The inlet and outlet are simulated by electrodes (brass or copper). As the geometry of the resulting flow net is independent of the absolute value of the filtration coefficient (conductivity – this follows from the Laplace equation) a variety of materials may be used for the conductor (e.g. tin foil sheets, electrolytes, varnished surfaces containing graphite powder, electroconductive paper). The flow net is determined by using a probe to find lines of equal potential (see Figure 5.4); a Wheatstone bridge is used with probe (4) moved along the conductor until the galvanometer G indicates zero.

Direct analogues using linear resistors and capacitors can be used for the analysis of groundwater resources.

Using *air* (or another fluid) instead of water in a scale model, or even in the prototype (e.g. in a duct), is not an analogue in the strict interpretation of the above definition, as both air and water flow are covered by the same equations as long as compressibility (and capillarity) effects are excluded.

The use of *aerodynamic models* – basically subcritical wind tunnels – for investigating problems governed by inertia and viscosity is straightforward and can offer some advantages (cost, measuring techniques) over

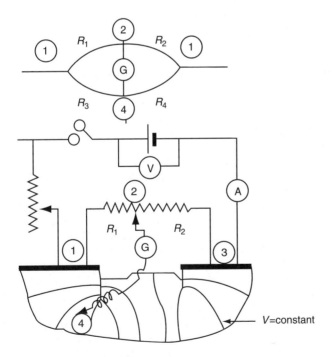

Figure 5.4 Electrical analogue (Novak and Čábelka (1981))

conventional hydraulic scale models; the main disadvantages of these models are the difficulty of making three-dimensional flow visible and in the difficulty of simulation of mixing flows.

The main problems in the use of aerodynamic modelling arise in cases where a free-water surface is involved. If the position of the free-water surface is known (or its exact level is not important) one can again use either a tunnel model with the flow cross-section inverted about the water surface, or, in a simpler way, substitute a plane ceiling for the free-water surface. If the position of the water surface or water depths are not known (e.g. this is the subject of the investigation), an iterative process for the placement of the (glass) plane ceiling has to be adopted. This second interesting technique of aerodynamic modelling of open-channel flow requires further elucidation of similarity criteria, which is best discussed together with the general discussion of physical modelling of open-channel flow in Chapter 7.

References

Ackers, P. (1987), Scale models – examples of how, why and when – with some ifs and buts. In *Topics in Hydraulic Modelling, Proceedings XXII Congress IAHR*, pp. 1–15 (eds J. A. Cunge and P. Ackers), Lausanne IAHR, Madrid.

ASCE (1982), Modelling hydraulic phenomena – a glossary of terms (ASCE Task Committee on Glossary of Hydraulic Modelling Terms), *Proc. ASCE J. Hydraulics. Div.*, 108(HY 7), pp. 845–852.

Barr, D. I. H. (1969), Method of synthesis – basic procedures for the new approach to similitude, *Water Power*, 21, April, 148–153, May, 183–188.

Barr, D. I. H. (1979), Echelon matrices in dimensional analysis, *Int. J. Mech. Eng. Education*, 7(2), 85–89.

Barr, D. I. H. (1983), A survey of procedures for dimensional analysis, *Int. J. Mech. Eng. Education*, 11(3), 147–159.

Ghetti, A. and D'Alpaos, L. (1977), Effect des forces de capillarité et de la viscosité dans les écoulements permanents examinées en modèle physique. In *Proceedings of 17th IAHR Congress*, Baden-Baden, pp. 389–396.

Hálek, V. (1965), *Methods of Analogy in Hydraulics, Hydraulic Research No. 3* (in Czech), Státní nakladatelství technické literatury, Prague.

Hay, D. (1988), Model – prototype correlation of hydraulic structures, keynote address. In *Proceedings. Int. Symp. on Model-Prototype Correlation of Hydraulic Structures*, Colorado, pp. 1–24 (ed. P. H. Burgi), ASCE, New York.

Karplus, W. J. and Soroka, W. W. (1959), *Analog Methods – Computations and Simulations*, 2nd Edn, McGraw-Hill, New York.

Kobus, H. (1980), Fundamentals. In *Hydraulic Modelling*, pp. 1–23, (ed. H. Kobus), Verlag Paul Parey/Pitman, Hamburg/London.

Kobus, H. and Koschitzky, H. P. (1991), Local surface aeration at hydraulic structures. In *Air Entrainment in Free-Surface Flows*, pp. 29–53, (ed. I. R. Wood) Hydraulic Structures Manual No. 4, IAHR, A. A. Balkema, Rotterdam.

Munson, B. R., Young, D. F. and Okiishi, T. H. (1998), *Fundamentals of Fluid Mechanics*, 3rd Edn, John Wiley & Sons, New York.

Novak, P. (1984), Scaling factors and scale effects in modelling hydraulic structures, General lecture. In *Proceedings of Symposium on Scale Effects in Modelling Hydraulic Structures*, Paper 0.3, pp. 1–5 (ed. H. Kobus), Technische Akademie, Esslingen.

Novak, P. and Čábelka, J. (1981), *Models in Hydraulic Engineering – Physical Principles and Design Applications*, Pitman, London.

Peakall, J. and Warburton, J. (1996), Surface tension in small hydraulic river models – the significance of the Weber number, *J. Hydrol (New Zealand)*, 35(2), 199–212.

Ranga Raju, K. G. and Garde, R. J. (1987), Modelling of vortices and swirling flows. In *Swirling Flow Problems at Intakes*, pp. 77–90 (ed. J. Knauss), Hydraulic Structures Design Manual IAHR, No. 1, A. A. Balkema, Rotterdam.

Schuring, D. J. (1977), *Scale Models in Engineering-Fundamentals and Applications*, Pergamon Press, Oxford.

Sharp, J. J. (1981), *Hydraulic Modelling*, Butterworth, London.

Selected bibliography

Allen, J. (1947), *Scale Models in Hydraulic Engineering*, Longmans, Green & Co. London.

Barenblatt, G. I. (1979), *Similarity, Self-similarity and Intermediate Asymptotics*, Plenum, New York.

Bear, J. (1960), Scales of viscous analogy models for groundwater studies, *Proc. ASCE J. Hydraulics Div.*, 86(HY2), 11–23.

Buckingham, E. (1915), Model experiments and the forms of empirical equations, *Trans. ASME*, 37, 263–296.

Cedergren, H. R. (1977), *Seepage, Drainage and Flownets*, 2nd Edn, John Wiley & Sons, Chichester.

Comolet, R. (1958), *Introduction a l'Analyse Dimensionelle et aux Problèmes de Similitude en Méchanique des Fluides*, Mason et Cie., Paris.

Escande, L. and Camichel, C. (1938), *Similitude Hydrodynamique et Technique des Modèles Reduits*, Publications Scientifiques et Techniques, *127*, Ministère de l'Air, Toulouse.

French, R. H. (1984),*Open Channel Hydraulics*, McGraw Hill, New York.

Ippen, A. T. (1968), Hydraulic scale models. In *Osborne Reynolds and Engineering Science Today* (eds. D. M. McDowell and J. D. Jackson), University Press, Manchester.

Isaacson, E. de St. Q. and Isaacson, M. de St. Q. (1975), *Dimensional Methods in Engineering and Physics*, Edward Arnold, London.

Ivicsics, L. (1975), *Hydraulic Models*, VITUKI, Budapest.

Kline, S. J. (1965), *Similitude and Approximation Theory*, McGraw-Hill, New York.

Langhaar, H. L. (1951), *Dimensional Analysis and Theory of Models*, John Wiley & Sons, New York.

Levi, J. J. (1960), *Modelling of Hydraulic Phenomena* (in Russian), Gosenergoizdat, Moscow.

Pawlowski, J. (1971), *Die Ahnlichkeitstheorie in der Physikalisch-technischen Forschung*, Springer Verlag, Berlin.

Prickett, T. A. (1975), *Modelling Techniques for Groundwater Evaluation, Advances in Hydroscience*, Academic Press, Oxford.

Proudovsky, A. M. (1984), General principles of approximate hydraulic modelling. In *Proceedings of Symposium on Scale Effects in Modelling Hydraulic Structures*, paper 0.3, pp. 1–5 (ed. H. Kobus), Technische Akademie, Esslingen.

Rayleigh, L. (1915), The principles of similitude, *Nature*, 95(2368), 66–68.

Reznjakov, A. B. (1959), *Metod Podobija*, Izdatelstvo Akademii Nauk Kazachskoj SSR, Alma Ata.

Sedov, L. I. (1959), *Similarity and Dimensional Methods in Mechanics*, Academic Press, New York.

Smetana, J. (1957), *Hydraulics* (in Czech), ČSAV, Prague.

Vasco Costa, F. (1982), Considerations of critical velocities in hydraulic modelling. In *Proceedings of the International Conference on the Hydraulic Modelling of Civil Engineering Structures*, Coventry, A.3, BHRA, Cranfield.

Vries, M. de (1986), *Hydraulic Scale Models*, International Institute for Hydraulic and Environmental Engineering, Delft.

Westrich, B. (1980), Air-tunnel models for hydraulic engineering. In *Hydraulic Modelling* (ed. H. Kobus), Paul Parey/Pitman, Hamburg/London.

Yalin, M. S. (1971), *Theory of Hydraulic Models*, Macmillan, London.

Zegzhda, A. P. (1938), *Theory of Similarity and Method of Computation of Hydraulic Models* (in Russian), Gosstrojizdat, Moscow.

Chapter 6

Tools and procedures

6.1 Laboratory installations

6.1.1 General laboratory installations

The planning, design and construction of a hydraulics laboratory – especially of a large one – can be a complicated procedure requiring specialized knowledge.

Covered hydraulic laboratories usually have a closed water circuit; open air installations, where water may be more quickly polluted, work with direct water supply without circulation.

The *closed water circuit* consists of an underground supply reservoir, a pumping station delivering water to high-head reservoirs (tanks), pipes discharging water from the reservoirs through discharge-measuring devices into flumes and models, and return channels delivering water back to the supply reservoir (where the water must be occasionally replaced if polluted).

The overhead tank must be placed sufficiently high above the laboratory floor and is usually fitted with a very long overfall edge to preserve an almost constant water level both during fluctuating water supply from the pumps and during changes in the discharge on the models. The overhead tank is usually permanent, built as part of the equipment of a hydraulic laboratory and frequently supplies several models. This solution is economical, but has the disadvantage that flows on separate models working simultaneously and supplied by a joint distribution pipe may influence each other. Therefore, separate tanks, or at least direct supply lines from the tank, are sometimes built for every permanent flume or larger models. It is also possible to supply individual models by pumps directly (without using an overhead tank); in that case a substantial model inlet tank is desirable.

In supply installations *without circulation* models are fed either by gravity or by water pumped from rivers, reservoirs, canals, lakes, etc., with water from the model discharged to the recipient.

Models built in the open should be protected against wind, rain and frost using windbreakers and temporary roofing.

The construction of models and work on them is greatly facilitated by light bridges or cranes. Apart from free areas used for temporary models and forming part of every hydraulic laboratory, a variety of permanent equipment (available commercially or built in-house) may be installed: fixed and tilting flumes, high-pressure tanks, flumes for rating instruments for velocity measurements, tanks for rating flow meters and specialized rigs (see Section 6.1.2).

Fixed hydraulic flumes, usually used for two-dimensional models of hydraulic structures, have vertical parallel lateral glass walls fitted into the reinforced concrete or steel supporting structure of the flume. The horizontal bed of the flume is usually of thick rolled steel sheet metal so that models are easily anchored or openings provided for piezometers. The part of the flume close to the inlet is sometimes higher than the rest to permit modelling of the upstream water level for gates and spillways. At the end of the flume a device for regulating the water level and a settling tank may be installed; a wave generator could be provided at either end. Above the lateral walls are rails for the longitudinal movement of measuring devices. Glass walls permit filming or photographing of the flow in planes parallel to them (and the use of a laser-Doppler anemometer – see Section 6.3.4). The usual dimensions of this type of hydraulic flume are: length 6–20 m, height 0.5–1.0 m (in the raised inlet part up to 2.0 m) and width 0.3–1.2 m. The discharge through the flumes does not usually exceed 200 L/s (according to the dimensions of its cross-section).

Flumes are usually placed on the floor of the laboratory, or occasionally set below its level, and covered by removable boards. Longer hydraulic flumes may be constructed as divisible with the outlet in the centre and feeding from both ends; naturally the whole may also serve as only one flume.

Flumes in which open-channel flow and wave phenomena are investigated must be considerably longer than flumes for two-dimensional models of hydraulic structures. Their length is usually more than a hundred times the depth of water, i.e. 25–100 m and more, and the width should be not less than five times the depth in order to minimize the influence of the lateral walls on the flow.

Hydraulic flumes with adjustable slope (tilting flumes) are used for studying the resistance of channels with various roughness, stability of movable beds, sediment transport, etc. They are mounted on a stiff supporting structure, which may be rotated around a horizontal pivot. The longitudinal slope of a tilting flume may be changed, usually by up to about 5%, with the aid of hydraulic or mechanical jacks, cogwheel segments, suspensions, etc. The width and height of these flumes are most frequently of the order of 0.50 m and 10–40 m, respectively; but due to structural complexities, smaller (although occasionally wider) flumes are often also in use. The horizontal pivot may be in the centre of the flume or, more frequently, near the inlet end to simplify its connection to an inlet tank. The outflow may, in this case, be through a telescopic pipe or chute. For sediment transport studies it is advantageous if the tilting flume is fitted with an independent circuit with an ejector for transporting the sediment back to the flume inlet, or at least a separate circuit for intermittent sediment measurement. Even though adjustable hydraulic flumes are expensive to install, they are essential for any hydraulic laboratory in which open-channel flow and sediment processes are investigated. The costly construction of the tilting flume can be avoided by placing an inclined false bed in a fixed horizontal flume or by levelling a sediment bed to the desired slope by traversing the length of the flume by a device with a regular vertical shift of a blade.

For the study of valves, bottom dam outlets, special energy dissipators, outlet jets, etc., *high-(pressure)-head installations* with independent water circuits are used. These may be a steel tower, 10–20 m high, in which a constant water surface may be maintained and water withdrawn at various heads, a tank placed high above the laboratory level, or a pressure tank of sufficiently large dimensions.

Towing tanks and rating flumes or circular *rating tanks* with stationary water are used mainly for the rating of instruments measuring velocity (current meters, etc.). Above the flumes a carriage with the attached rated instrument moves at even, adjustable and measured speeds between 0.02 and 7 m/s. The flume dimensions must be sufficiently large so that the walls do not influence the function of the rated device; its cross-section is usually almost square and if used for rating field current meters at least 1.5 m by 1.5 m and 100 m or more long. Appreciably smaller dimensions will be sufficient for rating laboratory instrumentation. If the flume is wide enough it may be used also to measure the resistance of towed objects (e.g. models of barges and ships).

Volumetric rating tanks are used for the rating of flow meters (notches, venturimeters, orifices, bends, etc.). They are large watertight tanks, usually underground, with straight perpendicular walls and accurate devices for the measurement of the water level. The relationship between the volume and water level in the tank is established by separate measurements.

6.1.2 Special laboratory installations

Aerated water flow and entrainment of air by a high-velocity flow with free surface are studied using special flumes with an adjustable slope of up to 45° (exceptionally even larger) (Reinauer and Lauber (1996)).

To investigate cavitation phenomena, special *cavitation tunnels* (see Figure 6.2) are used. These are hermetically sealed rigs with their own water circuit. In tunnels used for investigating flow around submerged bodies (turbine or pump impellers, ship propellers, etc.) the necessary low pressure is attained by exhausters reducing air pressure in the space above the inlet/outlet water surface and by placing the test section at a sufficient height above the downstream water level (Figure 6.1(a)). Another type of tank (Figure 6.1(b)) is used for simulating absolute pressure at structures with free surface flow (the Reynolds number should attain a value of around 10^6).

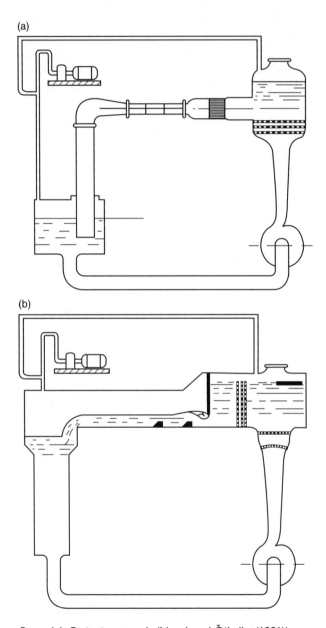

Figure 6.1 Cavitation tunnels (Novak and Čábelka (1981))

Studies of waves and models of coastal engineering works with tides and of estuaries with density-difference effects all require special laboratory equipment.

Wave flumes are often considerably longer, wider and/or deeper than normal laboratory flumes; sometimes they are also connected to a wind tunnel for the study of the interaction of wind effects on waves. *Wave generators* are usually in the form of a wedge, paddle or another body of suitable shape, the motion of which is controlled (through gearboxes) by variable-speed motors. Their controlled and programmed motion (often using sophisticated software) can generate regular, irregular and pseudo-random waves in wave flumes or in three-dimensional models. Irregular waves or waves of variable steepness can also be produced by wind super-imposed on waves produced by the regular movement of a wave generator. For pseudo-random waves the generator reacts to a random electronic signal filtered in accordance with the required energy spectrum of the waves. The pseudo-random sequence may be repeatable so that tests on different designs can be undertaken under the same conditions.

To simulate an oblique rather than frontal approach of waves to a structure or beach, wide *wave basins* with several wave generators, which may be operated with a gradual shift in their movement creating a random wave effect, are used. Wave basins with wave generators for multi-directional waves consisting of computer-controlled steering of segmented wave boards are used to investigate the effects of waves ranging from unidirectional monochromatic waves to mixed sea states with a combination of sea and swell with different principal approach directions and directional distribution functions (Gilbert and Huntington (1991), HR Wallingford (1998)).

Wave generators can also be installed in flumes with a current and the wave pattern superimposed on it. The inlet arrangement in this case is best designed by trial and error to achieve the desired combination of current and wave action in the test section; this should be sufficiently far removed from either end of the flume not to be affected by the boundary conditions.

A *wave absorber* must be provided at the end of the flume opposite the wave generator to prevent disturbing the generated waves and the test section by wave reflection. This usually takes the form of a sloping, rough (possibly porous) beach, but bales of steel wool, expanded metal, plastic fibre, etc., have also been used successfully. The best design is again achieved by undertaking preliminary experiments.

When studying the interaction of oscillatory flow with sediment move-ment or structures, *tunnels with pulsating flow* under pressure can also be used. It is important to realize that the vertical motion of orbiting particles cannot be reproduced correctly in this type of flow without a free water surface. Nevertheless, relevant results may be obtained, particularly when studying oscillatory flow in the boundary layer.

Tidal models, usually models of estuaries, need tide generators placed at their seaward boundary. In principle, the many different generator designs are of two types – gate-operated or pneumatic. The *weir-type tide generator* usually has one or several flap gates raised or lowered according to the tidal movement reproduced on the model and always discharging the excess inflow to waste (or the recirculating system). Modern servo-control techniques permit an easy reproduction of whole cycles of tides by following the movement of an eccentric cam or previously plotted curves of tidal movement.

Figure 6.2 is a schematic diagram of a *pneumatic tide generator*. A tide cam mechanism or automatic curve-reading apparatus or automated steering (1) is connected to a float and float-operated pot (2) through a comparing, stabilizing and amplifying circuit (3); the output from this circuit operates an air-control valve (6) through a servo-amplifier (4) and servomotor (5), which regulates the pressure in the tank (7) provided with an air bleed valve (8) and connected to the suction side of a fan (9).

In some cases it is possible to concentrate on the maximum ebb and flood flow rather than reproducing the whole tidal cycle. In this case steady-state conditions may be used and it may even be possible to simulate the whole cycle by means of a series of steady water levels.

In tidal models the reproduction of *density currents* may be important. If freshwater effects are negligible in the prototype then salt water is usually reproduced by using freshwater on the model, as using salt water could be costly and cause corrosion problems. If, however, the density difference and stratification effects are important, then this is controlled on the model by brine injection into the 'seawater' circulation. Salinity distribution in a model may also be maintained by extracting at the model periphery the surface layer of freshwater brought into the estuary by the modelled river. The use of saline water in the model may require a different zero setting for probes for water-level measurement using resistance to earth (see Section 6.3.2).

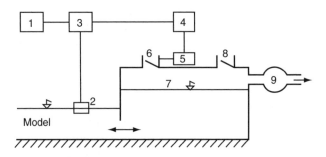

Figure 6.2 Tide generator (Novak and Čábelka (1981))

Large water areas simulated by, for example, distorted tidal models sometimes require the simulation of the geostrophic acceleration of flow. This can be achieved by using a number of rotating cylinders – *Coriolis simulators* or *Coriolis tops* – placed in the fluid; by utilizing the Magnus effect, the required additional acceleration on the model may thus be achieved (see also Chapter 12).

For research of density currents and turbulent boundary-layer flows, together with wave environment and possibly rainfall simulation, *total environmental simulators* provided with sophisticated instrumentation have been developed.

Model research of *ice phenomena* and their effects must be undertaken in special laboratories adapted for the investigation of the *thermal regime* and behaviour of rivers, reservoirs, lakes, etc., during frost, the study of measures for the elimination or reduction of difficulties caused by ice during the operation of hydraulic structures, and the study of the *structure and physical properties of ice* of various types. Investigations are usually carried out in covered and insulated concrete circulating flumes of rectangular cross-section and with a sufficiently long, straight working section fitted with observation windows. Water flowing in the flume is cooled by air (temperature $-10\,°C$ to $-20\,°C$), which is driven by a ventilator into the space between the water level and the ceiling of the flume. Special tanks placed in insulated rooms fitted with effective cooling devices for temperatures from $-15\,°C$ to $-25\,°C$ and smaller cooling cells in which the temperature may be reduced to $-30\,°C$ or even $-60\,°C$ are also used.

For debris-flow hazard assessment rheological models are required (see also Section 8.5.1). To study *debris flow* in the laboratory special equipment is sometimes used (Kaitna and Rickenmann (2007)) consisting of a rotating drum that allows surges of materials ranging from viscous slurries to granular flows to be generated and observed.

6.2 Physical models – types, construction, materials

The following text provides a brief outline of physical models. For further details, see, for example, Novak and Čábelka (1981).

6.2.1 Main types of physical models and their placing in the laboratory

Scale models in hydraulic engineering fall into two main groups: river and coastal engineering (including estuarine) models, and models of hydraulic structures. Combinations of the two types and various models of a special nature (e.g. aerodynamic models) may also be used.

River and estuary models are three-dimensional models of river reaches with or without hydraulic structures. Geometric similarity is often not possible, and thus models are usually built with distorted scales: inclined, narrowed, widened or, most frequently, as vertically distorted models with fixed or movable beds. Models with a fixed bed are used mainly for studies of water levels and flow patterns without investigations of sediment transport and local scour. Models with movable beds are used if sediment transport, scour and deposition are involved.

Models of structures (see Chapter 13) are usually geometrically similar to the prototype and can be three- or two-dimensional (sectional). Three-dimensional models are used to study complicated flow in order to find the hydraulically and economically most suitable solution, either for the whole layout or for individual parts of the structure. Two-dimensional models are used to investigate flow over spillways, stilling basins, etc., where the flow is either completely, or at least approximately, identical in all parallel (vertical) planes. A combination of three-dimensional models and two-dimensional sections of the main part of the structure is often used.

Models of rivers or hydraulic structures are placed either in specially constructed *temporary tanks* or watertight flumes on the laboratory floor or on a raised platform (of prefabricated slabs placed on steel beams supported by brick columns), or in *permanent hydraulic flumes* with perpendicular side walls (see Section 6.1.1). Temporary flumes have the advantage over permanent ones in that their walls may be suitably adapted to the outlines of the modelled river reach (reducing to a minimum both the material and space needed for the model) and provided with rails for instrument carriages. On the other hand, fixed facilities may provide a more cost-effective arrangement for a number of studies. A special laboratory space may be set aside for aerodynamic models.

Some 'semi-permanent' large flumes are used for special river and flood studies. For example, the EPSRC flood channel facility at HR Wallingford is a channel for studying the interaction between flows in a river channel and the flood plain (Knight and Sellin (1987)). The channel is 56 m long, 10 m wide, with a discharge capacity of up to 1.1 m³/s, and it has been used for a managed research programme over 15 years.

The following constraints apply to the choice of model type, its scales and its placing within the hydraulic laboratory:

(a) The limited space available in the laboratory influences the choice of the horizontal model scale.
(b) The model positioning must allow for its economic construction and operation, and for the installation of all auxiliary and measuring devices.
(c) The flow circulation must be resolved; this means checking whether it is possible to connect the inlet to the model to a sufficiently large laboratory distribution main without its operation being unduly influenced by

other models attached to the same line, and selecting a suitable method of measuring the model discharge. The outlet from the model to the laboratory sump should be as short as possible.

(d) Recycling has to be considered for experiments with sediment.

(e) Possible changes in the construction of the model must be borne in mind in the choice of the model site, method of construction and material, as well as access for transport of material, measurements, photographs, etc.

6.2.2 Inlets and outlets of models

The feeding installations of flumes and models in the laboratory normally include supply pipes (from pumps or overhead tank – see Section 6.1.1) with regulating valves, inlet tanks, discharge measurement and sometimes also other special devices (e.g. for sediment supply).

The *inlet tank*, separated from the model proper by a flow straightener, screen, damping grille or filter, must be sufficiently large. Various dampers and baffles or sharp-crested weirs are used to spread the concentrated flow at the inlet over the entire entry section of the model and/or direct it to simulate the influence of the upstream reach. Baffles are usually made either of wood, horsehair matting or perforated sheet metal, bricks or concrete blocks. Filters usually consist of a wire-mesh frame filled with gravel, a suitable choice of gravel grading ensuring the necessary velocity and flow distribution in the inlet section of the model.

The model *outlet* is used only exceptionally for discharge measurement (as the flow is usually directed from the outlet by a return channel into the supply reservoir (sump) of the laboratory with a minimum of head loss). It is provided with a device for setting or regulating, either manually or automatically, the downstream water level on the model according to the stage–discharge relationship. Manual control is suitable for experiments with a constant discharge, while automatic control systems are used for experiments with variable discharge or water level. A tailgate of the simple overflow type, more complicated venetian-blind-type gates, or various other gates and valves may be used for the downstream control of water levels.

Models with sediment transport and a sediment feeding device at the inlet may have a settling tank or sediment separator at the outlet. Settling tanks are used on mobile-bed models without independent sediment circulation; the sediment is usually removed from the tank by hand or conveyor belt. In models with automatic bed-load circulation, the water is separated from the sediment transported from the model in a separator, with sediment falling into a small container from which it can be transported hydraulically by an ejector and returned back to the model inlet, where the excess water is again separated.

6.2.3 Materials and model construction

The *material* used varies according to the type and aim of the model study. *Wood* (pinewood, cypress, larch, hardwood and marine plywood) is easily shaped; its great disadvantages are its non-transparency and its changes in volume and shape, due to warping and swelling, which are difficult to prevent fully even with good impregnation of a surface coating, especially if the model is operated for a long time. It is suitable mainly for the production of some less important parts where minor deformations are acceptable, for templates, temporary flumes, the shuttering of concrete parts, etc.

Metals, especially steel and non-ferrous metals, are used as sheets, plates, pipes, angles, bars or casts. Steel is used for the construction of flumes and for pipes, supports, simple model gates, etc. Non-ferrous metals are suitable for more sensitive parts of accurate models and measuring devices, model spillway surfaces, piezometers, etc., where corrosion or a protective coating would be harmful. Metals are easily formed, machined, cut, welded or soldered, and retain their shape and dimensions. Their non-transparency is a drawback.

Plastics can be shaped using various techniques and are well suited for pipes and thin-walled parts of all types of models, templates, models of structures, etc. Glass-fibre models are formed in a cold state by placing alternate layers of glass fibres and binding material on a previously prepared surface. If transparent walls are required, perspex, which can be shaped and pressed when heated, is used. The advantage common to all plastics is their malleability and ease of machining, which is similar to that of hardwood. They have the advantage over wood that they do not significantly change in volume or shape in water or damp conditions, and they are not subject to corrosion.

Complicated details of models requiring great accuracy are sometimes made of *wax or paraffin*, which are easily workable and do not deform in water. However, these materials are brittle, non-transparent and sensitive to temperature. The material is cast in forms and worked after cooling. With suitable additives that increase its brittleness, such material may be used to simulate the ice cover on rivers or canals and the movement of ice flows.

Cement-based mortar is very suitable for the construction of models of some structures and river reaches, as it is firm and, after hardening, changes neither in volume nor shape. For models of structures, certain parts can be fabricated separately and then assembled. In fixed-bed models of rivers and coastal regions a thin layer of mortar placed on a gravel–sand or brick foundation forms the firm bottom and banks. The mortar mixture is shaped with the aid of templates made of wood, metal or plastic. Permanent flumes can also be made of reinforced concrete. A disadvantage of concrete and mortar is their relatively great strength so that, after hardening, additional small changes are not easily made. This disadvantage can be partly overcome by using masonry saws or a special mixture that has the properties of cement mortar but hardens slowly.

Fine and slowly setting *plaster* can replace cement for various parts of models. It is more brittle and softer than concrete but can be worked after setting. Its instability in water is a disadvantage, but this can be reduced by the use of various additives (e.g. saltpetre). Hardened plaster does not change in volume or shape under damp conditions, but it does absorb water and should, therefore, be coated with paint.

Glass is suitable for the construction of transparent pipes, side walls of hydraulic flumes, inspection manholes and windows, small tanks, piezometer tubes, etc. For complicated shapes or easily broken parts it is often replaced by perspex.

Asphalt (bitumen) is used as a sealing material, mainly in river channels. *Rubber* is used as packing between pipe flanges and sections of metal flumes and models, for glass walls in hydraulic flumes, etc. *Putty, hemp,* and small plastic or rubber hose are also suitable for packing joints. For larger areas, sheets of plastic, varnished cloth or various types of coating are used as a sealing material.

Sediment in movable-bed models is simulated by sand, gravel, coal, fragmented hollow bricks or roof tiles, pumice, granular Bakelite or other plastics, treated hardwood, sawdust, etc., depending on the size and specific weight of the sediment required. For scour experiments it is sometimes useful to reduce the cohesion of the material used by, for example, treating it with quicklime. For studies of scour in rocks with steep sides in the plunge pools, low-cement-content mixtures of kaolin clay, chalk and sand may be used.

The *construction of models* is based on detailed drawings done to a scale chosen according to the complexity of the various parts and requirements for accuracy. For river models with a fixed bed, not only the general layout but also the channel cross-sections are drawn, the latter often in the actual scale of the model; templates cut according to these drawings may be divided into several parts. Sometimes it may be advantageous to use metal wire or plastic bars bent into the required shape on movable pegs instead of templates.

For river models with a movable bed a fixed bottom is constructed sufficiently below the bed and the movable part is modelled on it using removable templates; sometimes it is more suitable to model the bed using 'negative cross-sections' (i.e. inverted templates). Conventional surveying techniques are used for the setting out of the model and/or its temporary flume. The various parts and sections of the model must be easily positioned and correctly aligned; adjusting screws with metal plates as supports may be used. The construction of model structures in the flume consists of mounting its previously manufactured parts or templates and completing the masonry and modelling work. Hollow objects and those made of light materials must either be anchored to the bottom of the flume or sufficiently loaded to prevent floating when the flume has been filled.

To ensure similarity, the roughness of the model surfaces must be adjusted (smoothed or roughened) by painting (or by coating with glue) and, if required, sprinkling with graded sand or gravel of suitable size. Paving can be simulated by forming the concrete with a roller. Vegetation in the flood plains of river models is usually modelled later as part of the model-validating tests, using wood, plastics, wire mesh, expanded metal, horsehair matting, stones, etc.

6.3 Laboratory measuring methods and instrumentation

6.3.1 General

The instrumentation used is based mainly on mechanical, electrical or optical devices and methods; thermal and acoustic principles and methods are also important in measuring flow phenomena. A qualitative change from mainly mechanical instrumentation is due to the widespread use of electronics and modern measuring methods and techniques. Radio-isotopes, transistors, microchip technology, video, lasers, etc., have accelerated the development of new and more accurate measuring devices, and the use of computers, microprocessors and other data-processing techniques has become part of the operation of almost any hydraulic laboratory. Frequently, large models are controlled from special control rooms to which the readings from the model instrumentation are also transmitted for processing. When using a laser–Doppler anemometer, radiation techniques, etc., safety regulations have to be borne in mind during the design of experiments.

According to its purpose, hydraulic laboratory instrumentation may be roughly divided into devices for the measurement of water levels (steady or fluctuating) and movable bed levels (both below water and after termination of the experiment on the dry model), discharge, velocity (and its fluctuations), hydrodynamic pressures (and their fluctuation) and flow of mixtures (sediment transport and suspensions carried by the liquid, air entrained by the flow, etc.). Some of this instrumentation is commercially available, but often it is developed in-house, particularly in large laboratories, with plenty of scope for innovation.

Despite rapidly developing instrumentation techniques it is not necessary to use sophisticated measuring devices where a simple instrument will do. In the selection of instrumentation, the required accuracy is decisive, but cost and ease of operation must also be borne in mind. Thus, the choice of instrumentation is an inseparable part of the research methodology and the formulation of project objectives.

The following paragraphs give only a very brief summary of some of the more basic and frequently used methods and instrumentation. For a comprehensive text on laboratory measurement in fluid mechanics, including the treatment of measurement uncertainty and signal conditioning, discretization and analysis, see Goldstein (1996) and Tavoularis (2005). For flow measurement techniques in the field, see Herschy (1999).

6.3.2 Measurement of water levels and bed formation

6.3.2.1 Measurement of steady water levels

The simplest and most commonly used instrument for the measurement of steady water levels is a *point gauge*; the required information is read off the scale with an accuracy up to 0.1 mm (to facilitate its reading the gauge may also have a battery-operated liquid crystal digital display). For hydraulic flumes, gauges may be fitted on carriages, with vertical and transverse motion of the gauge. Indication of the correctly set gauge position is by a needle point, which may be straight (lowering of the gauge to the water level to be measured) or a hook with the point directed upwards (gauge approaching the water level from below). Contact of the point with the water level may be observed by eye or indicated electronically. Sometimes the gauge is fitted with an 'accuracy fork' (i.e. two points set apart at the maximum permitted error in the position of the measured water level). If both points are fitted with an electrical indicator the gauge must be set (or the water level regulated) so that always only one indicator lights up.

Water levels in river models can be read in glass or perspex containers connected to the model bed by flexible tubing.

When reading the water level in a reservoir, where an accuracy of about 2 mm will suffice, ordinary water gauges can be used.

6.3.2.2 Measurement of fluctuating water levels

Slowly fluctuating water levels may be measured in the same way as steady ones; however, the gauge would have to be constantly observed, and thus registration devices are most frequently used. Commercially produced, mechanically recording *float gauges* are usually unsuitable for use on models, and laboratories produce their own, often with a simple pen recorder attached. On large models, float gauges with transmitters and data registration from several dispersed points on a single recorder are used. HR Wallingford has developed an electronic float gauge based on measuring the distance between the head of a transducer and the magnetic field produced by a magnet mounted inside the float.

Slowly fluctuating water levels may also be measured using a gauge with a vibrating tip intermittently in contact with the water surface and automatically adjusted by a servo-mechanism (e.g. the *pointe vibrante* gauge developed by Neypric in Grenoble). HR Wallingford has developed a *water-level transmitter*, which has a sharp-pointed vertical probe driven by a servo-system that maintains the probe tip at a depth of about 0.1 mm below the surface. The impedance between the probe and a remote electrode forms one arm of a bridge network, which is balanced when the probe tip is at the correct depth. Any change in water level varies the immersion depth of the probe, causing an error signal to be produced from the bridge. The amplified signal applied to a servo-motor drives the probe to follow the water surface. The slider of a potentiometer coupled to the probe picks off a voltage proportional to the water level, which can be recorded in digital or analogue form. The accuracy of measurement is about 0.2–0.5 mm.

To measure *rapidly fluctuating water levels*, instruments with *capacitance gauges* are often used, where the length of the submerged part of the gauge is the measure of the instantaneous position of the water level. The gauge diameter should be less than 1 mm.

Twin-wire stainless-steel probes may also be used; the electrical conductivity between the wires is linearly related to their depth of submersion, and thus to the wave height. The accuracy of the instrument is about 0.5 mm. Another possibility is a resistance gauge, with either low (a thin wire) or high (about 50,000 ohm) resistance (non-conducting rod with a thin metal band).

For the measurement of *directional wave characteristics* in wave basins, Delft Hydraulics developed a special directional wave gauge; its principle is based on cross-correlation analysis of three mutually orthogonal components at one location. The gauge combines the two horizontal velocity components and the free-surface displacement in order to determine the directional-spreading distribution function. The output of the processing program consists of wave heights, the energy-density spectrum and a directional-spreading parameter as a function of frequency.

The position of the water level may also be ascertained by measuring the *hydrostatic pressure* (see Section 6.3.5). It may also be ascertained *photographically* by means of a camera placed either at an angle above the model or at the water level; photographs are taken through the glass wall of the model or flume on which a scale or a coordinate net is drawn. Strongly fluctuating water levels are best recorded using high-speed cameras. *Stereophotogrammetry* may be used to record the entire water level surface with local changes (near piers, etc.). The 'starry sky' procedure records wave orbits in models of harbours; it consists of photographing, with an exposure of one wave period, the reflection by the water surface of numerous points of light on the ceiling above the model.

The trace of the light reflection provides a good measure of the horizontal water-surface movements, which affect the movement of ships about their moorings.

6.3.2.3 Measurement of bed levels

After *draining the model*, the bed level may be measured using a normal point gauge. In three-dimensional models the point gauge is used to find contour lines on the movable bed; simultaneously, a white thread is placed on the contour line and the whole is then recorded photographically. Stereophotogrammetric methods may be used to measure the bed formation on larger models.

The bed levels may also be measured using profilographs, which are obtained by means of a wheel moving over the bed, fixed to a well-balanced arm of a recorder that makes a pantograph trace of the measured section to the required scale.

Measuring the bed level *under flowing water* during the experiment is relatively complicated because observation of the instruments is hindered by light diffraction. In addition, the approach of the tip of a gauge to the movable bed may disturb the scour formation. In simple experiments, thin rods fixed to a normal gauge can be used. For more accurate and sophisticated measurements, an optical instrument *without bed contact* is best. It consists of a fork fitted to a normal gauge, with both ends carrying a small light source and bent towards each other to form a right angle. The gauge is shifted vertically until the reflection of the light emitted by both sources, and observed through a tube close beneath the water surface, merges into one point on the movable bed. The device is then at a certain constant distance from that part of the bed surface.

In *bed-profile transmitters* (bed profilers) a beam of infrared radiation is reflected by the bed; alternatively, where the bed material is of very low reflectivity, a conductivity probe may be used. Both instruments maintain a constant distance (about 15 mm) above the bed.

An example of a *two-dimensional profiling system* that can be used with a variety of bed materials, both above and below water, is a profiler developed by HR Wallingford. The probe consists of a steel tube with a rack engaging the gear wheel of a servo-motor in the instrument carriage, which drives the probe up and down; at the bottom of the probe is a very lightweight 'finger' sensor, the position of which is measured optically. The whole system is computer controlled, with two screen displays – one to set the required parameter for the measurement and the second to display the position of the probe and the profile measurements. The system works with a resolution of ± 1 mm horizontally and ± 0.5 mm vertically.

6.3.3 Measurement and regulation of discharge

6.3.3.1 Measurement of discharge

The most accurate method of determining the discharge is to measure the liquid *weight or volume* delivered over a certain time into specially rated tanks, but this method is rarely used during model investigations.

Various *weirs and notches* built into fixed or mobile measuring tanks are the most common laboratory discharge measurement devices. Most frequently, a right-angled Thomson triangular (V-notch) weir is used; for larger discharges a rectangular or compound weir is more suitable, and for very small discharges a very narrow rectangular (slit) weir, a triangular weir with a small angle, or a proportional weir (with a linear relationship between the discharge and the overfall head) is used. The head above the weir crest is measured using a point gauge, usually situated in a small well connected to the measuring tank. The required tank sizes, the upstream position of the gauge for measuring the head above the crest, and details of the weir plates with appropriate discharge coefficients are given in standard specifications. If these are not followed, the tank (with inlet and baffles, etc.) must be rated before use and the rating checked from time to time. Measuring weirs need a spacious tank and cause loss of head, but they are simple, reliable and accurate.

Discharge can be measured by using devices based on the principle of *contracting the flow* and measuring the resulting pressure differences with a differential manometer. The most common instrumentation of this type can be a *venturimeter*, *orifices* or *nozzles* fitted into the pipe supplying the model.

Venturimeters cause relatively low head loss, but show low sensitivity for relatively small discharges (discharge is directly proportional to the square root of the pressure difference) and must be placed in a long straight pipe (if not rated at the place of use). According to specifications, for an orifice in a pipe of diameter D a straight length of at least $20D$ upstream and $5D$ downstream is required. The disadvantage of unequal sensitivity for greater ranges of discharges can be eliminated by installing batteries of parallel venturimeters of various diameters.

Nozzles and orifices are shorter than venturimeters but cause greater pressure losses. A shortened venturimeter with the contraction in the shape of a standard nozzle and the downstream expansion to a smaller pipe diameter than the original one combines the advantages of both. In all these devices great attention must be paid to pressure tappings; usually a number of openings (or a narrow slit) are connected to an annulus to which the manometer is fitted.

The *venturi-flume or the Parshall flume* (where a hydraulic jump is created by contraction) use the venturimeter principle for open-channel flow.

Bend discharge meters utilize the differences in hydrodynamic pressures on the concave and convex walls of (usually 90°) pipe bends. These meters (as well as venturimeters, nozzles and orifices, if placing and construction does not agree with specifications) must always be rated, and whenever possible the rating should be done under the actual operating conditions.

Watermeters are used only occasionally in a hydraulic laboratory. Small discharges are sometimes measured by means of rotating discharge meters (rotameters with a rotating float in a divergent glass pipe).

Induction (electromagnetic) discharge meters are based on the principle that, during the flow of conductive liquids between the poles of a magnet, an electromotive force arises that is directly proportional to the vector product of the intensity of the magnetic field and the velocity of movement of the conductor; accuracies of a few per cent are achievable by commercial instruments but *in-situ* calibration is desirable.

Under laboratory conditions, discharge may also be determined using *ultrasonic flow meters*, utilizing high-frequency pressure waves in Doppler and time-of-flight meters in pipes, and by the *dilution method*, especially in open-channel flow. In the latter, for a certain time a known and constant quantity of electrolyte (dye, isotopes) is added to measured flow in one cross-section, and the conductivity (colour, radiation intensity) recorded in another cross-section sufficiently distant from the dosing section for the measured and dosed liquid to mix completely. The discharge is then assessed on the basis of the mixing law. Alternatively, the salt-velocity method may be applied, during which the electrolyte is added either once in a slug or periodically. Dilution methods are used only exceptionally in the laboratory, and mostly only when other aims are pursued as well.

For methods for measuring velocities, from which discharge may be established, see Section 6.3.4.

6.3.3.2 Regulation of discharge

When modelling rivers and estuaries it may be necessary to change the inflow to the model with time. In this case, metal templates or *cams* simulating, for example, a modelled flood wave and moving at a constant speed given by the time scale can be used. The cams control the movement of a waste weir in the inlet measuring tank, thus changing the head on the fixed measuring weir.

Another method of discharge regulation involves maintaining a constant water level in the measuring tank (which is ensured by fixed weirs spilling to waste) by regulating the inflow to the model by the movement of the actual discharge weir (usually of the proportional type), which is controlled in the same way as above. The motor driving the inlet weir can also be remotely controlled *electronically* (e.g. simply by signals from a punched tape with a light source and a photo cell). The movement of the template or signals from an electronic control may also be used to control the movement of a valve, the characteristics of which are ascertained by independent measurements.

Software exists for *variable-discharge pumped systems* and control of hot water and/or saline injection (e.g. when simulating cooling water outfalls). Servo-control devices with feedback from a sensor to the control device may also be used, as appropriate.

6.3.4 Measurement of flow velocity

A simple method of measuring local velocities is by determining the velocity from the velocity head with the aid of various types of *Pitot tubes*. The magnitude of the velocity head h (for small velocities measured on an inclined or differential manometer) is found from the pressure difference at two or more points on the measuring device submerged in the flow. The most common type of device is the standard *Prandtl tube*, a Pitot tube shaped so that the coefficient in the equation $v = c\sqrt{(2gh)}$ is $c = 1$. The Prandtl tube measures velocity correctly only when used against the direction of flow with a deviation of up to $15°$. Spherical and cylindrical probes and gauges consisting of four or five Pitot tubes turned in various directions and suitable for measuring the velocity vector are also used. Pitot tubes, as well as other probes if differing from thoroughly tested types, must be rated before use and used only within the rated range. Probes with specially tested attachments are required for the measurement of high velocities (Schwalt and Hager (1993)).

A widely used velocity-measuring device working on a mechanical principle is the *current meter*. For model investigations it is important that its propeller should be as small and as sensitive as possible, two opposing demands that require careful construction and a contactless impulse transmitter. *Miniature current meters* of 4–12 mm diameter for use in the velocity range 0.03–3 m/s have been developed. For most laboratory types the method of registration permitting the measurement of both the mean and 'instantaneous' local velocities is based on the principle that, during the propeller rotation, the distance between a fixed electrode and the rotating propeller blade (or the cogs of a collector) changes, causing changes in electrical properties; the resulting impulses are recorded on an analogue or digital counter and/or plotter.

A small spring-loaded disc deflected by the pressure produced by the flowing liquid (i.e. a *miniature vane flow meter*) may also be used for velocity measurement. Accurate rating of the device before use is essential.

A general picture of the flow, the direction and magnitude of the velocities, and the turbulence characteristics may be gained by *flow-visualization* techniques.

Simple *photographic* methods are often quite cost-effective. The direction of flow is easily determined by means of cotton threads on a thin wire or by introducing grains of pumice, aluminium powder, etc., into the flow; in aerodynamic models burning sawdust or sparks are used as tracers of flow patterns. For two-dimensional flow it is also possible to judge the velocity at various points from the length of the recorded trace and the exposure time. In models with a free water surface, small floats carrying a light source may be used; the model is placed in a darkened room and the paths of the floats are recorded on the photographic plate by interrupted exposure (the exposure time and the intervals between exposures usually being 1 second). After the experiment the model is lit up and photographed on the same plate.

In the rather laborious *cinematographic* method, the flow pattern is usually made visible with the aid of an emulsion of vaseline oil dissolved in chlorobenzol and dyed with white oil colour. The mixture forms spherical particles of diameter of about 1–2 mm. From the difference of the coordinates of individual spheres on successive frames, recorded using a high-speed camera, the instantaneous velocity and other required parameters may be calculated. If the flow is recorded in two directions simultaneously (e.g. with the aid of a mirror inclined at 45°) then all velocity components may be recorded. The film record is usually assessed with using the corresponding reduction coefficients of length and depth and from a time scale determined from the film record of a light source connected to an alternating current of known frequency. High-speed videos, where the timing between frames is known, and digital image processing are also used.

Associated with flow-visualization techniques is the rapidly developing technique of *particle image velocimetry* (PIV), which uses a laser beam converted to a planar sheet of light and a camera producing an image of part of the illuminated flow. Three-dimensional flow vectors can be measured by recording particle images on two cameras.

A widespread method of measuring flow velocity and turbulence parameters is the *hot-wire and hot-film anemometry* (see e.g. Resch (1970)). Probes with an electrically heated thin wire or, preferably (for use in water), a heated metal film supported on a ceramic subplate are used on the principle that the rate of heat loss of the sensor heated to a higher temperature than that of the ambient fluid is proportional to the velocity of the medium at the point of measurement. This cooling results in a change in the resistance of the heated element, which becomes an indirect measure of the velocity. In a *constant-current anemometer* the change in resistance causes a change in voltage, which is measured and recorded. In a *constant-temperature anemometer* a feedback circuit keeps the temperature constant and the fluctuating current gives rise to fluctuating voltage, which is again the output measure. The instrument may be provided with a linearizer, which modifies the amplified output from the anemometer so that it is proportional to velocity. The linearized voltage is then recorded on an analogue recorder or sampled at set intervals and stored in digital form on tape or diskettes.

Thermistors can be used to measure very small velocities (several millimetres per second). Their disadvantages compared with a hot-wire probe are the difficult compensation of temperature influences and the large time constant.

While hot-wire and film anemometry require clean water free of suspended particles, the opposite is the case for instrumentation using a *laser beam* or *sound* for measuring local velocity and turbulence. One of the most widely used modern research tools is the *laser–Doppler anemometer (LDA)* (see e.g. Durst *et al.* (1976)), which is based on the fact that the crossing of two coherent light beams causes an interference pattern, which is displaced by the movement of scattered particles suspended in the fluid flow. The great advantage of LDA is that it is non-intrusive, i.e. there is no interference with the flow field (in fixed systems only). The anemometer consists of a light source (laser), a beam splitter and focusing lens (transmission optics), light-collecting optics, a photo-detector and a signal processor. A frequency shift of the light in one of the laser beams allows the measurement of both positive and negative values of the velocity component. The basic version measures velocities and turbulence in one direction only; more advanced instrumentation is available for simultaneous measurement in two and three directions. The LDA can be used to study the flow in glass-walled flumes, with all the instrumentation outside the flume, or on three-dimensional models, where an optic probe is placed in the flow (in this case the measurement point is typically 80–100 mm from the probe). The flow will probably need to be seeded, especially for a fibre-optic link system, which operates in a back-scatter mode, where the signals are weaker than for forward scatter, which can be used in a fixed LDA system. A certain amount of experimentation is needed to establish the correct seeding with neutrally buoyant particles; titanium dioxide particles approximately 1–2 μm in diameter are commonly used. A single-component system can measure turbulence intensity, a two-component system can measure turbulence intensities in two directions and one Reynolds stress term, and a three-component system measures all the Reynolds stress components (see Section 4.3.3).

The acoustic Doppler velocimeter (ADV) (e.g. Garcia *et al.* (2005), Muste *et al.* (2007), Nortek (1996)) uses acoustic sensing techniques to measure, with little disturbance, three velocity components of seeding particles in the flow as they pass through a remote sampling volume. The instrument has three main components: the measuring probe, the conditioning probe and the processing module. The acoustic sensor has three receiving transducers mounted on short arms around the transmitting transducer at 120° azimuth intervals and intercepting the transmit beam at 50–100 mm below the sensor. The cylindrical measuring volume is 3–9 mm long and has a diameter of about 6 mm; the ADV is able to record velocities up to 2.5 m/s at a sampling rate of about 25 Hz. The fairly robust and relatively inexpensive ADV can be used with different orientations of the measuring head to sample the flow up to 5 mm from a solid boundary. The accuracy quoted in the literature is 0.25% ±2.5 mm/s. The calibration (and operation) of the probe should be checked at regular intervals using the built-in checking system.

The *electromagnetic flow meter* (see Section 6.3.3) has been developed as a velocity meter for use both in the field and the laboratory. Two orthogonal pairs of electrodes in a single sensor give two-axis velocity measurements. In the laboratory version, discus and/or spherical sensors are available in sizes of 32 and 20 mm. The system consists of a sensor and electronics that drive the coil detecting signals and convert them into an analogue or digital output. The manufacturers quote an accuracy for the mean velocities of 1% ±5 mm/s and lower accuracy for instantaneous readings.

6.3.5 Measurement of hydrodynamic pressure

Hydrodynamic pressure is most frequently measured by means of simple glass or perspex *tube manometers* connected to a *piezometric opening* by rubber or plastic tubing. The manometers are usually placed next to each other on a panel with a grid showing the position of the piezometers so that the value of the pressure head on the model can be read directly. The accuracy of measurement (reading of meniscus) is increased by drawing a line on the panel behind the axis of the glass tube manometer, which should be at least 10–15 mm diameter to reduce the influence of surface tension (for a diameter of 10 mm the capillary elevation of the water in the tube is still about 3 mm) and/or by using inclined manometers. The details of the actual piezometer opening are very important: its axis must be perpendicular to the surface where the pressure is being measured, the orifice diameter must be relatively small (1–1.5 mm), its edge must not be too rounded and there must be no projection into the flow. The surface around the opening should be smooth up to a distance of at least 50 times the diameter of the piezometer.

Apart from the most frequent simple tube manometers, various *fluid manometers, micromanometers and differential manometers* with water, mercury, alcohol, etc., are used.

Membrane or other mechanical manometers, or manographs, are more suited for field work. Under laboratory conditions they are used only for measuring relatively high pressures, where the fluid manometer is less suitable.

Pressure transducers for measuring pressure fluctuations are important for investigations involving negative pressures, especially close to cavitation phenomena, and for the study of pressure fluctuations caused by strongly turbulent flow acting on hydraulic structures; pressure transducers may also be used to measure water depths. Most transducers are based on the action of pressure on an elastic membrane, which must be without (or with only a small) plastic deformation or hysteresis. Deformation of the membrane is most frequently transformed into changes in electrical resistance, capacitance or induction, which are measured (e.g. on an oscillograph). Tensometers can also be used as pressure transducers. All pressure transducers may also be used for measuring water depth.

6.3.6 Measurement of two-phase flow

The measurement of the hydraulic characteristics of mixtures of *solids and liquids* is important in particular in the study of the hydraulic transport of solids through pipelines and for the operation of such systems. The *discharge of the mixture* is measured by one of the methods used for the carrying fluid, most frequently by volumetric or weight measurement. Orifices, nozzles or venturimeters are also used. In homogeneous suspensions of fine particles, greater throttling may be permitted in the measuring section than for thick mixtures or mixtures with coarser particles. Bend meters have been found suitable, as have electromagnetic meters, which do not influence the flow of the mixture.

The *mean velocity of flow* is ascertained by measuring the discharge, or by salt-velocity and chemical-dilution methods. The *velocity of particles* in the flow of a mixture may be measured by the cinematographic method or with the aid of radioisotopes.

The important measurement of the *mixture concentration* is carried out by volumetric or weight measurement (at the delivery end of the pipe), either as continuous measurement, or as local values of the concentration by monitoring and recording the passage of waves or radiation through the mixture. The distribution of the concentration at various points in the flow can be determined by a special probe.

During work on *river models with a movable bed*, various types of sediment-dosing equipment are used (conveyors, screws or vibrators, taking either dry or wet material from a container with an adjustable opening). The grain distribution, geometric characteristics and specific weight of bed load and suspended sediment are usually determined before as well as after the experiment.

When working with a mixture of *air and liquid* two problems are of paramount interest, i.e. the quantity of air passing through the measuring cross-section and the size of the air bubbles. The *quantity of air* drawn into the water flow (e.g. by an aeration pipe) can easily be measured by standard means (e.g. by an orifice placed in the aeration vent). To measure the quantity of air entrained by the liquid in a conduit (air concentration), radiation may be used in the same way as for mixtures of solids and liquid. The measurement is based on the fact that gamma radiation is absorbed in the same way whether it passes through a mixture of two fluids (e.g. water–air) or two independent layers of the same fluid. A radiation source is attached to one side of the conduit and a Geiger counter fixed to the other. If this meter is rated by measuring the absorption of various layers of the liquid at rest (or flowing without aeration), the quantity of air entrained in the form of bubbles by the flow of liquid can be measured directly.

The quantity of air contained in a certain vertical of aerated water flow with a free water surface may be ascertained by measuring the hydrostatic pressure acting on the channel bed and comparing the specific weight of the mixture with the specific weight of liquid alone, or, for a known discharge of liquid without air, by measuring the discharge of the mixture (which can be determined from the depth of the aerated flow and the mean velocity found by integration from point measurements), or by special probes.

The *size of the air bubbles* and their velocity and direction are best measured by cinematographic or photographic methods.

6.4 Mathematical models – tools

Several software packages are available that permit the procedures mentioned in Chapters 2 and 3, and indeed many others, to be performed directly without the need for programming. All that is necessary is to call up the appropriate subroutine and input the necessary parameters; the output will be the required result. For example, the software will produce a spline function approximation from a set of data pairs (x_1, y_1), (x_2, y_2), $\ldots$, (x_n, y_n), while a numerical integration procedure will show that $\int_1^3 e^{-x} \sqrt{1 + 3x^2} \ln (1 + 2x) dx = 1.491938830$ (i.e. to nine decimal places). In addition, as many zeros as required can be found for functions like $f(x) = \tan hx - 2 \sin x$. Matrices with numerical entries can be manipulated to form sums and products, and when the matrices are $n \times n$ their determinants, eigenvalues and eigenvectors can be found, while large systems of linear algebraic equations can be solved. The Runge–Kutta (rk4) procedure and the Runge–Kutta–Fehlberg (rkf45) procedure will automatically integrate initial-value problems over an interval $a \leq x \leq b$ using a specified step length, after which the results can either be printed out or plotted. Furthermore, in the rkf45 procedure, the step length will be adjusted to maintain a prescribed accuracy throughout the interval of integration. These procedures can also be programmed to perform composite operations, such as using the output of one procedure as the input to a different one (e.g. typically, the output from an rkf45 procedure over an interval $a \leq x \leq b$ can be used to find its Fourier series representation over the same interval).

However, in addition to purely numerical software of the type just described, which will perform standard procedures on numerical data, there is another type of software called *computer algebra software*, which performs symbolic operations. Here, a 'symbolic operation' means that if, for example, an expression like $x/(1+x^3)$ is given as an input, the software is capable of simplifying it into its partial fraction form and giving as its output the result

$$\frac{x}{(1+x^3)} = \frac{x+1}{3(x^2+x+1)} - \frac{x}{3(x+1)}, \tag{6.1}$$

where the variable x has been treated as a symbol, and not as a number. Also, when given a function $f(x) = x + 1/x$, the software can expand it as a series about a prescribed point, e.g. $x = 4$ to a prescribed number of terms, so if an expansion up to the term in $(x-4)^3$ is required, the output will be

$$x+\frac{1}{x} = \frac{17}{4}+\frac{15}{36}(x-4)+\frac{1}{64}(x-4)^2 - \frac{1}{256}(x-4)^3 + O\left((x-4)^4\right) \tag{6.2}$$

where the last term indicates the order of the first term to be omitted from the expansion.

The term 'computer algebra software' is slightly misleading, because not only does such software perform symbolic algebraic operations like the ones just described, but it also manipulates matrices (treating their elements as symbols), solves systems of linear algebraic equations and inequalities, as well as performing a very wide range of other symbolic operations (e.g. integration and finding symbolic solutions of differential equations). By way of example, when such software is given the differential equation

$$y'' + 2y' + y = x \sin x, \tag{6.3}$$

it will produce the general solution

$$y(x) = C_1 e^{-x} + C_2 x e^{-x} - \frac{1}{2}x \cos x + \frac{1}{2}\cos x + \frac{1}{2}\sin x, \tag{6.4}$$

where C_1 and C_2 are arbitrary constants, and if it is also given the initial conditions $y(0) = 0$ and $y'(0) = 0$, it will produce the particular solution

$$y(x) = -\frac{1}{2}e^{-x} - \frac{1}{2}xe^{-x} - \frac{1}{2}x \cos x + \frac{1}{2}\cos x + \frac{1}{2}\sin x. \tag{6.5}$$

As an example of matrix multiplication and the expansion of determinants, when symbolic algebra software is given the matrices

$$\mathbf{A} = \begin{bmatrix} a & b \\ c & d \end{bmatrix} \quad \text{and} \quad \mathbf{x} = \begin{bmatrix} x_1 \\ x_2 \end{bmatrix},$$

it can produce algebraic results like

$$\mathbf{Ax} = \begin{bmatrix} ax_1 + bx_2 \\ cx_1 + dx_2 \end{bmatrix}, \quad \mathbf{x}^T\mathbf{Ax} = ax_1^2 + (b+c)x_1x_2 + dx_2^2, \quad \det[\mathbf{A}] = ad - bc$$

and

$$\det[\mathbf{A} - \lambda\mathbf{I}] = a\lambda^2 - (a+d)\lambda + ad - bc.$$

where the elements of $\mathbf{A}$, $\mathbf{x}$ and the parameter λ are all treated as symbols, and not as numbers.

Symbolic algebra software also allows numerical values to be assigned to symbols, and so it can produce numerical results that can be plotted using one of several different coordinate systems, such as Cartesian or polar coordinates. The output from such systems can also be combined with tables, text and numerical plots to produce documents in which the mathematics is displayed in standard printed form.

Of the three types of software mentioned in Chapters 2 and 3, only MAPLE® (a registered trademark of Waterloo Maple Inc.) is an example of purely symbolic algebra software. Its extremely large set of special functions and procedures, coupled with its ease of use, and its ability to be programmed and to allow symbols to be assigned numerical values, makes it extremely versatile and valuable for both symbolic and purely numerical work. MAPLE® has a good graphical output, as can be seen in Figure 3.14, which was produced by MAPLE®, and the software can be used for modelling and simulation. Furthermore, because text and diagrams can be combined in its output, along with mathematical expressions in standard form, it is capable of producing high-quality printed documents. The instruction manuals provided by MAPLE® are useful but not ideal when learning to take advantage of its full potential. Straightforward, and more detailed and very helpful, accounts of the capabilities and use of MAPLE® can be found in the books by Cornil and Testud (2001), Garvan (2002) and Heck (2003).

Initially, the other software mentioned in Chapters 2 and 3, namely MathCAD® and MATLAB®, was designed for accurate and flexible numerical computation with high-quality graphics and text output. They offer a wide range of standard numerical procedures that can be combined to form complex composite procedures. However, as the importance and power of computer algebra software have developed, the last decade has seen each of these software packages enhanced by having the capability of adding to them a large number of computer algebra procedures.

MathCAD® (a registered trademark of MathSoft Inc.) is designed specially for engineering use; it has many specialist add-on software procedures for use in different branches of engineering, and while its computer algebra capabilities are not as extensive as those of MAPLE®, they are more than adequate for most engineering purposes. MathCAD® allows tables to be printed out, and mathematics to be printed in standard form and combined with high-quality graphics and text to produce good-quality documents. The instruction manuals provided for MathCAD® are extensive and explain its many features in detail, while also providing examples of their use.

MATLAB® is somewhat different, because its structure is such that it performs its numerical computations by arranging results in a special way that makes use of matrices. MATLAB® has been designed to give highly accurate numerical results, and it has various supplementary software packages, called Toolboxes, one of which is the Symbolic Algebra Toolbox that gives it symbolic algebra capabilities, while another is the Partial Differential Equation Toolbox that performs very flexible and highly accurate finite-element calculations. Its graphical output is excellent, and the fact that it can be combined with text and equations in standard form enables it to produce documents of a very high quality. The instruction manuals provided for MATLAB® are valuable and extensive, as they list all of its capabilities, but they are not ideal when learning to use MATLAB®. Useful books explaining how to use MATLAB®, together with applications, include those by Biran and Breiner (1999), Knight (2000) and Part-Enander and Sjoberg (1999).

There are other powerful symbolic algebra packages, such as MATH-EMATICA® (a registered trademark of Wolfram Research Inc.), although only MAPLE® and MATLAB® software has been used in connection with Chapters 2 and 3. Chapters 7–13 deal with various computer packages appropriate for the specified applications.

6.5 Procedures during work with models

The following remarks refer mainly to applied research when models are used to solve design problems of engineering works.

Preparatory work for hydraulic modelling consists of two parts, theoretical and practical. In the theoretical preparation, the type of model to be used (physical, mathematical or hybrid), model scales and the degree of schematization are determined, and the content, extent and procedure of the research prepared. For physical models the practical preparation includes the choice of materials, the location, design, construction, link-up of the models with the water circuit of the hydraulic laboratory and instrumentation to be used.

The main aspects of the design of any physical model are the fulfilment of the appropriate criteria of similarity (see Chapter 5), while observing the limiting conditions, and taking into account the required accuracy of the results and the economy of the investigation. The model should be as large as possible for sufficient accuracy of results, but economy requires the smallest possible model that still satisfies the limiting similarity conditions and accuracy of results scaled up to the prototype. Experience gained from modelling similar problems is of great importance.

For mathematical (computational) models a survey of available computational models and software and the choice of the most appropriate one for the given problem – or the need to develop a new model – has to be made.

Field data, including topography, geology, hydrology, morphology and details of vegetation, must be carefully studied prior to any model investigations. The proposed operation of hydraulic structures during various construction stages as well as after completion should be noted.

Model studies should be conducted well ahead of construction, as their results must reach the client in time to be incorporated in the design. Models of construction stages should not be dismantled too soon, as they can contribute substantially to the solution of hydraulic problems encountered during the course of construction in the field.

The *relationship between investigator and client* should be one of partnership, particularly at the conceptual design stage, and close contact should be maintained throughout the investigation.

After agreeing the terms of reference, the researcher proposes the method of investigations, taking into account the existing data (field information), the availability of laboratory space and equipment, and/or of the necessary software, and the cost of the study. The consideration of the required accuracy of the results of investigation is an important element of the decision-making process at this stage.

The calibration and validation of any model is a vital part of the work; e.g. before 'running' river-engineering models their (adjustable) roughness and morphological features have to be calibrated.

In model investigations of engineering works, several alternative solutions suggested by the designing engineer and/or by the researcher are studied. Flow patterns, velocities, discharges, scour and other parameters are observed and measured. After analysis of the preliminary results the most suitable alternative is chosen and tested in detail. The results are processed qualitatively, graphically and/or numerically, but rarely analytically, because they are usually valid only for the investigated case and cannot be generalized. This does not mean, however, that a generally valid result could not be obtained from a series of similar experimental studies (e.g. for scour downstream of stilling basins).

After completing the experimental work and processing the results, a *final report* is prepared. This should include the original aims of the study, basic technical data, the results of a literature survey, the method of research used, a description of the experimental installations, procedure and results, as well as the scientific, technical and economical contribution of the conducted study. The final report of the investigation of a proposed hydraulic structure must contain not only the recommendations for a more effective design, as documented by graphs, photographs and tables, but also an analysis of the possible application of the obtained results to similar cases and the extent of their validity. In well-founded cases, and where technically possible, the final report should also recommend follow-up field investigations and a check of the agreement between the prototype and the model; for this it is advisable to propose the necessary measuring devices to be installed in the field.

References

Biran, A. and Breiner, M. (1999), *MATLAB 5 For Engineers*, 2nd Edn, Addison-Wesley, Reading, MA.

Cornil, J. M. and Testud, P. (2001), *An Introduction to MAPLE V*, Springer, New York.

Durst, F., Melling, A. and Greated, C. A. (1976), *Laser Systems in Flow Measurement*, Plenum, New York.

Garcia, C. M., Cantero, M. I., Nino, Y. and Garcia, M. H. (2005), Turbulence measurements with acoustic Doppler velocimeter, *J. Hydraulic Eng. ASCE*, 131(12), 1062–1073.

Garvan, F. (2002), *The MAPLE Book*, Chapman & Hall/CRC, London/Boca Raton, FL.

Gilbert, G. and Huntington, S. W. (1991), A technique for the generation of shortcrested waves in wave basins, *J. Hydraulic Res.*, 29(6), 789–799.

Goldstein, R. J. (ed.) (1996), *Fluid Mechanics Measurements*, 2nd Edn, Taylor and Francis, Philadelphia, PA.

Heck, A. (2003), *Introduction to Maple*, 3rd Edn, Springer, New York.

Herschy, R. W. (ed.) (1999), *Hydrometry – Principles and Practices*, 2nd Edn, Wiley, Chichester.

HR Wallingford (1998), *Multi-Element Wavemaker for Basins*, pp. 1–15, HR Wallingford, Wallingford.

Kaitna, R. and Rickenmann, D. (2007), A new experimental facility for laboratory debris flow investigation, *J. Hydraulic Res.*, 45(6), 797–810.

Knight, A. (2000), *Basics of MATLAB and Beyond*, Chapman & Hall/CRC, London/Boca Raton, FL.

Knight, D. W. and Sellin, R. H. J. (1987), The SERC flood channel facility, *J. Inst. Water Environ. Manage.*, 1(2), 198–204.

Muste, M. Vermeyen, T., Hotchkiss, R. and Oberg, K. (eds) (2007), Acoustic velocimetry for riverine environments, *J. Hydraulic Eng. ASCE (Special Issue)*, 133(12), 1297–1438.

Nortek (1996), *The Acoustic Doppler Velocimeter (ADV)*, Nortek, Vollen.

Novak, P. and Čábelka, J. (1981), *Models in Hydraulic Engineering – Physical Principles and Design Applications*, Pitman, London.

Part-Enander, E. and Sjoberg, A. (1999), *The MATLAB 5 Handbook*, Addison-Wesley, Reading, MA.

Reinauer, R. and Lauber, G. (1996), Steile Kanale im wasserbaulichen Versuchswesen. *Wasserbau*, 8, 121–124.

Resch, F. J. (1970), Hot film turbulence measurements in water flow, *Proc. ASCE J. Hydraulics Div.*, 96(HY3), 787–800.

Schwalt, M. and Hager, W. H. (1993), Ausmessen von Hochgeschwindigkeits-Strömungen, *Wasseer, Energie, Luft*, 85(7/8), 157–162.

Tavoularis, S. (2005), *Measurements in Fluid Mechanics*, Cambridge University Press, New York.

Chapter 7

Modelling of open-channel systems

7.1 Introduction

The purpose of the present chapter is to introduce the basic concepts and methods used in fixed boundary, open-channel system modelling.

In order to understand fully the developments in the present chapter, a number of basic notions regarding the classification and solution of partial differential equations (PDEs) should be mastered by the reader. It is strongly advised to study Chapter 2, more specifically Section 2.2 on the classification of PDEs, Sections 2.5 and 2.7 that deal with the initial- and boundary-conditions requirement, as well as Section 2.8 for an understanding of the method of characteristics. Reading Sections 2.10 and 2.12, where the shallow-water equations are covered in detail, also provides useful background reading.

The mathematical developments in this chapter make extensive use of linear algebra, a number of the basic aspects of which are recalled in the Appendix to Chapter 2.

Moreover, it is assumed that the reader is aware of the developments presented in Chapter 4; Sections 4.2–4.4 are necessary background reading. Further general reading in hydraulics and computational hydraulics is included in the references in Chapters 2–4. A deeper analysis of the open-channel-flow equations and the kinematic wave equation in non-prismatic channels can be found in Guinot (2008). For background reading on the physical modelling of open-channel flow (Section 7.5), see Chapter 5.

7.2 Mathematical description of open-channel processes

7.2.1 Governing assumptions – notation

Most existing models of open-channel flow are based on the following assumptions.

A1. The longitudinal dimension of the channel is much larger than its transverse (horizontal and vertical) dimensions. Consequently, the flow variables may be assumed to depend only on the longitudinal coordinate and to be homogeneous within a given channel cross-section. If this is not the case, an assumption is made to provide a relationship between the average value and the point variations of the variable over the cross-section (see Section 7.2.2 for such an example).

A2. The curvature of the streamlines is negligible in the horizontal and vertical planes. Consequently, the vertical and transverse components of the acceleration vector are negligible and the pressure field can be assumed to be hydrostatic within a given cross-section.

A3. The local slope of the channel is much smaller than unity. Therefore, the cosine of the angle between the bottom level and the horizontal is close to unity.

A4. The flow is turbulent. Consequently, the regular head loss is assumed to be proportional to the square of the flow velocity (see Sections 4.3.3 and 4.4.1 for more details and the justification of such an assumption on the basis of Reynolds' equations).

A5. In the range of pressure considered, the water can be assumed to be incompressible. Consequently, the (constant) water density does not appear in the final forms of the continuity and momentum equations.

In what follows, the channel is assumed to be prismatic, i.e. the shape of the cross-section is assumed to be constant with x. This has the particular consequence that the relationship between A and Y is identical for all x. The geometry of the channel is illustrated in Figure 7.1. The reader is referred to the list of notation in the preliminary pages of this book for a better understanding of the developments hereafter.

The open-channel-flow equations are derived from the two basic assumptions of conservation of mass and conservation of momentum (or energy).

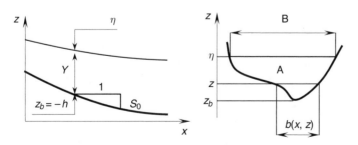

Figure 7.1 Definition of the geometry and variable notation

Background reading can be found in Section 4.3. Steady-state continuity, energy and momentum conservation are covered in Section 4.3.1, while non-uniform and unsteady flow in open channels are dealt with in Section 4.4.3.

7.2.2 The Saint Venant equations

7.2.2.1 The continuity equation

The continuity equation for open-channel flow is derived by writing the conservation of mass for a control volume extending from x to $x + dx$ between times t and $t + dt$. Conservation of mass implies the following equation

$$m(t + dt) - m(t) = F(x) - F(x + dx) \tag{7.1}$$

where $m(t)$ is the mass contained within the control volume at t and $F(x)$ (also known as the flux) is the mass that passes at x between t and $t + dt$. Note that m is the product of the density ρ and the volume of the slice. The volume of the slice is given by the product of $A(t)$ and dx. Consequently, the following equalities hold:

$$m(t) = \rho A(t) dx \tag{7.2a}$$

$$F(x) = \rho Q(x) dt \tag{7.2b}$$

Substituting equations (7.2a) and (7.2b) into equation (7.1) leads to

$$[\rho A(t + dt) - \rho A(t)] dx = [\rho Q(x) - \rho Q(x + dx)] dt \tag{7.3}$$

Noting that $A(t + dt) - A(t) = \partial A / \partial t\, dt$ and $Q(x) - Q(x + dx) = -\partial Q / \partial x\, dx$, simplifying by dt, dx and ρ leads to

$$\frac{\partial A}{\partial t} + \frac{\partial Q}{\partial x} = 0 \tag{7.4}$$

Note that the water density can be eliminated from the equation only because assumption (A5) allows the mass m to be written as the product of the volume $A\, dx$ and the (uniform) density ρ. Conversely, assumption (A5) is accountable for the simplicity of equation (7.2b) that allows the mass flux to be expressed as the product of the volume discharge and the (uniform) density ρ. Also note that, under the assumption of steady-state flow, equation (7.4) simplifies into equation (4.14), presented in Section 4.3.1.1.

7.2.2.2 The momentum equation

The momentum equation is obtained by applying Newton's second law of motion to the same slice of length dx as in Section 7.2.2.1. The momentum balance can be written as

$$M(t + dt) - M(t) = F(x) - F(x + dx) + Sdt \tag{7.5}$$

where $M(t)$ is the momentum of the fluid contained within the slice of length dx at the time t, $F(x)$ (also known as the momentum flux) is the amount of momentum transported by the flow over the time interval dt at the abscissa x, and S is the sum of the external forces exerted on the control volume between t and $t + dt$.

The momentum $M(t)$ is the product of the mass contained within the control volume and the average flow velocity

$$M(t) = \rho AV dx = \rho Q dx \tag{7.6}$$

where V is the average flow velocity over the cross-section. The momentum flux is defined as

$$F(x) = \int_{t}^{t+dt} \int_{A} \rho \tilde{u}^2 dA dt = \rho dt \int_{A} \tilde{u}^2 dA \tag{7.7}$$

where $\tilde{u}$ is the point value of the flow velocity. If the flow velocity is uniform over the entire cross-sectional area, then $\tilde{u} = V$ and equation (7.7) becomes

$$F(x) = \rho dt \int_{A} V^2 dA = \rho dt AV^2 = \rho \frac{Q^2}{A} dt \tag{7.8}$$

In practice, the non-uniform character of the velocity distribution over the cross-section is accounted for by a coefficient β

$$F(x) = \beta \rho \frac{Q^2}{A} dt \tag{7.9}$$

If the flow velocity is uniform over the entire cross-section, $\beta = 1$. Otherwise, $\beta > 1$.

The external forces applied to the control volume are the following:

1 Pressure forces exerted on the upstream and downstream sides of the control volume. The sum of the pressure forces on the upstream and downstream sides of the control volume is

$$P(x) - P(x + dx) = -\frac{\partial P}{\partial x} dx \tag{7.10}$$

Assumption (A2) of a hydrostatic pressure distribution leads to the following expression for the pressure force $P(x)$ exerted on the upstream edge of the control volume:

$$P(x) = \int_A p(x, z) dA = \int_{z_b}^{\eta} p(x, z) b(x, z) dz = \int_{z_b}^{\eta} \rho g(\eta - z) b(x, z) dz \tag{7.11}$$

Noting that the water density ρ can be taken out of the integral owing to assumption (A5), using the assumption of a prismatic cross-section allows equation (7.10) to be simplified as follows. The auxiliary variable $\xi = z - z_b$ is introduced, allowing equation (7.10) to be rewritten as

$$P(x) = \rho g \int_0^Y (Y - \xi) b(x, \xi + z_b) d\xi \tag{7.12}$$

In the case of a prismatic channel, b is a function of ξ alone and the dependence on x vanishes. Equation (7.12) is used in Section 7.2.2.3 to derive the non-conservation form of the equations.

2   Reaction of the bed. Assumption (A2) of negligible vertical accelerations implies that the vertical projection of the external forces exerted on the control volume is negligible. The external forces, the vertical component of which is not negligible, are (i) the weight of the control volume, and (ii) the reaction of the bottom (see Figure 7.2). Since the reaction of the bottom is exerted in the direction orthogonal to the bed of the channel, the following equality holds:

$$R_x = R_z S_0 \tag{7.13}$$

where R_x is the x-component of the bottom reaction. Moreover, as the sum of the vertical components of the forces is zero

$$R_z - mg = 0 \tag{7.14}$$

where g is the gravitational acceleration and R_z is the vertical component of the bed reaction. Consequently, the x-component of the bed reaction is given by

$$R_x = mg S_0 = \rho g A S_0 dx \tag{7.15}$$

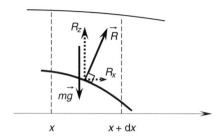

Figure 7.2 Definition of the weight of the control volume and the bed reaction. The dotted arrows indicate the horizontal and vertical components R_x and R_z of the bed reaction

3   Friction forces (see Section 4.4.1 for background reading on friction processes and their mathematical formulation). Friction is exerted in the direction tangential to the bed. From assumption (A3) of a nearly horizontal bed, friction can be assumed to be exerted in the x-direction only. The bed friction force F_b exerted on the control volume is proportional to the length of the control volume. It is usually written in the form

$$F_b = -mgS_e = -\rho g A S_e dx \tag{7.16}$$

where S_e is defined as the slope of the energy line (i.e. the friction-induced head loss per unit distance). Indeed, comparing equations (7.15) and (7.16) leads to the conclusion that S_e has the dimension of a slope. From assumption (A4) of a turbulent-flow regime, S_e is usually assumed to be proportional to the square of the flow velocity. The most widely used formulations are the Chezy, Manning and Strickler friction laws (see also equations (4.62) and (4.63)):

$$S_e = \frac{|V| V}{C^2 R} = \frac{|Q| Q}{A^2 C^2 R} \text{(Chezy)} \tag{7.17a}$$

$$S_e = n^2 \frac{|V| V}{R^{4/3}} = n^2 \frac{|Q| Q}{A^2 R^{4/3}} \text{(Manning)} \tag{7.17b}$$

$$S_e = \frac{|V| V}{K_{\text{Str}}^2 R^{4/3}} = \frac{|Q| Q}{A^2 K_{\text{Str}}^2 R^{4/3}} \text{(Strickler)} \tag{7.17c}$$

where C, K_{Str} and n are the so-called Chezy, Strickler and Manning's friction coefficients, and R is the hydraulic radius, defined as

$$R = \frac{A}{P'} \tag{7.18}$$

where P' is the wetted perimeter. Note that K_{Str} and n are the inverse of each other. All the formulae above may be written in the form $Q = C_{onv}S_e^{1/2}$, where C_{onv} is called the conveyance.

Substituting equations (7.6), (7.7), (7.10), (7.12) and (7.15) into equation (7.5), and simplifying by ρ, dt and dx, we obtain

$$\frac{\partial Q}{\partial t} + \frac{\partial}{\partial x}\left(\beta\frac{Q^2}{A} + \frac{P}{\rho}\right) = (S_0 - S_e)gA \qquad (7.19)$$

The quantity $\beta Q^2/A + P/\rho$ is known as the *impulse*.

7.2.2.3 Conservation and non-conservation form

Equations (7.5) and (7.20) can be rewritten in vector form as

$$\frac{\partial \mathbf{U}}{\partial t} + \frac{\partial \mathbf{F}}{\partial x} = \mathbf{S} \qquad (7.20)$$

where the vector variables $\mathbf{U}$, $\mathbf{F}$ and $\mathbf{S}$ are defined as

$$\mathbf{U} = \begin{bmatrix} A \\ Q \end{bmatrix}, \mathbf{F} = \begin{bmatrix} Q \\ \beta Q^2/A + P/\rho \end{bmatrix}, \mathbf{S} = \begin{bmatrix} 0 \\ (S_0 - S_e)gA \end{bmatrix} \qquad (7.21)$$

Equation (7.20) is known as the conservation form of the open-channel-flow equations. Another convenient form of equation (7.20) is the so-called non-conservation form

$$\frac{\partial \mathbf{U}}{\partial t} + \mathbf{A}\frac{\partial \mathbf{U}}{\partial x} = \mathbf{S} \qquad (7.22)$$

where the matrix $\mathbf{A}$ is defined as the Jacobian matrix of $\mathbf{F}$ with respect to $\mathbf{U}$:

$$\mathbf{A} = \frac{\partial \mathbf{F}}{\partial \mathbf{U}} \qquad (7.23)$$

Equation (7.22) is the vector extension of the scalar, first-order quasilinear equation (2.35) presented in Section 2.4. Note that equations (7.20) and (7.22) are equivalent only because the channel is assumed to be prismatic. Consequently, if $\mathbf{U}$ is a constant, both Q and A are constant and $\mathbf{F}$ is constant. If the channel is non-prismatic, the variations in the geometry of the cross-section must be accounted for in the space derivative of the impulse. In the case of a non-prismatic channel, the full expression of the derivative of $\mathbf{F}$ with respect to x is

$$\frac{\partial \mathbf{F}}{\partial x} = \left(\frac{\partial \mathbf{F}}{\partial \mathbf{U}}\right)_{x=Const}\frac{\partial \mathbf{U}}{\partial x} + \left(\frac{\partial \mathbf{F}}{\partial x}\right)_{U=Const} \qquad (7.24)$$

The extra derivative in equation (7.24) would lead to modifying the source term S in equation (7.22). The complete expression of the modified source term for a non-prismatic channel, provided in Guinot (2008), is rather complex and its detailed derivation is beyond the scope of the present text.

The expression of the Jacobian matrix **A** is given by

$$\mathbf{A} = \begin{bmatrix} 0 & 1 \\ c^2 - \beta V^2 & 2\beta V \end{bmatrix} \tag{7.25}$$

where the quantity c is defined as

$$c^2 = \frac{d(P/\rho)}{dA} \tag{7.26}$$

Note that the total derivative d is used in equation (7.26) because there is a one-to-one relationship between P and A. Indeed, A is an increasing function of the water depth Y, because an infinitesimal increase dY in the water depth generates an increase dA in the cross-sectional area given by

$$dA = BdY \tag{7.27}$$

As the width B of the free surface is strictly positive, the function $A(Y)$ is monotonically increasing. Moreover, an infinitesimal increase dY in the water depth induces an infinitesimal pressure increase $dp = \rho g dY$ over the entire channel cross-section. Therefore, the infinitesimal increase $d(P/\rho)$ is given by

$$d(P/\rho) = gAdY \tag{7.28}$$

As A is strictly positive, P/ρ is also a monotonically increasing function of Y, and the relationship between P/ρ and Y is also one-to-one. Consequently, P/ρ and A are related by a one-to-one relationship, and using the total derivative as in equation (7.26) is meaningful. Besides, substituting equations (7.27) and (7.28) into equation (7.26) leads to the following expression for c

$$c = \left(\frac{gA}{B}\right)^{1/2} \tag{7.29}$$

In most applications, the coefficient β is assumed equal to unity. In this case, the expression of **A** simplifies to

$$\mathbf{A} = \begin{bmatrix} 0 & 1 \\ c^2 - V^2 & 2V \end{bmatrix} \tag{7.30}$$

$\beta = 1$ is assumed to hold in what follows.

7.2.2.4 Alternative writing for the non-conservation form

The conservation and non-conservation forms (7.20) and (7.22) use the cross-sectional area A and the discharge Q as state variables. Unless the geometry is very simple, such variables may not be very convenient to use because they cannot be measured directly. For practical purposes the water depth Y (or the free-surface elevation η) and the mean flow velocity V are easier to use. This leads to defining a new vector variable $\mathbf{V} = [Y, V]^T$. Noting that $dA = B\,dY$ and $dQ = A\,dV + V\,dA = A\,dV + BV\,dY$ allows the variations in $\mathbf{U} = [A, Q]^T$ to be related to those in $\mathbf{V}$ by

$$
d\mathbf{V} = \left(\frac{\partial \mathbf{U}}{\partial \mathbf{V}}\right)^{-1} d\mathbf{U} = \begin{bmatrix} 1/B & 0 \\ -V/A & 1/A \end{bmatrix} d\mathbf{U} = \begin{bmatrix} 1/B\,dA \\ 1/A\,dQ - V/A\,dA \end{bmatrix}
$$

(7.31)

Multiplying equation (7.22) by $(\partial \mathbf{U}/\partial \mathbf{V})^{-1}$ yields

$$
\left(\frac{\partial \mathbf{U}}{\partial \mathbf{V}}\right)^{-1} \frac{\partial \mathbf{U}}{\partial t} + \left(\frac{\partial \mathbf{U}}{\partial \mathbf{V}}\right)^{-1} \mathbf{A} \frac{\partial \mathbf{U}}{\partial \mathbf{V}} \left(\frac{\partial \mathbf{U}}{\partial \mathbf{V}}\right)^{-1} \frac{\partial \mathbf{U}}{\partial x} = \left(\frac{\partial \mathbf{U}}{\partial \mathbf{V}}\right)^{-1} \mathbf{S}
$$

(7.32)

Using the definition (7.31) leads to

$$
\frac{\partial \mathbf{V}}{\partial t} + \mathbf{A}' \frac{\partial \mathbf{V}}{\partial x} = \mathbf{S}'
$$

(7.33)

where $\mathbf{A}'$ and $\mathbf{S}'$ are defined as

$$
\mathbf{A}' = \left(\frac{\partial \mathbf{U}}{\partial \mathbf{V}}\right)^{-1} \mathbf{A} \frac{\partial \mathbf{U}}{\partial \mathbf{V}} = \begin{bmatrix} V & A/B \\ g & V \end{bmatrix}
$$

(7.34a)

$$
\mathbf{S}' = \left(\frac{\partial \mathbf{U}}{\partial \mathbf{V}}\right)^{-1} \mathbf{S} = \begin{bmatrix} 0 \\ (S_0 - S_f)g \end{bmatrix}
$$

(7.34b)

The shallow-water equations (4.69) and (4.70) presented in Section 4.4.3 arise as a particular case of equation (7.33) when the channel is assumed to be horizontal, rectangular and infinitely wide (then $A/B = Y$) and when the motion is frictionless.

7.2.2.5 Characteristic form

The derivation and the characteristic form of scalar, first-order PDEs and their solution properties are covered in Section 2.8. The reader is referred to this section prior to reading what follows. The characteristic form of the vector equation (7.20) is obtained by rewriting equation (7.22) in the vector base formed by the eigenvectors of $\mathbf{A}$. This amounts to diagonalizing the matrix $\mathbf{A}$. To do so, equation (7.22) is multiplied by $\mathbf{K}^{-1}$, where $\mathbf{K}$ is the matrix formed by the right eigenvectors $\mathbf{r}_1$ and $\mathbf{r}_2$ of $\mathbf{A}$. As seen in Section 2.5.1, the eigenvalues and eigenvectors of $\mathbf{A}$ are

$$\lambda_1 = V - c, \lambda_2 = V + c \tag{7.35a}$$

$$\mathbf{r}_1 = \begin{bmatrix} 1 \\ V - c \end{bmatrix}, \mathbf{r}_2 = \begin{bmatrix} 1 \\ V + c \end{bmatrix} \tag{7.35b}$$

Note that the eigenvalues of $\mathbf{A}$ are real and distinct, which, from the definitions given in Chapter 2, indicates that the system (7.22) is hyperbolic. Consequently, it can be rewritten in the form of two characteristic equations, i.e. two first-order differential equations valid along two different trajectories in the (x, t)-plane. From equation (7.35b), the matrices $\mathbf{K}$ and $\mathbf{K}^{-1}$ are defined as

$$\mathbf{K} = \begin{bmatrix} 1 & 1 \\ V - c & V + c \end{bmatrix}, \mathbf{K}^{-1} = \frac{1}{2c} \begin{bmatrix} V + c & -1 \\ c - V & 1 \end{bmatrix} \tag{7.36}$$

Multiplying equation (7.22) by $\mathbf{K}^{-1}$, using the property $\mathbf{K}\mathbf{K}^{-1} = \mathbf{I}$ (where $\mathbf{I}$ is the identity matrix) leads to

$$\mathbf{K}^{-1} \frac{\partial \mathbf{U}}{\partial t} + \mathbf{K}^{-1} \mathbf{A} \mathbf{K} \mathbf{K}^{-1} \frac{\partial \mathbf{U}}{\partial x} = \mathbf{K}^{-1} \mathbf{S} \tag{7.37}$$

As mentioned in Chapter 2, the matrix $\mathbf{K}^{-1}\mathbf{A}\mathbf{K}$ is the expression of $\mathbf{A}$ in the basis of eigenvectors of $\mathbf{A}$. Consequently, it is diagonal. Equation (7.37) can be rewritten as

$$\frac{\partial \mathbf{W}}{\partial t} + \Lambda \mathbf{K}^{-1} \frac{\partial \mathbf{W}}{\partial x} = \mathbf{S}'' \tag{7.38}$$

where $\mathbf{W}$, Λ and $\mathbf{S}''$ are defined as

$$d\mathbf{W} = d\begin{bmatrix} W_1 \\ W_2 \end{bmatrix} = \frac{1}{2c} \begin{bmatrix} (V + c)dA - dQ \\ (c - V)dA + dQ \end{bmatrix} \tag{7.39a}$$

$$\Lambda = \begin{bmatrix} \lambda_1 & 0 \\ 0 & \lambda_2 \end{bmatrix} = \begin{bmatrix} V - c & 0 \\ 0 & V + c \end{bmatrix} \tag{7.39b}$$

$$\mathbf{S}'' = \frac{1}{2c} \begin{bmatrix} (S_e - S_0)gA \\ (S_0 - S_e)gA \end{bmatrix} \tag{7.39c}$$

The vector $\mathbf{W}$, defined in differential form by equation (7.39a), is called the vector of Riemann invariants. W_1 and W_2 are, respectively, the Riemann

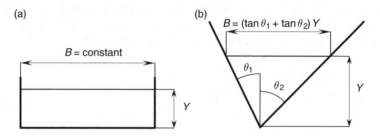

Figure 7.3 Definition of a rectangular (a) and a triangular (b) channel

invariants associated with the eigenvalues $\lambda^{(1)}$ and $\lambda^{(2)}$. Note that equation (7.38) is equivalent to the following set of differential equations

$$\frac{dW_1}{dt} = S_1'' \quad \text{for} \quad \frac{dx}{dt} = \lambda^{(1)} \tag{7.40a}$$

$$\frac{dW_2}{dt} = S_2'' \quad \text{for} \quad \frac{dx}{dt} = \lambda^{(2)} \tag{7.40b}$$

If the geometry of the channel is simple enough, equation (7.39a) can be integrated to provide an exact definition of the Riemann invariants W_1 and W_2. Particular expressions of the Riemann invariants are provided hereafter for the case of a rectangular and a triangular channel (Figure 7.3).

1   *Rectangular channel.* The width B is independent of Y and the following relationships hold:

$$A = BY \tag{7.41a}$$

$$Q = BYV \tag{7.41b}$$

$$c = (gY)^{1/2} \tag{7.41c}$$

Substituting equations (7.41a)–(7.41c) into equations (7.39a) and (7.39c) leads to

$$dW_1 = \frac{V+c}{2c}B\,dY - \frac{1}{2c}d(BYV) = \frac{V+c}{2c}B\frac{2c}{g}dc - \frac{B}{2c}(Y\,dV + V\,dY)$$
$$= \frac{V+c}{2c}B\frac{2c}{g}dc - \frac{B}{2c}\left(\frac{c^2}{g}dV + \frac{2c}{g}V\,dc\right) = \frac{Bc}{2g}d(2c - V) \tag{7.42a}$$

$$S_1'' = (S_e - S_0)\frac{gBY}{2c} = (S_e - S_0)\frac{Bc}{2} \tag{7.42b}$$

Note that, in deriving equations (7.42a) and (7.42b), the relationship $Y = c^2/g$ and its differential form $dY = 2c/gdc$ are used. Substituting equations (7.42a) and (7.42b) into equation (7.40a) and simplifying by Bc leads to

$$\frac{d}{dt}(V - 2c) = (S_0 - S_e)g \quad \text{for} \quad \frac{dx}{dt} = V - c \tag{7.43}$$

A similar reasoning leads to the following definition for the second Riemann invariant:

$$\frac{d}{dt}(V - 2c) = (S_0 - S_e)g \quad \text{for} \quad \frac{dx}{dt} = V - c \tag{7.44}$$

The Riemann invariants can then be redefined as

$$W_1 = V - 2c \tag{7.45a}$$
$$W_2 = V + 2c \tag{7.45b}$$

Note that the flow state can be determined uniquely from these two invariants, since

$$V = \frac{1}{2}(W_1 + W_2) \tag{7.46a}$$

$$c = \frac{1}{4}(W_2 - W_1) \tag{7.46b}$$

Using the relationship $Y = c^2/g$, $A = BY$ and $Q = AV$ allows the flow variables to be computed uniquely from the values of W_1 and W_2.

2   *Triangular channel.* The width B is proportional to the water depth Y. Denoting by θ_1 and θ_2 the angles between the left and right banks and the vertical, the following relationships hold:

$$A = \frac{By}{2}Y = \frac{1}{2}(\tan\theta_1 + \tan\theta_2)Y^2 \tag{7.47a}$$

$$Q = \frac{1}{2}(\tan\theta_1 + \tan\theta_2)Y^2 V \tag{7.47b}$$

$$c = \left(\frac{gY}{2}\right)^{1/2} \tag{7.47c}$$

Substituting equations (7.47a)–(7.47c) into equations (7.39a)–(7.39c) and simplifying as in the case of a rectangular channel leads to

$$\frac{d}{dt}(V - 4c) = (S_0 - S_e)g \quad \text{for} \quad \frac{dx}{dt} = V - c \tag{7.48a}$$

$$\frac{d}{dt}(V - 4c) = (S_0 - S_e)g \quad \text{for} \quad \frac{dx}{dt} = V - c \tag{7.48b}$$

Quite remarkably, the expression of the Riemann invariants does not depend on the slope of the embankments – just as it is independent of the width in the case of a rectangular channel. There, again, the flow state can be determined uniquely from these two invariants, since

$$V = \frac{1}{2}(W_1 + W_2) \tag{7.49a}$$

$$c = \frac{1}{8}(W_2 - W_1) \tag{7.49b}$$

Note that, if the flow is near to uniform conditions, $S_0 \approx S_f$ and $S'' \approx 0$. Equations (7.40) then simplify to

$$W_1 = \text{Const}_1 \quad \text{for} \quad \frac{dx}{dt} = V - c \tag{7.50a}$$

$$W_2 = \text{Const}_2 \quad \text{for} \quad \frac{dx}{dt} = V + c \tag{7.50b}$$

Thus, W_1 and W_2 are constant, or invariant, along the trajectories $dx/dt = V \pm c$, and hence the term 'Riemann invariant' used to designate them.

7.2.2.6 Physical interpretation of the characteristic form – flow regimes

The characteristic equations (7.40) and the particular cases (7.43)–(7.44), (7.48) and (7.50) can be interpreted as follows (see Section 2.5):

1   The space and time variations in the flow variables result from the propagation of two waves at speeds $V - c$ and $V + c$. The trajectories of these waves in the (x, t)-plane are called characteristics (see Figure 2.2).
2   The propagation speeds of the waves are obtained by superimposing the flow velocity V on the celerity $\pm c$ of the waves in still water (Figure 7.4). In a coordinate system moving at the same speed as the water molecules, the two waves travel at the same speed in opposite directions (see Figure 2.2).

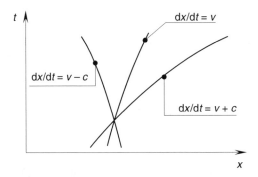

Figure 7.4 Definition of the characteristics in the (x, t)-plane. The lines $dx/dt = V \pm c$ are the characteristic lines. The line $dx/dt = V$ represents the trajectories of the water molecules

3   Along each of the two characteristics, a Riemann invariant can be defined, that obeys a first-order ordinary differential equation, also called a 'characteristic equation'. Each of the two Riemann invariants can be defined uniquely from the flow variables. Conversely, the flow variables can be determined uniquely provided that the two Riemann invariants are known.

4   In the case of uniform flow conditions, or in the case of frictionless motion over a horizontal bottom, both Riemann invariants are constant along the characteristic lines.

5   The domain of dependence and the domain of determinacy (see Section 2.7) of the solution are delimited by the two characteristics in the (x, t)-plane (see Figure 2.2(a–b) for the definition of the domain of dependence and the region of influence).

The ratio V/c is the Froude number Fr:

$$\mathrm{Fr} = \frac{V}{c} \tag{7.51}$$

Depending on the magnitude of the Froude number, the flow is said to be subcritical, critical or supercritical (Figure 7.5).

1   *Subcritical flow, $|Fr| < 1$*. In such a case, the flow velocity $|V|$ is smaller than the propagation speed c of the waves in still water and the two wave speeds $\lambda^{(1)}$ and $\lambda^{(2)}$ have opposite signs (Figure 7.5(a)). The two waves (and therefore the two Riemann invariants) propagate in

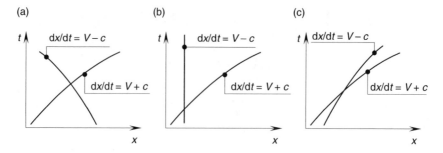

Figure 7.5 Definition of subcritical flow (a), critical flow (b) and supercritical flow (c)

opposite directions. The flow conditions in the channel are influenced by the points located upstream and downstream.

2   *Critical flow*, $|Fr| = 1$, *i.e. if* $|V| = c$. Then, one of the two characteristics is vertical in the (x, t)-plane (Figure 7.5(b)). In practice, critical conditions cannot be maintained over long channel reaches and are met only over very restricted regions of space that reduce to points.

3   *Supercritical flow*, $|Fr| > 1$. Under such conditions, $|V|$ is larger than c and both waves travel in the downstream direction (Figure 7.5(c)). Note that Figures 7.5(b–c) are drawn for a positive V.

7.2.2.7 Initial- and boundary-condition requirement

Consider a channel reach extending from $x = 0$ to $x = L$, over which the open-channel-flow equation (7.20) is to be solved for $t > 0$ (Figure 7.6). As inferred in Section 7.2.2.6 from the characteristic form (7.41), the solution at time t is determined uniquely provided that the two Riemann invariants W_1 and W_2 are known at all points of the segment $[0, L]$. The solution domain may be divided into three subregions.

Region A is defined in Section 2.4.1 as the domain of determinacy of the point located at the intersection of the characteristics $dx/dt = V + c$ passing at $(0, 0)$ and the characteristic $dx/dt = V - c$ passing at $(L, 0)$. As mentioned in Section 2.4.1, U can be computed uniquely in subregion A provided that the flow conditions (in the form $[A, Q]^T$, $[Y, V]^T$ or any combination of these) are known at all points of the segment $[0, L]$ at $t = 0$. In other words, initial conditions must be supplied for the flow variables over the entire computational domain.

Region B is located on the left-hand side of the characteristic $dx/dt = V + c$ passing at $(0, 0)$ in the (x, t)-plane. Consider a point M in this region (Figure 7.7). For U to be determined uniquely at M, W_1 and W_2 must be known. Several possibilities arise.

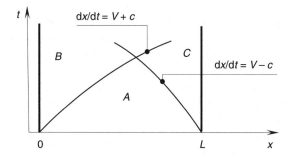

Figure 7.6 Solving the open-channel flow equations over a domain $[0, L] \times [0, t]$ in the (x, t)-plane. Definition of the three subregions of the solution domain. The bold lines indicate the domain boundaries

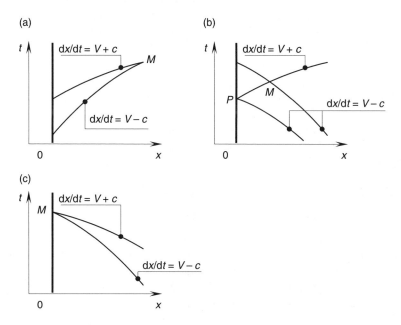

Figure 7.7 The various possible flow configurations at the left-hand boundary: (a) super-critical flow, entering the domain; (b) subcritical flow; (c) supercritical flow, leaving the domain

1   The flow is supercritical at the left-hand boundary, entering the domain (Figure 7.7(a)). In this case, both W_1 and W_2 are conditioned by the flow at the left-hand boundary, and the flow variables at M are not influenced by the flow inside the domain. Consequently, the knowledge of U at the boundary at all times suffices to the uniqueness of U at M.

2   The flow is subcritical at the left-hand boundary (Figure 7.7(b)). The characteristic $dx/dt = V - c$ passing at M comes from inside the domain. Consequently, the value of W_1 at M is determined by the flow conditions inside the domain. In contrast, the characteristic $dx/dt = V + c$ comes from the boundary. Therefore, the Riemann invariant W_2 must be prescribed at the boundary. Let P be the intersection between the characteristic $dx/dt = V + c$ and the boundary line $x = 0$. As the flow is subcritical at P, the characteristic $dx/dt = V - c$ that passes at P comes from inside the domain. Therefore, the Riemann invariant W_1 at P is determined entirely from the flow conditions inside the domain. The missing information on W_2 must be supplied as a boundary condition, in the form of a prescribed A, Q, Y, V or any relationship between two independent variables (e.g. a stage–discharge relationship).

3   The flow is supercritical, leaving the domain (Figure 7.7(c)). In this case, both characteristics leave the domain and the two Riemann invariants are entirely influenced by the flow conditions inside the domain. Even at the boundary, the flow conditions are entirely determined by what happens inside the domain. Therefore, no boundary condition is needed.

A similar reasoning for the right-hand boundary leads to the following general conclusion: at each boundary of the domain, the number of boundary conditions required to ensure the uniqueness of the solution is equal to the number of characteristics entering the domain.

7.2.3 The diffusive wave approximation

7.2.3.1 Assumptions – governing equation

The so-called diffusive wave approximation is a simplified version of the complete open-channel equations introduced in Section 7.2.2. It is obtained by neglecting the inertial terms $\partial Q/\partial t$ and $\partial(Q^2/A)/\partial x$ in the momentum equation (7.19). Introducing this simplification in the non-conservation form (7.22) and using the relationship $dA = B dY$ leads to

$$B\frac{\partial Y}{\partial t} + \frac{\partial Q}{\partial x} = 0 \qquad\qquad (7.52a)$$

$$\frac{\partial h}{\partial x} = S_0 - S_e \qquad\qquad (7.52b)$$

Equations (7.52a) and (7.52b) can be combined to provide a scalar equation in Q. This is done as follows. Differentiating equations (7.52a) and (7.52b) with respect to x and t respectively leads to

$$\frac{\partial B}{\partial x}\frac{\partial Y}{\partial t} + B\frac{\partial^2 Y}{\partial x \partial t} + \frac{\partial^2 Q}{\partial x^2} = 0 \tag{7.53a}$$

$$\frac{\partial^2 Y}{\partial x \partial t} = \frac{\partial}{\partial t}(S_0 - S_e) = -\frac{\partial S_e}{\partial t} \tag{7.53b}$$

Eliminating $\partial^2 h/\partial x \partial t$ from equations (7.53a) and (7.53b) leads to

$$b\frac{\partial S_f}{\partial t} + \frac{1}{b}\frac{\partial b}{\partial x}\frac{\partial Q}{\partial x} - \frac{\partial^2 Q}{\partial x^2} = 0 \tag{7.54}$$

The time derivative of S_e is rewritten as

$$\frac{\partial S_e}{\partial t} = \frac{\partial S_e}{\partial Q}\frac{\partial Q}{\partial t} + \frac{\partial S_e}{\partial Y}\frac{\partial Y}{\partial t} = \frac{\partial S_e}{\partial Q}\frac{\partial Q}{\partial t} - \frac{1}{B}\frac{\partial S_e}{\partial Y}\frac{\partial Q}{\partial x} \tag{7.55}$$

Substituting equation (7.55) into equation (7.54), and then using equation (7.52a) to eliminate the water depth Y from the resulting equation leads to

$$B\frac{\partial S_e}{\partial Q}\frac{\partial Q}{\partial t} + \left(\frac{1}{B}\frac{dB}{dY}\frac{\partial Y}{\partial x} - \frac{\partial S_e}{\partial h}\right)\frac{\partial Q}{\partial x} - \frac{\partial^2 Q}{\partial x^2} = 0 \tag{7.56}$$

This equation can be rewritten in the form of an advection–diffusion equation

$$\frac{\partial Q}{\partial t} + \lambda\frac{\partial Q}{\partial x} - D\frac{\partial^2 Q}{\partial x^2} = 0 \tag{7.57}$$

where D and λ are defined as

$$D = \left(B\frac{\partial S_e}{\partial Q}\right)^{-1} \tag{7.58a}$$

$$\lambda = \left(\frac{1}{B}\frac{dB}{dY}\frac{\partial Y}{\partial x} - \frac{\partial S_e}{\partial Y}\right)D = \frac{\frac{1}{B}\frac{dB}{dY}\frac{\partial Y}{\partial x} - \frac{\partial S_e}{\partial Y}}{B\frac{\partial S_e}{\partial Q}} \tag{7.58b}$$

Substituting the classical friction formulae (7.17a)—(7.17c) into (7.58), noting that $V = Q/A$ leads to

$$D = \frac{C^2 A^2 R}{2|Q|B} \quad \text{(Chezy)} \tag{7.59a}$$

$$D = \frac{A^2 R^{4/3}}{2n^2|Q|B} \quad \text{(Manning)} \tag{7.59b}$$

$$D = \frac{K_{Str}^2 A^2 R^{4/3}}{2|Q|B} \quad \text{(Strickler)} \tag{7.59c}$$

In most situations the energy-line slope S_e is a decreasing function of the water depth Y, and therefore $\partial S_e / \partial y < 0$. For a rectangular channel obeying the wide-channel approximation, $R \approx Y$ and $dB/dY = 0$. Then it is easy to check that, under the assumption of Manning's or Strickler's friction law, equation (7.58b) simplifies to

$$\lambda = \frac{3}{2} V \text{ (Chezy)} \tag{7.60a}$$

$$\lambda = \frac{5}{3} V \text{ (Manning, Strickler)} \tag{7.60b}$$

Once the equation (7.57) in Q has been solved, the water depth profile can be retrieved by integrating equations (7.52) from the initial or boundary conditions, and the flow state is then determined uniquely.

7.2.3.2 Solution properties, initial- and boundary-condition requirements

The advection–diffusion equation (7.57) can be rewritten using the characteristic form as

$$\frac{dQ}{dt} = D \frac{\partial^2 Q}{\partial x^2} \text{ for } \frac{dx}{dt} = \lambda \tag{7.61}$$

This equation is interpreted as follows: the discharge signal is transported at a speed λ given by equation (7.58b). At the same time it is subjected to diffusion, the intensity of which is determined by the coefficient D, the expression of which is given by equation (7.58a). Consequently, the discharge profile is smoothed out as it travels downstream. As the diffusion process occurs in the direction of both positive and negative x, a perturbation in the flow conditions at a given point is transmitted (although with some damping) in the direction of both positive and negative x.

In contrast with the full Saint Venant equations detailed in Section 7.2.2, the diffusive wave approximation always allows for the propagation (via diffusion) of the discharge signal in the upstream direction. This propagation mechanism, however, is due to diffusion, which is totally different from the wave propagation of two independent Riemann invariants at work in the full Saint Venant equations. In the diffusive wave approximation the distinction between subcritical and supercritical flow conditions is meaningless.

As equation (7.58) is a parabolic equation, its solution over a domain $[0, L]$ requires that the initial condition $Q(x, t = 0)$ be known over all the domain and that boundary conditions be specified at all times at each end of the domain (see Chapter 2 and, more specifically, Sections 2.2 and 2.5).

7.2.4 The kinematic wave approximation

7.2.4.1 Assumptions – governing equation

The so-called kinematic wave approximation is obtained as a further simplification of the diffusive wave approximation. If the channel slope is steep enough, or if the flow is very shallow, $\partial Y / \partial x$ may be neglected compared to the bottom slope S_0. In this case, the momentum equation (7.52b) is further simplified to

$$S_e = S_0 \tag{7.62}$$

If equation (7.62) holds, there is a one-to-one relationship between A and Q for a given x. The two components of $\mathbf{U}$ (or $\mathbf{V}$) not being independent of each other any more, assuming that equation (7.62) holds is equivalent to writing an equation in the form

$$Q = Q(A) \tag{7.63}$$

A particular consequence of this is that the flow conditions are known uniquely provided that one of the variables A and Q is known. It is then sufficient to write (and solve) a single scalar equation in A and Q over the domain of interest.

Substituting equation (7.63) into the continuity equation (7.4) leads to

$$\frac{\partial A}{\partial t} + \frac{\partial Q(A)}{\partial x} = 0 \tag{7.64}$$

It must be kept in mind that, in the general case, Q is not a function of A alone but also of the parameters that define the geometry of the channel (the bottom slope, the friction coefficient, etc.). Consequently, the derivative of the discharge may be expanded as

$$\frac{\partial Q}{\partial x} = \left(\frac{\partial Q}{\partial A}\right)_{x=\text{Const}} \frac{\partial A}{\partial x} + \left(\frac{\partial Q}{\partial x}\right)_{A=\text{Const}} \tag{7.65}$$

Substituting equation (7.65) into equation (7.64) yields the following equation in non-conservation form:

$$\frac{\partial Q}{\partial t} + \lambda \frac{\partial Q}{\partial x} = S \tag{7.66}$$

where λ and S are defined as

$$\lambda = \left(\frac{\partial Q}{\partial A}\right)_{x=\text{Const}} \tag{7.67a}$$

$$S = -\left(\frac{\partial Q}{\partial x}\right)_{A=\text{Const}} \tag{7.67b}$$

When the hydraulic properties of the channel are constant with x (i.e. if the friction coefficient, bottom slope and shape of the channel are identical all along the channel), the source term S is zero.

In a rectangular channel obeying the wide-channel approximation ($Y << B$, so that $R \approx Y$), it is easy to check that λ is given by

$$\lambda = 2V \text{ (Chezy)} \tag{7.68a}$$

$$\lambda = \frac{5}{3}V \text{ (Manning, Strickler)} \tag{7.68b}$$

7.2.4.2 Solution properties, initial- and boundary-condition requirements

As shown in Chapter 2, equation (7.66) is equivalent to the following characteristic formulation:

$$\frac{dQ}{dt} = S \quad \text{for} \quad \frac{dx}{dt} = \lambda \tag{7.69}$$

In contrast with the diffusive wave approximation, the kinematic wave approximation does not incorporate diffusion effects. Indeed, the source term S is only accountable for local, geometrical effects such as the variations in the bottom slope S_0, in the friction coefficient or in the geometry of the channel. No second-order derivative in A is incorporated in this term. Also note that, for a prismatic channel with constant bottom slope and friction coefficient, $S = 0$ and equation (7.69) reduces to

$$Q = \text{Const} \quad \text{for} \quad \frac{dx}{dt} = \lambda \tag{7.70}$$

Under such conditions, the discharge signal propagates downstream without attenuation. This is a fundamental difference from the diffusive wave approximation that allows for the smearing of the hydrograph as it travels downstream. Another fundamental difference from the diffusive wave approximation is the impossibility of the kinematic wave approximation to account for backwater effects, because there is only one wave propagating from upstream to downstream at the speed λ. As in the diffusive wave approximation, the kinematic wave approximation leaves no room for notions such as subcritical or supercritical flow, because there is only one wave, always travelling in the downstream direction.

The solution is determined uniquely provided that the initial condition is known at all points of the domain at $t = 0$ and that Q (or A) is known at all times at the upstream boundary of the domain. As the characteristics propagate in the downstream direction, the flow conditions at the downstream boundary of the solution domain are determined entirely by the flow conditions inside the domain. Consequently, no condition is needed at the downstream boundary.

7.2.5 Summary

Three main models are used in open-channel flow modelling: the Saint Venant equations, the diffusive wave approximation and the kinematic wave approximation. The solutions of these equations exhibit rather different behaviours, as illustrated in Figure 7.8.

The Saint Venant equations account for continuity, inertial effects, friction, and the influence of the bottom and the free-surface slope. The solution

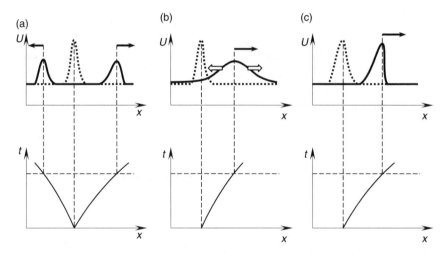

Figure 7.8 Behaviour of the solution for the Saint Venant, diffusive wave and kinematic wave models. (a) Saint Venant: propagation of two waves with possible deformation and attenuation. (b) Diffusive wave approximation: propagation of a single wave with diffusion in the upstream and downstream directions. (c) Kinematic wave approximation: propagation of a single wave without diffusion. Wave trajectories in the (x, t)-plane (bottom), schematic behaviour of the flow variables in the physical space (top). Dashed line: initial profile. Solid line: profile at $t > 0$. Solid arrows: wave propagation. Block arrows: diffusion

of these equations is composed of two waves, propagating at two different speeds $V + c$ and $V - c$ (Figure 7.8(a)). Because the propagation speeds are functions of the local flow conditions, the shape of the wave is most often altered during propagation. The presence of a source term in the equations allows for the damping of the waves via friction, and their local damping or amplification due to channel geometry variations. The Saint Venant equations allow backwater effects to be accounted for.

The diffusive wave approximation is a simplified form of the Saint Venant equations. It is based on the assumption that inertial effects can be neglected. It is therefore applicable to the modelling of slow transients with small flow velocities. The solution of the diffusive wave approximation is subjected to the combined influence of advection at a propagation speed different from that of the flow and diffusion of the discharge signal (Figure 7.8(b)). As diffusion acts in the direction of both positive and negative x, the diffusive wave approximation allows backwater effects to be accounted for. In contrast, it does not allow for the simulation of very rapid flows, where the flow regime may become supercritical.

The kinematic wave approximation results from a further simplification of the diffusive wave approximation. It is suitable for the simulation of shallow-water flows over steep slopes, where the slope of the energy line becomes equivalent to that of the channel bottom and where inertial effects can be neglected because the flow velocity remains reasonably small. The solution of the kinematic wave equation is made of a single wave travelling downstream at a speed different from that of the flow (Figure 7.8(c)). As there is only one wave, travelling in a single direction, the kinematic wave approximation does not allow backwater effects to be accounted for.

Note that in all three models, the propagation speed of the wave(s) is *not* equal to the flow velocity in the general case.

7.3 Computational models of open-channel flow

7.3.1 General

7.3.1.1 Numerical modelling as a three-step process

The three models presented in Section 7.2 provide a mathematical description of open-channel flow processes in the form of PDEs. Solving such PDEs exactly for real-world problems with arbitrary channel geometries and initial and boundary conditions is impossible. All the existing software packages for open-channel-flow modelling use numerical techniques to solve approximations of the governing PDEs, thus providing an approximation of the exact solution. The numerical solution process comprises three steps:

1   Derive approximations of the governing PDEs using discretization
     techniques such as the finite-difference, finite-volume of finite-element
     approach (see Section 3.9). In this operation the governing PDEs are
     transformed into systems of algebraic equations using standard interpo-
     lation (see Section 3.2) and/or integration (see Section 3.4) techniques.
     The discretization process consists of replacing the original problem,
     where the solution of the governing PDEs is a function of continu-
     ous time and space variables, with a much simpler problem, where
     the solution is sought only at predefined points at predefined times.
     Figure 7.9 illustrates the effect of dicretization on the representation of
     a river reach.
2   Solve the systems of algebraic equations obtained in the previous step
     using standard numerical techniques for root finding (see Section 3.3),
     integration (see Section 3.4) or matrix inversion (see Section 3.5).
3   Interpolate the so-obtained numerical solution onto the computational
     grid and convert it to flow variables that can easily be interpreted by
     the model's user.

Each of these steps induces approximations and, consequently, is a source
of inaccuracies in the solution process. A number of elementary precautions that allow the main sources of error to be minimized are mentioned
hereafter.

7.3.1.2 Discretization of the geometry

The discretization of the governing equations requires that the geometry of
the channel, as well as the initial and boundary conditions, be specified by
the modeller. In commercially available software packages, the specification
of the variations in $W(x, z)$ at all computational points is not needed. Rather,
the user is requested to provide the geometry at a number of key points
along the channel, the geometry being interpolated (in general linearly) at
the remaining computational points. The accuracy with which the geometry
is described in the computational model to a large extent influences the
accuracy of the modelling results. It is of particular importance that the

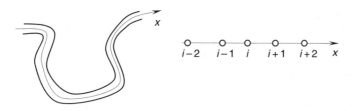

Figure 7.9 Discretization of a channel reach. Real-world channel (left), discretized channel
          (right)

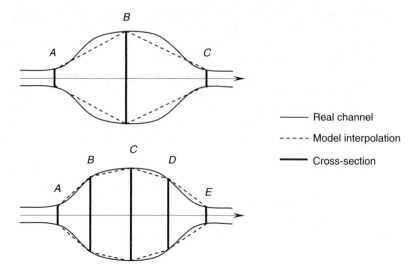

Figure 7.10 Two possible options for cross-section definition near a sudden widening and narrowing in a channel. Top: Option 1 (moderately accurate). Bottom: Option 2 (more accurate)

computational points be located in such a way that the variations in the channel geometry, such as the gradients of the river width in plan view and in a vertical cross-section, be captured as accurately as possible.

Figure 7.10 illustrates two possible options for the description of sudden channel-width variations.

Option 1 requires the definition of only three profiles, in comparison with five profiles in Option 2. However, Option 1 has the drawback that the interpolated geometry in the model leads to an underestimation of the volume available for plan storage within the reach [ABC]. From a mathematical point of view, Option 1 leads to an underestimate of the average value of B over the segment [ABC]. The consequence of underestimating the storage is an overestimate of the average propagation speed $c = (gA/B)^{1/2}$ of the waves in still water in the model compared to reality. A more physical interpretation of this is that, less storage being available in the modelled channel than in the real one, the modelled channel reacts faster than the real one to hydraulic transients, with a subsequent overestimation of the propagation speeds of the waves. In Option 2, the larger number of profiles allows the total storage in plan view to be better approximated than in Option 1.

Figure 7.11 illustrates two possible approaches to cross-section discretization. Although Option 1 (Figure 7.11, top) allows the total cross-sectional area of the reach to be represented correctly at full bank flow (because the area lost under the segment [AB] is compensated for by the area gained

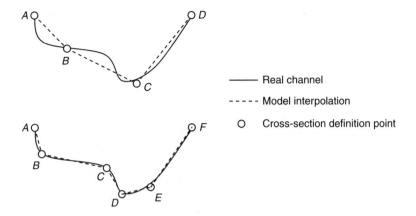

Figure 7.11 Two possible options to cross-section discretization. Top: Option 1 (moderately accurate). Bottom: Option 2 (more accurate)

above [BC]), it fails to provide a correct description of the variations in A and/or B for intermediate or small values of Y. Moreover, Option 1 leads to underestimating the wetted perimeter, with a subsequent underestimation of the energy slope S_e. In this respect, Option 2 (Figure 7.11, bottom) allows the variations in A, B and R with Y to be represented in a much more accurate fashion.

7.3.1.3 Numerical parameters

Cell size, time step. As mentioned in Section 3.9, the numerical solution of a PDE requires that a cell size Δx and a time step Δt be defined by the user. There exists a fundamental contradiction between time efficiency that requires that the number of computational operations be as small as possible (therefore requiring large values of Δx and Δt) and the accuracy of the numerical solution (which requires that Δx and Δt be kept as small as possible). In most engineering applications of modelling systems the choice of Δt and Δx is the result of a trade-off between these two contradictory requirements. The modeller should keep in mind that:

1   The larger Δt and Δx, the larger the truncation error (i.e. the difference between the real PDE to be solved and the approximate PDE that results from the discretization). Moreover, the truncation error is a function of both Δt and Δx. Consequently, reducing only Δt or Δx alone in order to reduce computational times allows only part of the truncation error to be removed. Substantial gains in model accuracy can be obtained only by decreasing both Δx and Δt.

2   The key factor in the performance of numerical schemes for hyperbolic problems (see Sections 7.3.2 and 7.3.3 for examples of such schemes) is the Courant number Cr, defined as

$$\text{Cr} = \frac{\lambda \Delta t}{\Delta x} \tag{7.71}$$

where λ is the propagation speed of the wave. Optimal solution quality is achieved when the Courant number remains reasonably close to unity. Consequently, reducing both Δx and Δt allows the quality of the numerical solution to be improved but only provided if the Courant number remains fairly constant (i.e. if Δx and Δt are reduced in the same ratio). Moreover, explicit numerical schemes are subjected to the so-called *Courant, Friedrichs and Lewy* (CFL) stability constraint $|\text{Cr}| \leq 1$ (see Section 3.9.5), which precludes the time step from exceeding a maximum permissible value.

3   Large time steps and/or cell sizes induce additional error in the interpolation of non-linear flow variables, with mass- or momentum-conservation problems as a possible consequence. Such an issue is easily illustrated by examining the equivalence between the variations in the channel cross-sectional area A and the water depth Y. For an infinitesimal variation dY in the water depth, the variation dA in the cross-sectional area is given by $dA = BdY$. Consequently,

$$\frac{\partial A}{\partial t} = B \frac{\partial Y}{\partial t} \tag{7.72}$$

Discretizing equation (7.72) between two consecutive time levels n and $n+1$ using a first-order approach leads to

$$\frac{A^{n+1} - A^n}{\Delta t} = B^{n+1/2} \frac{Y^{n+1} - Y^n}{\Delta t} \tag{7.73}$$

where the superscripts n and $n+1$ indicate the time levels at which the variable is sought. The superscript $n+1/2$ for B indicates that B is estimated in an average sense between the time levels n and $n+1$. The following estimates may be proposed

$$B^{n+1/2} = B^n \text{ (explicit)} \tag{7.74a}$$

$$B^{n+1/2} = B^{n+1} \text{ (implicit)} \tag{7.74b}$$

$$B^{n+1/2} = (B^n + B^{n+1})/2 \text{ (semi-implicit)} \tag{7.74c}$$

If B is not constant (i.e. if the channel is not rectangular), the three formulae above lead to three different estimates. Equation (7.74c) remains exact if the width of the channel varies linearly with Y, but in any other case it also fails to verify equation (7.73). In other words, the modelling result becomes dependent on the choice of the flow-state variables, which of course should not be the case. Even if the shape of the channel is defined as a piecewise linear function as illustrated in Figure 7.12, the estimate (7.74c) fails to provide the correct average value of B if Y varies over more than one interval where W is defined as a piecewise linear function. However, the smaller Δt, the smaller the difference between the estimates (7.74a)—(7.74c) and the more accurate the discretization of the equivalence (7.74).

As a consequence, the possible mass- and/or momentum-conservation errors that may arise from solving the equations in non-conservation form or for non-conservative variables are minimized when the cell size and time step are reduced.

Time- and space-centring coefficients. A number of the numerical schemes used in open-channel-flow modelling and presented in Sections 7.3.2–7.3.4 use time- and space-centring coefficients. Such coefficients may have a major influence on the behaviour of the numerical solution. As a broad rule, increasing the time-centring coefficient strengthens the implicit character of the numerical scheme and makes the solution more stable by adding numerical diffusion. However, the robustness of the computational process is increased at the expense of solution accuracy, with numerical diffusion smoothing out rapid transients. The reader is referred to Sections 7.3.2–7.3.4 for an analysis of the influence of such coefficients on the behaviour of the various schemes presented.

Iteration convergence criteria. The second step in the numerical solution process (see 7.3.1.1)) consists of solving the systems of algebraic equations obtained in the discretization step. Such systems being non-linear, their solution requires iterative procedures (see e.g. Newton–Raphson's method in Section 3.3, or Gauss–Seidel's method in Section 3.5). The iterative process is usually controlled using two types of parameters.

1   *Iteration stop criteria.* One or several iteration accuracy criteria may be defined, whereby the degree of accuracy of the solution is assessed. The iterations are stopped when the iteration accuracy criteria reach a predefined accuracy threshold, usually prescribed by the modeller. Typical iteration accuracy criteria are (i) the difference between the values taken by the flow variables at two successive iterations, and (ii) the residual of the system of equations to be solved.

2   *Maximum number of iterations.* In some situations, the iteration stop criteria defined by the modeller may prove to be too severe and to yield very time-consuming calculations, especially in the case of very fast transients, with strong changes in the flow variables within a given time step, inducing subsequent changes in the coefficients of the matrix systems to be solved. This is why a number of commercially available

modelling software packages allow a maximum number of iterations to be specified by the user. When the number of iterations made by the iterative procedure reaches the predefined maximum number of iterations, the iterations are stopped even if the iteration stop conditions specified in (1) above are not fulfilled.

Note that specifying iteration stop criteria that are too large may yield exactly the same problems regarding mass and/or momentum conservation as using too large a time step and/or cell size in the discretization step. As is usual in computational engineering, increased computational rapidity is often achieved at the expense of solution accuracy.

7.3.2 The method of characteristics

7.3.2.1 Treatment of internal points

The method of characteristics (MOC), initially applied by Courant, Isaacson and Rees (Courant *et al.* (1952)) to the equations of gas dynamics, uses the characteristic form (equations 7.40) of the Saint Venant equations. Except in the case of very simple channel geometry and initial and boundary conditions, where analytical solutions can be derived (Abbott (1966)), it is almost never possible to derive analytical solutions to the system (7.40) for real-world geometries or arbitrary initial and boundary conditions. The system (7.40) is integrated approximately using the finite-difference formalism (see Section 2.6.3 and Figure 2.7 for an outline of finite differences). It is shown in Section 3.9.5 how the characteristic form of a scalar PDE can be used to derive a solution to it. The present subsection generalizes the approach to the 2×2 set of equations (7.40).

The computational grid is defined as in Figure 7.12.

Space and time are discretized into computational points and time levels, respectively. The cell size $\Delta x_{i+1/2}$ is defined as the difference between

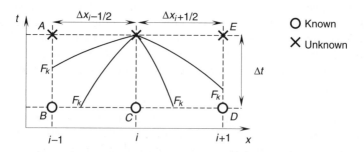

Figure 7.12 Definition of the method of characteristics. The four possible options for the foot F_k of the characteristic $dx/dt = \lambda_k$

the abscissae of the points i and $i+1$. The computational time step Δt is defined as the difference between the time levels n and $n+1$. The flow variables are assumed to be known at all points i at the time level n (circles in Figure 7.12). The purpose is to compute the flow variables at the time level n. Remember from Sections 7.2.2.5 and 7.2.2.7 that the vector $\mathbf{U}$ is known uniquely at any given point in the (x, t)-plane, provided that the two Riemann invariants W_1 and W_2 are known at this point. Computing the solution at the point i for the time level n then translates to computing the two Riemann invariants. This is achieved via numerical integration. Equations (7.40) are discretized as

$$\frac{W_{ki}^{n+1} - W_k(F_k)}{\Delta t} = S_k'''^{n+1/2}, \qquad k=1,2 \qquad (7.75)$$

where the subscript i and the superscript $n+1$ indicate that W_k is sought at the time level $n+1$ at the point i. The superscript $n+1/2$ denotes an average value between the time levels n and $n+1$. F_k is the foot of the kth characteristic $dx/dt = \lambda_k$, that is, the intersection of the characteristic $dx/dt = \lambda_k$ with one of the lines $t=t_n$ or $x=x_{i\pm1}$. Multiplying equation (7.75) by Δt and rearranging leads to

$$W_{ki}^{n+1} = W_k(F_k) + S_k'''^{n+1/2}\Delta t, \qquad k=1,2 \qquad (7.76)$$

The Riemann invariant W_k can be computed uniquely at $(i, n+1)$ provided that $W_k(F_k)$ and S_k'' can be estimated. This is done as follows.

1   *Estimating $W_k(F_k)$*. As F_k is, in general, not located at a computational point, the value of W_k must be interpolated. The most widely used interpolation techniques are the first-order (or linear) and second-order (or parabolic) interpolation methods. Note that Holly and Preissmann (1977) also proposed a two-point, third-order interpolation technique whereby both the solution variables and its space derivative are propagated simultaneously. A first-order interpolation technique leads to the following formulae

$$W_k(F_k) = \begin{cases} \dfrac{t_k - t_n}{\Delta t} W_{k i-1}^n + \left(1 - \dfrac{t_k - t_n}{\Delta t}\right) W_{k i-1}^{n+1} & \text{if } F_k \in [AB] \\[2ex] \dfrac{x_k - x_{i-1}}{\Delta x_{i-1/2}} W_{k i}^n + \left(1 - \dfrac{x_k - x_{i-1}}{\Delta x_{i-1/2}}\right) W_{k i-1}^n & \text{if } F_k \in [BC] \\[2ex] \dfrac{x_k - x_i}{\Delta x_{i+1/2}} W_{k i+1}^n + \left(1 - \dfrac{x_k - x_i}{\Delta x_{i+1/2}}\right) W_{k i}^n & \text{if } F_k \in [CD] \\[2ex] \dfrac{t_k - t_n}{\Delta t} W_{k i+1}^n + \left(1 - \dfrac{t_k - t_n}{\Delta t}\right) W_{k i+1}^{n+1} & \text{if } F_k \in [DE] \end{cases}$$

$$(7.77)$$

Noting that $t_{n+1} - t_n = \Delta t$, introducing the definition of the Courant number leads to

$$
W_k(F_k) = \begin{cases}
\frac{1}{\mathrm{Cr}_k} W^n_{k_{i-1}} + \left(1 - \frac{1}{\mathrm{Cr}_k}\right) W^{n+1}_{k_{i-1}} & \text{if } \mathrm{Cr}_k \geq 1 \\
\mathrm{Cr}_k W^n_{k_{i-1}} (1 - \mathrm{Cr}_k) W^n_{k_i} & \text{if } 0 \leq \mathrm{Cr}_k \leq 1 \\
-\mathrm{Cr}_k W^n_{k_{i+1}} + (1 + \mathrm{Cr}_k) W^n_{k_i} & \text{if } -1 \leq \mathrm{Cr}_k \leq 0 \\
-\frac{1}{\mathrm{Cr}_k} W^n_{k_{i+1}} + \left(1 + \frac{1}{\mathrm{Cr}_k}\right) W^{n+1}_{k_{i+1}} & \text{if } \mathrm{Cr}_k \leq -1
\end{cases}
\tag{7.78}
$$

where Cr_k is defined as

$$
\mathrm{Cr}_k = \begin{cases}
\lambda_k \dfrac{\Delta t}{\Delta x_{i+1/2}} & \text{if } \lambda_k \leq 0 \\
\lambda_k \dfrac{\Delta t}{\Delta x_{i-1/2}} & \text{if } \lambda_k \geq 0
\end{cases}
\tag{7.79}
$$

Note that for $\mathrm{Cr}_k = -1$, 0 and $+1$, the characteristic $dx/dt = \lambda_k$ passes at B, C and D, respectively, and equations (7.78) give the exact solution. When the absolute value of the Courant number is smaller than or equal to unity (second and third equations in system (7.78)), W_k can be determined directly from the known values of the flow variables at the time level n. Conversely, when the absolute value of the Courant number is larger than unity (first and fourth equations in system (7.78)), the estimate of $W_k(F_k)$ uses the unknown values of the flow variables at t_{n+1}. This makes the numerical scheme implicit.

2   *Estimating $S_k^{\prime\prime\prime+1/2}$*. In equations (7.75) and (7.76), $S_k^{\prime\prime\prime+1/2}$ should be seen as an average value of the right-hand side of equation (7.40) between the foot F_k of the characteristic and the point $(i, n+1)$. A widespread option is as follows:

$$
S_k^{\prime\prime\prime+1/2} = (1 - \theta) S_k^{\prime\prime}(F_k) + \theta S_{k_i}^{\prime\prime\prime+1}
\tag{7.80}
$$

where $S_k^{\prime\prime}(F_k)$ may be interpolated in the same way as W_k using equations (7.78). The parameter θ is a so-called implicitation parameter. For $\theta = 0$, $S_k^{\prime\prime}$ can be computed directly from the known values at the time level n, thus providing an explicit estimate of the source term. For $\theta > 0$, $S_k^{\prime\prime\prime+1/2}$ is a function of the unknown flow variables at the point i at the time level $n+1$.

3   *Estimating the wave celerities λ_k*. For equations (7.79) to be applicable, the wave celerities λ_k must be calculated numerically. The following formulae may be used:

$$\lambda_k = \lambda_{ki}{}^n \tag{7.81a}$$

$$\lambda_k = \frac{\Delta x_{i+1/2}\lambda_{ki-1}{}^n + \Delta x_{i-1/2}\lambda_{ki+1}{}^n}{\Delta x_{i-1/2} + \Delta x_{i+1/2}} \tag{7.81b}$$

$$\lambda_k = \begin{cases} [\mathrm{Cr}_k\lambda_{ki-1}{}^n + (1-\mathrm{Cr}_k)\lambda_{ki}{}^n](1-\theta) + \theta\lambda_{ki}{}^{n+1} & \text{if } \mathrm{Cr}_k \geq 0 \\ [(1+\mathrm{Cr}_k)\lambda_{k_i}{}^n - \mathrm{Cr}_k\lambda_{ki+1}{}^n](1-\theta) + \theta\lambda_{ki}{}^{n+1} & \text{if } \mathrm{Cr}_k \leq 0 \end{cases} \tag{7.81c}$$

Equation (7.81a) is the simplest option. It may lead to inaccuracies in the neighbourhood of critical points (i.e. near transitions from subcritical to supercritical flow), where one of the celerities λ_k changes sign. This drawback is eliminated to some extent in equation (7.81b), where λ_k at point i is estimated as the result of a linear interpolation between the points $i-1$ and $i+1$. In equation (7.81c), λ_k is estimated as the average between the foot of the characteristic and the point $(i, n+1)$. While equations (7.81a) and (7.81b) provide explicit estimates for λ_k, equation (7.81c) assumes an implicit character via the implicitation coefficient θ.

7.3.2.2 Treatment of boundary points

The treatment of supercritical flow is straightforward, as it involves prescribing both Riemann invariants (i.e. both A and Q, or both Y and V) for a supercritical inflow and none for a supercritical outflow. Consequently, for the sake of clarity and conciseness only the treatment of a left-hand boundary under subcritical conditions is detailed hereafter. The treatment of right-hand boundaries may be inferred from symmetry considerations.

Remember that, under subcritical conditions, the characteristics $dx/dt = \lambda_1$ and $dx/dt = \lambda_2$ leave and enter the domain, respectively. Consequently, $W_1(F_1)$ and $S_{11}^{\prime\prime n+1/2}$ can be determined by applying equation (7.76) for $i = k = 1$

$$W_{1_1}^{n+1} = W_1(F_1) + S_1^{\prime\prime n+1/2}\Delta t \tag{7.82}$$

where the Riemann invariant W_1 at the foot F_1 of the first characteristic and the average source term are estimated according to equations (7.78) and (7.80) for $i = k = 1$:

$$W_1(F_1) = \begin{cases} -\mathrm{Cr}_1 W_{k2}^n + (1+\mathrm{Cr}_1) W_{k1}^n & \text{if } -1 \leq \mathrm{Cr}_1 \leq 0 \\ -\dfrac{1}{\mathrm{Cr}_1} W_{k2}^n + \left(1 + \dfrac{1}{\mathrm{Cr}_1}\right) W_{k2}^{n+1} & \text{if } \mathrm{Cr}_1 \leq -1 \end{cases} \tag{7.83a}$$

$$S_1^{\prime\prime n+1/2} = (1-\theta)S_1^{\prime\prime}(F_1) + \theta S_{11}^{\prime\prime n+1} \tag{7.83b}$$

The missing piece of information is provided in the form of a boundary condition. In practice, the following types of boundary condition are used.

1 *Prescribed water level η or depth Y*. Prescribing η or Y is equivalent to prescribing A and B. Consequently, the celerity $c = (gA/B)^{1/2}$ of the waves in still water is known. For instance, in the case of the rectangular and triangular channels presented in Section 7.2.2.5 (see Figure 7.3), one has from equations (7.41c) and (7.47c)

$$c_b = (gY_b)^{1/2} \text{ (rectangular channel)} \tag{7.84a}$$

$$c_b = (gY_b/2)^{1/2} \text{ (triangular channel)} \tag{7.84b}$$

where the subscript b indicates the prescribed value at the boundary. As the celerity c of the waves in still water is a combination of the two Riemann invariants W_1 and W_2, prescribing $c = c_b$ is equivalent to prescribing a known relationship between W_1 and W_2. Taking again the example of the rectangular and triangular channels, one obtains from equations (7.46b) and (7.49b)

$$W_{21}^{n+1} - W_{11}^{n+1} = 4c_b \text{ (rectangular channel)} \tag{7.85a}$$

$$W_{21}^{n+1} - W_{11}^{n+1} = 8c_b \text{ (triangular channel)} \tag{7.85b}$$

Equations (7.82) and (7.85) form a 2×2 system of algebraic equations that can be solved uniquely for W_1^{n+1} and W_2^{n+1}.

2 *Prescribed velocity V*. Prescribing a velocity V_b at the left-hand boundary is equivalent to prescribing a combination of W_1 and W_2. It can be seen from equations (7.46a) and (7.49a) that for a rectangular or triangular channel

$$W_{11}^{n+1} + W_{21}^{n+1} = 2V_b \tag{7.86}$$

There, again, equations (7.82) and (7.86) form a 2×2 system that can be solved uniquely for W_{11}^{n+1} and W_{21}^{n+1}.

3 *Prescribed discharge Q*. Prescribing a discharge Q_b at the left-hand boundary is equivalent to prescribing the product of A and V, which in turn is equivalent to prescribing a combination of W_1 and W_2. In the case of the rectangular and triangular channels, substituting equations (7.41) and (7.47) into equations (7.46b) and (7.49b), respectively, yields

$$Q_b = BY_bV_b = \frac{B}{g}c_b^2V_b \quad \text{(rectangular channel)} \tag{7.87a}$$

$$= \frac{B}{32g}\left(W_{21}^{n+1} - W_{11}^{n+1}\right)^2\left(W_{11}^{n+1} + W_{21}^{n+1}\right)$$

$$Q = \frac{1}{2}(\tan\theta_1 + \tan\theta_2)Y_b^2V_b$$

$$= \frac{1}{2}(\tan\theta_1 + \tan\theta_2)\left(\frac{2c_b^2}{g}\right)^2V_b \quad \text{(triangular channel)} \tag{7.87b}$$

$$= \frac{\tan\theta_1 + \tan\theta_2}{2^{12}g^2}\left(W_{21}^{n+1} - W_{11}^{n+1}\right)^4\left(W_{11}^{n+1} + W_{21}^{n+1}\right)$$

4   *Prescribed stage–discharge relationship.* Such a relationship can be written in the form

$$f(Q_b, \eta_b) = 0 \tag{7.88}$$

From equations (7.84) and (7.87), it is easy to see that equation (7.88) can also be rewritten in the form

$$f_2(W_{11}^{n+1}, W_{21}^{n+1}) = 0 \tag{7.89}$$

7.3.2.3 Algorithmic aspects

From a practical point of view, the calculation process that allows the flow variables at the time level $n + 1$ to be computed from the known variables at the time level n and the boundary conditions at t_{n+1} is a three-step process.

1   Sweep the domain from $i = 1$ to $i = M$ (the number of calculation points in the computational domain). For each i compute the Riemann invariants W_{ki}^n, $k = 1, 2$, from the known flow variables at the time level n.

2   Use equations (7.76)–(7.81) to compute the Riemann invariants W_{ki}^{n+1} at the internal points (from $i = 2$ to $i = M - 1$) and solve equations (7.82)–(7.89) to compute the Riemann invariants W_{k1}^{n+1} and W_{kM}^{n+1} at the domain boundaries. The sequence is the following: for each internal point i, (i) determine the wave celerities using one of equations (7.81), (ii) use equation (7.79) to compute the Courant number for each wave, (iii) estimate the Riemann invariant at the foot of each characteristic using equation (7.78) and the average source terms $S_k^{'''n+1/2}$ using equation (7.80), and (iv) use equation (7.76) to compute the Riemann invariants W_k at the next time level.

3   Use the so-obtained values of W_{ki}^{n+1} to compute the flow variables (U_i^{n+1} or V_i^{n+1} depending on the choice of state variables made).

The number of operations involved in step 2 is strongly dependent on the options chosen for the determination of the celerities λ_k and the average source term $S_k^{'n+1/2}$ as well as the numerical values chosen for the computational time step and cell size. Two options may be considered:

Option 1: *purely explicit method.* A purely explicit calculation at internal points is possible provided that the following conditions are satisfied: (i) the computational time step is restricted in such a way that the absolute value of the Courant number Cr_k for both waves ($k = 1$ and $k = 2$) remains smaller than unity for all i, (ii) the implicitation parameter θ is set to zero in the estimate (7.80) of the source term, and (iii) equation (7.81a) or (7.81b) is used in the estimate of the wave celerities. It is easy to check that this combination of options is indeed the only one that allows the Riemann invariants at the time level $n + 1$ to be computed directly from the known values at the time level n.

Option 2: *implicit method.* Any other choice for the estimate of the wave speed, source term or a time step leading to Courant numbers larger than unity leads to relating several unknown values of the Riemann invariants within the same equation. In this case, the calculation procedure becomes iterative. Then, steps 2 and 3 of the above algorithm must be repeated until convergence. In the first iteration, an initial guess must be provided for the unknown values of W_k, λ_k and S_k'' at the time level $n + 1$. The most commonly used option consists of using the current value at the known time level.

In this option, the computational rapidity of the method depends to a large extent on the convergence criteria and the maximum number of iterations defined by the user (see Section 7.3.1.3 – Iteration convergence criteria). Specifying severe convergence criteria allows the convergence of the solution to be increased, but at the expense of computational rapidity.

7.3.2.4 Behaviour of the numerical solution

The degree of accuracy of the numerical solution obtained using the MOC is basically conditioned by two factors: (i) the nature of the interpolation method used, and (ii) the numerical value of the Courant number. The influence of these two parameters is briefly discussed hereafter.

First-order interpolation methods are essentially monotone. In other words, whatever the numerical values of the Riemann invariants that are used as a basis for interpolation in equations (7.78), the result of the interpolation always lies between the minimum and maximum values used as a basis for the interpolation. Consequently, no undershooting or overshooting is possible, and first-order interpolation methods tend to introduce a smoothing in the solutions by adding numerical diffusion, an effect also known as 'amplitude error' (Cunge *et al.* (1980)). In contrast, second- and higher-order interpolation techniques are essentially non-monotone, with the consequence that the result of the interpolation may lie outside the initial range of the variables. From a practical point of view, this results in oscillations in the neighbourhood of strong gradients and steep fronts in the flow variables. This effect, also called 'phase error' (Cunge *et al.* (1980)), is the result of numerical dispersion.

As mentioned in Section 7.3.2.1, particular values of the Courant number such as 0 and ± 1 yield the exact solution because in such cases the foot of the characteristic is located at a computational point and the interpolation formulae give the exact solution. Since the celerities of the two waves are different in the general case, no value can be found for Δt to make the Courant number equal to unity for both waves at the same time. Consequently, at least one of the two waves is either subjected to damping or under/overshooting in the interpolation process. When a second-order interpolation is used, numerical dispersion is usually the strongest when the Courant number is around 1/2. When a first-order interpolation is used, numerical diffusion is zero for $Cr = 0$, around its maximum for $Cr = 1/2$, vanishes again for $Cr = 1$ and increases with Cr for values of Cr larger than unity. Such considerations, however, refer to the amount of numerical diffusion introduced within a single time step. When the computational time step is decreased, the number of time steps needed to simulate a given period increases, with the consequence that the numerical smoothing process is repeated a larger number of times. Although the amount of numerical diffusion brought within a given time step decreases with Δt, this may result in a strong degradation of the quality of the numerical solution when Δt tends to zero (Guinot (1998)).

A final recommendation concerns the use of the interpolation equations (7.78). Each of these four equations is to be used within a well-defined range for the Courant number. Attempting to use any of these equations outside its range of applicability yields numerical instabilities in the calculation of the corresponding Riemann invariant, with the entire numerical solution becoming unstable as a final result. In particular, when the explicit MOC (given by the second and third equations in system (7.78)) is used, the computational time step should be kept sufficiently small for the absolute value of the Courant number of both waves to remain smaller than or equal to unity.

7.3.3 Preissmann's scheme

7.3.3.1 Treatment of internal points

The Preissmann scheme (Preissmann (1961), Preissmann and Cunge (1961a), (1961b)) uses two time levels and two points in space in the discretization of the governing equations. Although the scheme may be used with equal success to solve the conservation form (equation 7.20), the non-conservation forms (equations 7.22 or 7.33) or the characteristic form (equation 7.40) of the Saint Venant equations, most industrial implementations concern the non-conservation form of the equations. The computational grid is defined as in Figure 7.13.

The space and time derivatives of U are estimated as

$$\frac{\partial U}{\partial x} \approx (1 - \theta) \frac{U_{i+1}^n - U_i^n}{\Delta x_{i+1/2}} + \theta \frac{U_{i+1}^{n+1} - U_i^{n+1}}{\Delta x_{i+1/2}} \tag{7.90a}$$

$$\frac{\partial U}{\partial t} \approx (1 - \psi) \frac{U_i^{n+1} - U_i^n}{\Delta t} + \psi \frac{U_{i+1}^{n+1} - U_{i+1}^n}{\Delta t} \tag{7.90b}$$

Substituting equations (7.90) into equation (7.22) and rearranging gives

$$\mathbf{a}_{i+1/2}^{n+1/2} \mathbf{U}_i^{n+1} + \mathbf{b}_{i+1/2}^{n+1/2} \mathbf{U}_{i+1}^{n+1} = \mathbf{c}_{i+1/2}^{n+1/2} \tag{7.91a}$$

$$\mathbf{a}_{i+1/2}^{n+1/2} = \frac{1 - \psi}{\Delta t} \mathbf{I} - \frac{\theta}{\Delta x_{i+1/2}} \mathbf{A}_{i+1/2}^{n+1/2} \tag{7.91b}$$

$$\mathbf{b}_{i+1/2}^{n+1/2} = \frac{\psi}{\Delta t} \mathbf{I} + \frac{\theta}{\Delta x_{i+1/2}} \mathbf{A}_{i+1/2}^{n+1/2} \tag{7.91c}$$

$$\mathbf{c}_{i+1/2}^{n+1/2} = \mathbf{S}_{i+1/2}^{n+1/2} + \left[\frac{1 - \psi}{\Delta t} \mathbf{I} + \frac{1 - \theta}{\Delta x_{i+1/2}} \mathbf{A}_{i+1/2}^{n+1/2} \right] \mathbf{U}_i^n + \left(\frac{\psi}{\Delta t} \mathbf{I} - \frac{1 - \theta}{\Delta x_{i+1/2}} \mathbf{A}_{i+1/2}^{n+1/2} \right) \mathbf{U}_{i+1}^n \tag{7.91d}$$

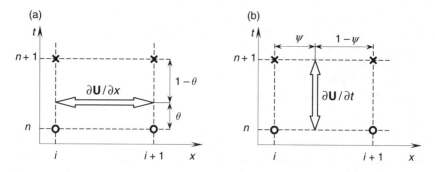

Figure 7.13 Preissmann's scheme. Definition of the discretization of the space derivative (a) and the time derivative (b). Circles: known values at the time level n. Crosses: unknown values at the time level $n + 1$

where $\mathbf{I}$ is the identity matrix and the subscript and superscript $i+1/2$ and $n+1/2$ indicate that $\mathbf{A}$ and $\mathbf{S}$ are to be estimated in an average sense over the 'box' delimited by $(i,\ n)$, $(i+1,\ n)$, $(i+1,\ n+1)$ and $(i,\ n+1)$ in the $(x,\ t)$-plane. Equation (7.91a) yields a 4×4 system that must be solved for the unknowns A and Q at the points i and $i+1$ at the time level n. Substituting the expression (7.25) for the matrix $\mathbf{A}$ into equations (7.91b)–(7.101d) leads to

$$
\mathbf{a}_{i+1/2}^{n+1/2} =
\begin{bmatrix}
\dfrac{1-\psi}{\Delta t} & -\dfrac{\theta}{\Delta x_{i+1/2}} \\[2ex]
\dfrac{\theta(u^2-c^2)_{i+1/2}^{n+1/2}}{\Delta x_{i+1/2}} & \dfrac{1-\psi}{\Delta t} - \dfrac{2\theta\, u_{i+1/2}^{n+1/2}}{\Delta x_{i+1/2}}
\end{bmatrix}
\tag{7.92a}
$$

$$
\mathbf{b}_{i+1/2}^{n+1/2} =
\begin{bmatrix}
\dfrac{\psi}{\Delta t} & \dfrac{\theta}{\Delta x_{i+1/2}} \\[2ex]
\dfrac{\theta(c^2-u^2)_{i+1/2}^{n+1/2}}{\Delta x_{i+1/2}} & \dfrac{\psi}{\Delta t} + \dfrac{2\theta\, u_{i+1/2}^{n+1/2}}{\Delta x_{i+1/2}}
\end{bmatrix}
\tag{7.92b}
$$

$$
\mathbf{c}_{i+1/2}^{n+1/2} =
\begin{bmatrix} 0 \\ (S_0 - S_e)gA \end{bmatrix}_{i+1/2}^{n+1/2}
+
\begin{bmatrix}
\dfrac{1-\psi}{\Delta t} & -\dfrac{1-\theta}{\Delta x_{i+1/2}} \\[2ex]
\dfrac{(1-\theta)(u^2-c^2)_{i+1/2}^{n+1/2}}{\Delta x_{i+1/2}} & \dfrac{1-\psi}{\Delta t} - \dfrac{2\theta\, u_{i+1/2}^{n+1/2}}{\Delta x_{i+1/2}}
\end{bmatrix}
$$
$$
\times
\begin{bmatrix} A \\ Q \end{bmatrix}_i^n
+
\begin{bmatrix}
\dfrac{\psi}{\Delta t} & -\dfrac{1-\theta}{\Delta x_{i+1/2}} \\[2ex]
\dfrac{(1-\theta)(u^2-c^2)_{i+1/2}^{n+1/2}}{\Delta x_{i+1/2}} & \dfrac{\psi}{\Delta t} - \dfrac{2\theta\, u_{i+1/2}^{n+1/2}}{\Delta x_{i+1/2}}
\end{bmatrix}
\begin{bmatrix} A \\ Q \end{bmatrix}_{i+1}^n
$$
$$
\tag{7.92c}
$$

Substituting equations (7.92) into the vector equation (7.91a) leads to the following two scalar, algebraic equations

$$
D_{i+1/2}^{(1)} A_i^{n+1} + E_{i+1/2}^{(1)} Q_i^{n+1} F_{i+1/2}^{(1)} A_{i+1}^{n+1} + G_{i+1/2}^{(1)} Q_{i+1}^{n+1} = H_{i+1/2}^{(1)},\ i = 1, \ldots,\ M-1
\tag{7.93a}
$$

$$
D_{i+1/2}^{(2)} A_i^{n+1} + E_{i+1/2}^{(2)} Q_i^{n+1} F_{i+1/2}^{(2)} A_{i+1}^{n+1} + G_{i+1/2}^{(2)} Q_{i+1}^{n+1} = H_{i+1/2}^{(2)},\ i = 1, \ldots,\ M-1
\tag{7.93b}
$$

Equations (7.93) provide $2M - 2$ equations in $\left[A_i^{n+1},\ Q_i^{n+1}\right]$, for a total number of unknowns equal to $2M$. The missing two pieces of information must be taken from the boundary conditions.

7.3.3.2   Treatment of boundary points

The treatment of boundary points is examined only for subcritical flow configurations. In this case, exactly one boundary condition must be supplied at each end of the channel. For the sake of conciseness, only the treatment of the left-hand boundary ($i = 1$) is detailed hereafter.

1   *Prescribed water level η or depth Y.* Prescribing η or Y is equivalent to prescribing a known value A_b for A at the point $i = 1$:

$$A_1^{n+1} = A_b \tag{7.94}$$

2   *Prescribed velocity.* Prescribing a known velocity V_L at the left-hand boundary is achieved by specifying that the following relationship should hold:

$$V_L A_1^{n+1} - Q_1^{n+1} = 0 \tag{7.95}$$

3   *Prescribed discharge.* Prescribing a known discharge Q_L at the left-hand boundary yields the straightforward condition:

$$Q_1^{n+1} = Q_b \tag{7.96}$$

4   *Prescribed stage–discharge relationship.* A stage–discharge relationship may be expressed in the form:

$$Q_1^{n+1} = f(A_1^{n+1}) \tag{7.97}$$

As the function f is non-linear in the general case, solving equation (7.97) usually requires iterative procedures. A widespread approach consists of linearizing the relationship (7.97) between the known time level n and the unknown time level $n + 1$:

$$
\begin{aligned}
Q_1^{n+1} \approx Q_1^n + (A_1^{n+1} - A_1^n)\frac{dQ}{dA} &= Q_1^n + (A_1^{n+1} - A_1^n)\frac{dQ}{d\eta}\left(\frac{dA}{d\eta}\right)^{-1} \\
&= Q_1^n + \frac{A_1^{n+1} - A_1^n}{B}\frac{dQ}{d\eta}
\end{aligned}
\tag{7.98}
$$

Rearranging equation (7.98) yields

$$B^{n+1/2}Q_1^{n+1} - \left(\frac{dQ}{d\eta}\right)^{n+1/2} A_1^{n+1} = B^{n+1/2}Q_1^n - \left(\frac{dQ}{d\eta}\right)^{n+1/2} A_1^n \tag{7.99}$$

where $dQ/d\eta$ and B must be estimated in an average sense between the time levels n and $n + 1$.

7.3.3.3 Algorithmic aspects

The coefficients D to H in equations (7.93) involve averaged values of V, c and S_e between the points i and $i+1$ and the time levels n and $n+1$. The following, general approximation is proposed :

$$V_{i+1/2}^{n+1/2} = (1-\theta')\left[(1-\psi')V_i^n + \psi'V_{i+1}^n\right] + \left[(1-\psi')V_i^{n+1} + \psi'V_{i+1}^{n+1}\right]\theta'$$
$$(7.100a)$$

$$c_{i+1/2}^{n+1/2} = (1-\theta')\left[(1-\psi')c_i^n + \psi'c_{i+1}^n\right] + \left[(1-\psi')c_i^{n+1} + \psi'c_{i+1}^{n+1}\right]\theta'$$
$$(7.100b)$$

$$S_{e_{i+1/2}}^{n+1/2} = (1-\theta')\left[(1-\psi')S_{e_i}^n + \psi'S_{e_{i+1}}^n\right] + \left[(1-\psi')S_{e_i}^{n+1} + \psi'S_{e_{i+1}}^{n+1}\right]\theta'$$
$$(7.100c)$$

Another approach consists of computing V, c and S_e using interpolated values of A and Q, that is,

$$V_{i+1/2}^{n+1/2} = V\left(A_{i+1/2}^{n+1/2}, Q_{i+1/2}^{n+1/2}\right) = Q_{i+1/2}^{n+1/2}/A_{i+1/2}^{n+1/2} \qquad (7.101a)$$

$$c_{i+1/2}^{n+1/2} = c\left(A_{i+1/2}^{n+1/2}, Q_{i+1/2}^{n+1/2}\right) = \left[gA_{i+1/2}^{n+1/2}/B\left(A_{i+1/2}^{n+1/2}\right)\right]^{1/2} \qquad (7.101b)$$

$$S_{e_{i+1/2}}^{n+1/2} = S_e\left(A_{i+1/2}^{n+1/2}, Q_{i+1/2}^{n+1/2}\right) \qquad (7.101c)$$

$$A_{i+1/2}^{n+1/2} = (1-\theta')\left[(1-\psi')A_i^n + \psi'A_{i+1}^n\right] + \left[(1-\psi')A_i^{n+1} + \psi'A_{i+1}^{n+1}\right]\theta'$$
$$(7.101d)$$

$$Q_{i+1/2}^{n+1/2} = (1-\theta')\left[(1-\psi')Q_i^n + \psi'Q_{i+1}^n\right] + \left[(1-\psi')Q_i^{n+1} + \psi'Q_{i+1}^{n+1}\right]\theta'$$
$$(7.101e)$$

The most frequently used options consist of using $\theta' = \psi' = 1/2$, or $\theta' = \theta$ and $\psi' = \psi$. A similar approach may be used for the estimate of B and $dQ/d\eta$ in equation (7.99).

The solution algorithm between the time levels n and $n+1$ can be broken down into the following sequence.

1   For each interval $[i, i+1]$, compute the coefficients in equations (7.93) using the estimates (7.100) or (7.101). In the first iteration, the unknown variables at t_{n+1} are assumed to be equal to their previous value at t_n. Incorporate the boundary conditions using equations (7.94)–(7.99) depending on the nature of the boundary condition. If equation (7.99) is used, B and $dQ/d\eta$ are taken equal to their values at t_n for the first iteration.

2   Solve the $2M \times 2M$ system of algebraic equations for A_i^{n+1} and Q_i^{n+1} $(i = 1, \ldots, M)$. As shown in Cunge *et al.* (1980) and Abbott and Minns (1998), the vector of unknowns can be arranged in such a way that

the system to be solved is pentadiagonal. Standard matrix-inversion techniques may be used, one of the most efficient ones for such narrow bandwidth systems being the so-called 'double-sweep algorithm' (Abbott and Minns (1998), Cunge *et al.* (1980)).

3   Use the solutions A_i^{n+1} and Q_i^{n+1} $(i = 1, \ldots, M)$ to update the coefficients in equations (7.93) and (7.99) using equations (7.100) or (7.101) depending on the option retained.

Steps 2 and 3 must be repeated until convergence. In practice, three or four iterations prove to be sufficient in most situations.

7.3.3.4 Behaviour of the numerical solution

The parameter ψ influences the centring of the estimate of the time derivative with respect to x. If ψ is set to 0, the time derivative is estimated using the numerical solution at the point i, which yields an increased stability of the waves with negative celerities and makes the waves with positive celerities less stable. Conversely, setting ψ to 1 gives full weight to the point $i + 1$, thus enhancing the stability of the solutions travelling with positive wave speeds and decreasing the stability of solutions travelling at negative celerities. As solving the Saint Venant equations under subcritical conditions implies dealing with two waves travelling in opposite directions, optimal stability for both waves is achieved for $\psi = 1/2$. Departing from the symmetrical value $\psi = 1/2$ increases the stability of one wave, while decreasing the stability of the other, with the risk of making the entire solution unstable. For this reason it is strongly advised to use $\psi = 1/2$ in standard channel-flow-modelling applications.

Increasing the implicitation parameter θ increases the stability of the numerical solution. For the standard configuration $\psi = 1/2$, values of θ smaller than 1/2 yield unconditionally stable solutions (i.e. the numerical solution is unstable whatever the numerical value of the Courant number), while values of θ larger than or equal to 1/2 yield stable solutions. When $\theta = \psi = 1/2$, the analytical solution is obtained for the waves, the absolute value of the Courant number of which is unity. If the absolute value of the Courant number is different from unity, numerical dispersion appears and the solution exhibits spurious oscillations near steep fronts and rapidly varying flow variables. Increasing the value of θ leads to increased numerical diffusion, leading to a more stable solution with a stronger damping of rapid transients.

The Preissmann scheme in its original version is not equipped to deal with transcritical flow simulations (e.g. a channel with supercritical inflow and subcritical outflow), because in such configurations the number of equations to be solved does not match the number of unknowns (Meselhe and Holly (1997)). Recent developments (Johnson *et al.* (2002)) allow the Preissmann scheme to be used for the simulation of critical points and hydraulic jumps by combining equations (7.93) appropriately depending on the direction of propagation of the waves.

7.3.4 Abbott–Ionescu's scheme

7.3.4.1 Treatment of internal points

The Abbott–Ionescu scheme uses a staggered, regular grid (Abbott and Ionescu (1967)). The water depth Y (or the cross-sectional area A) is sought at the points i, $i+2$, $i+4$, etc., while the flow velocity V (or the discharge Q) is sought at the points $i-1$, $i+1$, $i+3$, etc. (Figure 7.14). A consequence of the staggered character of the scheme is that the derivatives of Y (or A) and V (or Q) with respect to x cannot be approximated at the same points. Therefore, the governing equations must be rewritten in such a way that the space derivative of only one of the variables appears in each equation.

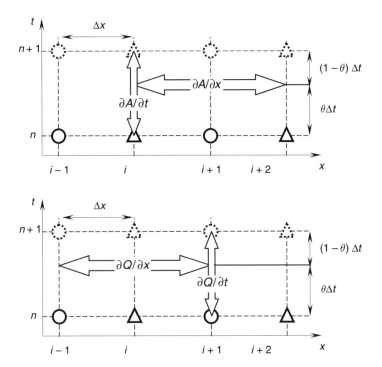

Figure 7.14 Definition of the Abbott–Ionescu scheme. Derivatives of Y or A (top), derivatives of V or Q (bottom). Triangles: computational points for A or Y. Circles: computational points for Q or V. Solid markers: known values at time level n. Dashed markers: unknown values at time level $n+1$

Several variants may be proposed for this scheme.

1 *Original scheme* (Abbott and Ionescu (1967)). The scheme was originally presented for the solution of the non-conservation form (equation 7.33), which can be rewritten as

$$B\frac{\partial Y}{\partial t} + BV\frac{\partial Y}{\partial x} + A\frac{\partial V}{\partial x} = 0 \tag{7.102a}$$

$$BV\frac{\partial Y}{\partial t} + A\frac{\partial V}{\partial t} + (c^2 + V^2)B\frac{\partial Y}{\partial x} + 2Q\frac{\partial V}{\partial x} = (S_0 - S_e)gA \tag{7.102b}$$

The derivative $\partial V/\partial x$ can be eliminated by multiplying equation (7.102a) by $2V$ and subtracting the resulting equation from equation (7.102b). The derivative $\partial Y/\partial x$ can be eliminated by multiplying equations (7.102a) and (7.102b) by $c^2 + V^2$ and V, respectively, and subtracting them from each other. The following system of equations is obtained:

$$-BV\frac{\partial Y}{\partial t} + A\frac{\partial V}{\partial t} + (c^2 + V^2)B\frac{\partial Y}{\partial x} = (S_0 - S_e)gA \tag{7.103a}$$

$$-gA\frac{\partial Y}{\partial t} + Q\frac{\partial V}{\partial t} + (QV - c^2A)\frac{\partial V}{\partial x} = (S_0 - S_e)gQ \tag{7.103b}$$

Equation. (7.103a) is discretized at the point $i + 1$ using the following estimates of the derivatives:

$$\frac{\partial Y}{\partial t} \approx \frac{1}{2}\left(\frac{Y_i^{n+1} - Y_i^n}{\Delta t} + \frac{Y_{i+2}^{n+1} - Y_{i+2}^n}{\Delta t}\right) \tag{7.104a}$$

$$\frac{\partial V}{\partial t} \approx \frac{V_{i+1}^{n+1} - V_{i+1}^n}{\Delta t} \tag{7.104b}$$

$$\frac{\partial Y}{\partial x} \approx (1-\theta)\frac{Y_{i+2}^n - Y_i^n}{2\Delta x} + \theta\frac{Y_{i+2}^{n+1} - Y_i^{n+1}}{2\Delta x} \tag{7.104c}$$

Equation (7.103b) is discretized at the point i using the following estimates for the derivatives:

$$\frac{\partial Y}{\partial t} \approx \frac{Y_i^{n+1} - Y_i^n}{\Delta t} \tag{7.105a}$$

$$\frac{\partial V}{\partial t} \approx \frac{1}{2}\left(\frac{V_{i-1}^{n+1} - V_{i-1}^n}{\Delta t} + \frac{V_{i+1}^{n+1} - V_{i+1}^n}{\Delta t}\right) \tag{7.105b}$$

$$\frac{\partial V}{\partial x} \approx (1-\theta)\frac{V_{i+1}^n - V_{i-1}^n}{2\Delta x} + \theta\frac{V_{i+1}^{n+1} - V_{i-1}^{n+1}}{2\Delta x} \tag{7.105c}$$

Substituting equations (7.104) and (7.105) into equations (7.103a) and (7.103b), respectively, yields a system in the form

$$D_{i+1}^{(1)} Y_i^{n+1} + E_{i+1}^{(1)} V_i^{n+1} F_{i+1}^{(1)} Y_{i+2}^{n+1} = G_{i+1}^{(1)}, \quad i = 1, \ldots, M-1 \qquad (7.106a)$$

$$D_i^{(2)} V_{i-1}^{n+1} + E_i^{(2)} Y_i^{n+1} F_i^{(2)} V_{i+1}^{n+1} = G_i, \quad i = 1, \ldots, M-1 \qquad (7.106b)$$

2 *Solution in pseudo-conservation form* (DHI (2005)). In some commercially available software packages such as Mike 11 (DHI (2005)), the following set of equations is solved:

$$B\frac{\partial Y}{\partial t} + \frac{\partial Q}{\partial x} = 0 \qquad (7.107a)$$

$$\frac{\partial Q}{\partial t} + \frac{\partial}{\partial x}\left(\frac{Q^2}{A}\right) + gA\frac{\partial Y}{\partial x} = (S_0 - S_e)gA \qquad (7.107b)$$

The continuity equation is discretized around the point i using the following estimates:

$$\frac{\partial Y}{\partial t} \approx \frac{Y_i^{n+1} - Y_i^n}{\Delta t} \qquad (7.108a)$$

$$\frac{\partial Q}{\partial x} \approx (1-\theta)\frac{Q_{i+1}^n - Q_{i-1}^n}{2\Delta x} + \theta\frac{Q_{i+1}^{n+1} - Q_{i-1}^{n+1}}{2\Delta x} \qquad (7.108b)$$

The momentum equation is discretized around the point $i+1$ using the following estimates:

$$\frac{\partial Q}{\partial t} \approx \frac{Q_{i+1}^{n+1} - Q_{i+1}^n}{\Delta t} \qquad (7.109a)$$

$$\frac{\partial(Q^2/A)}{\partial x} \approx (1-\theta)\frac{\dfrac{Q_{i+1}^n Q_{i+1}^{n+1}}{A_{i+2}^n} - \dfrac{Q_{i+1}^n Q_{i+1}^{n+1}}{A_i^n}}{2\Delta x} + \theta\frac{\dfrac{Q_{i+1}^n Q_{i+1}^{n+1}}{A_{i+2}^{n+1}} - \dfrac{Q_{i+1}^n Q_{i+1}^{n+1}}{A_i^{n+1}}}{2\Delta x}$$

$$(7.109b)$$

$$\frac{\partial A}{\partial x} \approx (1-\theta)\frac{A_{i+2}^n - A_i^n}{2\Delta x} + \theta\frac{A_{i+2}^{n+1} - A_i^{n+1}}{2\Delta x} \qquad (7.109c)$$

which yields a system similar to (7.106). Note that using the value of Q at the point $i+1$ alone in equation (7.109b) does not yield a full discretization of the term $\partial(Q^2/A)/\partial x = Q^2\partial(1/A)/\partial x + 2V\partial A/\partial x$. Equation (7.109b) is actually an approximation for the quantity $Q^2 \partial(1/A)/\partial x$. Consequently, the term $2V \partial A/\partial x$ is absent from the discretized momentum equation.

Denoting by M the total number of points in the computational grid, M flow variables must be computed at the time level $n+1$ (remember that only one flow variable, either Y or Q, must be computed at each computational point). However, the momentum or continuity equations can be written only for the points $i=2$ to $i=M-1$ because the discretization uses three adjacent points in space. Consequently, one boundary condition must be specified at each end of the domain.

7.3.4.2 Treatment of boundary points

As mentioned in Cunge et al. (1980), prescribing the water level $\eta(t)$ or the discharge $Q(t)$ at the domain boundaries is straightforward. The only necessary condition is that the grid should be designed in such a way that boundaries where the discharge is to be prescribed are materialized by a Q-computational point, and boundaries where the water level η (or depth Y or cross-sectional area A) is to be prescribed are materialized by a Y-computational point.

Difficulties arise when the boundary condition is a combination of both Y and Q (or V), as is the case with stage–discharge relationships. In essence, a stage–discharge relationship $Q=Q(\eta)$ or $Q=Q(Y)$ implies that both Q and Y are known at the same point. The staggered nature of the Abbott–Ionescu scheme, however, makes this impossible. This difficulty may be eliminated by interpolating linearly the rating curve between two successive Y-computational points, as proposed by Verwey (1971). Another option (DHI (2005)) consists of defining the point M as a Y-computational point and writing a simplified continuity equation for this point:

$$\frac{Y_M^{n+1} - Y_M^n}{\Delta t} A_M^{n+1/2} = \frac{1}{2} \left(\frac{Q_{M-1}^n - Q_b^n}{\Delta x} + \frac{Q_{M-1}^{n+1} - Q_b^{n+1}}{\Delta x} \right) \tag{7.110}$$

where Q_b is the discharge at the point M. Q_b^n is known from the previous time step, but Q_b^{n+1} is still to be determined. The system is closed using a linearization of the rating curve:

$$Q_b^{n+1} = a^{n+1/2} Y_M^{n+1} + b^{n+1/2} \tag{7.111}$$

7.3.4.3 Algorithmic aspects

Whatever the option retained for the treatment of internal points, the system to be solved is a tri-diagonal system (Abbott and Minns (1998), Cunge et al. (1980)) that can be solved using standard inversion techniques such as the double-sweep algorithm.

The solution sequence is iterative. At each iteration, the following steps must be made:

1   Compute the coefficients in the discretized equations. In equations (7.103) and (7.107), the average value of the free-surface width

B between the time levels n and $n+1$ must be estimated at the Y- and V-points. The following estimates may be used:

$$B_i^{n+1/2} = (1 - \theta)B(Y_i^n) + \theta B(Y_i^{n+1})(Y - \text{point}) \tag{7.112a}$$

$$B_{i+1}^{n+1/2} = \frac{1 - \theta}{2}\left[B(Y_i^n) + B(Y_{i+2}^n)\right] + \frac{\theta}{2}\left[B(Y_i^{n+1}) + B(Y_{i+2}^{n+1})\right](V - \text{point}) \tag{7.112b}$$

The hydraulic radius that is necessary in the calculation of the slope of the energy line may be interpolated in the same way. At the beginning of the first iteration, the solution at the time level $n+1$ is not yet known and the known value at the time level n must be used instead.

2 Solve the four-diagonal system (7.106) using any standard matrix-inversion technique. One of the fastest techniques known to date is the double-sweep algorithm (Abbott and Minns (1998), Cunge *et al.* (1980)).

3 Update the coefficients as in step 1 using the newly computed solution at the time level $n+1$ and repeat steps 2 and 3 until convergence.

7.3.4.4 Behaviour of the numerical solution

For $\theta = \frac{1}{2}$, the Abbott–Ionescu scheme is exactly centred in time and does not induce any numerical diffusion. However, it is dispersive and yields unphysical oscillations in the numerical solution, irrespective of the value of the Courant number. In contrast with the Preissmann scheme, the Abbott–Ionescu scheme is dispersive even for $\text{Cr} = 1$.

Just as for the Preissmann scheme, increasing the value of θ for the Abbott–Ionescu scheme strengthens the implicit character of the scheme and introduces numerical diffusion, the effect of which is to make the numerical solution more stable, at the expense, however, of accuracy.

7.3.5 Modelling of structures and junctions

The equations and algorithms detailed so far deal only with single reaches, over which the Saint Venant equations are applicable at all points. The equations are not applicable across structures (sills, weirs, gates, bridges with local section narrowing, etc.). Standard modelling practice consists of breaking down the channel into two elementary channels, one upstream and the other downstream of the structure (Figure 7.15a). From an algorithmic point of view, the classical-flow equations remain applicable within each of the reaches and the structure is considered as a boundary for each of the elementary channels, hence the term 'internal boundary'.

A similar problem arises in the modelling of branched-channel networks, where the confluence of a main channel with a tributary may be represented as an internal boundary between three elementary channels (Figure 7.15b).

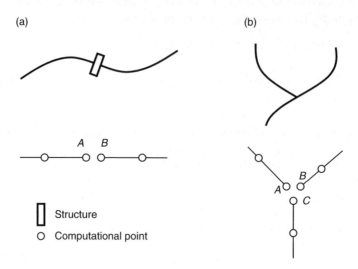

(a)   (b)

A   B

☐ Structure

O Computational point

Figure 7.15 Representation of (a) structures and (b) channel branching as internal boundary conditions. Top: physical situation. Bottom: algorithmic representation in
the solution algorithm

Consider first the case of a structure (Figure 7.15a). If the flow is subcritical in the channel, one characteristic enters the domain at each of the points
A and B, and therefore one boundary condition must be prescribed at each
of these points. Consequently, two equations must be provided in order to
close the system. These two equations are the continuity equation across the
structure and the formula for the head loss across the structure:

$$Q_A = Q_B + Q_{out} \tag{7.113a}$$

$$\eta_A + \frac{V_A^2}{2g} = \eta_B + \frac{V_B^2}{2g} + \Delta H \tag{7.113b}$$

where Q_{out} is the discharge lost (e.g. by overspilling in the case of lateral
weirs, or discharge diversion across irrigation gates, etc.) and ΔH is the
head loss across the structure. In most cases this system is linearized with
respect to Q and Y and incorporated in the matrix system to be solved.

Consider now the case of a junction (Figure 7.15b). For the sake of
simplicity, only three channels are represented in Figure 7.15b, but the
considerations hereafter remain valid for an arbitrary number of channels
joining at the same point. Under subcritical flow, one characteristic leaves
each of the elementary channels at the points A, B and C, while one characteristic enters each of the channels. Consequently, the number of internal

boundary conditions needed to close the system is equal to the number of channels at the junction. The first, obvious, condition is continuity:

$$\sum_{k}^{J} \varepsilon_k Q_k = 0 \tag{7.114}$$

where J is the number of channels at the junction (in Figure 7.15b, $J = 3$), Q_k is the discharge in the kth channel connected to the junction and ε_k is a topological indicator, which is equal to $+1$ if the junction is considered as an upstream boundary to the channel k and to -1 if the junction is considered to be the downstream boundary of the channel. The remaining $J - 1$ conditions are provided in the form of an assumption of equal water levels across the junctions:

$$\eta_1 = \eta_2 = \cdots = \eta_J \tag{7.115}$$

Equation (7.115) yields $J - 1$ independent equations, which, together with equation (7.114), suffice to close the system. Note that the assumption of equal heads may also be used:

$$\eta_1 + \frac{V_1^2}{2g} = \eta_2 + \frac{V_2^2}{2g} = \cdots = \eta_J + \frac{V_J^2}{2g} \tag{7.116}$$

Note, however, that the option (7.116) introduces an extra complication in the treatment of boundary conditions compared to the straightforward condition (7.115) because of the non-linearity introduced by the term V_k^2.

From a general point of view, internal boundary conditions involve the determination of both A and Q (or Y and V) at the same boundary points. Consequently, the treatment of structures and junctions is easier to implement algorithmically when the solution scheme is collocated (i.e. when the two independent flow variables A and Q or Y and V are computed at the same points, as in the MOC and in Preissmann's scheme) than when the numerical scheme is staggered (e.g. the Abbott–Ionescu scheme).

7.3.6 Modelling of dry beds and small depths

The treatment of small depths and dry beds is often a source of inaccuracy, if not numerical instability, in the numerical solution of the open-channel-flow equations. Three main issues are described hereafter: (i) problems arising from the discretization of naturally shaped channels, (ii) the non-uniqueness of the relationship between the discharge and the water depth in the discretized equations, and (iii) unphysical computational results induced by coarse dicretizations.

The first difficulty lies in that, in natural channels, the channel width $B(Y)$ tends to zero when the depth Y tends to zero. In most computational codes, the equations are solved for Y and the continuity equation is discretized as

$$B_i^{n+1/2} \frac{Y_i^{n+1} - Y_i^n}{\Delta t} = -\left(\frac{\partial Q}{\partial x}\right)_i^{n+1/2} \tag{7.117}$$

where the superscript $n + 1/2$ denotes the average value of the variable between the time levels n and $n + 1$. Solving equation (7.117) for Y_i^{n+1} gives

$$Y_i^{n+1} = Y_i^n - \frac{\Delta t}{B_i^{n+1/2}} \left(\frac{\partial Q}{\partial x}\right)_i^{n+1/2} \tag{7.118}$$

The free-surface width $B_i^{n+1/2}$ may be approximated using explicit, implicit or semi-implicit equations such as (7.74) or (7.112). All options, even the semi-implicit ones, lead to initialize $B_i^{n+1/2}$ with the value B_i^n. In the case of a wave flowing on an initially dry bed, $Y_i^n = 0$, and consequently $B_i^n = 0$. Then the calculation with equation (7.118) fails owing to division by zero. Note that similar problems may occur even if the initial depth is not zero. If Y_i^n is very small, B_i^n is also very small. Consequently, $B_i^{n+1/2}$, which is initialized as $B_i^{n+1/2} = B_i^n$ in semi-explicit procedures, is very small at the beginning of the iterative process. This may result in a dramatic overestimation of the variations in the water depth as computed by equation (7.118). For negative $\partial Q/\partial x$, Y_i^{n+1} is strongly overestimated, and so is the resulting value of $B_i^{n+1/2}$ in the next iteration. In turn, overestimating the value of $B_i^{n+1/2}$ leads to an underestimate of Y_i^{n+1} at the following iteration. If the computational time step is too large, the iterative process may converge very slowly. If the maximum number of iterations allowed by the user is too small, convergence may not be achieved, with the consequence that the discretized quantity $B\partial y/\partial t$ may fail to provide a correct approximation of $\partial A/\partial t$. Repeating this problem over several time steps generally leads to severe mass-conservation problems, if not to numerical instability.

A solution to this problem consists of modifying the shape of the cross-section artificially so that the bed does not become dry for $Y = 0$. To do so, an artificial, triangular slot of base width W_s (to be specified by the modeller) is created in the bottom (Figure 7.16a). The width of the slot is usually a few centimetres. In the calculations, any value of η smaller than the bottom level z_b is reset to z_b, which guarantees that the free-surface width never becomes zero. This option, however, introduces a strong bias in the calculation of the cross-sectional area when $z < z_s$. For this reason the so-called 'area-preserving slot' (Figure 7.16b) is sometimes preferred. The area-preserving slot is constructed as the superimposition of two triangular slots.

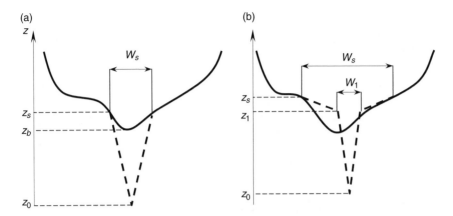

Figure 7.16 Artificial slot in the river bottom for dry bed computations. Simple, triangular slot (a) and area-preserving triangular slot (b). Solid line: real cross-sectional profile. Dashed line: modified cross-sectional profile. The size of the slot is exaggerated compared to its actual dimensions

A second problem associated with small depths is the appearance of oscillations near wetting and drying fronts due to the non-uniqueness of the relationship between the discharge and the depth in the discretized equations. The non-uniqueness of the relationship can be illustrated by considering a horizontal channel, where the depth is small enough for friction to be assumed reasonably to be the dominant phenomenon, so that the inertial terms can be neglected. The only terms remaining in the momentum equation are the term $c^2 \partial A / \partial x$ on the left-hand side and the term $-gAS_e$ on the right-hand side. Introducing the derivative of Y via equation (7.27) and using the definition (7.29) for c allows the momentum equation to be simplified to

$$\frac{\partial Y}{\partial x} = -S_e \tag{7.119}$$

For the sake of simplicity, the channel is assumed to be rectangular and the wide-channel approximation $R \approx Y$ is assumed to hold in equations (7.17). Substituting, for example, equation (7.17b) into equation (7.119) and using the wide-channel approximation and the relationship $V = Q/(BY)$ leads to

$$\frac{\partial Y}{\partial x} = -\frac{n^2}{B^2} \frac{|Q| Q}{Y^{10/3}} \tag{7.120}$$

This equation can be rewritten as

$$Q = \varepsilon \frac{B}{n} Y^{5/3} \left| \frac{\partial Y}{\partial x} \right|^{1/2} \tag{7.121}$$

where $\varepsilon = \pm 1$, depending on the sign of $\partial Y / \partial x$. Discretizing equation (7.121) between two adjacent computational points i and $i+1$ leads to (the superscript indicating the time level is omitted for the sake of generality)

$$Q = \frac{\varepsilon B}{n} y_{i+1/2}^{5/3} \left(\frac{Y_{i+1} - Y_i}{\Delta x_{i+1/2}} \right)^{1/2} \tag{7.122}$$

In the general case, $Y_{i+1/2}^{5/3}$ is estimated as a linear combination of Y_i and Y_{i+1}, such as

$$Y_{i+1/2} = (1 - \psi) Y_i + \psi \ Y_{i+1} \tag{7.123}$$

where the centring coefficient Y is between zero and unity. Substituting equation (7.123) into equation (7.122) leads to

$$Q = \frac{\varepsilon B}{n} [(1 - \psi) Y_i + \psi \ Y_{i+1}]^{5/3} \left(\frac{Y_{i+1} - Y_i}{\Delta x_{i+1/2}} \right)^{1/2} \tag{7.124}$$

When Q is positive, i is the upstream point and $i+1$ is the downstream point. A straightforward function variation analysis indicates that Q as given by equation (7.124) is not a monotonic function of Y_{i+1} for a given Y_i, unless ψ is set to zero. Conversely, for a negative discharge, Q is a monotonic function of the downstream point Y_i only if ψ is set to 1. The non-uniqueness of the relationship between the discharge and the downstream depth may lead to artificial oscillations in the computed flow variables during the (iterative) numerical solution process. Repeating the process along several time steps may lead to solution instability.

The solution to this problem, suggested by Cunge et al. (1980), consists of setting Y to 0 for positive values of Q and to unity for negative values of Q. In other words, the conveyance $B/n \ Y^{5/3}$ should be estimated using only the upstream point. This numerical stabilization procedure is known as 'conveyance upwinding'. It should be used only for small depths, more classical estimates being used when the depth becomes larger than a given threshold value that is specified by the modeller.

The third problem associated with small depths is the possibly unphysical computational results triggered by the coarseness of the computational grid. Consider a channel where the flow equations are solved using, for example, Preissmann's scheme. The continuity equation is discretized as in equation (7.90) with $\psi = 1/2$:

$$\frac{A_{i+1}^{n+1} + A_i^{n+1}}{2\Delta t} + \theta \frac{Q_{i+1}^{n+1} - Q_i^{n+1}}{\Delta x} = \frac{A_{i+1}^n + A_i^n}{2\Delta t} + (1 - \theta)\frac{Q_{i+1}^n - Q_i^n}{\Delta x} \quad (7.125)$$

If the flow is initially uniform, $A_i^n = A_{i+1}^n = A_0$ and $Q_i^n = Q_{i+1}^n = Q_0$. Assume now that a wave travelling in the direction of positive x reaches the point i at t_{n+1}, modifying slightly the discharge at t_{n+1} by a quantity ΔQ and A by the quantity ΔA, while the discharge at the point $i + 1$ remains unchanged because the wave has not had the time to reach it between the time levels n and $n + 1$. Then, $A_i^{n+1} = A_0 + \Delta A$, $Q_i^{n+1} = Q_0 + \Delta Q$ and $Q_{i+1}^{n+1} = Q_0$. Equation (7.125) then gives

$$A_{i+1}^{n+1} = A_0 + 2\theta \Delta t \frac{\Delta Q}{\Delta x} - \Delta A \quad (7.126)$$

If A_0 is small and the conveyance of the channel is low (owing to high roughness, narrow free-surface width or a combination of both), even a small value ΔQ may lead to a large variation ΔA and A_{i+1}^{n+1} may drop below the initial value A_0 without any physical reason, thus creating artificial oscillations in the computed water depth and discharge. A_{i+1}^{n+1} may even become negative for some combinations of the initial water depth, geometry, hydraulic and numerical parameters. This undesirable behaviour can be prevented by keeping Δx sufficiently small to ensure the positiveness of the quantity $2\theta \Delta t \Delta Q/\Delta x - \Delta A$. Note that increasing the value of θ also has a stabilizing effect on the solution.

7.3.7 Elementary precautions in boundary-condition definition

7.3.7.1 Usual combinations of boundary conditions

The most widely used combinations of boundary conditions in natural river modelling are: (i) prescribed discharge at the upstream boundary and prescribed water level at the downstream boundary; (ii) prescribed discharge at the upstream boundary and a prescribed stage–discharge relationship at the downstream boundary.

When the channel system to be modelled is subjected to a strong backwater influence, as is the case in estuary modelling or in the modelling of coastal streams, the influence of the tide precludes the definition of one-to-one stage–discharge relationships. If the upstream boundary of the river system is too close to the sea, measuring the discharge reliably via water-level measurements becomes impossible. In such cases, water levels must be prescribed as functions of time at both ends of the river.

It should be noted, however, that some types of boundary conditions should be used with care, while some others should be avoided or they will lead to instability in the numerical solution. Typical examples are given in the following subsections.

7.3.7.2 Prescribed outflowing discharge

Although the theory of characteristics allows a discharge to be prescribed at the downstream end of the channel system (in which case it is a prescribed outflowing discharge), the modeller should be aware that prescribing exactly the desired values of the outflowing discharge may not always be possible. More specifically, the prescribed value of the discharge should be such that (i) no more water is taken from the channel than can actually be supplied physically, and (ii) the prescribed value of the discharge must be such that it does not yield a supercritical outflow at the downstream boundary.

Consider the downstream end of a channel where a discharge Q_b is to be prescribed. For the sake of simplicity of the analysis, the channel is assumed to be rectangular and horizontal, and friction is assumed to be negligible. Writing the invariance of the positive Riemann invariant $V + 2c$ between the channel and the boundary yields

$$V_0 + 2c_0 = V_b + 2c_b \tag{7.127}$$

where the subscripts 0 and b denote the values inside the channel and at the boundary, respectively. Inserting the equality $Q_b = A_b V_b = B/g\ c_b^2 V_b$ leads to

$$Q_b = \frac{B}{g}(V_0 + 2c_0 - 2c_b)c_b^2 \tag{7.128}$$

A simple variation analysis indicates that Q_b is an increasing function of c_b for $0 \le c_b \le (V_0 + 2c_0)/3$ and a decreasing function of c_b for $c_b \ge (V_0 + 2c_0)/3$ (Figure 7.17a). Q_b is zero for $c_b = 0$ and $c_b = (V_0 + 2c_0)/2$, and takes its maximum value Q_{max} of Q_b for $c_b = (V_0 + 2c_0)/3$, with

$$Q_{max} = \frac{B}{g}\left(\frac{V_0 + 2c_0}{3}\right)^3 \tag{7.129}$$

The Froude number Fr_b at the boundary is given by

$$Fr_b = \frac{V_b}{c_b} = \frac{V_0 + 2c_0}{c_b} - 2 \tag{7.130}$$

Fr_b is a decreasing function of c_b that is equal to unity for $c_b = (V_0 + 2c_0)/3$ (Figure 7.17b).

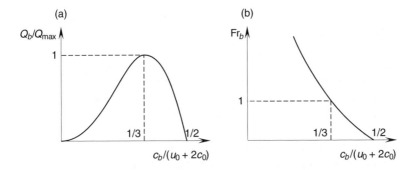

Figure 7.17 Variations in the discharge Q_b and the Froude number Fr_b at the downstream boundary as a function of the celerity c_b of the waves in still water

The values of Q_b that can be prescribed are those for which the Froude number at the boundary is smaller than unity. It stems from the above analysis of variations that all values of Q_b between 0 and Q_{max} can be prescribed. However, it is not possible to prescribe values of Q_b larger than Q_{max}, which is the maximum value of the discharge that can be physically taken out at the downstream end of the channel.

7.3.7.3 Prescribed water depth

From the analysis carried out in Section 7.3.7.2, it is easy to infer that prescribing any arbitrary value Y_b at the downstream or upstream end of a channel is not always possible.

Considering downstream boundaries and using the relationship $c = (gY)^{1/2}$ for a rectangular channel, it is easy to see from equation (7.130) that only values of Y_b that correspond to $c_b/(V_0 + 2c_0) = 1/3$ can be prescribed. In other words, the minimum meaningful value Y_{ds} for Y_b at a downstream boundary is given by

$$Y_{ds} = \frac{1}{9g}(V_0 + 2c_0)^2 \tag{7.131}$$

A similar analysis leads to the following formula for the minimum meaningful upstream boundary condition Y_{us}:

$$Y_{us} = \frac{1}{9g}(V_0 - 2c_0)^2 \tag{7.132}$$

Attempting to prescribe a water depth smaller than Y_{us} at an upstream boundary or smaller than Y_{ds} at a downstream boundary will result in an erroneous (if stable at all) numerical solution.

7.3.7.4 Stage–discharge relationship

Stage–discharge relationships should be used with great care. The following rules should be used.

1   Stage–discharge relationships where Q is an increasing function of η should never be used at upstream boundaries.
2   Stage–discharge relationships where Q is a decreasing function of η should never be used at downstream boundaries.

Consider an increasing function $Q(\eta)$ implemented at the upstream boundary of the channel. Assume that the initial flow conditions in the channel satisfy steady state. Then, if for some reason (a change in the downstream boundary conditions, operation of a gate within the channel, etc.) a perturbation propagating in the channel system reaches the boundary and triggers a change $\Delta\eta$ in the water level, two possibilities arise.

1   If $\Delta\eta$ is positive, a positive variation ΔQ arises in the discharge due to the increasing nature of the function $Q(\eta)$. Owing to continuity, this triggers a new, positive variation $\Delta\eta$ in the elevation of the free surface at the next computational time step. This variation in turn triggers a new, positive variation ΔQ in the discharge at the upstream boundary. The positive feedback thus created leads to solution instability.
2   If $\Delta\eta$ is negative, the increasing nature of the function $Q(\eta)$ yields a negative variation ΔQ in the discharge. This in turn generates another negative variation $\Delta\eta$, with an associated negative ΔQ. The process is repeated until the depth at the upstream boundary becomes zero.

Reasoning along the same lines allows rule 2 to be established. Note that if an increasing $Q(\eta)$ relationship is prescribed at a downstream boundary or a decreasing $Q(\eta)$ relationship is prescribed at an upstream boundary, $\Delta\eta$ and ΔQ have opposite signs, which has a stabilizing effect on the solution.

7.4 Special applications

7.4.1 Quasi-two-dimensional models

7.4.1.1 Application fields – governing equations

Quasi-two-dimensional models are developed for floodplain modelling (Cunge (1975), Cunge *et al.* (1980)). Floodplains mostly act as storage compartments, where the water flows very slowly compared to the typical flow velocities in the main channel system. This is because the water depth (and consequently the hydraulic radius) is also much smaller in the floodplain than in the channel system, with an increased influence of friction.

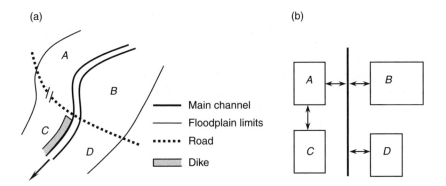

Figure 7.18 Definition of a quasi-two-dimensional model. Real-world situation (a), schematization in the quasi-two-dimensional model (b)

Representing the floodplain directly as part of the channel cross-section would lead to violating assumption (A1) that the flow variables (and, more specifically, the flow velocity) can be considered uniform over a given cross-section. The floodplain is represented using so-called 'cells' or 'storage units' (Figure 7.18). The cells may exchange water with each other and with the main channel, thus providing a simplified yet acceptably accurate representation of the flow processes.

The definition and parameterization of the links between the channel and the cells, as well as the links between the storage cells, must incorporate the existing exchange patterns. Consider the example of a main road crossing a floodplain, with a dike isolating the right-hand floodplain from the main channel downstream of the road, as sketched in Figure 7.18a. Assume that the continuity of the floodplain on the right-hand side of the river is ensured by a bridge or by culverts that allow the domains A and C to communicate, while such communication is disabled on the left-hand side of the channel between the subregions B and D. In the quasi-two-dimensional model a link must be defined between A and C, while there is no such communication between B and D. The impossibility of the main channel to exchange water with the cell C is accounted for by the absence of a link in the model.

7.4.1.2 Governing equations

The flow between the cells and the channel network obeys continuity:

$$\frac{\partial V_{S_i}}{\partial t} = -\sum_{j}^{i} Q_{i,j}$$

(7.133)

where V_{Si} is the volume stored in the cell i and $Q_{i,j}$ is the discharge flowing from the cell i to the cell j. The summation in equation (7.133) is carried out for all the neighbouring cells j of i. Note that an additional discharge may be present in equation (7.133) to account for the exchange between the cell i and the channel network. The integral of q_i with respect to the x-coordinate represents the amount of water that flows from the cell i into the channel system and has the dimension of a discharge. Note that equation (7.133) is usually solved in non-conservation form using the elevation η_i in the cell as a dependent variable

$$A_i \frac{\partial \eta_i}{\partial t} = -\sum_j Q_{i,j}(\eta_i, \eta_j) \tag{7.134}$$

where A_i is the plan view area of the cell i. Various options are available for the exchanges between the floodplain cells. A classical-resistance formula may be applied, such as

$$Q_{i,j} = \varepsilon_{i,j} K_{i,j} W_{i,j} Y_{i,j} \left(\frac{|\eta_i - \eta_j|}{L_{i,j}} \right)^{1/2} \tag{7.135a}$$

$$Y_{i,j} = \frac{y_i + y_j}{2} \tag{7.135b}$$

$$\varepsilon_{i,j} = \text{sgn}(\eta_i - \eta_j) \tag{7.135c}$$

where $K_{i,j}$, $L_{i,j}$, $W_{i,j}$ and $Y_{i,j}$ are, respectively, the average friction coefficient, the distance, the average cell width and average water depth between the centroids of the cells i and k. In equation (7.135b), $Y_{i,j}$ is computed assuming that the free-surface elevation varies linearly between the cells i and j.

When the cells are separated by structures or topographical singularities such as levees, roads, etc., a weir stage–discharge relationship may provide a more realistic description of the flow:

$$Q_{i,j} = \varepsilon_{i,j} \mu_{i,j}^{(1)} (\eta_{us} - z_b)[(\eta_{us} - z_b)2g]^{1/2} \tag{7.136a}$$

$$Q_{i,j} = \varepsilon_{i,j} \mu_{i,j}^{(2)} (\eta_{ds} - z_b)[(\eta_{us} - \eta_{ds})2g]^{1/2} \tag{7.136b}$$

$$\eta_{us} = \max(\eta_i, \eta_j) \tag{7.136c}$$

$$\eta_{ds} = \min(\eta_i, \eta_j) \tag{7.136d}$$

where z_b is the elevation of the crest of the weir and $\mu_{i,j}^{(1)}$ and $\mu_{i,j}^{(2)}$ are discharge coefficients to be used under free-flowing and flooded conditions, respectively.

7.4.1.3 Algorithmic aspects

The system formed by the flow equations (7.134) between the cells is usually linearized and solved using an implicit approach

$$A_i^{n+1/2} \frac{\eta_i^{n+1} - \eta_i^n}{\Delta t} = -\sum_j Q_{i,j}(\eta_i^{n+1}, \eta_j^{n+1}) \qquad (7.137)$$

or a semi-implicit approach

$$A_i^{n+1/2} \frac{\eta_i^{n+1} - \eta_i^n}{\Delta t} = -\sum_j \theta Q_{i,j}(\eta_i^{n+1}, \eta_j^{n+1}) + (1 - \theta)Q_{i,j}(\eta_i^n, \eta_j^n) \qquad (7.138a)$$

$$A_i^{n+1/2} \frac{\eta_i^{n+1} - \eta_i^n}{\Delta t} = -\sum_j Q_{i,j}(\theta \eta_i^{n+1} + (1 - \theta)\eta_i^n, \theta \eta_j^{n+1} + (1 - \theta)\eta_j^n)$$

$$(7.138b)$$

Note that equations (7.130) and (7.138) converge to the same solution when the computational time step tends to zero. The system formed by the set of equations (7.137) or (7.138) is non-linear and may be solved using standard, iterative techniques for system inversion (see Chapter 3 and Section 3.5). The linearization of equations (7.138a) and (7.138b) requires comparable effort. Note that $A_i^{n+1/2}$ is an average value for A_i between the time levels n and $n+1$ that must be updated within the iterative process.

In natural cells, the plan-view area A_i usually tends to zero when η_i tends to the bottom level. This triggers similar problems to those described in Section 7.3.6. Namely, an exaggerated sensitivity of the model's response to small variations in the water level in the neighbouring cells. It may be necessary to modify the law $A(\eta)$ artificially in such a way that $A(z_b) \neq 0$, while preserving the storage volume in the cell.

7.4.2 Two-dimensional models

7.4.2.1 Application fields – governing equations

Although two-dimensional models for estuary modelling are treated in detail in Chapter 11 (see Section 11.3), their use in floodplain modelling, as well as the governing equations and the basic solution techniques, are outlined here.

Although allowing two-dimensional flow patterns to be reproduced to some extent via main-channel–cell and intercell exchange, quasi-two-dimensional models may fail to provide an accurate representation of the actual flow patterns in situations where (i) the flow patterns cannot be assumed to be one-dimensional, and (ii) the inertial terms cannot be

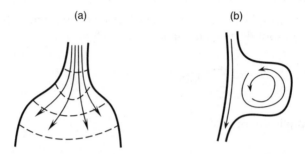

Figure 7.19 Typical situations where the one- and quasi-two-dimensional approaches are invalid. (a) Sharp transient propagating into a channel with strong geometry variations. (b) Recirculating flow in a lateral storage pocket due to momentum diffusion

neglected any more in the momentum equations. In such cases the quasi-two-dimensional approach becomes insufficient. To give but a few examples, this is the case when a sharp transient (e.g. a flash flood) propagates into a valley or floodplain with very strong variations in channel geometry, such as sudden widenings or narrowings (Figure 7.19a), or when the flow in the main channel creates a swirl in an expansion zone as a result of lateral momentum diffusion (Figure 7.19b). The swirl that appears as a result of momentum diffusion cannot be reproduced using classical exchange laws in the form of equations (7.135) and (7.136).

In such cases, a two-dimensional description of the flow processes is needed. Classical two-dimensional models retain assumptions (A2)–(A5) introduced in Section 7.2.1, while assumption (A1) is 'relaxed' into the hypothesis of a uniform velocity distribution over the vertical. These assumptions lead to the following vector equation in conservation form:

$$\frac{\partial \mathbf{U}}{\partial t} + \frac{\partial \mathbf{F}_x}{\partial x} + \frac{\partial \mathbf{F}_y}{\partial y} = \mathbf{S} \tag{7.139}$$

where the conserved variable $\mathbf{U}$, the x- and y-fluxes $\mathbf{F}_x$ and $\mathbf{F}_y$ and the source term $\mathbf{S}$ are given by

$$\mathbf{U} = \begin{bmatrix} Y \\ Yu \\ Yv \end{bmatrix}, \mathbf{F}_x = \begin{bmatrix} Yu \\ Yu^2 + gY^2/2 \\ Yuv \end{bmatrix}, \mathbf{F}_y = \begin{bmatrix} Yv \\ Yuv \\ Yv^2 + gY^2/2 \end{bmatrix},$$
$$\mathbf{S} = \begin{bmatrix} 0 \\ (S_{0,x} - S_{e,x})gY \\ (S_{0,y} - S_{e,y})gY \end{bmatrix}$$

$$\tag{7.140}$$

where $S_{0,x}$ and $S_{0,y}$ are the bottom slope in the x- and y-direction, respectively, and $S_{e,x}$ and $S_{e,y}$ are the friction slopes in the x- and y-direction, respectively. $S_{e,x}$ and $S_{e,y}$ are assumed to obey similar laws to the classical one-dimensional friction laws presented in Section 7.2.1, with the difference that the hydraulic radius in a two-dimensional context is equal to the water depth Y. For instance, Manning's friction law may be extended to two dimensions of space as follows:

$$S_{e,x} = n^2 \frac{(u^2 + v^2)^{1/2} u}{Y^{4/3}} \tag{7.141a}$$

$$S_{e,y} = n^2 \frac{(u^2 + v^2)^{1/2} v}{Y^{4/3}} \tag{7.141b}$$

Some commercially available packages incorporate the effects of momentum diffusion and wind-induced stress by modifying the source term $\mathbf{S}$ as follows:

$$\mathbf{S} = \begin{bmatrix} 0 \\ (S_{0,x} - S_{e,x})gY + D\frac{\partial^2 (Yu)}{\partial x^2} + \frac{\tau_x}{\rho} \\ (S_{0,y} - S_{e,y})gY + D\frac{\partial^2 (Yv)}{\partial y^2} + \frac{\tau_y}{\rho} \end{bmatrix} \tag{7.142}$$

where D is a momentum diffusion coefficient and τ_x and τ_y are the wind-induced stress components, usually modelled using classical turbulent drag formulae:

$$\tau_x = C_D (u_w^2 + v_w^2)^{1/2} u_w \tag{7.143a}$$

$$\tau_y = C_D (u_w^2 + v_w^2)^{1/2} v_w \tag{7.143b}$$

where C_D is the wind-drag coefficient and the subscript w indicates the wind velocity. A number of formulae for C_D are available in the literature (Charnock (1995), Geernaert *et al.* (1986), Large and Pond (1981), Luyten *et al.* (1999), Smith and Banke (1975), Wu (1969)).

7.4.2.2 Behaviour of the solution – boundary conditions

Equation (7.139) can also be rewritten in non-conservation form as

$$\frac{\partial \mathbf{U}}{\partial t} + \mathbf{A}_x \frac{\partial \mathbf{U}}{\partial x} + \mathbf{A}_y \frac{\partial \mathbf{U}}{\partial y} = \mathbf{S} \tag{7.144}$$

where A_x and A_y are, respectively, the Jacobian matrices of F_x and F_y with respect to U. From equation (7.140),

$$A_x = \begin{bmatrix} 0 & 1 & 0 \\ c^2 - u^2 & 2u & 0 \\ -uv & v & u \end{bmatrix}, \quad A_x = \begin{bmatrix} 0 & 0 & 1 \\ -uv & v & u \\ c^2 - u^2 & 0 & 2v \end{bmatrix} \quad (7.145)$$

The right-hand side of equation (7.144) is said to be a hyperbolic PDE because any linear combination of the matrices A_x and A_y has real, distinct eigenvalues. The hyperbolic nature of the equations allows a characteristic formulation to be derived (Daubert and Graffe (1967), Gerritsen (1982), Guinot (2008)). Two characteristic surfaces can be identified in the (x, y, t) space (Figure 7.20). The first surface (S_1) is reduced to a curved line of equations $(dx/dt = u, \ dy/dt = v)$; the second surface (S_2) is conical and expands from the first surface at a speed c in all directions of space. The domain of dependence of the solution is contained within the extension of the characteristic surface (S_2) to negative times.

As shown in Guinot (2008), the vector equation (7.144) can be rewritten in the following characteristic form:

$$\frac{d}{dt}(u_\xi - 2c) = S_1 \quad \text{for} \quad \frac{d\xi}{dt} = u_\xi - c \quad (7.146a)$$

$$\frac{d}{dt}v_\xi = S_2 \quad \text{for} \quad \frac{d\xi}{dt} = u_\xi \quad (7.146b)$$

$$\frac{d}{dt}(u_\xi + 2c) = S_3 \quad \text{for} \quad \frac{d\xi}{dt} = u_\xi + c \quad (7.146c)$$

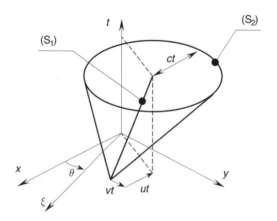

Figure 7.20 Definition of the characteristic surfaces in the (x, y, t) space

where ξ is the coordinate along the axis that makes an angle θ with the x-axis (Figure 7.20), and u_ξ and v_ξ are the components of the velocity vector in the directions parallel and orthogonal to ξ, respectively. The expressions for S_p ($p = 1, 2, 3$) can be found in Daubert and Graffe (1967), Gerritsen (1982) or Guinot (2008). Equations (7.146) are another way of saying that the two-dimensional shallow-water equations are invariant by rotation. The two characteristic lines $dx/dt = V \pm c$ of the one-dimensional Saint Venant equations are the intersection of the two-dimensional characteristic surface (S_2) with the plane ξ. The characteristic relationships in the one-dimensional Saint Venant equations may be seen as the one-dimensional restriction of equations (7.146a) and (7.146c).

The additional characteristic relationship (7.146b) (as compared to the two characteristic relationships in the one-dimensional Saint Venant equations) induces a dependence of the number of boundary conditions to be prescribed, not only on the flow regime but also on the flow direction. Four possibilities arise, which are discussed here assuming that the ξ-axis is oriented in the direction normal to the boundary (positive outwards).

1   *Supercritical flow entering the domain*. The three characteristics (7.146) enter the domain. The behaviour of the flow in the immediate neighbourhood of the boundary is entirely determined by the boundary conditions, and three independent flow variables must be prescribed.
2   *Subcritical flow entering the domain*. The two characteristics $d\xi/dt = u_\xi - c$ and $dx/dt = u_\xi$ enter the domain, while the characteristic $d\xi/dt = u_\xi + c$ leaves the domain. Two boundary conditions are needed, one of which must be the transverse velocity.
3   *Subcritical flow leaving the domain*. The characteristic $d\xi/dt = u_\xi - c$ enters the domain, and the characteristics $d\xi/dt = u_\xi$ and $d\xi/dt = u_\xi + c$ leave the domain. Only one boundary condition is needed.
4   *Supercritical flow leaving the domain*. All three characteristics leave the domain and the solution at the boundary is entirely determined by the flow state inside the domain. Consequently, no boundary condition is needed.

7.4.2.3 Numerical solution

Many approaches are available for the solution of the two-dimensional flow equations. As a broad rule, finite-difference-based techniques use structured grids (either Cartesian or curvilinear), that make the discretization of the spatial derivatives easier, while finite-volume- and finite-element-based methods use unstructured grids. The first attempts to solve the two-dimensional shallow-water equations were based on the two-dimensional method of characteristics on Cartesian grids (Daubert and Graffe (1967), Katopodes (1977), Katopodes and Strelkoff (1979)).

Most commercially available software packages solve the governing equations using implicit schemes, which allows large time steps to be used without any stability constraint on the Courant, Friedrichs and Lewy (CFL) number. This implies the solution of large, non-linear systems, which is the reason why the non-conservation form (equation 7.144), which can be easily linearized, is solved in most cases.

Time splitting (also known as operator splitting, see Strang (1968) for an overview of the technique) is often used in the solution of the two-dimensional equations. Two main approaches may be used for time splitting.

1   *Dimensional splitting.* Dimensional splitting, also called 'alternate directions', is used when the equations are to be solved using Cartesian grids. In this case, equation (7.139) is rewritten in the following two forms:

$$\frac{\partial \mathbf{U}}{\partial t} + \frac{\partial \mathbf{F}_x}{\partial x} = \mathbf{S} - \frac{\partial \mathbf{F}_y}{\partial y} \tag{7.147a}$$

$$\frac{\partial \mathbf{U}}{\partial t} + \frac{\partial \mathbf{F}_y}{\partial y} = \mathbf{S} - \frac{\partial \mathbf{F}_x}{\partial x} \tag{7.147b}$$

Equations (7.147a) and (7.147b) are solved successively, the result of the previous being used as an initial condition for the solution of the next. If the solution method is explicit, solving successively equations (7.147a) and (7.147b) is sufficient. If the solution method is implicit, the right-hand side terms in equations (7.147) must be discretized in a semi-implicit way, thus leading to an iterative procedure. Iterations are stopped when the difference between the solution obtained after two successive solutions of equation (7.147b) falls below a predefined threshold that is specified by the modeller. This solution technique is often referred to as the 'alternate directions implicit' (ADI) technique.

Alternate directions have the advantage of simplicity because they imply the solution of one-dimensional equations. They lead to the construction of tri-diagonal to pentadiagonal matrix systems *s*, for which standard inversion techniques are available. This makes them easy to implement.

2   *Operator splitting.* In most commercially available packages using unstructured grids the governing equations are broken into several parts, each of which expresses different physical processes. For instance, equations (7.139) may be solved in the following sequence:

$$\frac{\partial U}{\partial t} + \frac{\partial F_x^{(a)}}{\partial x} + \frac{\partial F_y^{(a)}}{\partial y} = 0 \tag{7.148a}$$

$$\frac{\partial U}{\partial t} + \frac{\partial F_x^{(p)}}{\partial x} + \frac{\partial F_y^{(p)}}{\partial y} = 0 \tag{7.148b}$$

$$\frac{\partial U}{\partial t} + \frac{\partial F_x^{(d)}}{\partial x} + \frac{\partial F_y^{(d)}}{\partial y} = 0 \tag{7.148c}$$

$$\frac{\partial U}{\partial t} = S \tag{7.148d}$$

where the superscripts (a), (p) and (d) represent the advection, propagation and diffusion part of the flux function, respectively:

$$\mathbf{F}_x^{(a)} = \begin{bmatrix} uY \\ u^2Y \\ uvY \end{bmatrix}, \mathbf{F}_x^{(p)} = \begin{bmatrix} 0 \\ gY^2/2 \\ 0 \end{bmatrix}, \mathbf{F}_x^{(d)} = \begin{bmatrix} 0 \\ -\partial(uY)/\partial x \\ 0 \end{bmatrix} \tag{7.149a}$$

$$\mathbf{F}_y^{(a)} = \begin{bmatrix} vY \\ uvY \\ v^2Y \end{bmatrix}, \mathbf{F}_y^{(p)} = \begin{bmatrix} 0 \\ 0 \\ gY^2/2 \end{bmatrix}, \mathbf{F}_y^{(d)} = \begin{bmatrix} 0 \\ 0 \\ -\partial(vY)/\partial y \end{bmatrix} \tag{7.149b}$$

In a first-order time-splitting approach, equations (7.148) are solved sequentially, each step using the result of the previous one as an initial condition for the solution of the next time step. In contrast with alternate directions, this procedure is not iterative. The interest of the time-splitting approach is that different numerical techniques may be used for the solution of equations (7.148a)–(7.148d), each of these techniques being best suited for the solution of the relevant part of the equations. This is the case of the advection part (equation (7.148a)) of the governing equations, for which specific upwinding techniques (such as the SUPG, finite-element method, see Hervouët (2007)) providing optimal advective gradient discretization are available.

Each of the above time-splitting techniques has advantages and drawbacks. Their common advantage is programming simplicity. A drawback of the alternate-directions approach is the accurate representation of the flow patterns near steep topographical gradients. When the ADI approach is used, too large a computational time step or an insufficiently converged iteration sequence may lead to unphysical solutions such as larger velocities in the floodplain than in the main channel, or velocity fields and wetting fronts unnaturally aligned with the main grid directions. When large time steps are used, the operator-splitting approach may lead to erroneous solutions in that the final solution at the end of the sequence (equation (7.148)) may not satisfy the weak form of equation (equation (7.139)). Although the consequences of this may remain unnoticed in most standard engineering applications, they may induce solution inaccuracy (e.g. wrongly located bores, inaccurate hydraulic jump height computation) in the case of more specific applications such as sharp transients.

7.4.2.4 Mesh design

The finite-element or finite-volume solution of the two-dimensional shallow-water equations on unstructured grids has gained considerable popularity in the engineering community over the last decade. Commercially available packages are supplied with grid-generation programs that allow arbitrary geometries to be represented accurately. Although largely facilitated by increasingly sophisticated and user-friendly packages, two-dimensional grid generation for refined hydraulic studies remains time-consuming; it cannot be fully automated and still requires considerable operating time. Depending on the complexity of the geometry and the requirements of the modelling study in terms of accuracy, the design of the mesh may represent up to 50% of the time devoted to a two-dimensional hydraulic study. As a consequence, the technician in charge of grid generation and the hydraulic engineer in charge of the supervision bear considerable responsibility for the quality of the modelling results. The present subsection is devoted to elementary guidelines for two-dimensional grid design.

As in the one-dimensional case, the most influential factor in the quality of the numerical solution of the two-dimensional shallow-water equations is the Courant number, which expresses the ratio of the distance covered by the wave within a time step to the size of the computational cell. Stability analyses of numerical techniques for solving the two-dimensional shallow-water equations (Soares-Frazão and Guinot (2007)) have led to the following two-dimensional generalization of the Courant number:

$$\mathrm{Cr}_{i,j} = \frac{(u_{i,j} + c_i)w_{i,j}\Delta t}{A_i} \tag{7.150}$$

where A_i and c_i are, respectively, the plan-view area and the propagation speed of the waves in still water for the cell i, $u_{i,j}$ is the flow velocity in the direction orthogonal to the jth interface of the cell i, $w_{i,j}$ is the width of the jth interface of the cell i and Δt is the computational time step. The Courant number $\mathrm{Cr}_{i,j}$ as defined in equation (7.150) may be interpreted as the ratio of the area covered by the fastest of the waves in the direction normal to the jth interface of the cell i within a time step Δt to the plan-view area of the cell. Note that, in most situations (except special applications such as dambreak flood wave modelling, see Section 7.5.4, or the refined modelling of overspilling over dikes or weirs), the flow velocity is small compared to the propagation speed of the waves in still water, and equation (7.150) may be approximated as

$$\mathrm{Cr}_{i,j} \approx \frac{c_i w_{i,j} \Delta t}{A_i} \tag{7.151}$$

For the accuracy of the numerical solution to be optimal, the Courant number must be as isotropic and homogeneous as possible over the computational domain. This is the case if (i) the mesh is mostly isotropic and (ii) the mesh is coarser in regions where $(u^2 + v^2)^{1/2} + c$ is large and finer in regions where $(u^2 + v^2)^{1/2} + c$ is small. Strongly distorted meshes may yield flow velocity fields that are abnormally aligned with the main directions of the grid.

7.4.3 Three-dimensional models

As three-dimensional models are covered in detail in Section 11.4 on estuary modelling, only a short description of their underlying principles and applications to river modelling is given here.

Three-dimensional models are needed in applications where the assumptions of negligible vertical accelerations and/or uniform velocity fields over the vertical do no longer hold. Such applications include, but are not limited to, the refined modelling of flow patterns near structures such as gates, bridges, sills and weirs, and culverts through levees. Significant vertical accelerations lead to invalidate the assumption of a hydrostatic-pressure distribution. One of the best known consequences of the invalidity of hydrostatic-pressure distribution is the development of undular bores near mobile jumps, with the maximum amplitude of the waves exceeding the amplitude of the bore computed from the Saint Venant equations by a factor of up to 2 (Cunge *et al.* (1980)). Note that the non-uniform character of the velocity field in three dimensions of space and the non-hydrostatic nature of the pressure field are two different issues (although the latter may be a consequence of the former). Some three-dimensional models use the assumption of hydrostatic-pressure distributions over the vertical, while some others do not.

Three-dimensional models are based on the three-dimensional Navier–Stokes equations (equations 4.3–4.5). Note, however, that these equations can be solved uniquely for u, v and w only if the pressure field $p\,(x, y, z)$ is known. The pressure is usually formulated as

$$p(x, y, z) = (\eta - z)\rho g + p^* \tag{7.152}$$

where the first term on the right-hand side of equation (7.152) accounts for hydrostatic factors and p^* is the so-called 'excess pressure' that accounts for non-hydrostatic effects. Note that introducing the notation in (7.152) allows the z-momentum equation to be written in exactly the same way as the x- and y-momentum equations by balancing the vertical volume force ρg with the vertical gradient of the hydrostatic part of the pressure as given by equation (7.152). The so-called 'kinematic boundary condition' allows η to be related to the flow velocity components as

$$\frac{\partial \eta}{\partial t} + u(x, y, \eta)\frac{\partial \eta}{\partial x} + v(x, y, \eta)\frac{\partial \eta}{\partial y} = w(x, y, \eta) \tag{7.153}$$

while the atmospheric pressure is used as a boundary condition at the free surface: $p(x, y, \eta) = p_{\text{atm}}$. Note that integrating equation (7.153) over the vertical leads to an equation for η

$$\frac{\partial \eta}{\partial t} + \frac{\partial}{\partial x}\int_{z_b}^{\eta} u\,dz + \frac{\partial}{\partial x}y\int_{z_b}^{\eta} v\,dz = 0 \tag{7.154}$$

Standard solution techniques use time splitting (see e.g. Casulli (1999), Stelling and Zijlema (2003)). A typical solution sequence is:

1   Solve the momentum equations without the excess pressure term:

$$\frac{\partial u_i}{\partial t} + \sum_{j=1}^{3}\frac{\partial}{\partial x_j}(u_i u_j) = \sum_{j=1}^{3}v\frac{\partial^2 u_i}{\partial x_j}, \quad i = 1, 2, 3 \tag{7.155}$$

where the subscripts i and j indicate the vector components ($i = 1, 2, 3$ for the x-, y- and z-direction, respectively). This step yields intermediate values for the velocity components. These are used as a starting point in the second step of the process.

2   Solve the continuity equation and the momentum equation with the excess pressure term only:

$$\sum_i \frac{\partial u_i}{\partial x_i} = 0 \tag{7.156a}$$

$$\frac{\partial u_i}{\partial t} + \frac{1}{\rho}\frac{\partial p^*}{\partial x_i} = 0, \quad i = 1, 2, 3 \tag{7.156b}$$

Equations (7.156a) and (7.156b) are solved for the excess pressure p^* and the velocity components $u_i(i = 1, 2, 3)$.

Steps 1 and 2 are usually part of an iterative, implicit or semi-implicit procedure, which must be repeated until the excess pressure field p^* and the velocity components u_i converge.

The equations may be discretized in the horizontal plane using structured or unstructured grids, but the vertical discretization is a structured, multi-layer discretization. A widespread discretization technique called the 'sigma coordinate approach' (see Section 11.4) consists of fixing the number of discretization points (or layers) over the vertical and to distributing them equally between the bottom and the free surface. The vertical locations of the computational points thus move vertically when the elevation of the free surface changes.

7.4.4 Dambreak flood-wave modelling

The salient feature of dambreak-generated flood waves from a hydraulic point of view is the presence of supercritical flow regions and discontinuous flows. The breaking of a dam is usually modelled as the instantaneous removal of the wall (or part of it), thus making the initial water level discontinuous. The upstream water level is the initial elevation of the free surface in the dam immediately before the failure occurs. The downstream water depth (if any) is usually taken from uniform, steady-state simulation under the average discharge being released by the dam. The discontinuity in the free-surface elevation yields an infinite acceleration over a time interval of zero length, resulting in a discontinuous velocity profile near the front of the wave. The thus-generated flood wave can usually be separated into three regions (Figure 7.21). The first region consists of a backward wave propagating into the reservoir at a celerity c. This wave connects the reservoir to region 2, where the flow is strongly supercritical. Region 3 is the front of the wave, the steepness of which depends on the intensity of friction and the initial water level downstream of the dam. In the case of zero friction, region 3 is infinitely narrow and the flow is discontinuous across it, which is the definition of a shock wave. The reader interested in a detailed analysis of shock-wave formation and propagation in open

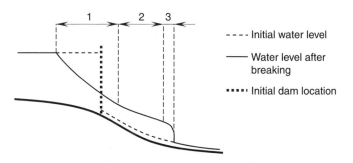

Figure 7.21 The three regions in a dambreak flood wave

channels may refer to Cunge *et al.* (1980), Guinot (2008) or Stoker (1957). More theoretical background may be found in Lax (1957), Stoker (1957) or Whitham (1974). Analytical solutions to the dambreak problem with zero and non-zero downstream depth can be found in Stoker (1957).

When a shock wave is present in the solution, the classical, differential form of the one- or two-dimensional open-channel equations ceases to be valid because the time and space derivatives become locally infinite. Only the weak form of the equations is applicable, and the so-called jump relationships (also called the Rankin–Hugoniot conditions) must be used across the shock. In the direction of propagation of the shock, such conditions may be written as follows:

$$\mathbf{F}_1 - \mathbf{F}_2 = (\mathbf{U}_1 - \mathbf{U}_2)c_s \qquad (7.157)$$

where c_s is the speed of the shock and the subscripts 1 and 2 denote the values of $\mathbf{F}$ and $\mathbf{U}$ on the left- and right-hand sides of the shock. Equation (7.134) is obtained from a balance over a control volume that contains the shock in the limit case where the size of the control volume and the time interval over which the balance is carried out tend to zero. It is a direct application of the conservation form (equations (7.20) and (7.139)). Note that, when the solution is discontinuous, the conservation, non-conservation and characteristic forms of a hyperbolic system of conservation laws cease to be equivalent. Solving the non-conservation or non-characteristic form of the equations without any specific treatment of discontinuities may result in incorrect shock location and/or amplitude calculation.

This stems from the above-mentioned considerations that dambreak modelling studies can be carried out using only specific modelling software packages for which two essential basic requirements are: (i) the capability to deal with transcritical flow; and (ii) a conservative solution of the governing equations that allows for the treatment of weak solutions, thus enabling the accurate computation of the location and amplitude of shocks such as hydraulic jumps and/or moving bores. For instance, the Preissmann scheme in its original form satisfies condition (ii) but fails to fulfil condition (i) (Meselhe and Holly (1997)). However, a new formulation of the scheme similar to the flux-splitting formalism (Johnson *et al.* (2002)) allows both conditions to be satisfied, but does not appear to have been implemented in industrial packages to date. The Abbott–Ionescu scheme in its original form fails to satisfy condition (ii), while the solution in pseudo-conservation form additionally fails to satisfy condition (i) because of the missing term $2u\partial Q/\partial x$ in the differentiation of the term Q^2/A (see Section 7.3.4.1).

Dambreak simulations are increasingly carried out using two-dimensional software packages that solve the two-dimensional shallow-water equations presented in Section 7.5.2. Most software packages use finite-volume or finite-element techniques to solve the governing equations on unstructured grids. Note that finite-element techniques deal with the weak form of the governing equations (see Section 3.5.7), which raises the issue of the non-uniqueness of weak solutions (see e.g. LeVeque (1990) for an outline of the problem). Finite-volume techniques are essentially conservative, at least in their explicit implementations, and do not exhibit such problems. Note that even a fully conservative method solving the correct weak form of the equations may fail to provide the correct solution when the hyperbolic part of the equations is not solved within a single algorithmic step (see Section 7.5.2.3). Moreover, the reader should be aware that, although classically presented as essentially conservative, most implicit finite-volume-based methods presented in the literature do not ensure exactly the conservation of mass and momentum because they are based on a linearized version of the governing equations (i.e. on the non-conservation form). Although the non-conservation form of the equations may be discretized in such a way that it is made conservative (see Cunge *et al.* (1980), Guinot (2008)), this is not always the case in the wide range of methods available in the literature.

7.5 Physical models of open-channel flow

7.5.1 General

As mentioned earlier (see Chapter 1), physical (scale, hydraulic) models of open-channel flow have been in use for well over a hundred years. In fact, models of rivers were among the first models to be conceived for solving practical engineering problems. Equally, however, the sphere of river engineering, particularly where fixed boundaries were involved, saw the early introduction and rise of computational modelling. This, as well as the substantial space requirements and the cost of hydraulic models of river systems, resulted in their relative decline during the second half of the 1900s. Nevertheless, they retain their role as a basic research tool and for modelling flow in open channels with complicated geometries and river–structure interaction. As interest in environmental processes in open channels grew, physical modelling of open-channel flow went through a renaissance and firmly remains an important modelling discipline – particularly as part of hybrid modelling.

As morphological processes and models with sediment transport are introduced in Chapter 8, this section deals only with models with fixed boundaries.

The following text gives only a brief introduction to the subject; for further reading, see, for example, de Vries (1986), French (1984), Knauss (1980) and Novak and Čábelka (1981).

7.5.2 Governing equations

The equations used for the design and operation of hydraulic models of open-channel flow are based on the equations dealt with in Sections 4.2–4.4 and 7.2 and on the procedures outlined in Chapter 5.

From the Saint Venant equations for unsteady flow (equations (4.66) and (4.67)), the Darcy–Weisbach equation (equation (4.36)), the equation for local head losses (equation (4.65)), bearing in mind that the difference in total heads H between two sections is

$$\Delta H = \Delta z + \Delta y + \Delta \left(\frac{V^2}{2g} \right) = \Delta \left(\frac{\lambda l V^2}{2gR} \right) + \Delta \left(\frac{\xi V^2}{2g} \right) \tag{7.158}$$

and utilizing the procedure outlined in Section 5.5.2, we can establish six equations for nine scale variables: M_z, M_b, M_l, M_h, M_λ, M_R, M_ξ, M_t, M_Q or M_v, where

M_z – scale of height (e.g. of channel bed) above datum

M_b – scale of width

M_l – scale of length

M_h – scale of depth

M_λ – scale of friction coefficient

M_R – scale of hydraulic radius

M_ξ – scale of local losses coefficient

M_t – scale of time

M_Q – scale of discharge

M_v – scale of velocity

$$M_z = M_h \tag{7.159}$$

$$M_v = M_h^{1/2} \text{ (Froude law)} \tag{7.160}$$

and thus

$$M_Q = M_A M_v = M_h M_b M_h^{1/2} = M_h^{3/2} M_b \tag{7.161}$$

$$M_R = M_l M\lambda \tag{7.162}$$

$$M_\xi = 1 \tag{7.163}$$

$$M_R = \Phi(M_h, M_h/M_b, h) \tag{7.164}$$

$$M_t = M_l/M_v = M_l M_h^{-1/2} \tag{7.165}$$

The above equations assume $M_g = 1$ and apply to a *distorted model* ($M_l \neq M_b \neq M_h$) of one-directional unsteady flow. With *six* equations for *nine* variables there are *three* degrees of freedom, and we can thus choose *three* variables.

For steady non-uniform flow, M_t and equation (7.165) (essentially the identity of Strouhal numbers) are redundant, leaving *five* equations for *eight* variables.

For uniform flow, there are no local losses (due to a change in cross-section or direction of flow) and thus equation (7.163) and M_ξ are also redundant. There is also no need for the scale above datum (e.g. of the channel bed, M_z) to be equal to the scale of depth (i.e. the model can be tilted with a suitable adjustment in roughness and equation (7.159) can be omitted). Finally, the necessity of choosing the discharge scale M_Q strictly according to the Froude law of similarity does not generally arise, and it may be changed slightly (say by 20%) if it is found to be necessary (e.g. during model verification tests; see also the following sections). Therefore, in this case we have only two equations (equations (7.162) and (7.164)) for *seven* variables (M_z, M_h, M_b, M_l, M_Q, M_λ, M_R), and *five* degrees of freedom.

The above is a general case; in practice it is nearly always useful to have only one horizontal scale, i.e. $M_b = M_l$ reducing the choice for unsteady and steady non-uniform flow to two scales (or to four for uniform flow). It has to be realized that there can be only one time scale for the water movement in the model resulting (from equation (7.165)) in the scale of the speed of vertical movements (e.g. of the water level) (see also equation 5.33):

$$M_w = M_h/M_t = M_h^{3/2} M_l^{-1} \tag{7.166}$$

There is also the need for a single horizontal scale for the case of a general (three-dimensional) unsteady flow, i.e. $M_l = M_b$.

The laboratory space, the pumping capacity and the type of problem to be studied in the model determine which scales are chosen in the model design; the parameters selected are usually M_l and M_h or M_l and M_Q.

In non-distorted models $M_b = M_l = M_h = M_R = M_z$ and the Froude law applies ($M_Q = M_h^{5/2}$). Thus, only one parameter, usually the length scale M_l, need be chosen.

The scale of the hydraulic radius is not only dependent on the depth and width scales but also on the absolute value of the depth modelled (equation (7.164)). Only for non-distorted models are the scales of depth and hydraulic radius equal; they are also approximately equal for wide channels. With $M_R = M_h = M_l$ it follows from equation (7.162) that $M_\lambda = 1$; therefore, the friction and local head-loss coefficients in the model and prototype are identical. The fact that the proportion of friction and local head losses is identical in the model and prototype can be important in river- and flood-flow modelling.

Recalling equations (4.37), (4.61) and (4.64a)–(4.64c), it follows that for fully developed turbulent flow (independent of the Reynolds number – see Section 7.4.3)

$$M_\lambda = \Phi(M_R, M_k) \qquad (7.167)$$

For *non-distorted* models ($M_R = M_h = M_l$ and $M_\lambda = 1$) it follows from equations (4.64) that

$$M_k = M_R = M_l \qquad (7.168)$$

For *distorted models*, equation (7.167) can be approximated from equation (4.64) as $M_\lambda = (M_m/M_R)^{1/3}$; combining this expression with equation (7.162) and using $M_R \approx M_h$ gives

$$M_k = M_h^4 M_l^{-3} \qquad (7.169)$$

Thus, the relative roughness in a distorted model is always larger than in the prototype.

It must be remembered, however, that the roughness size 'k' is really a concept using an 'equivalent uniform roughness size' giving the same friction loss as the actual roughness. In reality, the coefficient λ and the friction head loss will depend not only on the roughness size but also on the shape and distribution of roughness elements, and the head loss can only be properly determined by conducting verification experiments on the scale model.

7.5.3 Boundary conditions and scale effects

The equations presented in Section 7.4.2 are sufficient for the design and operation of open-channel-flow models on the assumption that the appropriate boundary conditions are observed in order to avoid, or at least minimize, the scale effect.

(a) The most important boundary condition is given by the need to avoid the influence of viscosity and to have fully turbulent flow in the model (see Sections 4.4 and 5.8.2), i.e.

$$\mathrm{Re}_m > \mathrm{Re}_{sq} \qquad (7.170)$$

The above condition should really apply for the smallest discharge used in the model. However, sometimes this is not practical; in this case it is obvious that for smaller discharges than the one satisfying the above condition the model Reynolds number Re_m must at least be large enough to prevent laminar flow, i.e. $Re_m > Re_{cr}$; in this case some scale effects for the lower flow are inevitable.

Just as for pipe flow, where an equation for Re_{sq} was established on the basis of Nikuradse's experimental work, an equation for open channels can be derived from similar work by, for example, Zegzhda (see Novak and Čábelka (1981)). Equation (4.52) can be written as:

$$Re_{sq} = \left(\frac{vD}{v}\right)_{sq} = 400\frac{r}{k\lambda^{1/2}} = 400\frac{R}{k\lambda^{1/2}} \qquad (7.171)$$

Therefore,

$$Re_{sq} = \left(\frac{vR}{v}\right)_{sq} = 200\frac{r}{k\lambda^{1/2}} = 100\frac{R}{k\lambda^{1/2}} \qquad (7.172)$$

The corresponding equation for open-channel flow is

$$Re_{sq} = \left(\frac{vR}{v}\right)_{sq} = 130\frac{r}{k\lambda^{1/2}} = 650\frac{R}{k\lambda^{1/2}} \qquad (7.173)$$

The difference in the coefficients in equations (7.172) and (7.173) reflects the shape factor of the conduit flow.

Another way of checking that the model has fully turbulent flow is to compare the components of the denominator in equation (4.61) $C = 18\log[6R/(\delta'/7 + k/2)]$ and ensure that $\delta'/7 << k/2$, where δ' is given by equation (4.31) ($\delta' = 11.6v/V^* = 11.6v/v(\lambda/8)^{1/2}$).

(b) The model should be large (deep) enough to avoid surface-tension effects (i.e. the influence of the Weber number). This was discussed in Section 5.3.2. Unfortunately, the guidance here is not as clearly defined as for the limiting values of the Reynolds number; it is, however, generally accepted that the depth of flow in river models should not fall below 0.03 m to avoid surface-tension effects, and that for a depth less than 0.015 m considerable scale effects may result.

(c) One condition concerning the Froude law (equation (7.161)) has already been mentioned. The observance of this law is essential when modelling non-uniform open-channel flow with an interaction with submerged structures (e.g. involving flow over submerged obstacles such as groynes) or through bridges, and it is preferable to adhere to it in all conditions. However, a relatively minor deviation of up to 20% is permissible for overall uniform flow conditions if this does not substantially alter the flow pattern, and may be required as a result of verification experiments (when the same overall effect cannot easily be achieved by the adjustment of roughness) or because of limitations in the pumping capacity. It is, of course, necessary to observe the condition of identical flow regimes in both the model and the prototype.

(d) Equation (7.164) shows that in a distorted model with significant water-level changes there always will be a scale effect. To minimize this, the distortion should be as small as possible and consistent with other boundary conditions. In practice, unless dictated by other conditions, a limit should be placed on the distortion M_l/M_b of about 5 ($M_l/M_b < 5$). It should be noted, however, that a distorted model satisfies the condition of a sufficiently large Reynolds number (equation (7.170)) more easily than does a non-distorted model, due to its large relative roughness (equation (7.169)).

(e) Equation (7.163) indicates that in cases where the more detailed flow pattern is important, rather than (or as well as) an overall friction loss, there should be a limit on the size of individual 'roughness' elements representing the overall roughness. The condition often used is $R/k > 5$, although good results have been obtained for greater roughness sizes.

(f) In a distorted model, the coefficient of local energy loss ξ due to a change in the cross-section is often greater in the model than in the prototype. In this case, equation (7.163) ($M_\xi = 1$) can be interpreted as applying to the sum of local losses, and a significant difference in local losses due to changes in channel section can be compensated for by a reduction in the losses due to a change in direction (or, ultimately, by the adjustment of roughness and the friction loss).

7.5.4 Aerodynamic modelling of open-channel flow

The basis of aerodynamic modelling of open-channel flow is the conversion of free-water surface flow to flow in a pressurized conduit by substituting a smooth surface for the air–water interface. If the position of the free-water surface is known beforehand, then the procedure is quite simple and the model can be used to investigate the flow patterns in a channel or, for example, in the approaches to a barrage, power station, navigation lock, etc. (see also Chapter 13). If, however, we are dealing with non-uniform flow, where the position of the water level is not known beforehand (and is, in fact, the objective of the investigation), the procedure is much more complicated.

The conditions for the simulation of flow in a hydraulically rough channel have been discussed in the previous sections. If a free-water surface is replaced by a smooth plate and the flow is pressurized, with the pressure head simulating the free surface, the Froude number becomes irrelevant, and the only conditions remaining are the similarity of conduit shapes, the relative roughness and that the Reynolds number in the model should be in the hydraulically rough region ($Re_m > Re_{sq}$). In principle, we could use any Newtonian fluid in this type of model; the use of air instead of water is mainly a matter of convenience, economics and technical suitability, as pumping costs and space requirements are substantially reduced and the need for a circuit with inlet and outlet tanks is avoided. If air is used, it is, of course, essential to limit the air-flow velocity to values where compressibility effects are negligible (i.e. limit of the Mach number); this value is about 50 m/s, a condition that is easily satisfied in practice.

The main advantage of aerodynamic modelling of open-channel flow is that the models can be much smaller than normal free-surface hydraulic models which are governed by the relationship of the Froude numbers in the model and the prototype, and a permissible minimum Reynolds number for the model as the necessary value of the Reynolds number may be achieved by an increase in velocity (and thus also discharge) of air without regard to the Froude number. (It is necessary to remember, however, that the coefficient of kinematic viscosity for air at temperature 20 °C and atmospheric pressure is about 15×10^{-6} m^2/s, i.e. about fifteen times greater than the coefficient of viscosity for water.)

The introduction of a smooth plane surface (e.g. a glass plate) instead of the free surface distorts the boundary conditions; a number of studies have dealt with the influence of this surface on the deformation of the velocity field. It follows that air models of open-channel flow are particularly useful wherever this deformation does not substantially influence the total flow picture and where the water level in the prototype does not deviate significantly from a plane surface.

Theoretical and experimental studies aimed at establishing the type of free-surface flow suitable for study in air models were carried out in the former USSR by Averkiev (1957) and Ljatcher and Proudovskij (1959, 1971). They investigated the extent of the reversal of flow behind a sudden expansion (see Figure 7.22(a). On the assumption that $M_\lambda = 1$, Averkiev found that the percentage error in the relative length of the vortex $l/(b_2 - b_1)$ due to the use of a fixed plate instead of a free surface depended on the geometric

(a)   (b)

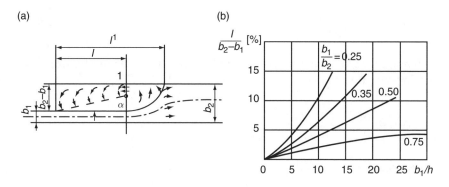

Figure 7.22(a) (b) Scope of the aerodynamic modelling of open-channel flow

parameters b_1/b_2 and b_1/h, as shown in Figure 7.22(b) (where h is the depth of flow simulated by the distance between the smooth plate and the bed). Accepting an error of 5% in the length of the vortex, the aerodynamic model may be used for all open channels where $b_1/b_2 > 0.75$ and $b_1/h < 30$; limiting the error in the vortex size to 10% would mean $b_1/b_2 > 0.5$ and $b_1/h < 23$, etc.

Thus, although in fact the value of $l/(b_2 - b_1)$ is also a function of the Froude number (for very low values its influence is insignificant), it is evident that there is a wide range of situations where the aerodynamic models of open-channel flow may be used with reasonable confidence.

As the Froude law is no longer applicable and as the model discharge may be chosen arbitrarily (as long as $\text{Re}_m > \text{Re}_{sq}$ and $V < 50\,\text{m/s}$), the scale of the velocities is given by

$$M_v = \frac{Q_p A_m}{Q_m A_p} = \frac{M_Q}{M_b M_h} = \frac{M_Q}{M_l M_h} \tag{7.174}$$

(This scale is valid for mean cross-sectional velocities as well as for local velocities.)

The scale of the friction coefficient for a distorted model is again given by equation (7.162), but it is necessary to take into account that the hydraulic radius in the model can no longer be approximated by the depth (for a wide channel), as in this case $R = h/2$ and $M_R = 2M_h$. The scale of the friction coefficient then becomes

$$M_\lambda = \frac{M_R}{M_l} = 2\frac{M_h}{M_l} \tag{7.175}$$

From the Darcy–Weisbach equation the pressure changes between two sections of a pressurized model are given by

$$\Delta p = \rho_m \lambda_m l_m \frac{V_m^2}{2R_m} \tag{7.176}$$

Noting that, for the difference in water levels above datum between two sections in prototype,

$$\Delta h_p = \frac{\Delta b_p}{\rho_p g} = \lambda_p l_p \frac{V_p^2}{2gR_p} \tag{7.177}$$

we get for $\Delta p_p/\Delta_m = M_p$

$$M_p = M_\rho M_\lambda M_l M_v^2 M_R^{-1} \tag{7.178}$$

Substituting from equations (7.174) and (7.175) leads to

$$M_p = \frac{M_\rho M_Q^2}{M_l^2 M_h^2} \tag{7.179}$$

The difference in the water levels above datum between two prototype sections $\Delta h = \Delta p_p/(\rho g)$ will be simulated in the model by the difference in elevation of the cover as $\Delta h_m = \Delta h_p/M_h$. Thus, from equation (7.174) we get the condition

$$\Delta h_m = \frac{\Delta h_p}{M_h} = \frac{\Delta p_p}{\rho_p g M_h} = \frac{M_p \Delta p_m}{\rho_p g M_h} = \frac{M_Q^2 \Delta p_m}{\rho_m g M_l^2 M_h^2} \tag{7.180}$$

If the difference in water levels in the prototype is initially not known, we proceed by iteration and gradually adjust the position of the model cover until equation (7.175) is satisfied.

There is sufficient evidence that observation of the above criteria gives good results even for detailed flow analysis and head losses on distorted aerodynamic models of river flow up to a distortion of about 2.5. When only total losses in a river stretch are required for qualitative studies (e.g. pressure losses for an arbitrary position of the water surface, i.e. a glass cover), very good results have been reported for much greater distortions (e.g. 20). With a suitable distortion it certainly is possible to achieve the same distribution of velocities as for an open channel in the lower part of the cross-section (below the zone of maximum velocity); the distortion in the velocity distribution produced by the cover could be partially offset by placing the cover in a slightly more elevated position than would correspond to the water level.

For equal roughness on the model bed and cover, the relationship between λ, Re and R/k is given by Zegzhda's experiments, but for different roughness on the bed and on the cover the resulting value of λ could deviate substantially. The total frictional resistance coefficient for a cross-section of varying roughness on the bottom and covering plate may be sufficiently accurately calculated as the arithmetical mean of the resistance coefficients for even roughness over the entire section and the same depth.

In aerodynamic models, apart from the pressures that are measured by means of manometers connected to normal piezometric openings, velocities are usually measured using small pressure head (Pitot) tubes or hot-wire anemometers, and flow paths are studied using various tracers (smoke, sparks, etc.) or light threads. The models are usually placed on tables or trestles with access from below for the insertion of velocity probes and pressure tappings. The morphological features are usually modelled using thin plasticine templates, with a small amount of linseed oil added to prevent cracking, or sometimes layers of plywood have been used. Due to the small size of the models, great care and accuracy are required in their construction. The cover is usually made of glass or perspex (5–10 mm thick) and the contact between the cover and the model has to be carefully sealed (e.g. with putty).

Air may be either driven into the model or exhausted from it by fans discharging $1,000–10,000\,\text{m}^3/\text{h}$ at a pressure of about 500–600 mm of water; air discharge is usually measured by means of an orifice. An arrangement where the outlet from the model is connected to the suction side of the fan is used more frequently, as this allows an easy introduction of tracers at the model inlet. In this case it is necessary to check the model carefully and, if necessary, prevent the bending of the model cover by supporting it at intervals by thin rods that act as distance pieces; for this reason a combination of driving air into the model and extracting it at the end is sometimes used so that atmospheric pressure is maintained approximately in the middle of the model.

Because of their small dimensions and the possibility of rapid adjustment, aerodynamic models of open-channel flow are particularly well suited for preliminary studies of various alternatives of the layouts of complex low-head hydraulic structures, intakes, river training schemes, etc. The 'optimum' solution is then often tested in greater detail using a conventional hydraulic model, but in less important cases it is possible to base the final design on the aerodynamic model alone. Reliable quantitative results may be achieved, especially for river studies where there is a fixed bed and predominantly two-dimensional flow.

7.6 Case studies

7.6.1 The Vidourle river

The purpose of this section is to describe the steps in the construction of a one- and two-dimensional free-surface-flow model. In order to better illustrate the differences between the approaches, the one- and two-dimensional models are built for the same river reach. The selected site is a reach of the Vidourle river that has been used as an application example in previous river-model calibration studies (Guinot and Cappelaere (2009)). The topography of the bed is shown in Figure 7.23.

The flow in this part of the river is intermittent. This behaviour stems from: (i) the highly seasonal rainfall pattern, with intense events concentrated mainly in autumn and spring; and (ii) the strongly karstified underlying geological layers, combined with a bed made of gravel and stones that allows large seepage rates from the river. As a consequence, the discharge pattern is a succession of short peaks, occurring mainly between autumn and spring, separated by long zero-discharge periods. The river bed is made

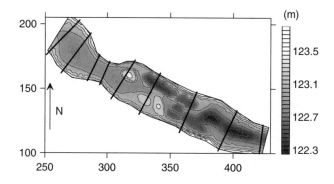

Figure 7.23 Bathymetry of the river reach used for the comparison between a one-dimensional model and a two-dimensional model. Bold lines: the location of the cross-sections in the one-dimensional model. Contour line spacing: 5 cm

of gravel and small stones, with sizes ranging typically from 1 to 5 cm. As a consequence of the highly contrasted flow regime, the bathymetry of the bed is not stable over time, with local depressions and island-moving from year to year along the reach. The strong variability in the river bathymetry allows the differences between one- and two-dimensional model outputs to be illustrated strikingly.

7.6.1.1 One-dimensional model

As mentioned in Section 7.3, the geometry of the river in a one-dimensional model is described using cross-sectional profiles located upstream and downstream of the major changes in cross-section geometry (see Section 7.3.1.2 for a justification). The locations of the transverse profiles are shown in plan view in Figure 7.23. The variability in the geometry is illustrated by Figure 7.24(a)–(b), which shows the cross-sectional profiles for the second, fourth, sixth and eighth sections.

The one-dimensional model is built using the Mike 11 modelling package (DHI (2005)). Like all one-dimensional models, the software uses precomputed tables for the free-surface width, cross-sectional area and conveyance. The actual free-surface width, cross-sectional area and energy slope are updated at every iteration within each computational time step using an interpolation between the entries of the table. The free-surface elevation is used as the input variable for the interpolations. The standard option in the software package is that the free-surface width, cross-sectional area and conveyance tables are composed of 20 entries defined automatically between the bottom and top levels of the cross-sectional area. However,

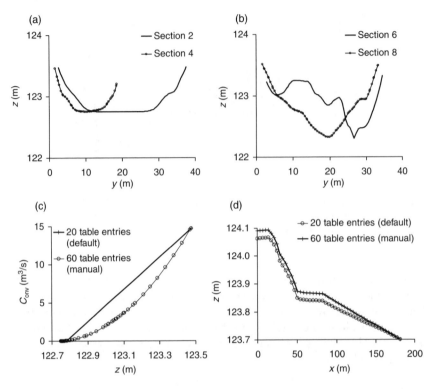

Figure 7.24 Input data, processed data and model results for the one-dimensional modelling package. (a) Cross-sectional profiles for cross-sections 2 and 4. (b)
Cross-sectional profiles for sections 6 and 8. (c) Processed conveyance data
for the default value (20 interpolation levels) and for the manually adjusted,
60 interpolation levels. (d) Longitudinal profiles computed by the software for
20 and 60 interpolation levels

the user of the software is entitled to specify a different number of interpolation levels if the default value is deemed too coarse. Figure 7.24(c)
shows the conveyance curves computed for the fourth cross-section for
the default value of 20 table entries and a manually adjusted table with
60 interpolation levels. As illustrated in the figure, the default option
yields an artificially linear variation between the last-but-one and last
points in the curve. Computing the conveyance curve using 60 interpolation levels yields a much smoother (and more accurate) conveyance
curve.

Figure 7.24(d) shows the steady-state longitudinal profiles obtained for
a discharge $Q = 31\,\mathrm{m^3/s}$ and a downstream water level $z = 123.7\,\mathrm{m}$,

which corresponds to full bank flow. The only difference between the two longitudinal profiles shown in the figure is the number of entries used in the conveyance table. As shown in the figure, the levels at the upstream end of the model differ by 3 cm. This difference represents more than 7.5% of the water-level difference computed between the upstream and downstream boundaries of the domain, which illustrates the importance of the discretization of the geometry to the quality of the output of a one-dimensional model.

7.6.1.2 Two-dimensional model

The two-dimensional model solves the two-dimensional shallow-water equations using a finite-volume approach on unstructured grids. A variety of numerical solvers and interpolation schemes are available for the solution of the equations (Guinot (2003), Guinot and Soares-Frazão (2006), Lhomme and Guinot (2007), Soares-Frazão and Guinot (2007)). The computational grid is made of triangular cells with sizes ranging from 10 to 1 m (see Figure 7.25 for a global view and a zoomed-in view of the central part of the mesh). The elevations of the cells are interpolated from measurements at points scattered all over the modelled area.

The modelling results for an upstream discharge $Q = 31 \, \text{m}^3/\text{s}$ and a downstream water level $z = 123.7 \, \text{m}$ are shown in Figure 7.26. The

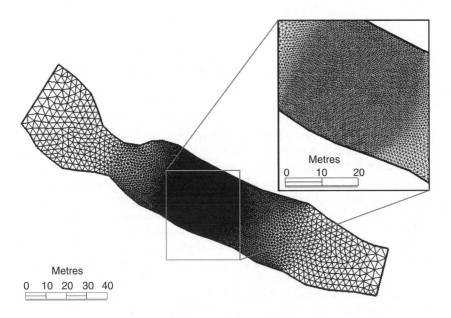

Figure 7.25 Two-dimensional shallow-water model. Plan view of the computational grid

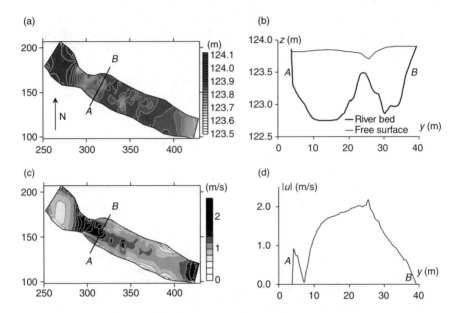

Figure 7.26 Simulation results for the two-dimensional shallow-water model. Free-surface elevation in plan view (a) and along the cross-section [AB] (b). The norm of the flow velocity in plan view (c) and along the cross-section [AB] (d)

water level is shown in plan view in Figure 7.26(a). Figure 7.26(b) shows a cross-sectional view of the river bed and the free surface along the segment [AB]. Figures 7.26(c) and 7.26(d) show the plan view and cross-sectional distributions, respectively, of the norm of the flow velocity.

The results of the two-dimensional simulation clearly contradict the leading assumption of uniform flow variables over a cross-section that is behind classical, one-dimensional river modelling. As illustrated by the cross-sectional free-surface profile in Figure 7.26(b), the variation in the free-surface elevation computed by the two-dimensional model along the segment [AB] is 19 cm (for an average flow depth of 80 cm), while the roughly triangular flow velocity profile (Figure 7.26(d)) varies between 0 and 2.5 m/s across the section. A close inspection of the vector velocity field reveals that the peak in the norm of the velocity near the point A is due to a swirl that appears close to the right bank (Figure 7.27). The numerical integration of the velocity and unit-discharge profiles along [AB] yield a value $\beta = 1.2$ for Boussinesq's coefficient in the momentum equation (7.19) for a one-dimensional model, in contrast with the value $\beta = 1.0$ used in most applications.

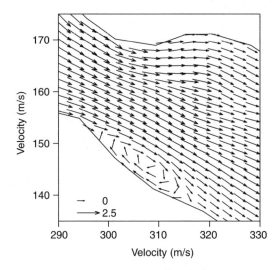

Figure 7.27 Zoomed-in view of the swirl appearing near the right bank

The present application shows that one- and two-dimensional free-surface flow models provide different kinds of information. One-dimensional river models are best suited for long channels with smooth, uniform or slowly varying geometries. They do not allow the variations in the flow variables to be represented in detail in the presence of local geometric effects such as sudden narrowings/widenings, highly variable flow depths and/or velocity distributions. Local effects, such as dead zones, flow diversion around banks or shallow flow zones, etc., can be represented accurately only by two- or three-dimensional models.

Although more complete than one-dimensional models in their description of the physics of the flow, two- and three-dimensional models require more time and computational effort. Setting up and calibrating the one-dimensional model presented in Section 7.6.2 is only a matter of a few hours. Setting up and calibrating the two-dimensional model presented in Section 7.6.3 demands at least twice or three times as much time (and human labour). Moreover, this time ratio is valid only provided that the topographical data have been acquired appropriately, thus allowing semi-automatic mesh generation. Reconstructing any missing topographical and/or geometric information requires the modeller's knowledge, experience and judgement.

This is why one-dimensional open-channel models still represent the vast majority of hydraulic models in operation, while two- or three-dimensional free-surface flow modelling still remains an activity for skilled and experienced users.

7.6.2 The river Dargle flood-management scheme

A hydraulic model was built at HR Wallingford to study the impact on the town of Bray of a proposed flood risk-management scheme on the river Dargle in the Republic of Ireland (Éire).

A distorted model with scales $M_l = 100$ and $M_h = 50$ was used. The river model extended to the sea and was run for a number of tidal levels.

The model (Figure 7.28) was used to provide data on flood levels, flow velocities and inundated areas in order to assess the performance of a number of different flood-mitigation measures (regrading of the bed, realignment of the river channel, flood defence walls, debris trap). As there was concern about the impact of the proposed scheme on trees within the flood plain, and also to aid the assessment of the visual impact, the trees were represented in the model by shrub twigs.

Figure 7.28 The river Dargle model (courtesy of HR Wallingford)

A physical model was used (in preference to a mathematical one) as it was believed that it would give a better representation of complex flow interactions within an urban environment and that it would be a useful tool in presenting to the general public the scheme and its impact. During the outline design stage of the project a one-dimensional numerical model was used for a longer reach of the river to help develop the scheme options.

7.6.3 Entrance to Ústí harbour and regulation of the Labe (Elbe)

The entrance to the harbour at Ústí and the regulation of the adjoining section of the river Labe (Elbe) was studied at the T. G. Masaryk Water Research Institute in Prague.

The harbour entrance had a tendency for silting up after *high* discharges, mainly from sediments washed down from upstream and from tributaries (i.e. not from the river bed itself), and the design for improvement was finalized after a study on a fixed-bed undistorted hydraulic model of scale $M_l = 60$, with the feeding of crushed fruit stones at its inflow to simulate the sediment (see Figure 7.29), and modelling discharges from $600\,\mathrm{m^3/s}$ up to a 50-year flood $(4,000\,\mathrm{m^3/s^1})$.

Figure 7.29 Model of the entrance to Ústí harbour (courtesy of VÚV-TGM, Prague)

In contrast, modelling the river regulation required emphasis at *low* discharges from $70 \, \text{m}^3/\text{s}$, when the depths of flow for a substantial reach of the river downstream of the harbour are so low that navigation on this important international waterway has to be severely restricted or abandoned altogether. The aim of the study was to increase the depths by at least 100 mm by means of longitudinal dykes or by groynes (without interfering with navigation). The river bed in the considered reach of about 2.5 km downstream of the harbour is composed of rock with some very coarse sediment. Computation showed that more than 50% of the total head losses was due to local losses rather than friction caused by sudden substantial changes in bed level (macro-roughness) and changes in river width.

To study the options for increasing the flow depth, a distorted fixed-bed hydraulic model at scales $M_h = 25$ and $M_l = 65$ with a bed-slope scale of $M_S = 1$ (the model was tilted about its downstream end) was built; this rather unusual arrangement was a compromise between the accuracy of water-level measurements in the model, the distortion of river cross-sections, the laboratory space and the preservation of the longitudinal profile macro-roughness. As a 100 mm prototype is represented in the model only by 4 mm, it was felt that this might cast doubts on the conclusions drawn from the model, despite the great care taken in its operation. Therefore, further studies were carried out on an aerodynamic model, with scales of $M_h = 130$ and $M_l = 350$ (i.e. with the same distortion as used in the hydraulic model).

The main advantage of the aerodynamic model was that it could be arbitrarily inclined (without having to account for the correction of the datum and the possibility of using sufficiently large discharges and mean velocities so that the pressure differences (corresponding to water levels could be measured with sufficient accuracy. Thus, with a flow of air of 30 l/s representing a discharge of $70 \, \text{m}^3/\text{s}$, a change in water level by 100 mm results in a pressure change of 46 mm water column (see equation (7.17)), which could easily and accurately be measured. The river bed in the model was constructed using a mixture of two-thirds plasticine and one-third gypsum, covered by a special paint (which could be peeled away) with sand grains for roughness. The cover was of plastic sheets, also covered with sand grains, so that not too large roughness elements had to be placed on the bed. The inlet was connected to the delivery and the outlet to the suction of air blowers so that there was atmospheric pressure approximately in the middle of the model. The model was calibrated using the known water levels at $70 \, \text{m}^3/\text{s}$. Figure 7.30 shows the model with the cover, without roughness, in place. Using the procedure outlined in Section 7.4.4, the effectiveness of various regulation measures could be assessed.

The results from both the hydraulic and aerodynamic models confirmed that neither longitudinal dykes nor groynes achieved the required result (increase of water depth by 100 mm for low river flows) unless the river width was narrowed to levels unacceptable to navigation. The only

Figure 7.30 Aerodynamic model of the Elbe (Labe) regulation (courtesy of VÚV-TGM, Prague)

temporary solution was dredging in critical localities, and the only satisfactory permanent solution was the construction of a low-head barrage (see also Section 13.5.2). For further details, see Novak (1967) and Libý and Novak (2002).

References

Abbott, M. B. (1966), *An Introduction to the Method of Characteristics*, Thames and Hudson/Elsevier, London/New York.

Abbott, M. B. and Ionescu, F. (1967), On the numerical computation of nearly-horizontal flows, *J. Hydraulic Res.*, 5(2), 97–117.

Abbott, M. B. and Minns, A. W. (1998), *Computational Hydraulics*, 2nd Edn, Aldershot.

Averkiev, A. G. (1957), *Methodology of Study of Open Channel Flow on Pressurised Models* (in Russian), Gosenergoizdat, Moscow.

Casulli, V. (1999), A semi-implicit finite difference method for non-hydrostatic, free-surface flows, *Int. J. Numer. Methods Fluids*, 30, 425–440.

Charnock, H. (1995), Wind stress on a water surface, *Quart. J. Royal Meteorological Soc.*, 81, 639–640.

Courant, R., Isaacson, and Rees (1952), On the solution of nonlinear hyperbolic differential equations by finite differences, *Comm.Pure Appl. Math.*, 5, 243–255.

Cunge, J. A. (1975), *Two-Dimensional Modelling of Floodplains, Unsteady Flow in Open Channels*, Chapter 17, Water Resources Publications, Fort Collins, CO.

Cunge, J.A., Holly, F. M. Jr and Verwey, A. (1980), *Practical Aspects of Computational River Hydraulics*, Pitman, London.

Daubert, A. and Graffe, O. (1967), Quelques aspects des écoulements presque horizontaux à deux dimensions en plan et non permanents, *La Houille Blanche*, 8, 847–860.

DHI (2005), *Mike 11, Reference manual*, DHI Water and Environment, Hørsholm.

French, R. H. (1984), *Open Channel Hydraulics*, McGraw Hill, New York.

Geernaert, G. L., Katsaros, K. B. and Richter, K. (1986), Variation of the drag coefficient and its dependence on sea state, *J. Geophysical Res.*, 91, 7667–7679.

Gerritsen, H. (1982), *Accurate Boundary Condition Treatment in Shallow Water Flow Calculations*, PhD Thesis, University of Twente, The Netherlands.

Guinot, V. (2003), *Godunov-Type Schemes. An Introduction for Engineers*, Elsevier, Amsterdam.

Guinot, V. (2008), *Wave Propagation in Fluids. Models and Numerical Techniques*, ISTE/Wiley, London/Hoboken, NJ.

Guinot, V. and Cappelaere, B. (2009), Sensitivity analysis of 2D steady-state shallow water flow. Application to free-surface flow model calibration, *Adv. Water Res.*, 32, 540–560.

Guinot, V. and Soares-Frazão, S. (2006), Flux and source term discretization in two-dimensional shallow water models with porosity on unstructured grids, *Int. J. Numer. Methods Fluids*, 50, 309–345.

Hervouët, J. M. (2007), *Hydrodynamics of Free Surface Flows*, John Wiley, Chichester.

Holly, F. M. Jr and Preissmann, A. (1977), Accurate transport calculation in two dimensions, *J. Hydraulic Eng.*, ASCE, 11, 103.

Johnson, T., Baines, M. J. and Sweby, P. K. (2002), A box scheme for transcritical flow, *Int. J. Numer. Methods Eng.*, 55, 895–212.

Katopodes, N. D. (1977), *Unsteady Two-Dimensional Flow Through a Breached Dam by the Method of Characteristics*, PhD thesis, University of California.

Katopodes, N. D. and Strelkoff, T. (1979), Two-dimensional shallow water-wave models, *J. Eng. Mech. Div.*, ASCE, 105, 317–334.

Knauss, J. (1980) River models with fixed bed. In *Hydraulic Modelling* (ed. H. Kobus), Paul Parey/Pitman, Hamburg/London.

Large, W. G. and Pond, S. (1981), Open ocean momentum flux measurements in moderate to strong winds, *J. Phys.Oceanogr.*, 11, 324–336.

Lax, P. D. (1957), Hyperbolic systems of conservation laws II. *Comm.Pure Appl. Math.*, 13, 217–237.

LeVeque, R. J. (1990), *Numerical Methods for Conservation Laws*, Birkhauser, Boston, MA.

Lhomme, J. and Guinot, V. (2007), A general approximate-state Riemann solver for hyperbolic systems of conservation laws with source terms, *Int. J. Numer. Methods Fluids*, 53, 1509–1540.

Libý, J. and Novak, P. (2002), *Model Investigations of Improvement of Navigation Conditions on the Lower Elbe (Labe) between Střekov and Prostřední Žleb*, T. G. Masaryk Water Research Institute, Prague, p. 40.

Ljatcher, V. M. and Proudovskij, A. M. (1959), *Some Problems of Aerodynamic Modelling of Open Channel Flow – New Methods and Instrumentation for Research of River Processes*, Izd. ANSSSR, Moscow (in Russian).

Ljatcher, V. M. and Proudovskij, A. M. (1971), *Investigation of Open Channels on Pressurised Models*, Energija, Moscow (in Russian).

Luyten, P. J., Jones, P. E., Proctor, R., Tabor, A., Tett, P. and Wild-Allen, K. (1999), COHERENS – A coupled hydrodynamical-ecological model for regional and mathematical models of the North Sea, MUMM report, Management unit of the Mathematical Models of the North Sea.

Meselhe, E. and Holly, F. M. Jr (1997), Invalidity of the Preissmann scheme for transcritical flow, *J. Hydraulic Engineering, ASCE*, 123, 605–614.

Novak, P. (1967), Model similarity and training of rivers with large channel irregularities, *Proc. 11th Congress IAHR*, Fort Collins, CO, Vol. 1, paper A47, pp. 379–388.

Novak, P. and Čábelka, J. (1981), *Models in Hydraulic Engineering – Physical Principles and Design Applications*, Pitman, London.

Preissmann, A. (1961), Propagation des intumescences dans les canaux et rivières, *Prococeedings, Premier Congrès de l'Association Française du Calcul*, Grenoble, September 1961.

Preissmann, A. and Cunge, J. A. (1961a), Calcul des intumescences sur machines électroniques, *Proceedings of the 9th IAHRE Conference*, Dubrovnik, 1961.

Preissmann, A. and Cunge, J. A. (1961b), Calcul du mascaret sur machine électronique, *La Houille Blanche*, 5, 588–596.

Smith, S. D. and Banke, E. G. (1975), Variation of the sea surface drag coefficient with wind speed, *Quart.J. Meteorolog. Soc.*, 101, 665–673.

Soares-Frazão, S. and Guinot, V. (2007), An eigenvector-based linear reconstruction scheme for the shallow water equations on two-dimensional unstructured meshes, *Int. J. Numer. Methods Fluids*, 53, 23–55.

Stelling, G. and Zijlema, M. (2003), An accurate and efficient finite-difference algorithm for non-hydrostatic free-surface flow with application to wave propagation, *Int.J. Num. Meth. Fluids*, 43, 1–23.

Stoker, J. J. (1957), *Water Waves*, Interscience, New York.

Strang, G. (1968), On the construction and comparison of difference schemes, *SIAM. J. Numer. Anal.*, 5, 506–517.

Verwey, A. (1971), *Mathematical Model for Flow in Rivers with Realistic Bed Configurations*, Report Series No. 12, International Courses in Hydraulic and Sanitary Engineering, Delft.

Vries, M. de (1986), *Hydraulic Scale Models*, International Institute for Hydraulic and Environmental Engineering, Delft.

Whitham, G. B. (1974), *Linear and Nonlinear Waves*, Wiley-Interscience, New York.

Wu, J. (1969), Wind stress and surface roughness at air-sea interface, *J. Geophys. Res.*, 74, 444–455.

Chapter 8

Environmental modelling of open-channel systems

8.1 Introduction

The present chapter deals with models for dissolved substances' transport, water quality and sediment movement in open channels.

Chapters 2–4 are necessary background reading for a correct understanding of the mathematical, numerical and physical notions developed hereafter. Moreover, the reader is assumed to have mastered the basis of open-channel modelling presented in Chapter 7.

In Section 8.2, the physical processes and governing equations for solute transport are presented. In most existing solute-transport models the solute concentrations are assumed to be small enough for the solute concentration field not to interact with the flow, which makes the mathematical formulation and solution techniques rather straightforward. In contrast, morphological processes, which are dealt with in Section 8.3, involve an interaction (also called coupling) between the transport and the hydraulic processes, which has consequences on the model behaviour, solution techniques and algorithmic aspects. Section 8.4 deals with water-quality modelling and Section 8.5 is devoted to physical modelling of morphological processes.

8.2 Computational models of transport of dissolved matter

8.2.1 Physical processes and governing equations for one-dimensional transport

8.2.1.1 Assumptions

Solving a solute-transport problem involves computing the concentration in the dissolved substance at all points at all times of the open-channel system. Most existing models for one-dimensional transport of dissolved substances in open-channel systems are based on the following assumptions:

(1) The typical ratio of longitudinal to transverse channel dimensions is large enough for the flow and transport processes to be considered one-dimensional (see also Section 7.2.1 for a justification and illustration of this assumption). Consequently, the concentration in the dissolved substance is assumed to be homogeneous within a given cross-section of the channel.
(2) The range of concentration of the dissolved substance is small enough to have no influence on water density and thus on the flow field. In other words, open-channel hydrodynamics are not influenced by the transport process, and the flow field in the channel system may be determined independently of the solution of the transport problem. This means, in particular, that previously computed flow fields may be used for solute-transport studies.
(3) The main processes at work in solute transport are (i) advection, (ii) diffusion, (iii) hydrodynamic dispersion and (iv) degradation processes. The mathematical formulation of such processes is examined hereafter.
(4) The dissolved substance is transported at the same velocity as the water molecules, and therefore the advection velocity is the flow velocity u.
(5) Molecular diffusion is assumed to obey Fick's law, whereby the diffusion flux is assumed to be proportional to the concentration gradient and the so-called molecular diffusion coefficient.
(6) Hydrodynamic dispersion may be modelled using a Fickian diffusion-based law where the diffusion coefficient is proportional to the average flow velocity. The arguments militating for such a formulation will not be detailed hereafter. The underlying idea to representing hydrodynamic dispersion using a diffusion law is very similar to the idea that leads to the eddy viscosity concept in the representation of Reynolds stress for turbulent flows (see Section 4.3.3).

8.2.1.2 Mathematical formulation in conservation form

The governing equation for the solute concentration C is derived directly from a mass balance. The method used to derive the mass-balance (also called continuity) equation is detailed in Section 7.2.2.1 and will not be presented here. Applying the reasoning of Section 7.2.2.1 to the mass M of solute per unit length of channel, the following conservation equation is obtained:

$$\frac{\partial M}{\partial t} + \frac{\partial F}{\partial x} = S \tag{8.1}$$

where F is the flux (i.e. the amount of dissolved substance that passes at a given cross-section per unit time) and S is the source term that results from source or non-point source pollutions, as well as from lateral flow diversions or inflows. The expression of the various terms in equation (8.1) is examined hereafter.

Expression for M. The mass M of solute per unit length of river may be expressed as the product of the concentration and the volume per unit length of river. As the volume per unit length of river is the cross-sectional area A, one has $M = AC$.

Expression for F. The flux F expresses the combined effects of advection, molecular diffusion and hydrodynamic dispersion

$$F = F_a + F_m + F_d \tag{8.2}$$

where the advection flux F_a, the molecular diffusion flux F_m and the dispersive flux F_d are given by

$$F_a = \int_A uC\,dA = \int_A u\,dAC = QC \tag{8.3a}$$

$$F_m = \int_A -D_m\frac{\partial C}{\partial x}dA = -D_m A\frac{\partial C}{\partial x} \tag{8.3b}$$

$$F_d = \int_A -D_d\frac{\partial C}{\partial x}dA = -D_d A\frac{\partial C}{\partial x} = -\alpha_L uA\frac{\partial C}{\partial x} = -Q\alpha_L\frac{\partial C}{\partial x} \tag{8.3c}$$

where D_m and D_d are the molecular diffusion and dispersion coefficients, respectively, and α_L is the so-called longitudinal dispersivity $D_d = \alpha_L u$. Note that in deriving equations (8.3a)–(8.3c), the assumption of a constant C over the channel cross-section is used to take C out of the integrals. In most cases, F_m is negligible compared to F_d and is omitted from the final expression of F. Substituting equations (8.3a)–(8.3c) into equation (8.2) and neglecting F_m gives

$$F = \left(C - \alpha_L\frac{\partial C}{\partial x}\right)Q \tag{8.4}$$

Expression for S. The source term may be expressed as

$$S = S_d + S_{in} + S_{out} \tag{8.5}$$

where S_d, S_{in} and S_{out} are the source terms that account for the influence of degradation, lateral inflow of possibly contaminated water and lateral outflow, respectively. These fluxes are usually formulated as follows:

$$S_d = -k_d M = -k_d AC \tag{8.6a}$$

$$S_{in} = \frac{q + |q|}{2}C_{in} \tag{8.6b}$$

$$S_{out} = \frac{q - |q|}{2}C \tag{8.6c}$$

where k_d is the linear degradation constant, q is the laterally inflowing discharge per unit length and C_{in} is the concentration in the inflowing water. Note that if q is positive (inflow), S_{out} is zero, while S_{in} is zero if q is negative, hence the presence of the terms $(q \pm |q|)/2$ in equations (8.6b) and (8.6c).

Final expression in conservation form. Substituting the expressions for M, F and S into the governing equation (8.1) yields the final expression in conservation form

$$\frac{\partial(AC)}{\partial t} + \frac{\partial}{\partial x}\left[\left(C - \alpha_L \frac{\partial C}{\partial x}\right)Q\right] = -k_d AC + \frac{q + |q|}{2}C_{in} + \frac{q - |q|}{2}C \qquad (8.7)$$

8.2.1.3 Alternative expressions

The conservation form (equation 8.7) is not the easiest possible formulation of the transport equation, because the quantity AC is not directly measurable, while C is. For this reason, the non-conservation and characteristic forms of equation (8.7), where C is the dependent variable, are often preferred.

Non-conservation form. The non-conservation form is obtained by expanding the derivatives in equation (8.7)

$$A\frac{\partial C}{\partial t} + C\frac{\partial A}{\partial t} + C\frac{\partial Q}{\partial x} - \alpha_L \frac{\partial C}{\partial x}\frac{\partial Q}{\partial x} + Q\frac{\partial}{\partial x}\left(C - \alpha_L \frac{\partial C}{\partial x}\right)$$
$$= -k_d AC + \frac{q + |q|}{2}C_{in} + \frac{q - |q|}{2}C \qquad (8.8)$$

For convenience, we recall that the liquid-continuity equation is

$$\frac{\partial A}{\partial t} + \frac{\partial Q}{\partial x} = q \qquad (8.9)$$

Noting that the second and third terms in equation (8.8) are nothing but the left-hand side of equation (8.9) multiplied by C, then rearranging and dividing by A, equation (8.8) is rewritten as

$$\frac{\partial C}{\partial t} + \left(u - \frac{\alpha_L}{A}\frac{\partial Q}{\partial x}\right)\frac{\partial C}{\partial x} - u\frac{\partial}{\partial x}\left(\alpha_L \frac{\partial C}{\partial x}\right) = -k_d C + \frac{q + |q|}{2A}(C_{in} - C)$$

$$(8.10)$$

Neglecting the term $\alpha_L/A \partial Q/\partial x$ (which is justified in most real-world applications) and assuming a constant α_L allows equation (8.10) to be simplified further

$$\frac{\partial C}{\partial t} + u\frac{\partial C}{\partial x} - u\alpha_L\frac{\partial^2 C}{\partial x^2} = -k_d C + \frac{q + |q|}{2A}(C_{in} - C) \qquad (8.11)$$

which is the form solved in most models for solute transport in open channels.

Characteristic form. From the developments presented in Chapter 7, it follows that equation (8.11) may be rewritten in characteristic form as follows:

$$\frac{dC}{dt} = u\alpha_L \frac{\partial^2 C}{\partial x^2} - k_d C + \frac{q + |q|}{2A}(C_{in} - C) \qquad \text{for } \frac{dx}{dt} = u \qquad (8.12)$$

8.2.2 Initial and boundary conditions in one dimension – branched and looped networks

8.2.2.1 Initial and boundary conditions for a single reach

Equation (8.10) is parabolic (see Section 2.2). The conditions for the existence and uniqueness of parabolic equations (see Chapter 2) require that (i) the initial condition (in the present case the value of C at $t = 0$) be known at all points of the computational domain and (ii) a condition be prescribed at each boundary of the domain. When a single reach is to be modelled, classical boundary condition types are:

(1) *Prescribed concentration.* Such boundary conditions, also known as 'Dirichlet conditions' (see Section 2.5.3), are usually prescribed at the upstream boundary of channel models.
(2) *Prescribed flux or prescribed concentration gradient.* Such boundary conditions, also known as 'Neumann conditions' (see Section 2.5.4), are more often used at the downstream boundaries of channel systems.

8.2.2.2 Internal boundary conditions for branched and looped networks

As mentioned in Chapter 7, the junctions between several reaches in a channel network are considered as internal boundaries (see Section 7.3.5). At such boundaries, mass conservation must hold:

$$\sum_{k}^{J} \varepsilon_k F_k = 0 \qquad (8.13)$$

where J is the number of junctions, F_k is the flux at the boundary of the branch k connected to the junction, and ε_k is a topological indicator that is equal to $+1$ if the junction is considered as the upstream boundary of the branch and to -1 if the junction is the downstream boundary of the branch k. From the definition of F, equation (8.13) becomes

$$\sum_{k}^{J} \varepsilon_k \left(C_k - \alpha_{Lk} \frac{\partial C_k}{\partial x} \right) Q_k = 0 \tag{8.14}$$

where Q_k is the liquid discharge at the boundary of the kth branch (note that the discharges Q_k satisfy the liquid-continuity condition (7.114)). As evaluating the derivative of the concentration at a model boundary is rather complicated from an algorithmic point of view, many commercially available software packages approximate the flux at the boundary as the advection flux, thus simplifying equation (8.14) to

$$\sum_{k}^{J} \varepsilon_k Q_k C_k = 0 \tag{8.15}$$

The solution is made unique by requiring in addition that the concentration C_k be the same for all branches for which the junction is an upstream boundary. This assumption, known as the 'perfect mixing' assumption, leads to the following formula:

$$C_{us} = \frac{\displaystyle\sum_{ds} Q_k C_k}{\displaystyle\sum_{us} Q_k} \tag{8.16}$$

where C_{us} is the concentration used as an upstream boundary in all the branches for which the junction is an upstream boundary, and the summation indexes 'ds' and 'us' indicate the branches for which the junction is a downstream boundary and the branches for which the junction is an upstream boundary, respectively.

8.2.3 Numerical techniques for one-dimensional models

The numerical techniques for the solution of the one-dimensional transport-reaction equation (8.7), (8.11) or (8.12) can be classified into: (i) unsplit techniques, whereby all the terms in the equation are solved in a single step; and (ii) operator-splitting techniques, whereby the advection, diffusion and reaction parts of the equations are treated sequentially within a given time step. The broad outlines of these methods are provided hereafter.

8.2.3.1 Unsplit techniques

These techniques use a simultaneous discretization of all the terms in the transport-degradation equation. In many cases the non-conservation form (8.11) of the equation is solved because the solute concentration arises as a 'natural' variable in this form. The concentration is easily measured in the field, which makes a direct comparison with the model results straightforward. The derivatives in equation (8.11) are usually approximated using implicit or semi-implicit techniques in order to avoid any numerical stability problem. For instance, one may choose to discretize the advection and diffusion parts using the Crank–Nicholson technique:

$$\frac{\partial C}{\partial x} \approx (1 - \theta_1)\frac{C_{i+1}^n - C_{i-1}^n}{2\Delta x} + \theta_1 \frac{C_{i+1}^{n+1} - C_{i-1}^{n+1}}{2\Delta x} \tag{8.17a}$$

$$\frac{\partial^2 C}{\partial x^2} \approx (1 - \theta_2)\frac{C_{i+1}^n - 2C_i^n + C_{i-1}^n}{\Delta x^2} + \theta_2(1 - \theta_2)\frac{C_{i+1}^{n+1} - 2C_i^{n+1} + C_{i-1}^{n+1}}{\Delta x^2} \tag{8.17b}$$

where θ_1 and θ_2 are implicitation parameters between 0 and 1. θ_1 and θ_2 are usually equal, but other options are possible. The derivative with respect to time and the terms in C in the equation (8.11) are approximated as

$$\frac{\partial C}{\partial t} \approx \frac{C_i^{n+1} - C_i^n}{\Delta t} \tag{8.18a}$$

$$-k_d C + \frac{q + |q|}{2A}(C_{in} - C) \approx -k_d C_i^{n+1} + \left(\frac{q + |q|}{2A}\right)_i^{n+1/2}(C_{in} - C_i^{n+1}) \tag{8.18b}$$

Substituting equations (8.17) and (8.18) into equation (8.11) yields

$$A_i C_{i-1}^{n+1} + B_i C_i^{n+1} + D_i C_{i+1}^{n+1} = E_i^n \tag{8.19}$$

where the coefficients A_i, B_i, D_i and E_i are given by

$$A_i = -\frac{\theta_1}{2}Cr - \theta_2 F \tag{8.20a}$$

$$B_i = 1 + 2\theta_2 F + \left(k_d + \frac{q + |q|}{2A}\right)\Delta t \tag{8.20b}$$

$$D_i = \frac{\theta_1}{2}Cr - \theta_2 F \tag{8.20c}$$

$$E_i^n = \left[\frac{1 - \theta_1}{2}Cr + (1 - \theta_2)F\right]C_{i-1}^n + [1 - 2(1 - \theta_2)F]C_i^n$$
$$+ \left[-\frac{1 - \theta_1}{2}Cr + (1 - \theta_2)F\right]C_{i+1}^n + \frac{q + |q|}{2A}C_{in}\Delta t \tag{8.20d}$$

where Cr and F are the Courant number and the diffusive Courant number, respectively, and are defined as

$$Cr = \frac{u\Delta t}{\Delta x} \tag{8.21a}$$

$$F = \frac{u\alpha_L \Delta t}{\Delta x^2} \tag{8.21b}$$

For a reach with M computational points, $M - 2$ equations (8.19) can be written for the points $I = 2, 3, \ldots, M - 1$. The missing two equations are provided by the boundary conditions. Note that, owing to the presence of the diffusion term, one boundary condition is needed at each end of the reach.

8.2.3.2 Operator-splitting techniques

In these techniques the various parts of the transport equation are treated in a sequence. This allows, in particular, the most appropriate (or accurate) method to be selected for, for example, the advection part of the equation. A typical time-splitting sequence for the solution of equation (8.11) is the following.

(1) Solve the advection part of the equation over the computational time step Δt using the values of C at the time level n as an initial condition.

$$\frac{\partial C}{\partial t} + u\frac{\partial C}{\partial x} = 0 \tag{8.22a}$$

$$\frac{dC}{dt} = 0 \qquad \text{for} \quad \frac{dx}{dt} = u \tag{8.22b}$$

Equations (8.22a) and (8.22b) are the non-conservation and characteristic forms of the advection equation, respectively. Any standard method, such as the Crank–Nicholson, Preissmann or upwind schemes, may be used to solve equation (8.22a), while the method of characteristics (MOC) is best suited to the solution of the characteristic form (equation (8.22b)).

Step (1) provides a first estimate $C_i^{n+1,A}$ of the solution at the time level $n + 1$. The superscript A indicates that this estimate is the result of the advection part only.

(2) The values $C_i^{n+1,A}$ at the computational points are used as a starting point for the solution of the diffusion part of the equation over the time step Δt:

$$\frac{\partial C}{\partial t} - u\alpha_L \frac{\partial^2 C}{\partial x^2} = 0 \tag{8.23}$$

This yields a new estimate $C_i^{n+1,D}$ of the solution, where the superscript D indicates that the solution is the result of the diffusion part of the equation.

(3) The values $C_i^{n+1,D}$ are used as initial conditions to compute the remaining part of the equation over the computational time step Δt:

$$\frac{\partial C}{\partial t} = -k_d C + \frac{q + |q|}{2A}(C_{in} - C) \tag{8.24}$$

8.2.4 Two-dimensional transport of dissolved matter

8.2.4.1 Assumptions – mathematical formulation

The assumptions behind two-dimensional models for the transport of dissolved substances are identical to assumptions (2)–(6) for one-dimensional transport. Assumption (1), valid only for one-dimensional flow configurations, must be replaced with the following assumption:

(7) The typical ratio of the vertical to horizontal dimensions of the physical domain is small enough for the flow to be considered nearly horizontal and the flow velocity homogeneous over the vertical (see Section 7.4.2 on two-dimensional free-surface-flow modelling). The concentration in the dissolved substance is also assumed to be homogeneous over the vertical.

Transposing the reasoning of Section 8.2.1 to a two-dimensional flow configuration leads to the following conservation form of the two-dimensional transport equation:

$$\frac{\partial(YC)}{\partial t} + \frac{\partial}{\partial x}\left[\left(C - \alpha_L \frac{\partial C}{\partial x}\right)Yu\right] + \frac{\partial}{\partial y}\left[\left(C - \alpha_L \frac{\partial C}{\partial y}\right)Yv\right] = -k_d YC \tag{8.25}$$

where u and v are the x- and y-velocities, respectively, and Y is the water depth. The non-conservation form of the transport equation is obtained by expanding the derivatives in equation (8.25)

$$Y\frac{\partial C}{\partial t} + C\frac{\partial Y}{\partial t} + Yu\frac{\partial}{\partial x}\left(C - \alpha_L \frac{\partial C}{\partial x}\right) + \left(C - \alpha_L \frac{\partial C}{\partial x}\right)\frac{\partial}{\partial x}(Yu)$$

$$+ Yv\frac{\partial}{\partial v}\left(C - \alpha_L \frac{\partial C}{\partial v}\right) + \left(C - \alpha_L \frac{\partial C}{\partial v}\right)\frac{\partial}{\partial v}(Yv) = -k_d YC \tag{8.26}$$

Equation (8.26) can be simplified using the liquid-continuity equation obtained from the first component of the vector equation (7.140):

$$\frac{\partial Y}{\partial t} + \frac{\partial}{\partial x}(Yu) + \frac{\partial}{\partial y}(Yv) = 0 \tag{8.27}$$

Multiplying equation (8.27) by C, subtracting the resulting equation from equation (8.26) and dividing by Y leads to

$$\frac{\partial C}{\partial t} + \left[u - \frac{\alpha_L}{Y}\frac{\partial}{\partial x}(Yu)\right]\frac{\partial C}{\partial x} + \left[v - \frac{\alpha_L}{Y}\frac{\partial}{\partial y}(Yv)\right]\frac{\partial C}{\partial y} =$$
$$u\alpha_L\frac{\partial^2 C}{\partial x^2} + v\alpha_L\frac{\partial^2 C}{\partial y^2} - k_d C \tag{8.28}$$

Assuming, as in Section 8.2.1, that $\alpha_L/Y\partial(Yu)/\partial x$ is negligible compared to u and that $\alpha_L/Y\partial(Yv)/\partial y$ is negligible compared to v leads to the following writing for the non-conservation form of the transport equation:

$$\frac{\partial C}{\partial t} + u\frac{\partial C}{\partial x} + v\frac{\partial C}{\partial y} = u\alpha_L\frac{\partial^2 C}{\partial x^2} + v\alpha_L\frac{\partial^2 C}{\partial y^2} - k_d C \tag{8.29}$$

The characteristic form of the transport equation is obtained directly from equation (8.29):

$$\frac{dC}{dt} = u\alpha_L\frac{\partial^2 C}{\partial x^2} + v\alpha_L\frac{\partial^2 C}{\partial y^2} - k_d C \quad \text{for} \quad \begin{cases} \dfrac{dx}{dt} = u \\[2mm] \dfrac{dy}{dt} = v \end{cases} \tag{8.30}$$

8.2.4.2 Numerical techniques

The two-dimensional transport-degradation equation is usually solved using time-splitting procedures. As mentioned in Section 7.4.2.3, two main types of time splitting are available:

(1) *Process, or operator, splitting.* This option is used when optimal accuracy is needed in the solution of the advection, diffusion and/or degradation terms. Each of the advection, diffusion and degradation processes is then modelled using the most appropriate technique. A classical process-splitting framework consists of solving sequentially the

advection, diffusion and degradation parts of the equations. Assuming that the conservation form (equation (8.25)) is to be solved, the solution sequence is:

$$\frac{\partial (YC)}{\partial t} + \frac{\partial}{\partial x}(YuC) + \frac{\partial}{\partial y}(YvC) = 0 \tag{8.31a}$$

$$\frac{\partial (YC)}{\partial t} - \frac{\partial}{\partial x}\left[\left(\alpha_L \frac{\partial C}{\partial x}\right) Yu\right] - \frac{\partial}{\partial y}\left[\left(\alpha_L \frac{\partial C}{\partial y}\right) Yv\right] = 0 \tag{8.31b}$$

$$\frac{\partial (YC)}{\partial t} = -k_d YC \tag{8.31c}$$

Equation (8.31a) is solved over the computational time step Δt using the concentration field at the time level n as an initial condition. The solution of equation (8.31a) is used as an initial condition to solve equation (8.31b) over the computational time step. The resulting solution is used as an initial condition to solve equation (8.31c) over the computational time step. The solution obtained after solving equation (8.31c) is the final solution at the time level $n + 1$.

(2) *Dimensional splitting.* This option can be used when the transport equation is to be solved on structured grids (e.g. Cartesian grids). It is interesting because many high-order numerical schemes for transport modelling (in particular schemes for advection modelling) are formulated for one-dimensional equations. Robust and accurate numerical techniques, such as those presented in Section 8.2.3, can then be readily used for the solution of the two-dimensional equations, with optimal tuning of the numerical parameters leading to optimal accuracy of the numerical solution. Applying explicit dimensional splitting to the non-conservation form (equation (8.29)) yields the following computational sequence:

$$\frac{\partial C}{\partial t} + u\frac{\partial C}{\partial x} - u\alpha_L \frac{\partial^2 C}{\partial x^2} = -\frac{k_d}{2}C \tag{8.32a}$$

$$\frac{\partial C}{\partial t} + v\frac{\partial C}{\partial y} - v\alpha_L \frac{\partial^2 C}{\partial y^2} = -\frac{k_d}{2}C \tag{8.32b}$$

In this sequence equation (8.32a) is solved over one computational time step. The solution of equation (8.32a) is used as an initial condition to solve equation (8.32b). The result is the final value at the time level $n + 1$.

8.3 Computational models of morphological processes

8.3.1 Introduction – assumptions

Morphological processes represent an important engineering field of hydraulics as well as a vast research topic with many pending questions. Compared to classical free-surface flow modelling, morphological modelling introduces two extra difficulties: (i) the need to account for the coupling between sediment and flow dynamics, and (ii) the difficulty of deriving models that accurately describe sediment motion.

Prior to reading this section, the reader is referred to Section 4.6.3 on the basics of sediment mechanics. Many models are available for the description of morphological processes in open channels. The simplest existing model is the so-called Saint Venant–Exner model. The present section is devoted to a short presentation of this model. For the sake of clarity, the equations are derived for a rectangular, prismatic channel, i.e. $b(x, z) = B(x) = $ constant, and for a uniform sediment grain-size distribution.

The uniform grain-size distribution assumption is a very limiting assumption because, as shown by Cunge *et al.* (1980), non-uniform sediment grain-size distributions yield different mobility properties for sediment motion, leading to armouring, variable bed erodability, etc. Ideally, morphological simulations should be carried out taking non-uniform sediment grain-size distributions into account. The purpose of the present section is not to give a detailed description of the theory of non-uniform sediment motion but to present the dynamics of morphological flows as well as basic considerations on their modelling.

The Saint Venant–Exner model presented hereafter is based on the following assumptions.

> Assumption 1: both the flow and sediment-transport processes are one-dimensional. Such an assumption is classical in the field of open-channel flow modelling (see Chapter 7). This means, in particular, that the sediment is eroded or removed in a uniform fashion across a given cross-section.

> Assumption 2: the flow pattern can be viewed as the superimposition of two layers (Figure 8.1): (i) a bottom layer flowing over the river bed, referred to as the bedload layer; and (ii) an upper layer, where the sediment is transported in the form of a suspension, referred to as the suspended-transport layer. The thicknesses of the bedload and transport layers are hereafter denoted by Y_b and Y_s.

> Assumption 3: the sediment velocity in the suspended-transport layer is equal to the flow velocity V, while it is equal to a lower velocity V_b in the bedload layer.

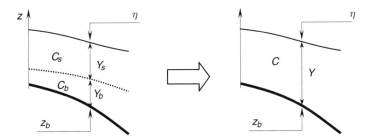

Figure 8.1 Definition of and notation for the Saint Venant–Exner equations. Two-layer flow system (left) modelled using a single-layer approximation (right)

Assumption 4: the sediment concentrations C_b and C_s in the bedload and suspended-transport layer are constant. The density ρ_b of the sediment is uniform. The sediment concentration in both layers is assumed to be small enough for the density of the (sediment + water) mixture to be constant.

Assumption 5: once deposited on the bottom, the sediment instantaneously takes a zero velocity. However, the bottom may not be entirely made of sediment but may include immobile water. The porosity of the bed is assumed to be uniform, and is denoted by ϕ hereafter.

Assumption 6: erosion or sediment deposition occur uniformly over the cross-section at a given point, i.e. the bottom level rises or falls in a uniform fashion at all points of a cross-section.

Assumption 7: the pressure distribution is hydrostatic over both the bedload and the suspended-transport layer.

Assumption 8: the bed elevation varies slowly compared to the total depth, i.e. $\partial z_b/\partial t << \partial(Y_b + Y_s)/\partial t$.

Assumption 9: the average sediment concentration over the vertical in the bedload and the suspended-transport layers is assumed to be very small compared to the sediment concentration in the bed.

The two-layer system may be approximated using an equivalent, single-layer system, as shown on the right-hand side of Figure 8.1.

8.3.2 Governing equations

Only the broad lines of the derivation of the continuity and momentum equations are given hereafter. The detailed principle of their derivation can be found in Chapter 7 and will not be recalled here. The Saint Venant–Exner equations are obtained from conservation considerations.

8.3.2.1 Liquid continuity

The liquid-continuity equation is written under assumption 9 by summing the continuity equation for each of the bedload and suspended-transport layers. Under the assumption of a constant width B, the following equation is obtained:

$$\frac{\partial}{\partial t}(BY_b + BY_s) + \frac{\partial}{\partial x}(BY_b u_b + BY_s u_s) = 0 \tag{8.33}$$

Simplifying by the constant width B and introducing the average flow velocity V

$$V = \frac{Y_b u_b + Y_s u_s}{Y_b + Y_s} = \frac{Y_b}{Y} u_b + \frac{Y_s}{Y} u_s \tag{8.34}$$

Substituting equation (8.34) into equation (8.33) and simplifying by the constant width B allows equation (8.33) to be rewritten as

$$\frac{\partial Y}{\partial t} + \frac{\partial q}{\partial x} = 0 \tag{8.35}$$

where $q = YV$ is the unit discharge. From the point of view of liquid continuity, the single-layer system sketched on the right-hand side of Figure 8.1 is equivalent to the two-layer system, provided that the flow velocity V is defined as in equation (8.34).

8.3.2.2 Momentum balance

According to assumption (4), the sediment concentration in the bedload and suspended-transport layers is very small. Therefore, the density of the water–sediment mixture in both layers can be considered constant, and equal to that of water. Consequently, the momentum equation takes the same form as the momentum equation in the Saint Venant model (see Chapter 7). It is recalled that the equation is

$$\frac{\partial Q}{\partial t} + \frac{\partial}{\partial x}\left(\beta\frac{Q^2}{A} + P\right) = (S_0 - S_e)gA \tag{8.36}$$

The reader may refer to Chapter 7 for the details of the derivation, and the meaning and detailed expression of the terms in equation (8.36). In the case of sediment transport, the formulation of the friction terms needs to be adapted to account for the friction shear stress, thus yielding a modified expression for S_e. In the case of a rectangular, prismatic channel of width B and $\beta = 1$, equation (8.36) simplifies to

$$\frac{\partial q}{\partial t} + \frac{\partial}{\partial x}\left(\frac{q^2}{Y} + \frac{g}{2}Y^2\right) = (S_0 - S_e)gY \tag{8.37}$$

Note, however, that, since the bed level z_b is variable, it must be considered as a flow variable, and the bed slope $S_0 = -\partial z_b/\partial x$ cannot be considered independent of time. Consequently, equation (8.37) is rewritten as

$$\frac{\partial q}{\partial t} + \frac{\partial}{\partial x}\left(\frac{q^2}{Y} + \frac{g}{2}Y^2\right) + gY\frac{\partial z_b}{\partial x} = -S_e gY \tag{8.38}$$

8.3.2.3 Sediment balance

Applying the assumption of mass conservation to the sediment included in both the bedload and the suspended-transport layer yields a conservation equation in the form

$$\frac{\partial M}{\partial t} + \frac{\partial Q_s}{\partial x} = 0 \tag{8.39}$$

where M is the mass of sediment per unit length of channel and Q_s is the sediment mass flux (i.e. the mass of sediment passing at a given x per unit time). The variation of M is expressed as the sum of three terms

$$\frac{\partial M_s}{\partial t} = \frac{\partial M_r}{\partial t} + \frac{\partial M_b}{\partial t} + \frac{\partial M_s}{\partial t} \tag{8.40}$$

where $\partial M_r/\partial t$, $\partial M_b/\partial t$ and $\partial M_s/\partial t$ are the variations in the mass of sediment stored under the (immobile) bed level, in the bedload layer and in the suspended-transport layer, respectively. As the bed has a porosity p, a variation dz_b in the bed level yields a variation

$$dM_r = (1-p)\rho_s B dz_b = d[(1-p)\rho_s B z_b] \tag{8.41}$$

because z_b is assumed to vary uniformly across the cross-section and because $B = $ constant can be incorporated in the differential. A variation dY_b in the depth of the bedload layer yields a variation

$$dM_b = C_b B \, dY_b = d[C_b B Y_b] \tag{8.42}$$

Conversely, one has

$$dM_s = C_s B \, dY_s = d[C_s B Y_s] \tag{8.43}$$

In the same way, the sediment mass flux can be written in the form

$$Q_s = Q_b + Q_s = Y_b u_b B C_b + Y_s u_s B C_s \qquad (8.44)$$

Substituting equations (8.40)–(8.44) into equation (8.39) and dividing by the constant width B gives

$$\frac{\partial}{\partial t}[(1-p)\rho_s z_b + C_b Y_b + C_s Y_s] + \frac{\partial}{\partial x}(Y_b u_b C_b + Y_s u_s C_s) = 0 \qquad (8.45)$$

An average sediment concentration may be defined as

$$C = \frac{Y_b C_b + Y_s C_s}{Y_b + Y_s} = \frac{Y_b}{Y} C_b + \frac{Y_s}{Y} C_s \qquad (8.46)$$

Then, equation (8.45) may be rewritten as

$$\frac{\partial}{\partial t}(YC) + (1-p)\rho_s \frac{\partial z_b}{\partial t} + \frac{\partial q_s}{\partial x} = 0 \qquad (8.47)$$

where $q_s = Q_s/B$ is the unit sediment discharge. Assumption (9) allows the first time derivative in equation (8.47) to be neglected. Assumptions (4) and (5), of constant ρ_s and ϕ, allow equation (8.47) to be simplified to

$$\frac{\partial z_b}{\partial t} + \frac{\partial}{\partial x}\left(\frac{q_s}{(1-p)\rho_s}\right) = 0 \qquad (8.48)$$

As mentioned in Section 4.6.3, the sediment flux is most often expressed as

$$q_s = \rho_s(\Delta g)^{1/2} d^{3/2}\phi \qquad (8.49)$$

where d is the mean grain diameter, Δ is the relative density difference between the sediment and water (i.e. the ratio of sediment to water), and ϕ is the so-called 'sediment-transport parameter'. In classical sediment-transport models, sediment transport cannot occur unless the bed shear stress is larger than a given threshold. An example of such a law is Meyer–Peter and Müller's model (see equation (4.100)), which can be rewritten in the form

$$\begin{aligned} q_s &= \rho_s \operatorname{sgn}(V) d^{3/2}(\Delta g)^{1/2}\left[\max\left(\frac{4V^2}{\Delta g d} - 0.188, 0\right)\right]^{3/2} \\ &= \rho_s \operatorname{sgn}(q) d^{3/2}(\Delta g)^{1/2}\left[\max\left(\frac{4}{\Delta g d}\frac{q^2}{Y^2} - 0.188, 0\right)\right]^{3/2} \end{aligned} \qquad (8.50)$$

where $\operatorname{sgn}(\cdot)$ is the sign function, equal to -1 if the argument is negative and to $+1$ if the argument is positive. Note that the $\operatorname{sgn}(\cdot)$ function allows

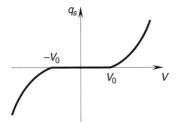

Figure 8.2 Definition of the Meyer–Peter and Müller model

both positive and negative flow velocities to be accounted for and that the $\max(\cdot)$ operator yields a zero value for q_s if the absolute value of V is smaller than the threshold value $V_0 = (0.47\Delta gd)^{1/2}$ (Figure 8.2).

8.3.2.4 Vector form of the Saint Venant–Exner equations

The liquid continuity, momentum and sediment-balance equations (equations (8.35), (8.38) and (8.48)) cannot be recast completely in conservation form because the term $gY\,\partial z_b/\partial x$ cannot be integrated in the general case. The non-conservation form is used for the vector writing of the governing equations

$$\frac{\partial \mathbf{U}}{\partial t} + \mathbf{A}\frac{\partial \mathbf{U}}{\partial x} = \mathbf{S} \tag{8.51}$$

where $\mathbf{U}$, $\mathbf{A}$ and $\mathbf{S}$ are given by

$$\mathbf{U} = \begin{bmatrix} Y \\ q \\ z_b \end{bmatrix},\ \mathbf{A} = \begin{bmatrix} 0 & 1 & 0 \\ c^2 - V^2 & 2V & c^2 \\ \dfrac{1}{(1-\phi)\rho_s}\dfrac{\partial q_s}{\partial Y} & \dfrac{1}{(1-\phi)\rho_s}\dfrac{\partial q_s}{\partial q} & 0 \end{bmatrix},\ \mathbf{S} = \begin{bmatrix} 0 \\ -gYS_e \\ 0 \end{bmatrix} \tag{8.52}$$

Note that in equation (8.52) the two derivatives on the third line of $\mathbf{A}$ are obtained from equation (8.50) as

$$\frac{1}{\rho_s}\frac{\partial q_s}{\partial Y} = -V\frac{1}{\rho_s}\frac{\partial q_s}{\partial q} \tag{8.53a}$$

$$\frac{1}{\rho_s}\frac{\partial q_s}{\partial q} = 12\,\mathrm{sgn}(q)\left(\frac{d}{\Delta g}\right)^{1/2}\frac{q}{Y^2}\left[\max\left(\frac{4}{\Delta gd}\frac{q^2}{Y^2} - 0.188, 0\right)\right]^{1/2} \tag{8.53b}$$

Substituting equations (8.53) into equation (8.52) leads to the following expression for **A**:

$$\mathbf{A} = \begin{bmatrix} 0 & 1 & 0 \\ c^2 - V^2 & 2V & c^2 \\ \dfrac{-V}{(1-\phi)\rho_s}\dfrac{\partial q_s}{\partial q} & \dfrac{1}{(1-\phi)\rho_s}\dfrac{\partial q_s}{\partial q} & 0 \end{bmatrix} \tag{8.54}$$

8.3.3 Wave-propagation speeds

8.3.3.1 General

The wave-propagation speeds λ of the solution are the eigenvalues of the matrix **A**. They satisfy the condition $|\mathbf{A} - \lambda\mathbf{I}| = 0$, that is,

$$\lambda^3 - 2V\lambda^2 + \left[V^2 + \frac{1+\theta-\phi}{1-\phi}c^2\right]\lambda + \frac{\theta}{1-\phi}Vc^2 = 0 \tag{8.55}$$

No analytical solution can be found for the roots of equation (8.55) in the general case. However, it is possible to solve equation (8.55) numerically for particular values of the flow variables. As an example, Figure 8.3 shows the variations in the dimensionless ratio λ/c, with the Froude number for the sediment and flow parameters given in Table 8.1. Figure 8.3 is plotted by using various values for the average velocity V and computing the corresponding value of $\mathrm{Fr} = V/c$.

As illustrated by Figure 8.3, there is a clearly visible asymptotic behaviour for the wave-propagation speeds at large values of the Froude number. The

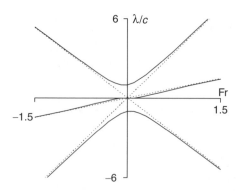

Figure 8.3 Variations in the Saint Venant–Exner wave-propagation speeds with the Froude number. Sediment-transport formula used: Meyer–Peter and Müller

Table 8.1 Parameters for the graph in Figure 8.3.

Symbol	Meaning	Value
d	Sediment grain size	10^{-2} m
g	Gravitational acceleration	9.81 m/s^2
Y	Total mixture depth	1 m
ϕ	Sediment porosity	0.25
ρ_s	Sediment density	3×10^3 kg/m^3

analysis of the wave-propagation speed is carried out in the neighbourhood of zero velocities and for asymptotic values of the Froude number hereafter.

8.3.3.2 Limit case 1: zero sediment discharge

If the average velocity V is smaller than the threshold value V_0, then $\theta = 0$ and equation (8.55) simplifies to

$$\lambda^3 - 2V\lambda^2 + (V^2 + c^2)\lambda = 0 \tag{8.56}$$

the roots of which are

$$\lambda^{(1)} = V - c \tag{8.57a}$$

$$\lambda^{(2)} = 0 \tag{8.57b}$$

$$\lambda^{(3)} = V + c \tag{8.57c}$$

This may be interpreted as follows. If the sediment unit discharge is zero, the bottom elevation z_b is constant. Therefore, the channel bed does not move, and hence the zero wave-propagation speed $\lambda^{(2)}$. The remaining two wave-propagation speeds $\lambda^{(1)}$ and $\lambda^{(3)}$ are the classical celerities of the Saint Venant system. Therefore, the wave-propagation properties of the Saint Venant equations are retrieved as a particular case of the Saint Venant–Exner system.

8.3.3.3 Limit case 2: large Froude numbers

Dividing equation (8.57) by c^3 yields the following equality:

$$\left(\frac{\lambda}{c}\right)^3 - 2\mathrm{Fr}\left(\frac{\lambda}{c}\right)^2 + \left[\mathrm{Fr}^2 - \left(1 + \frac{\theta}{1-\phi}\right)\right]\frac{\lambda}{c} + \frac{\theta}{1-\phi}\mathrm{Fr} = 0 \tag{8.58}$$

Equation (8.58) is an equation of the type $f(x) = 0$. A simple variation analysis shows that equation (8.58) has only one root if $\mathrm{Fr}^2 > 7/4 + \theta/(1-\phi)$, because in such a case the derivative $f'(x)$ is always non-zero. In most situations with ordinary sediment density, porosity and grain size, however, this is not the case, and equation (8.58) has three roots.

It can be shown that, for large values of V, θ becomes very large compared to unity and that q can be approximated as

$$\theta \approx \frac{24}{\Delta} Fr^2 \gg 1 \tag{8.59}$$

Then equation (8.58) becomes equivalent to

$$\left(\frac{\lambda}{c}\right)^3 - 2Fr\left(\frac{\lambda}{c}\right)^2 + \left(1 - \frac{24}{(1-\phi)\Delta}\right)Fr^2\frac{\lambda}{c} + \frac{24}{(1-\phi)\Delta}Fr^3 = 0 \tag{8.60}$$

Noting that $\lambda/c = Fr$ is a root of equation (8.60), equation (8.60) can be rewritten as

$$\left(\frac{\lambda}{c} - Fr\right)\left[\left(\frac{\lambda}{c}\right)^2 - Fr\frac{\lambda}{c} - \frac{24Fr^2}{(1-\phi)\Delta}\right] = 0 \tag{8.61}$$

which leads to the following eigenvalues for $\mathbf{A}$

$$\frac{\lambda_1}{c} = \left\{\frac{1}{2} - \left[1 + \frac{96}{(1-\phi)\Delta}\right]^{1/2}\right\}Fr \tag{8.62a}$$

$$\frac{\lambda_2}{c} = Fr \tag{8.62b}$$

$$\frac{\lambda_3}{c} = \left\{\frac{1}{2} + 2\left[1 + \frac{96}{(1-\phi)\Delta}\right]^{1/2}\right\}Fr \tag{8.62c}$$

Note that $\lambda_1 + \lambda_3 = \lambda_2$. Also note that $\lambda_1 < V - c$ and $\lambda_3 > V + c$.

8.3.3.4 Interpretation

The following conclusions may be drawn from the present analysis.

(1) The wave-propagation speeds λ_1 and λ_3 of the Saint Venant–Exner system are very different from those of the original Saint Venant equations. With the parameters in Table 8.1, the ratio between the wave speeds of the two models may be as large as 4. Consequently, solving the Saint Venant–Exner equations using explicit techniques implies that much shorter computational time steps should be used than when the Saint Venant equations alone are to be solved.

(2) The asymptotic expressions (equations (8.62a) and (8.62b)) indicate that the wave-propagation speed is lower when the sediment is denser and when the porosity is smaller. This can be expected because larger values of Δ tend to reduce the volumetric sediment discharge q_s/ρ_s and because a smaller porosity allows more sediment to be stored (or removed) at a given place for a given variation in the bottom level.

(3) The ratio of the speed λ_2 of the central wave to the speeds λ_1 and λ_3 of the other two waves is between 0 and, typically, 1/6. Consequently, when solving the Saint Venant–Exner equations numerically, there is at least one wave with a Courant number significantly different from unity. Unity is well known to be the Courant number value for which most numerical techniques for wave propagation are optimally accurate (Cunge *et al.* (1980). This means that at least one of the waves is bound to be poorly resolved in the numerical solution. Indeed, two options are possible: either (i) the time step and cell size are chosen such that the Courant numbers of the first and third waves are close to unity, thus yielding a Courant number for the central wave significantly smaller than unity; or (ii) the Courant number of the central wave is kept close to unity, thus yielding Courant number values much larger than unity for the first and third waves. Determining the most appropriate computational time step and cell size is a matter of determining which of the three waves is of particular importance in the representation of the morphological transients.

8.3.4 Initial and boundary conditions

Assume that the Saint Venant–Exner equations with Meyer–Peter and Müller's formula are to be solved over a domain $[0, L]$ for $t > 0$. The wave-propagation speed analysis in Section 8.3.4 allows the following conclusions to be derived for solution existence and uniqueness.

For a point M, the domain of dependence of which is located within the interval $[0, L]$ (see Figure 8.4), the solution is unique provided that the initial condition at the feet A, B and C of the three characteristics $dx/dt = \lambda_k$ ($k = 1$, 2, 3) is known. If this is the case, a characteristic relationship can be written along each of the characteristic segments [AM], [BM] and [CM]. The solution of the resulting 3×3 system is unique.

For a point located next to the upstream boundary (such as the point M' in Figure 8.4), only one characteristic (the characteristic $dx/dt = \lambda_1$) travels from within the domain. The missing two pieces of information must be supplied in the form of two upstream boundary conditions.

For a point located near the downstream boundary (such as the point M'' in Figure 8.4), the characteristics $dx/dt = \lambda_1$ travels from the downstream boundary into the domain, while the remaining two characteristics $dx/dt = \lambda_2$ and $dx/dt = \lambda_3$ come from within the domain. Consequently, only one boundary condition is needed downstream.

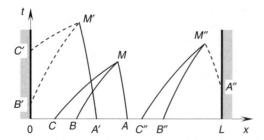

Figure 8.4 Definition sketch of the characteristics in the Saint Venant–Exner model

There exist situations where prescribing a downstream boundary condition (e.g. a boundary condition in terms of hydrodynamic variables) yields a second boundary condition because of the specific formulation for the sediment discharge q_s. Consider the downstream boundary condition $q(L, t) = 0$. Such a condition is equivalent to prescribing a zero velocity, $V(L, t) = 0$. Then, from Meyer–Peter and Müller's formula, the sediment discharge is necessarily zero at the boundary, $q_s(L, t) = 0$. This additional condition in q_s should not be considered as an independent boundary condition, and it should not be inferred that two conditions (i.e. one for the flow and one for the sediment discharge) can be prescribed at the downstream boundary. Indeed, obtaining $q_s(L, t) = 0$ as a particular consequence of the boundary condition $V(L, t) = 0$ is not the same as prescribing $q_s(L, t) = 0$, irrespective of the flow conditions at the downstream boundary.

8.3.5 Numerical techniques

The impossibility to derive exact, analytical expressions for the wave celerities in the Saint Venant–Exner model makes numerical methods based on exact expressions of the Riemann invariants, such as the method of characteristics (see Section 7.3.2), difficult to use. One of the earliest industrial implementations of the coupled solutions of the flow and sediment-transport equations described in the literature is based on Preissmann's scheme (Preissmann (1961), Preissmann and Cunge (1961a,b)) (see also Section 7.3.3). Preissmann's scheme is described in Section 7.3.3 and the principle of the scheme will not be recalled here. It is simply recalled that the derivatives of $\mathbf{U}$ with respect to time and space are approximated as

$$\frac{\partial \mathbf{U}}{\partial x} \approx (1 - \theta) \frac{\mathbf{U}_{i+1}^n - \mathbf{U}_i^n}{\Delta x_{i+1/2}} + \theta \frac{\mathbf{U}_{i+1}^{n+1} - \mathbf{U}_i^{n+1}}{\Delta x_{i+1/2}} \qquad (8.63a)$$

$$\frac{\partial \mathbf{U}}{\partial t} \approx (1 - \psi) \frac{\mathbf{U}_i^{n+1} - \mathbf{U}_i^n}{\Delta t} + \psi \frac{\mathbf{U}_{i+1}^{n+1} - \mathbf{U}_{i+1}^n}{\Delta t} \qquad (8.63b)$$

Substituting these approximations into equation (8.51) yields a vector equation in the form

$$a_{i+1/2}^{n+1/2} U_i^{n+1} + b_{i+1/2}^{n+1/2} U_{i+1}^{n+1} = c_{i+1/2}^{n+1/2} \tag{8.64a}$$

$$a_{i+1/2}^{n+1/2} = \frac{1-\psi}{\Delta t} I - \frac{\theta}{\Delta x_{i+1/2}} A_{i+1/2}^{n+1/2} \tag{8.64b}$$

$$b_{i+1/2}^{n+1/2} = \frac{\psi}{\Delta t} I + \frac{\theta}{\Delta x_{i+1/2}} A_{i+1/2}^{n+1/2} \tag{8.64c}$$

$$c_{i+1/2}^{n+1/2} = S_{i+1/2}^{n+1/2} + \left[\frac{1-\psi}{\Delta t} I + \frac{1-\theta}{\Delta x_{i+1/2}} A_{i+1/2}^{n+1/2} \right] U_i^n$$
$$+ \left(\frac{\psi}{\Delta t} I - \frac{1-\theta}{\Delta x_{i+1}} A_{i+1/2}^{n+1/2} \right) U_{i+1}^n \tag{8.64d}$$

where I is the identity matrix and A is given by equation (8.54). Substituting equation (8.54) into equations (8.64b)–(8.64d) leads to a system of equations in the form

$$D_{i+1/2}^{(k)} A_i^{n+1} + E_{i+1/2}^{k)} Q_i^{n+1} + F_{i+1/2}^{(k)} z_{b_i}^{n+1} + G_{i+1/2}^{(k)} A_{i+1}^{n+1}$$
$$+ H_{i+1/2}^{(k)} Q_{i+1}^{n+1} + I_{i+1/2}^{(k)} z_{b_{i+1}}^{n+1} = J_{i+1/2}^{(k)}, \quad \begin{cases} i = 1, \ldots, M-1 \\ k = 1, 2, 3 \end{cases} \tag{8.65}$$

where M is the number of computational points in the domain. Writing equation (8.65) for $k = 1, 2, 3$ for $i = 1$ to $M-1$ yields a system with $3M - 3$ equations for $3M$ unknowns. The missing three equations are provided in the form of boundary conditions. As mentioned in Section 8.3.3, two boundary conditions must be specified at the upstream end of the reach and one boundary condition must be prescribed at the downstream end.

8.4 Models of water-quality processes

8.4.1 Physical processes and governing equations

The basics of water-quality modelling are presented in Chapter 4 (Section 4.6.4). As mentioned in Section 4.6.4, water quality may be described in terms of physical, chemical and biological indicators. Modelling of temperature and dissolved chemical transport are described in Section 8.2. The present section is devoted to modelling of biological processes. It is recalled from Section 4.6.4 that two biological processes are usually taken into account in water-quality modelling, namely: (i) the biological oxidation of organic matter, described in terms of biological oxygen demand (BOD); and (ii) self-purification processes, described in terms of dissolved oxygen (DO).

The variations of the BOD in water are accounted for by the concentration L of organic matter in the water. Assuming that organic matter is transported at the velocity of the flow and subjected to the same hydrodynamic dispersion processes as any dissolved chemical substance, L obeys an equation similar to (8.7), where the concentration C is replaced with L, and the degradation constant k_d of the chemical is replaced with an oxidation constant K_1:

$$\frac{\partial(AL)}{\partial t} + \frac{\partial}{\partial x}\left[\left(L - \alpha_L \frac{\partial L}{\partial x}\right)Q\right] = -K_1 AL + \frac{q+|q|}{2}L_{in} + \frac{q-|q|}{2}L \quad (8.66)$$

where A is the cross-sectional area, Q is the liquid discharge, q is the liquid discharge of point or non-point sources, L_{in} is the concentration of organic matter in incoming flow and α_L is the longitudinal dispersion coefficient.

The governing equation for DO is somewhat similar to the BOD equation, with the difference that the kinetic degradation term $-K_1 L$ in equation (4.115) is replaced with the term $K_1 L - K_2 D$ in the so-called 'sag-curve equation' (4.118). This leads to the following equation:

$$\frac{\partial(AD)}{\partial t} + \frac{\partial}{\partial x}\left[\left(D - \alpha_L \frac{\partial D}{\partial x}\right)Q\right] = (K_1 L - K_2 D)A$$
$$+ \frac{q+|q|}{2}D_{in} + \frac{q-|q|}{2}D \quad (8.67)$$

The conservation forms (equations (8.66)–(8.67)) of the equations may be rewritten in characteristic form as

$$\frac{dL}{dt} = u\alpha_L \frac{\partial^2 L}{\partial x^2} - K_1 L + \frac{q+|q|}{2A}(L_{in} - L) \quad \text{for} \quad \frac{dx}{dt} = u \quad (8.68a)$$

$$\frac{dD}{dt} = u\alpha_L \frac{\partial^2 D}{\partial x^2} + K_1 L - K_2 D + \frac{q+|q|}{2A}(L_{in} - L) \quad \text{for} \quad \frac{dx}{dt} = u \quad (8.68b)$$

Equations (8.68a) and (8.68b) are coupled via the term $K_1 L$.

The main approaches for solving the transport-reaction equations (equations (8.66)–(8.68)) are described in Section 8.2.

8.4.2 Initial and boundary conditions

The initial and boundary conditions needed to ensure solution existence and uniqueness are identical to those needed to solve a classical chemical-transport problem. In order to solve equations (8.66)–(8.67) or equations (8.68) over a domain $[0, L]$ for all $t > 0$, the initial conditions $D(x, 0)$ and $L(x, 0)$ must be known for all x, $0 \le x \le L$. Owing to the parabolic terms that account for the hydrodynamic dispersion process, one condition must be specified at each boundary of the solution domain for D (or $\partial D/\partial x$) and L (or $\partial L/\partial x$) at all times $t > 0$.

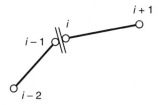

Figure 8.5 Structure treatment in computational models for water quality. The double bar represents the structure, and the points $i-1$ and i are considered as internal boundaries

As recalled from Section 4.6.4, hydraulic structures with overfalls or jumps may be considered as point sources in terms of dissolved oxygen (see Chapter 13). Option 1: the structure is treated as an internal boundary (Figure 8.5). The computational points on the upstream and downstream sides of the structure are denoted by $i-1$ and i, respectively. The point $i-1$ is a downstream boundary for the reach $[i-2,\ i-1]$ and the point i is an upstream boundary for the reach $[i,\ i+1]$. Prescribing a relationship between D and L at the points $i-1$ and i allows the system of equations to be closed:

$$L_i^{n+1} = L_{i-1}^{n+1} \tag{8.69a}$$

$$D_i^{n+1} = f\left(D_{i-1}^{n+1}, Q_{i-1}^{n+1}, Q_i^{n+1}\right) \tag{8.69b}$$

because equations (8.69) replace the transport equation that would normally be solved between the points $i-1$ and i if those were ordinary points in the reach.

8.5 Physical models of morphological processes

8.5.1 General

River engineering studies usually deal with river-training structures designed to create favourable geometric and kinematic flow conditions for various purposes and/or to provide adequate flood protection. Studies of scour and bed-level changes caused by structures (e.g. bridge piers, cofferdams, intakes) are also frequent topics for open-channel flow models with movable beds (see also Chapter 13). The physical (scale) modelling of morphological processes will obviously be closely connected with models of open-channel flow with fixed boundaries (see Section 7.4) and incorporate the procedures outlined there with the additional conditions governing the introduction of a movable bed required to simulate sediment transport.

The following text gives only a brief introduction to the subject; for further reading, refer to, for example, Gehrig (1980), Novak and Čábelka (1981), de Vries (1993) and Yalin (1971).

8.5.2 Governing equations

Equations (7.159)–(7.167) involving 11 variables (M_z, M_b, M_l, M_h, M_R, M_ξ, M_λ, M_t, M_v (or M_Q), M_w and M_k) deal basically with the modelling of water movement. To these have to be added equations for the similarity of the beginning of sediment motion, sediment transport and the time scale of morphological processes.

8.5.2.1 Beginning of sediment motion

The condition for similarity of incipient sediment motion can be derived from the well-known Shields diagram giving the relationship between the square of the densimetric Froude number (see Section 5.8) in the form $\mathrm{Fr_d}^2 = U^{*2}/gd\Delta$ (see equations (4.96) and (4.99a); also, $1/\mathrm{Fr_d}^2$ is the flow parameter ψ – see Section 4.6.3) and the Reynolds number in the form $\mathrm{Re}^* = U^*d/\nu(\Delta = (\rho_s - \rho)/\rho$, where ρ_s is the sediment-specific mass and d is the sediment diameter.

In general, we have to consider both the densimetric Froude number and the Reynolds number introducing two new variables d and $\rho_s(\Delta)$. The scale of the sediment size is, of course, closely related to the roughness scale M_k, i.e. $M_k = f(M_d)$. The 'equivalent' roughness size k (see Section 4.4.1) in a channel with a sediment bed and transport will be a function of the overall channel morphology and sediment size and grading. If, for simplicity, we use $k \cong d$ (or, more accurately, d_{90}), i.e. $M_k = M_d$, then equations (7.168) and (7.169) will also apply.

For $U^{*2} = \tau_0/\rho = gRS$ (equations (4.38) and (4.39)), where S is the channel/water surface/energy line slope, using $M_R = M_b$, $M_S = M_h/M_l$ and for $M_v = 1$ we get from $M_{\mathrm{Fr}} = 1$, $M_{\mathrm{Re}} = 1$ and equation (7.169) the following three equations:

$$M_b M_l^{-3/2} M_d^{-1/2} M_\Delta^{-1/2} = 1 \tag{8.70}$$

$$M_b M_l^{-1/2} M_d = 1 \tag{8.71}$$

$$M_l^3 M_b^{-4} M_d = 1 \tag{8.72}$$

There are four variables in the above three equations, and thus there is only one degree of freedom and a *distorted* model is *needed* (usually M_l is chosen). Should it not be necessary to consider roughness (e.g. in a short model the water surface slope may not be important), equation (8.72) is redundant and we have two degrees of freedom and *may* thus choose a *non-distorted* model. Finally, if the Reynolds number effect can be neglected (equation (8.71) redundant), we also can choose two variables (or even three on short models using only equation (8.70)).

Examining more closely the last case (equation (8.63) redundant), as is the case for fully turbulent flow and Fr_d constant (see equation (4.99a)) – i.e. for $Re^* > 400$ (at $Re^* = 400$ the effect of viscosity is negligible and at $Re^* = 1,000$ it disappears completely) – and for a *non-distorted model* $(M_h = M_l)$, from equation (8.70):

$$M_\Delta M_d = M_l \tag{8.73}$$

If M_d is also equal to the length scale M_l then

$$M_\Delta = 1 \tag{8.74}$$

i.e. the model sediment-specific mass must be the same as that of the prototype.

For *distorted* models $(M_h < M_l)$, from equations (8.70) and (8.72)

$$M_\Delta = (M_l / M_h)^2 \tag{8.75}$$

and from equations (8.70) and (8.71)

$$M_h^3 = M_\Delta M_l^{3/2} \tag{8.76}$$

Thus

$$M_\Delta = M_l^{3/5} \tag{8.77}$$

i.e. $M_\Delta > 1$ (the model must have lighter material than the prototype).

8.5.2.2 Sediment transport

From equations (4.100)–(4.103) expressing the correlation between the sediment-transport and flow parameters, it is evident that the scale of the sediment transport is

$$M_{qs} = f(M_h, M_l, M_\Delta, M_d, M_\lambda) \tag{8.78}$$

Using, for example, equation (4.100) and the procedure outlined in Chapter 5 results in the simple equation

$$M_{qs} = M_h^3 M_l^{-3/2} M_\Delta \tag{8.79}$$

(a different form of equation (8.79) would result if equations (4.101)–(4.103) were used as the basis, because q_s varies with different powers of the velocity V – see Section 4.6.3).

For the derivation of model scales for suspended-sediment transport and concentration we can start with equations (4.105)–(4.109), all of which involve the dimensionless Rouse number (the ratio of sediment fall velocity w and the shear velocity U^*). From this criterion we can state

$$M_w M_{U^*}^{-1} = 1 \tag{8.80}$$

Substituting for $U^{*2} = gRS$ and using for w equation (4.98), equation (8.79) leads again to

$$M_d M_\Delta = M_R M_S = M_b^2 M_l^{-1} \tag{8.81}$$

For the sediment fall velocity with the time scale given by the Froude law, equation (7.166) $\left(M_w = M_b^{3/2}/M_l\right)$ must also apply. Combining equations (7.166) and (8.69) leads to the condition of an undistorted model $(M_b = M_l)$.

8.5.2.3   Time scale of morphological processes

The time scale of morphological processes will not be the same as that for water movement as it will depend on the speed of sediment transport. It can be determined from corresponding bedload volumes transported in the prototype and on the model. This involves the use of the ratio of the void fraction M_p (which depends on the grain sieve curves of the prototype and the model bedload material). The morphological time scale will be given by

$$M_{ts} = M_{vol} M_p M_{qs}^{-1} M_b^{-1} \tag{8.82}$$

Thus, using equation (8.79) for q_s

$$M_{ts} = \frac{M_l M_b M_h M_p}{\frac{M_b^3 M_b}{M_l^{3/2} M_\Delta}} = M_l^{5/2} M_\Delta M_p M_b^{-2} \tag{8.83}$$

Considering that equation (8.83) involves the void fraction ratio (which often is considered as unity), which together with the required knowledge of the prototype bedload transport rate introduces considerable uncertainties, it is best to determine the morphological time scale, if at all possible, by reproducing on the model a known prototype sequence of discharges resulting in a known sediment deposition (or scour) (see the case study in Section 8.6.3).

8.5.3 Boundary conditions and scale effects

(a) The boundary conditions stated in Chapter 7 for fixed-bed, open-channel flow models will apply also for movable-bed models. In this connection it is important to note that it is much easier to adjust the roughness of the channel in a fixed-bed model than in one with a movable bed. Sometimes the differing demands of reproducing correctly the roughness and the morphology of the channel and that of reproducing sediment transport can be reconciled by using lightweight material to simulate transport on a fixed bed or a sand/gravel bed of appropriate roughness.

(b) If lightweight material is used (coal, plastics, sawdust, etc.), the shape of the particles may be quite different from natural sand and gravel; this can introduce scale effects, which may at least partly be offset by varying the other relevant parameters.

(c) It has already been mentioned that ignoring viscous effects requires $Re^* > 400$. Sometimes this is considered to be rather strict, and a limiting value of about 70 or even lower may be acceptable without major scale effects (see Figure 8.6). An interesting possibility is also offered by the fact that the same constant value of $Fr_d^2 = 0.056$ occurs approximately also at $Re^* = 3.5$. This, in fact, represents a value when the forces acting on a sediment particle are caused mainly by its frontal resistance (as would be the case in the prototype) and when tangential stresses are negligible. This therefore presents a possibility of modelling a limited range of discharges and simulating the incipient motion of fine sediment by using coarser material.

(d) Using a distorted-scale model with a movable bed, the condition of not exceeding the natural slope of bed material under water is relevant. This is particularly important when studying the bank-slope stability

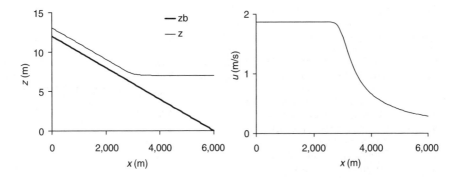

Figure 8.6 Side view of the channel bed and free surface (left); velocity profile along the channel (right)

and/or the extent of local scour (see also Section 4.6.3); the condition of $M_l/M_h < 5$ is, therefore, also valid for these reasons (usually an appreciably smaller distortion is used).

(e) Similarity of bedload transport requires that not even the smallest grain of a mixture on the model should move in suspension (unless this is the case in the prototype). This condition requires the sediment fall velocity w to be greater than the shear velocity U^*, i.e. a Rouse number > 1.

8.6 Case studies

8.6.1 Transport of a dissolved chemical in a channel

8.6.1.1 Test case description

The purpose of this test case is to illustrate the typical behaviour of solutions of the transport-degradation reaction presented in Section 8.2. The following problem is considered: a river of length L, the discharge of which can be considered constant at the time scale of the study, discharges into a lake. The water is initially free from any contamination. The purpose of the study is to assess the consequences of an industrial pollution occurring at the upstream boundary of the channel. At $t = 0$, the concentration of the dissolved chemical at the upstream boundary of the river $(x = 0)$ rises instantaneously to the constant value C_{us}. The chemical is subjected to advection at the flow velocity, hydrodynamic dispersion in the river and degradation. The parameters of the test case are summarized in Table 8.2.

8.6.1.2 Simulation results

The velocity field is obtained by solving the steady-state backwater curve equation over the domain using the downstream water depth Y_{ds} given in

Table 8.2 Parameters for the case study.

Symbol	Meaning	Value
C_0	Initial concentration of dissolved chemical	0 g/l
C_{us}	Concentration at the upstream boundary	1 g/l
k_d	Chemical degradation constant	0 (Option 1), 10^{-4}/s^1 (Option 2)
L	Length of the computational domain	6×10^3 m
n_M	Manning's friction coefficient	2.5×10^{-2}
q	Unit discharge in the river	$2 \, \text{m}^2/\text{s}$
S_0	River-bed slope	2×10^{-3}
Y_{ds}	Downstream water depth	7 m
α_L	Longitudinal dispersivity	5 m
Δt	Computational time step	20 s
Δx	Cell width	60 m

Table 8.2 as a boundary condition. The resulting water level and velocity field over the solution domain are shown in Figure 8.6.

The characteristic form (equation (8.12)) of the transport equation is solved using the parameters in Table 8.2. Note that two options are considered regarding the behaviour of the dissolved chemical. In Option 1, the chemical is assumed to be fully conservative (i.e. there is no degradation). In Option 2, the degradation constant is $k_d = 10^{-4}/\text{s}$, which corresponds to a half-life of approximately 2 hours. Such a value for the degradation constant may be considered as an extreme case, corresponding to a highly reactive chemical. Therefore, Options 1 and 2 can be considered as the boundaries of a wide range of possible situations, the behaviour of most contaminants being somewhere between these two extreme behaviours.

The computed concentration profiles at various times are shown for both options in Figure 8.7. As the transport-degradation problem is linear, the plots are drawn for the dimensionless ratio of the concentration to the concentration at the upstream boundary (C/C_{us}).

The profile for Option 1 shows no damping of the concentration signal. This could be expected because, in the absence of degradation, $C = $ constant is a solution of the characteristic equation (8.12) when C is constant at the upstream boundary. The concentration profiles are drawn every 20 minutes in the figure. The dramatic decrease in the advection velocity in the deeper part of the channel ($x > 4{,}000\,\text{m}$) is easily deduced from the smaller distance between the profiles after $t = 2{,}400\,\text{s}$. The negative velocity gradient around $x = 4{,}000\,\text{m}$ is also responsible for the compression of the contamination front. Indeed, it can be observed that for $t = 1{,}200\,\text{s}$ the front is smeared over 1,200 m, while the smearing occurs within only 900 m at $t = 2{,}400\,\text{s}$.

In the presence of degradation the concentration decreases as the contaminant travels downstream. When the contaminant enters the deeper part of the channel ($x > 4{,}000\,\text{m}$) its degradation properties remain unchanged, but

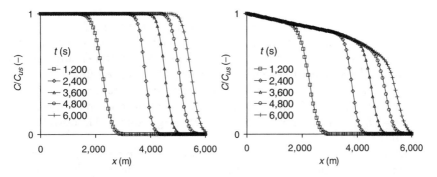

Figure 8.7 Simulated concentration profiles in the channel at various times. Left: no degradation (Option 1). Right: $k_d = 10^{-4}/\text{s}^1$ (Option 2)

a given distance is covered within a longer time than in the shallower part of the channel ($x < 4,000$ m). This is why the concentration decreases faster with distance in the deeper part of the channel than in the shallower part. This explains the change in the slope of the envelope of the concentration profiles around $x = 4,000$ m.

It must be stressed that the present case study is given only for illustration purposes. In the case of a river entering a lake, the flow and transport patterns become two-dimensional (if not three-dimensional) at the mouth of the river. Attempting to model contaminant transport in such a situation using a one-dimensional description of the flow and transport processes would obviously lead to an oversimplification of the problem and would inevitably generate incorrect modelling results.

8.6.2 Arve diversion

The Arve river and valley upstream of Chamonix are affected by landslides. These and the resulting sediment transport in the Arve are threatening the village downstream, could raise the river bed up to Chamonix, and affect it by clogging its banks by fine sediments for 100 km as far as Geneva. The solution for these problems consisted of:

- construction of a water and sediment intake and an 800 m-long Arve torrent-diversion tunnel to stop further undermining of the landslide base;
- provision of a deposition area for the transported material by a sediment-retaining barrage to store $20,000$ m^3 of sediment and to control the area between the diversion intake and the tunnel outlet;
- improvement of the Arve channel capacity.

To study the above measures, two hydraulic models were built at the Sogreah laboratory, Grenoble:

- A large model, scale $M_l = 27$, of a reach of the torrent with a movable bed and bedload transport, the diversion intake, a vortex shaft (height 40 m and diameter 3 m), a dissipation chamber and tunnel, with all the structures made of perspex. The diverted discharge is up to 27 m^3/s and all sediments smaller than 0.3 m are diverted into the tunnel. Figure 8.8 shows the intake and entrance to the shaft of the model, with the deposition of sediment larger than 0.3 m downstream of the intake.
- A model, scale $M_l = 45$, of the sediment-retaining barrage simulating the sediment transport and torrential mudflow, with the model extending beyond the downstream village. Figure 8.9 shows the sediment-retaining barrage of the model.

Figure 8.8 The diversion intake in the Arve model, $M_l = 27$ (courtesy of Sogreah, Grenoble)

Figure 8.9 The sediment-retaining barrage in the Arve model, $M_l = 45$ (courtesy of Sogreah, Grenoble)

Both undistorted models were operated according to Froude law and the principles outlined in Section 8.5. The boulders and gravel of the prototype were represented on the model by gravel and sand to geometrical scale. A special feature of the models was the representation of the mudflow. A mixture of kaolin and additives was used to give the appropriate rheological properties, and was calibrated according to the actual observed mudflow.

8.6.3 Danube confluence

The Danube below Bratislava undergoes a substantial change in slope resulting in a network of branches and difficult navigation conditions at low discharges in a river with substantial sediment transport. This situation was only improved to a substantial degree in 1992 by the construction of the Gabčíkovo hydroelectric plant and a long navigation canal. In 1953, the T. G. Masaryk Water Research Institute in Prague was asked to carry out a study of a particularly difficult reach of the Danube at the confluence with a strong branch – the Denkpal – with the aim of improving the situation by river-training measures and changes at the confluence. The actual sediment transport at various discharges was not known, but detailed surveys of the river bed and its changes over three years were available.

A hydraulic model of the confluence (Novak (1966)), to scales of $M_h = 100$ and $M_l = 300$, and using coal with a specific gravity 1.38, $d_{90} = 2.4\,\text{mm}$ and $d_s = 1.15\,\text{mm}$ was used; the slope of the river channel was given by the length and depth scales $M_S = 1/3$.

To determine the sediment rating discharge in the fairly short model, the bed was set according to prototype for a certain date and the model operated with the corresponding (constant) discharge for 30 minutes while dosing an estimated sediment volume at entry and collecting sediment at exit; the procedure was repeated until both sediment discharges were equal. In the same way, sediment discharges were obtained for other water discharges. To prove the model, the actual sequence of discharges in the Danube (according to the daily readings at the gauging station in Bratislava) was reproduced on the model for the period of three years, and the resulting bed forms compared with those recorded in the prototype. The time scale for sediment transport was determined from equation (8.83) to be about $M_{ts} = 1,000$, and after a few runs was adjusted to $M_{ts} = 1,200$ for best results (see Section 8.5.2.3). Comparing this result with the conventional time scale for water movement according to the Froude law (equation (5.32c)), $M_t = M_l M_h^{-1/2} = 30$, it can be seen that the use of lightweight material to simulate sediment transport and the distorted scales resulted in speeding up the channel-forming process, and that three years were represented by 21 hours of operations on the model.

Figure 8.10 Distorted model of the Danube confluence (courtesy of VÚV-TGM, Prague)

The same sequence of discharges for the period of three years was then used to study the effect of various river-training measures to find the best solution (see Figure 8.10). The proposed measures, when carried out in the prototype, resulted in a bed configuration that on the whole closely corresponded to the one predicted from the model. Small differences – apart from other uncertainties – could be attributed to the fact that, although the actual flow duration curve for the period used on the model closely resembled the actual one, the actual sequence of discharges was, of course, not the same as the one used on the model.

References

Cunje, J. A., Holly, F. M. Jr and Verwey, A. (1980), Practical aspects of river computational Hydraulics, Pitman, London.

Gehrig, W. (1980), River models with movable bed. In *Hydraulic Modelling* (ed. H. Kobus), Paul Parey/Pitman, Hamburg/London.

Novak, P. (1966), A study of river regulation with extensive bed load movement, *Proc. Golden Jubilee Symposia (Model and Prototype Conformity)*, CWPRS Poona, 1, pp. 9–13.

Novak, P. and Čábelka, J. (1981), *Models in Hydraulic Engineering – Physical Principles and Design Applications*, Pitman, London.

Preissmann, A. (1961), Propagation des intumescences dans les canaux et rivières, *Proceedings of Premier Congrès de l'Association Française du Calcul*, Grenoble, September 1961.

Preissmann, A. and Cunge, J. A. (1961a), Calcul des intumescences sur machines électroniques, *Proceedings of the 9th IAHRE Conference*, Dubrovnik, 1961.

Preissmann, A. and Cunge, J. A. (1961b), Calcul du mascaret sur machine électronique, *La Houille Blanche*, 5, 588–596.

Vries, M. de (1993), *Use of Models for River Problems*, UNESCO, Paris.

Yalin, M. S. (1971),*Theory of Hydraulic Models*, Macmillan, London.

Chapter 9

Modelling of closed-conduit flow

9.1 Introduction

The present chapter deals with modelling of steady-state and transient hydraulics in pressurized pipe networks. The case of free-surface flow in closed conduits is dealt with separately in Chapter 10.

Prior to reading this chapter, for a correct understanding of the mathematics of pipe transients the reader should be aware of the notions presented in Chapter 2 (more specifically Sections 2.1–2.5).

In Chapter 3, Section 3.3 is needed for an understanding of a number of system inversion techniques, and Section 3.6 provides the necessary background for the development of the characteristic form of the equations for transients (Section 9.3). Section 3.9.5 provides an introduction to numerical methods for the solution of hyperbolic partial differential equations.

In Chapter 4, the reader should be aware of Sections 4.3 and 4.4. In particular, Section 4.4.1 is an indispensable prerequisite to the understanding of the head-loss formulae used in Section 9.2.

Two configurations are dealt with in the present chapter: (i) pressurized, quasi-steady flow in networks, to be simulated over large time scales; and (ii) fast transients (known as waterhammer), the typical time scale of which is, at most, a few seconds. These two aspects are dealt with in separate sections hereafter.

9.2 Computational models of quasi-steady closed-conduit flow

9.2.1 General – assumptions

Pipe networks are schematized as nodes connected by pipes (Figure 9.1). From a topological and hydraulic point of view, a node is needed at any place where the geometric properties (diameter) or hydraulic parameters (roughness coefficient) of the pipe change. Any point where water is abstracted from the network (in the form of a known demand or in the form of a discharge–head-loss relationship) or supplied to the network by a

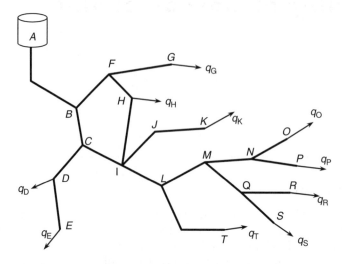

Figure 9.1 Definition sketch for a water-supply network

tank or a pump also needs to be indicated by a node. In the network shown in Figure 9.1, the presence of the node J is justified only if the pipe [IJ] has a different diameter or is made from another material than that of the pipe [JK], or if a structure such as a diaphragm or a pump is located at this node. The node D is needed along the line [CDE] because there is a demand at D.

Networks are usually divided into loops and branches. A loop is a set of pipes that form a closed contour (e.g. [BCIHF] in Figure 9.1). A branch is a set of pipes where an upstream and a downstream end can be clearly identified (e.g. [ILMNP] in Figure 9.1). While steady-state flow in branched networks can be easily calculated (the discharges in all the pipes are simply computed by summing all the demands from the downstream to the upstream end of the branch), the calculation of steady flow in looped networks is more computationally intensive because the direction of the flow cannot be guessed *a priori* and may even change depending on the variations in the boundary conditions.

Models for quasi-steady flow in pipe networks rely on the following assumptions:

(1) The time scales involved are large. Consequently, the fast transients originating from the operation of valves, pumps, tanks or any other device can be neglected at such time scales. Therefore, compressibility effects are negligible and the density of water is assumed to be constant.

(2) Devices such as pumps, valves, controlled tanks, etc., satisfy known relationships between the discharge and head, or head loss.

9.2.2 Governing equations

9.2.2.1 Conservation of mass

Conservation of mass imposes that, at each node in the network, the total amount of water flowing to the node (or leaving the node) be zero. Combined with the assumption of incompressible water, this leads to the following equation for volume discharge:

$$\sum_{p=1}^{P_n} Q_p = q_n \tag{9.1}$$

where P_n is the number of pipes connected to the node n, Q_p is the discharge in the pth pipe connected to the node, and q_n is the demand at the node n. In equation (9.1), the discharge Q_p is taken as positive if the water is flowing toward the node and as negative if the water flows away from the node. By definition, the demand q_n is positive if the water is being consumed.

Equation (9.1) is also known as the 'law of nodes'.

9.2.2.2 Conservation of energy

Conservation of energy states that, in the absence of friction and any dissipation mechanism, the hydraulic head is invariant between two consecutive nodes on a pipe. In real-world situations, dissipation mechanisms yield a head loss that can be classified as either regular or singular. Note that such notions are fully developed in Chapter 4 and only the broad lines are recalled here. The reader is strongly advised to refer to Section 4.4.1 for a review of the main friction laws available in the literature.

Regular head losses are due to friction against the pipe walls. They yield a head loss per unit length of pipe. The head loss per unit length is classically expressed in the form of Darcy–Weisbach's law

$$\frac{\mathrm{d}H}{\mathrm{d}x} = \lambda \frac{L}{D} \frac{|V| V}{2g} = \lambda \frac{L}{2gDA^2} |Q| Q = \frac{8}{\pi^2 g} \lambda \frac{L}{D^5} |Q| Q \tag{9.2}$$

where D and L are the diameter and length of the pipe, respectively, λ is the Darcy–Weisbach friction factor, g is the gravitational acceleration, and Q and V are the discharge and average flow velocity in the pipe, respectively. The absolute value in equation (9.2) accounts for the dependence of the head loss per unit length on the direction of the flow. For a pipe of length

L with a uniform diameter D, integrating equation (9.2) over the length of the pipe gives

$$H_{us} - H_{ds} = \lambda \frac{L^2}{D} \frac{|V| V}{2g} = \lambda \frac{L^2}{2gDA^2} |Q| Q = \frac{8}{\pi^2 g} \lambda \frac{L^2}{D^5} |Q| Q \qquad (9.3)$$

where the subscripts 'us' and 'ds' denote the upstream and downstream nodes of the pipe. Many formulae for λ are available in the literature. A widely used friction model is Colebrook's formula, where the friction factor is defined implicitly via (see also equation (4.54))

$$\lambda^{-1/2} = -2 \log \left(\frac{\kappa}{3.7D} + \frac{2.51}{Re} \lambda^{-1/2} \right) \qquad (9.4)$$

where κ is the roughness height of the pipe (κ/D is called the relative roughness) and Re is the dimensionless Reynolds number, $Re = VD/\nu$, where ν is the kinematic viscosity of the fluid. Although equation (9.4) is implicit and solving for λ may seem time-consuming at first sight, using equation (9.4) in a recursive way allows λ to be determined quite accurately after a few iterations (3–4 iterations are usually sufficient to converge with a relative precision of 10^{-4}). Alternatively, equation (4.56), providing an explicit equation for λ, can be used with sufficient accuracy, particularly with large Reynolds numbers ($Re > 10^5$).

Note that for a totally smooth pipe ($\kappa/D = 0$), Colebrook's equation simplifies to

$$\lambda^{-1/2} = -2 \log \left(\frac{2.51}{Re} \lambda^{-1/2} \right) \qquad (9.5)$$

Also note that, when the Reynolds number becomes very large, equation (9.4) can be approximated as

$$\lambda = \frac{1}{4 \log^2 \left(\frac{\kappa}{3.7D} \right)} \qquad (9.6)$$

In such a case, λ is independent of the value of the Reynolds number (and thus of V), and the head loss is strictly proportional to the square of the flow velocity, which is the case when the flow is fully turbulent (i.e. $Re > Re_{sq}$). Equations (9.5) and (9.6) represent two asymptotic behaviours for λ.

The variation in λ with the Reynolds number Re and the relative roughness κ/D are usually plotted on a Moody diagram (Figure 9.2). The asymptotic law for totally smooth pipes, as given by equation (9.5), is plotted as the bold line in the figure, while the other lines represent the

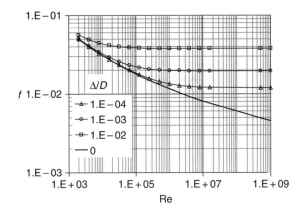

Figure 9.2 Part of the Moody diagram for the friction law. The diagram is plotted for $Re > 2 \times 10^3$ (the lower limit of turbulent flow)

variations in f for three different values of the relative roughness. The horizontal part of the curves corresponds to the asymptotic behaviour given by equation (9.6).

Singular head losses are due to sudden changes in the flow direction (e.g. bends), in pipe diameter (sudden or rapid widenings or narrowings) or to singularities such as valves, diaphragms, etc. Energy loss occurs due to the turbulent dissipation in the swirls created by the singularities. The corresponding head loss formulae are classically assumed to be proportional to the square of the flow velocity (or liquid discharge). The singularity may be considered as a pipe of zero length (usually referred to as a link) over which the following formula is applicable:

$$H_{us} - H_{ds} = \alpha \, |Q| \, Q \qquad (9.7)$$

where the subscripts 'us' and 'ds' denote the nodes connected to the upstream and downstream sides of the singularity, respectively.

9.2.2.3 Boundary conditions

In a network with P pipes connecting N nodes, there are P unknown discharges and N unknown heads. As N equations (9.1) can be written and P equations (9.3) can be written for the head losses along the pipes, the number of equations matches the number of unknowns provided that the demands q_n are known at all the nodes in the network. However, prescribing only demands at the nodes of the network is not sufficient to guarantee the existence and uniqueness of the solution, for the following reasons.

(1) If the nodal demands are the only boundary conditions prescribed in the network, at least one of them is redundant, because under steady state the total amount of water in the network is constant, and therefore

$$\sum_{n=1}^{N} q_n = 0 \qquad (9.8)$$

Consequently, the N equations (9.3) are equivalent to only $N-1$ independent equations.

(2) The set of equations (9.3) provides relationships between the discharge and the difference between the heads at both ends of a pipe. Consequently, the solution of the set of equations (9.3) is *a priori* independent of the reference level (datum) fixed for the heads.

For these two reasons, at least one of the boundary conditions must involve the hydraulic head at one of the nodes. Examples of such conditions are prescribed heads (when the node is connected to a tank with a fixed water level), or prescribed relationships between head and discharge (e.g. the characteristic of a pump or a group of pumps). A general equation for a head–discharge boundary condition at the node n is

$$b_n(q_n, H_n) = 0 \qquad (9.9)$$

where b_n is a known function for the boundary condition. Note that head–discharge relationships should be handled with care, as detailed in the paragraph 'Modelling precautions' in Section 9.2.3. Also note that such relationships can be modelled by connecting a node with a fixed head to the node n and placing a pump or a singularity with a known head–discharge relationship between them. Moreover, the boundary condition (9.9) is also applicable to fixed-head boundary conditions.

9.2.3 Numerical techniques and solution methods

9.2.3.1 Vector writing

The laws of nodes (9.1), the head-loss laws (9.3) and (9.7) and the boundary conditions (9.8) may be written in vector form as

$$\mathbf{F}(\mathbf{X}) = 0 \qquad (9.10)$$

where $\mathbf{F}$ and $\mathbf{R}$ are the function and unknown vectors, respectively, defined as

$$\mathbf{F} = [c_1, \ldots, c_N, h_1, \ldots, h_P, b_1, \ldots, b_R]^T \tag{9.11a}$$

$$\mathbf{X} = [H_1, \ldots, H_N, Q_1, \ldots, Q_P, q_1, \ldots, q_R]^T \tag{9.11b}$$

where N is the number of nodes in the network, P is the number of pipes, and R is the number of nodes where head–discharge relationships in the form of equation (9.9) are prescribed. The functions c_n $(n = 1, \ldots, N)$ account for the continuity equations, or laws of nodes (equation 9.1):

$$c_n(\mathbf{X}) = \sum_{p=1}^{P_n} Q_p - q_n \tag{9.12}$$

while the functions h_p $(p = 1, \ldots, P)$ account for the head loss relationships (equations (9.3) and (9.7)):

$$h_p = H_{n_1} - H_{n_2} - \frac{8}{\pi^2 g} \frac{L_p^2}{D_p^5} \lambda_p \left| Q_p \right| Q_p \text{ (pipe)} \tag{9.13a}$$

$$h_p = H_{n_1} - H_{n_2} - \alpha_p \left| Q_p \right| Q_p \text{ (singularity)} \tag{9.13b}$$

where n_1 and n_2 are the node numbers that correspond to the upstream and downstream ends of the pipe (or link) p. The friction factor λ_p in equation (9.13a), being computed using Colebrook's formula, is a function of the discharge Q_p.

9.2.3.2 Solution techniques

The vector equation (9.10) is non-linear and must be solved using iterative techniques. A widely used method is the Newton–Raphson method, an extension of Newton's method presented in Section 3.3. Like the original Newton method, Newton–Raphson's algorithm is based on a local linearization of the vector equation (9.10):

$$\mathbf{F}(\mathbf{X}^{(m+1)}) - \mathbf{F}(\mathbf{X}^{(m)}) = \mathbf{J} \cdot [\mathbf{X}^{(m+1)} - \mathbf{X}^{(m)}] \tag{9.14}$$

where the superscript between parentheses indicates the iteration number and $\mathbf{J} = \partial \mathbf{F} / \partial \mathbf{X}$ is the Jacobian matrix of $\mathbf{F}$ with respect to $\mathbf{X}$. The purpose is to find $\mathbf{X}^{(m+1)}$ such that equation (9.10) be satisfied. Substituting this condition into equation (9.14) leads to the following equivalent conditions:

$$\mathbf{J} \cdot \mathbf{X}^{(m+1)} = \mathbf{J} \cdot \mathbf{X}^{(m)} - \mathbf{F}(\mathbf{X}^{(m)}) \tag{9.15a}$$

$$\mathbf{X}^{(m+1)} = \mathbf{X}^{(m)} - \mathbf{J}^{-1} \cdot \mathbf{F}(\mathbf{X}^{(m)}) \tag{9.15b}$$

Although equations (9.15a) and (9.15b) are equivalent, equation (9.15b) implies that the Jacobian matrix $\mathbf{J}$ be inverted at each iteration. Inverting $\mathbf{J}$ exactly is extremely time-consuming, and it is often preferred to solve equation (9.15a) directly, for which iterative methods, such as conjugate gradients, or any such iterative approach, are very efficient.

The solution technique thus involves two nested loops: (i) the iterative matrix-inversion method used to solve equation (9.15a); and (ii) the recursive application of equation (9.15a) or (9.15b). A convergence, or iteration stop criterion, must be defined for each of these two loops. Convergence parameters are usually of two kinds:

(1) A first parameter, often called 'precision', is the threshold level under which the equation to be solved can be considered verified. At each iteration, $\mathbf{F}(\mathbf{X}^{(m+1)})$ is computed. The solution is considered satisfactory if the following condition is fulfilled:

$$\left| c_n(\mathbf{X}^{(m+1)}) \right| \le \varepsilon_c, \quad n = 1, \ldots, N \tag{9.16a}$$

$$\left| b_p(\mathbf{X}^{(m+1)}) \right| \le \varepsilon_b, \quad p = 1, \ldots, P \tag{9.16b}$$

$$\left| b_n(\mathbf{X}^{(m+1)}) \right| \le \varepsilon_b, \quad n = 1, \ldots, B \tag{9.16c}$$

where ε_b, ε_c and ε_h are predefined convergence criteria (either absolute or relative).

(2) In some cases, however, one or several convergence criteria may be too small for equations (9.16) to be satisfied within a reasonable number of iterations. In order to avoid time-consuming iterations that may fail to converge, it is customary to specify a maximum permissible number of iterations, after which the recursive application of equations (9.15) is stopped, even if equations (9.16) are not satisfied.

9.2.3.3 Modelling precautions

The numerical solution of the vector equation (9.10) may fail if a number of conditions are not met during model building. A number of classical sources for failure are given hereafter (the list is non-exhaustive). Some commercially available packages may be sufficiently well programmed to check for such mistakes prior to starting the calculation procedures. However, this is not always the case and the user of a network-simulation package is strongly advised to check whether any of the following conditions are verified in the case of a failure.

(1) There is no boundary condition (equation (9.9)) involving the hydraulic head in the model. In this case, the system is not closed because the

number of equations does not match the number of unknowns (see Section 9.2.2.2).

(2) Two nodes A and B with different prescribed heads are connected using pipes, all with a zero length and/or singularities with nil head-loss coefficients. In such a case, summing the equations (9.13a) and (9.13b) between A and B yields an equation of the form

$$H_A - H_B = 0 \tag{9.17}$$

As $H_A \neq H_B$, equation (9.17) has no solution.

(3) One or several loops is/are made of pipes with a zero head-loss coefficient. This may result in a zero determinant for the Jacobian matrix J in equations (9.14) and (9.15). If a Newton–Raphson-based algorithm is used, the system (9.10) cannot be solved because J cannot be inverted.

(4) The head–discharge relationship (equation (9.9)) is decreasing. Then the iterative solution sequence will be unstable, for the following reason. Assume first that q_n is negative (this corresponds to water being pumped into the system, point A in Figure 9.3). If, from one iteration to the next, q_n decreases (i.e. it remains negative but becomes larger in absolute value, point B in Figure 9.3), then the head H_n at the node n decreases (point C in Figure 9.3). This in turns triggers a new decrease in q_n at the next iteration due to friction (point D), which yields a new increase in H (point E). Repeating this positive-feedback iteration after iteration induces instability. The problem is exactly the same if q_n is positive (water being pumped out of the network at the node n). The only permissible relationship is an increasing H–q relationship. This is the

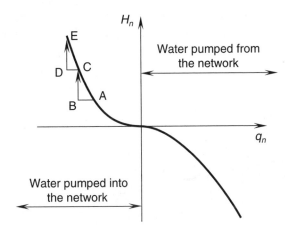

Figure 9.3 Instability triggered by a decreasing H–q relationship at a boundary node

case with pumps (H decreases with the discharge $-q$ pumped into the network) and orifice or head-loss formulae at outlets (the outflowing discharge q increases with the head H at the outlet).

9.3 Computational models of pipe transients

9.3.1 Introduction – assumptions

Pipe transients, also known as waterhammer, are often associated with the names of Joukowski (1898) and Allievi (1903), who provided the first mathematically correct derivation of the governing equations. The derivation of the waterhammer equations has been widely published (Fox (1989), Jaeger (1933), (1977), Swaffield and Boldy (1993), Wylie and Streeter (1977)).

Pipe transients arise as a consequence of rapid variations in the flow conditions in pressurized networks. Due to the (small) elasticity of the pipe material and the (small) compressibility of the fluid, such variations in the pressure and/or flow velocity propagate in the pipes at very high speeds. A typical wave celerity for waterhammer is 1,000 m/s. The actual value of c will depend on the pipeline diameter, wall thickness and material, and on the liquid bulk modulus K, which in turn is dependent on the entrained-air content (see equation (4.78)). This large contrast between the wave-propagation speed and the flow velocity has the consequences that: (i) pipe transients occur over very small time scales (at most a few seconds); and (ii) as the energy of the transient is dissipated only by friction or via singular head losses, the pressure wave may propagate over long distances with very little dissipation, because the flow velocity is small compared to the wave-propagation speed.

Waterhammer may lead to considerable pressure variations (see e.g. Section 9.3.3.3) within very short times, which may induce considerable damage to the pipes, junctions or installations, be it in the form of overpressure and/or cavitation induced by the pressure dropping below the vapour pressure of water in the pipe.

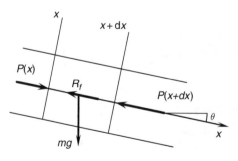

Figure 9.4 Definition sketch and notation for the derivation of the governing equations

The notation used in the present section is standard (see the list of Main Symbols given at the front of this book). The angle of the axis of the pipe with the horizontal is denoted by θ (Figure 9.4). The x-axis is the symmetry axis of the pipe. For the sake of clarity and conciseness, the governing equations and their characteristic form are derived for a pipe with a constant cross-sectional area. More complete expressions involving pipes with variable cross-sectional areas are presented in Guinot (2008).

Assumptions:

(1) The water is compressible and the pipe is deformable. Both the fluid and pipe material are assumed to remain in the elastic domain, i.e. the relative variations in the density ρ of water and the cross-sectional area A of the pipe are proportional to the variations in the pressure. A particular consequence of this is that the propagation speed of the pressure waves in still water is constant.
(2) The pressure p is assumed to be uniform over the cross-sectional area of the pipe.
(3) Both the water and pipe are only slightly deformable. Therefore, the variations in the pressure force Ap exerted on the cross-sectional area of the pipe are mainly due to the variations in the pressure, $d(Ap) \approx A dp$ and the variations in the mass flux are mainly due to those in the liquid discharge, $d(\rho Q) \approx \rho dQ$.
(4) The variations in the pressure p may reach several million pascals, while the typical order of magnitude of the velocity is a few metres per second. Consequently, the momentum discharge is much smaller than the pressure force and may be neglected in the momentum equation.
(5) Friction is accounted for by steady-state, turbulent friction laws such as Darcy–Weisbach's equation (9.2).

9.3.2 Governing equations

In view of the above statements, only a broad outline as necessary for the following sections is given here; for further details, see Chapter 7.

9.3.2.1 Continuity equation

Consider the control volume between the abscissas x and $x + dx$ in Figure 9.4. Conservation of mass imposes that the variation in the mass stored in the control volume between times t and $t + dt$ is due to the difference between the mass entering the control volume at x and the mass leaving the volume at $x + dx$. The mass stored in the control volume is equal to $\rho A dx$ and the amount of water passing at a given point over the time interval dt is given by $\rho Q dt$. Using these expressions for mass balance yields the following equality:

$$(\rho A dx)(t + dt) - (\rho A dx)(t) = (\rho Q dt)(x) - (\rho Q dt)(x + dx) \tag{9.18}$$

In the limit of infinitesimal dt and dx, the two members of equation (9.18) become

$$(\rho A dx)(t + dt) - (\rho A dx)(t) = \frac{\partial}{\partial t}(\rho A dx)\,dt = dt dx \frac{\partial}{\partial t}(\rho A) \tag{9.19a}$$

$$(\rho Q dt)(x) - (\rho Q dt)(x + dx) = -\frac{\partial}{\partial x}(\rho Q dt)dx = -dt dx \frac{\partial}{\partial x}(\rho Q) \tag{9.19b}$$

Substituting equations (9.19) into equation (9.18) and simplifying by dt/dx yields the continuity equation for a compressible fluid:

$$\frac{\partial}{\partial t}(\rho A) + \frac{\partial}{\partial x}(\rho Q) = 0 \tag{9.20}$$

9.3.2.2 Momentum equation

The momentum equation is obtained by applying Newton's second law of motion to the control volume of Figure 9.4:

$$\frac{\partial M}{\partial t} = Q_M(x) - Q_M(x + dx) + F \tag{9.21}$$

where F is the projection onto the x-axis of the sum of the forces exerted onto the control volume, M is the momentum of the fluid contained in the control volume and Q_M is the momentum discharge, i.e. the amount of momentum carried at the flow velocity that passes at a given point per unit time. F, M and Q_M are given by

$$F = P(x) - P(x + dx) + mg \sin \theta - R_x \tag{9.22a}$$

$$M = \rho Q dx \tag{9.22b}$$

$$Q_M = \rho Q u = \rho A u^2 = \rho Q^2 / A \tag{9.22c}$$

$$m = \rho A g dx \tag{9.22d}$$

where m is the mass of the fluid contained in the control volume, P is the pressure force exerted onto the cross-section of the pipe and R_x is the force exerted onto the fluid owing to friction (see Figure 9.4). From Assumption (2),

P is equal to the product of the cross-sectional area A and the pressure p. Moreover, the friction force R_x is related to the slope of the energy line via

$$R_x = \rho A g S_e dx \tag{9.23}$$

where S_e is the slope of the energy line, given by equation (9.2). Substituting equations (9.22) and (9.23) into equation (9.21) leads to

$$\frac{\partial}{\partial t}(\rho Q dx) = \left(\frac{\rho Q^2}{A}\right)(x) - \left(\frac{\rho Q^2}{A}\right)(x + dx)$$
$$+ (Ap)(x) - (Ap)(x + dx) + \rho Ag(\sin\theta - S_e)dx \tag{9.24}$$

Introducing the derivative of $\rho Q^2/A$ and Ap with respect to x and simplifying by dx leads to

$$\frac{\partial}{\partial t}(\rho Q) + \frac{\partial}{\partial x}\left(\frac{\rho Q^2}{A} + Ap\right) = (\sin\theta - S_e)\rho g A \tag{9.25}$$

Using Assumption (4) allows the momentum discharge $\rho Q^2/A$ to be neglected compared to Ap, and equation (9.25) simplifies to

$$\frac{\partial}{\partial t}(\rho Q) + \frac{\partial}{\partial x}(Ap) = (\sin\theta - S_e)\rho g A \tag{9.26}$$

9.3.2.3 Vector writing: conservation and non-conservation form

The continuity equation (9.20) and the momentum equation (9.26) can be recast in vector conservation form as

$$\frac{\partial \mathbf{U}}{\partial t} + \frac{\partial \mathbf{F}}{\partial x} = \mathbf{S} \tag{9.27a}$$

$$\mathbf{U} = \begin{bmatrix} \rho A \\ \rho Q \end{bmatrix}, \quad \mathbf{F} = \begin{bmatrix} \rho Q \\ Ap \end{bmatrix}, \quad \mathbf{S} = \begin{bmatrix} 0 \\ (\sin\theta - S_e)\rho g A \end{bmatrix} \tag{9.27b}$$

The non-conservation form of equations (9.27) is written as

$$\frac{\partial \mathbf{U}}{\partial t} + \mathbf{A}\frac{\partial \mathbf{U}}{\partial x} = \mathbf{S}' \tag{9.28}$$

where the general definitions for the matrix **A** and the source term **S'** are (see Chapter 7)

$$A = \frac{\partial F}{\partial U} \qquad (9.29a)$$

$$S' = S - \left(\frac{\partial F}{\partial x}\right)_{U=Const} \qquad (9.29b)$$

The diameter of the pipe being constant, $(\partial F/\partial x)_{U=Const}$ and $S' = S$. From equations (9.27), the expression for **A** is

$$A = \begin{bmatrix} 0 & 1 \\ c^2 & 0 \end{bmatrix} \qquad (9.30a)$$

$$c^2 = \frac{d(Ap)}{d(\rho A)} \qquad (9.30b)$$

From Assumption (1) of a linear dependence between the pressure, pipe cross-sectional area and water density (see Section 9.3.1), the speed of sound c as defined in equation (9.30b) is independent of the pressure and the Jacobian matrix **A** is constant.

Equations (9.27) are not the preferred form to deal with the waterhammer equations because the conserved variables ρA and ρQ are not directly measurable. The flow variables that are directly accessible to measurement are the pressure p and the discharge Q. Equations (9.27) can be transformed into equations involving only the derivatives of p and Q by noting from equation (9.30b) and Assumption (3) that

$$d(\rho A) = \frac{1}{c^2}d(Ap) \approx \frac{A}{c^2}dp \qquad (9.31a)$$

$$d(\rho Q) \approx \rho dQ \qquad (9.31b)$$

Substituting equations (9.31) into (9.28) leads to an expression of the form

$$\frac{\partial V}{\partial t} + B\frac{\partial V}{\partial x} = R \qquad (9.32)$$

with

$$V = \begin{bmatrix} p \\ Q \end{bmatrix}, \quad B = \begin{bmatrix} 0 & \rho c^2/A \\ A/\rho & 0 \end{bmatrix}, \quad R = \begin{bmatrix} 0 \\ (\sin\theta - S_e)gA \end{bmatrix} \qquad (9.33)$$

Another possibility is to use the flow velocity V rather than the discharge Q for the second component of the variable vector **V**. This leads to the following definition for **B**, **R** and **V**:

$$\mathbf{V} = \begin{bmatrix} p \\ V \end{bmatrix}, \quad \mathbf{B} = \begin{bmatrix} 0 & \rho c^2 \\ 1/\rho & 0 \end{bmatrix}, \quad \mathbf{R} = \begin{bmatrix} 0 \\ (\sin\theta - S_e)A \end{bmatrix} \qquad (9.34)$$

9.3.2.4 Characteristic form

The characteristic form is obtained by transforming equation (9.28) or equation (9.32) into a vector equation where the matrix **A** becomes diagonal. As shown in Chapter 7, this can be done by introducing the matrix **K** of the eigenvectors of **A** and left-multiplying equation (9.28) by $\mathbf{K}^{-1}$:

$$\mathbf{K}^{-1}\frac{\partial \mathbf{U}}{\partial t} + \mathbf{K}^{-1}\mathbf{A}\mathbf{K}\mathbf{K}^{-1}\frac{\partial \mathbf{U}}{\partial x} = \mathbf{K}^{-1}\mathbf{S}' \qquad (9.35)$$

The matrix $\mathbf{K}^-\mathbf{A}\mathbf{K}$ is diagonal. The terms on the diagonal are the eigenvalues of **A**. Equation (9.35) can be rewritten as

$$\frac{\partial \mathbf{W}}{\partial t} + \mathbf{\Lambda}\frac{\partial \mathbf{W}}{\partial x} = \mathbf{S}'' \qquad (9.36)$$

with the following definitions of **W**, **Λ** and **S**″:

$$d\mathbf{W} = \mathbf{K}^{-1}\,d\mathbf{U} \qquad (9.37a)$$

$$\mathbf{\Lambda} = \mathbf{K}^{-1}\mathbf{A}\mathbf{K} \qquad (9.37b)$$

$$\mathbf{S}'' = \mathbf{K}^{-1}\mathbf{S}' = \mathbf{K}^{-1}\mathbf{S} \qquad (9.37c)$$

W is called the vector of Riemann invariants because, as shown hereafter, its components are invariant along some particular trajectories in the (x, t)-plane, called 'characteristics'. It is easy to check that the eigenvalues λ_p and eigenvectors $\mathbf{r}_p (p = 1, 2)$ of **A** are

$$\lambda_1 = -c, \qquad \lambda_2 = +c \qquad (9.38a)$$

$$\mathbf{r}_1 = \begin{bmatrix} 1 \\ -c \end{bmatrix}, \qquad \mathbf{r}_2 = \begin{bmatrix} 1 \\ +c \end{bmatrix} \qquad (9.38b)$$

leading to the following expressions for **K** and $\mathbf{K}^{-1}$:

$$\mathbf{K} = \begin{bmatrix} 1 & 1 \\ -c & c \end{bmatrix}, \quad \mathbf{K}^{-1} = \frac{1}{2c}\begin{bmatrix} c & -1 \\ c & 1 \end{bmatrix} \qquad (9.39)$$

Substituting equations (9.38) into equations (9.37) leads to

$$dW = \frac{1}{2c} \begin{bmatrix} cd(\rho A) - d(\rho Q) \\ cd(\rho A) + d(\rho Q) \end{bmatrix}$$

$$\approx \frac{1}{2c} \begin{bmatrix} d(Ap)/c - \rho dQ \\ d(Ap)/c + \rho dQ \end{bmatrix} \approx \frac{1}{2c} \begin{bmatrix} A/c\, dp - \rho dQ \\ A/c\, dp + \rho dQ \end{bmatrix} \tag{9.40a}$$

$$\Lambda = \begin{bmatrix} -c & 0 \\ 0 & c \end{bmatrix} \tag{9.40b}$$

$$S'' = \frac{1}{2c} \begin{bmatrix} -(\sin\theta - S_e)\rho g A \\ (\sin\theta - S_e)\rho g A \end{bmatrix} \tag{9.40c}$$

Substituting equations (9.40) into (9.36) and simplifying by $1/(2c)$ yields the following two characteristic equations:

$$\rho\frac{dQ}{dt} - \frac{A}{c}\frac{dp}{dt} = (\sin\theta - S_e)\rho g A \quad \text{for} \quad \frac{dx}{dt} = -c \tag{9.41a}$$

$$\rho\frac{dQ}{dt} + \frac{A}{c}\frac{dp}{dt} = (\sin\theta - S_e)\rho g A \quad \text{for} \quad \frac{dx}{dt} = +c \tag{9.41b}$$

Assumptions (1) and (3) allow equations (9.41) to be transformed into

$$\frac{d}{dt}\left(\rho Q - \frac{Ap}{c}\right) = (\sin\theta - S_e)\rho g A \quad \text{for} \quad \frac{dx}{dt} = -c \tag{9.42a}$$

$$\frac{d}{dt}\left(\rho Q + \frac{Ap}{c}\right) = (\sin\theta - S_e)\rho g A \quad \text{for} \quad \frac{dx}{dt} = +c \tag{9.42b}$$

or

$$\frac{d}{dt}(\rho c u - p) = (\sin\theta - S_e)\rho c g \quad \text{for} \quad \frac{dx}{dt} = -c \tag{9.43a}$$

$$\frac{d}{dt}(\rho c u + p) = (\sin\theta - S_e)\rho c g \quad \text{for} \quad \frac{dx}{dt} = +c \tag{9.43b}$$

Note that, for frictionless flow in a horizontal pipe, equations (9.43) become

$$\rho c u - p = \text{Const}_1 \quad \text{for} \quad \frac{dx}{dt} = -c \tag{9.44a}$$

$$\rho c u + p = \text{Const}_2 \quad \text{for} \quad \frac{dx}{dt} = -c \tag{9.44b}$$

Then W_1 and W_2 are said to be invariant along their respective characteristic lines, and hence the term 'Riemann invariants'.

9.3.3 Behaviour of analytical solutions – initial and boundary conditions

9.3.3.1 General

The characteristic forms (equations (9.42)–(9.44)) lead to the following conclusions about the behaviour of the analytical solutions of the waterhammer equations.

(1) The state of the flow at any point M in the (x, t)-plane is determined by the flow state in the domain of dependence delimited by the characteristics $dx/dt = -c$ and $dx/dt = +c$ (domain [ABM] in Figure 9.5).

(2) Conversely, a perturbation in the flow state at the point M triggers variations in the flow at the points contained within the domain of influence delimited by the characteristics $dx/dt = -c$ and $dx/dt = +c$ (domain [A′B′M] in Figure 9.5).

(3) The speed of sound c being independent from the flow state, the characteristics are straight lines in the (x, t)-plane.

(4) The flow at M is determined uniquely at a given point provided that $W_1 = \rho cu - p$ and $W_2 = \rho cu + p$ are known. Indeed,

$$u = \frac{W_1 + W_2}{2\rho c} \tag{9.45a}$$

$$p = \frac{W_2 - W_1}{2} \tag{9.45b}$$

then $Q = Au$ can be computed and all the flow variables are known.

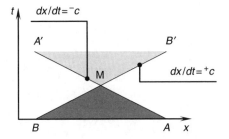

Figure 9.5 Waterhammer. Definition sketch for the propagation speeds of the waves in the (x, t)-plane. Domain of influence (light-grey area), domain of dependence (dark-grey area)

9.3.3.2 Initial and boundary conditions

Assume that the waterhammer equations are to be solved over the domain $[0, L]$ for all times $t > 0$ (Figure 9.6). Three possibilities arise depending on the location of the point $M(x, t)$ in the phase space.

(1) If the domain of dependence is included in the segment $[0, L]$ (point M and domain of dependence [AB] in Figure 9.6), the knowledge of the initial condition is sufficient to compute the solution at M. The invariant W_1 can be computed from the known flow state at the point A and the invariant W_2 can be computed from the known initial condition at the point B. The invariants W_1 and W_2 at M can be determined by integrating equations (9.43) between the feet A and B of the characteristics and the point M:

$$(W_1)_M = (W_1)_A + \int_{t_A}^{t_M} S'' dt \tag{9.46a}$$

$$(W_2)_M = (W_2)_B + \int_{t_B}^{t_M} S'' dt \tag{9.46b}$$

thus enabling the calculation of u, Q and p at M via equations (9.45).

(2) If part of the domain of dependence comprises the left-hand boundary (point M′ and domain of dependence [A′B′] in Figure 9.6), the knowledge of the initial state alone is not sufficient to compute the solution at M. Indeed, W_2 cannot be computed at the point M from the initial condition because the characteristic $dx/dt = +c$ passing at M does not intersect the line $t = 0$. The earliest time at which W_2 can be known corresponds to the point B′ in the phase space (see Figure 9.6). The missing information must be supplied in the form of a boundary condition

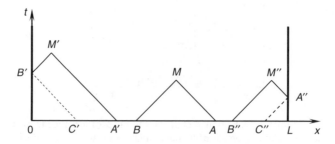

Figure 9.6 Waterhammer equations. Definition sketch for initial and boundary conditions

involving the velocity (or discharge) and/or the pressure. The general form of such a boundary condition is

$$f_L(p_{B'}, u_{B'}, t) = 0 \qquad (9.47)$$

where f_L is a known function of both the unknown pressure $p_{B'}$ and velocity $u_{B'}$. From the definitions of W_1 and W_2, one has

$$(W_1)_{B'} = \rho c u_{B'} - p_{B'} \qquad (9.48a)$$

$$(W_2)_{B'} = \rho c u_{B'} + p_{B'} \qquad (9.48b)$$

Equations (9.47) and (9.48) must be solved for $p_{B'}$, $u_{B'}$ and $(W_2)_{B'}$. To do so, the system must be closed by finding $(W_1)_{B'}$. This is done using the characteristic relationship (9.43a) between the foot C' of the characteristic $dx/dt = -c$ and the boundary point B':

$$(W_1)_{B'} = (W_1)_{C'} + \int_{t_{C'}}^{t_{B'}} S'' dt \qquad (9.49)$$

The calculation sequence is as follows. In a first step, $(W_1)_{B'}$ is computed using equation (9.49). The second step consists of solving equations (9.47)–(9.48a) for $p_{B'}$ and $u_{B'}$. In a third step, $(W_2)_{B'}$ is computed using equation (9.48b).

(3) The treatment of the right-hand boundary is similar to that of the left-hand boundary. A boundary condition is needed at the right-hand end of the domain to ensure the existence and uniqueness of the solution. Such a boundary condition is written in the form of a known relationship between the pressure and velocity at the point A'':

$$f_R(p_{A''}, u_{A''}, t) = 0 \qquad (9.50)$$

and the following relationships are available:

$$(W_1)_{A''} = \rho c u_{A''} - p_{A''} \qquad (9.51a)$$

$$(W_2)_{A''} = \rho c u_{A''} + p_{A''} \qquad (9.51b)$$

$$(W_2)_{A''} = (W_2)_{C''} + \int_{t_{C''}}^{t_{A''}} S'' dt \qquad (9.51c)$$

The calculation sequence at the right-hand boundary consists of (i) computing $(W_2)_{A''}$ from the known value at the point C'' using equation (9.51c), (ii) solving equations (9.50) and (9.51b) for $p_{A''}$ and $u_{A''}$, and (iii) computing $(W_1)_{A''}$ using equation (9.51a).

9.3.3.3 Joukowski's formula

Equations (9.46) allow the derivation of the so-called Joukowski formula. Consider a pressure wave across which the pressure and velocity variations are denoted by Δp and Δu, respectively. If the wavefront is narrow, the time interval needed to cross the wave is negligible and the integral in equations (9.46) can be neglected. If the wave moves at the speed $+c$, equation (9.46a) gives

$$\Delta p = \rho c \Delta u \tag{9.52}$$

while for a wave moving at a speed $-c$, equation (9.46b) gives

$$\Delta p = -\rho c \Delta u \tag{9.53}$$

Joukowski's formula provides a relationship between the magnitude of the pressure variation and that of the flow velocity by summarizing equations (9.52) and (9.53) under the same equation

$$|\Delta p| = \rho c |\Delta u| = \frac{\rho c}{A} |\Delta u| Q \tag{9.54}$$

Equation (9.54) may be used to estimate the maximum possible pressure variations caused by the sudden operation of valves, pumps, turbines, etc., in a pipe network.

9.3.4 The method of characteristics

9.3.4.1 General

The principle of the method of characteristics (MOC) for waterhammer simulations is the same as for the Saint Venant equations (see Sections 2.8 and 7.3.2), but with the difference that the celerity of the pressure waves is fixed and is independent of the local value of the flow variables. The MOC consists of solving the characteristic form (equation (9.43)) of the equations approximately. Space is discretized into computational points (Figure 9.7) at which the solution is to be computed for predefined times. The distance Δx between two adjacent points is usually called the 'cell width', while the difference Δt between two successive computational times is called the

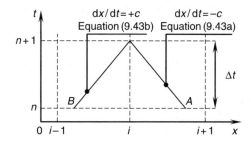

Figure 9.7 Definition sketch for the method of characteristics (MOC)

'computational time step'. A flow variable U computed at the point i at the time level n is denoted by U_i^n.

In the MOC, the characteristic equations (9.43) are integrated approximately between the foot of each characteristic and the point $(i, n+1)$ where the solution is sought in the phase space. The invariant W_1 at the foot A and the invariant W_2 at the foot B are computed from the known values at the points $(i-1, n)$, (i, n), $(i+1, n)$, etc., using interpolation formulae. The treatment of internal points and boundary conditions is dealt with in separate sections hereafter.

9.3.4.2 Treatment of internal points

Assume that the pipe is discretized using M computational points. The present section describes the solution technique for internal points, i.e. for $i = 2, 3, \ldots, M-1$. The technique consists of applying equations (9.46) with $S'' = 0$, where the point M is the point $(i, n+1)$ in the phase space:

$$\rho c u_i^{n+1} - p_i^{n+1} = \rho c u_A - p_A + S_1'' \Delta t \tag{9.55a}$$

$$\rho c u_i^{n+1} + p_i^{n+1} = \rho c u_B + p_B + S_2'' \Delta t \tag{9.55b}$$

The expressions for p_i^{n+1} and u_i^{n+1} are obtained from the sum and difference between equations (9.55)

$$p_i^{n+1} = \frac{p_A + p_B}{2} + \frac{\rho c}{2}(u_B - u_A) + \frac{S_2'' - S_1''}{2} \Delta t \tag{9.56a}$$

$$u_i^{n+1} = \frac{u_A + u_B}{2} + \frac{p_B - p_A}{2\rho c} + \frac{S_1'' + S_2''}{2\rho c} \Delta t \tag{9.56b}$$

The pressure and flow velocity are interpolated at the points A and B. The formulae for a linear interpolation over an irregular computational grid are

$$U_A = \begin{cases} (1+1/Cr_1)U_{i+1}^n - U_{i+1}^{n+1}/Cr_1 & \text{if } Cr_1 \leq -1 \\ (1+Cr_1)U_i^n - Cr_1 U_{i+1}^n & \text{if } Cr_1 \geq -1' \end{cases} \quad U = p, u \qquad (9.57a)$$

$$U_B = \begin{cases} (1-Cr_2)U_i^n + Cr_2 U_{i-1}^n & \text{if } Cr_2 \leq 1 \\ (1-1/Cr_2)U_{i-1}^n + U_{i-1}^{n+1}/Cr_2 & \text{if } Cr_2 \geq 1' \end{cases} \quad U = p, u \qquad (9.57b)$$

where the Courant numbers Cr_1 and Cr_2 are defined as

$$Cr_1 = -\frac{c\Delta t}{\Delta x_{i+1/2}} \qquad (9.58a)$$

$$Cr_2 = \frac{c\Delta t}{\Delta x_{i-1/2}} \qquad (9.58b)$$

where $\Delta x_{i-1/2}$ and $\Delta x_{i+1/2}$ are the width of the cells $i - 1/2$ (between the points $i - 1$ and i) and $i + 1/2$ (between the points i and $i + 1$), respectively.

(1) If the absolute values of the Courant numbers Cr_1 and Cr_2 are smaller than unity, substituting the interpolation formulae (9.57) into equations (9.56) yields the explicit formulae

$$p_i^{n+1} = \frac{Cr_2 p_{i-1}^n + (2 + Cr_1 - Cr_2)p_i^n - Cr_1 p_{i+1}^n}{2}$$

$$+ \frac{\rho c}{2}\left[Cr_2 u_{i-1}^n - (Cr_1 + Cr_2)u_i^n + Cr_1 u_{i+1}^n\right] + \frac{S_2'' - S_1''}{2}\Delta t \qquad (9.59a)$$

$$u_i^{n+1} = \frac{Cr_2 u_{i-1}^n + (2 + Cr_1 - Cr_2)u_i^n - Cr_1 u_{i+1}^n}{2}$$

$$+ \frac{1}{2\rho c}\left[Cr_2 p_{i-1}^n - (Cr_1 + Cr_2)p_i^n + Cr_1 p_{i+1}^n\right] + \frac{S_1'' + S_2''}{2\rho c}\Delta t \qquad (9.59b)$$

with

$$S_1'' = (1 + Cr_1)S'''_i - Cr_1 S'''_{i+1} \qquad (9.60a)$$

$$S_1'' = (1 - Cr_2)S'''_i + Cr_2 S'''_{i-1} \qquad (9.60b)$$

(2) If Cr_1 and Cr_2 are larger than unity, substituting equations (9.57) into equations (9.56) leads to a set of implicit relationships:

$$p_i^{n+1} - \frac{(1+1/Cr_1)p_{i+1}^{n+1} + (1-1/Cr_2)p_{i-1}^{n+1}}{2}$$

$$+ \frac{\rho c}{2}\left[(1+1/Cr_1)u_{i+1}^{n+1} - (1-1/Cr_2)u_{i-1}^{n+1}\right] + \frac{S_2'' - S_1''}{2}\Delta t \quad (9.61a)$$

$$= -\frac{p_{i+1}^n/Cr_1 + p_{i-1}^n/Cr_2}{2} + \frac{\rho c}{2}\left(\frac{u_{i-1}^n}{Cr_2} - \frac{u_{i+1}^n}{Cr_1}\right)$$

$$u_i^{n+1} - \frac{(1+1/Cr_1)u_{i+1}^{n+1} + (1-1/Cr_2)u_{i-1}^{n+1}}{2}$$

$$+ \frac{(1+1/Cr_1)p_{i+1}^{n+1} - (1-1/Cr_2)p_{i-1}^{n+1}}{2\rho c} \quad (9.61b)$$

$$= \frac{Cr_2 u_{i-1}^n - Cr_1 u_{i+1}^n}{2} + \frac{Cr_1 p_{i+1}^n + Cr_2 p_{i-1}^n}{2\rho c} + \frac{S_1'' + S_2''}{2\rho c}\Delta t$$

Many options are available for the implicit discretization of the source term S''. One could propose the following two formulae:

$$S_1'' = S_2'' \approx \rho g c(\sin\theta - S_{ei}^{n+1}) = \rho g c\left(\sin\theta - \frac{\lambda L^2}{2g}\left|u_i^{n+1}\right|u_i^{n+1}\right)$$

(fully implicit) $\qquad\qquad$ (9.62a)

$$S_1'' = S_2'' \approx \rho g c\left(\sin\theta - \frac{\lambda L^2}{2gD}\left|u_i^n\right|u_i^{n+1}\right) \text{ (semi-implicit)} \qquad (9.62b)$$

Equation (9.62a) is not used in practice because it introduces a second-degree term in the sought variable u_i^{n+1}. The option (9.62b) is preferred. It yields a system of equations of the form

$$A_i^{(1)}p_{i-1}^{n+1} + B_i^{(1)}p_i^{n+1} + C_i^{(1)}p_{i+1}^{n+1} - \rho c A_i^{(1)}u_{i-1}^{n+1} - \rho c C_i^{(1)}u_{i+1}^{n+1} = D_i^{(1)} \quad (9.63a)$$

$$A_i^{(2)}u_{i-1}^{n+1} + B_i^{(2)}u_i^{n+1} + C_i^{(2)}u_{i+1}^{n+1} - \frac{A_i^{(2)}}{\rho c}p_{i-1}^{n+1} - \frac{C_i^{(2)}}{\rho c}p_{i+1}^{n+1} = D_i^{(2)} \quad (9.63b)$$

with

$$A_i^{(1)} = A_i^{(2)} = \frac{1 - Cr_2}{2Cr_2} \qquad\qquad (9.64a)$$

$$B_i^{(1)} = 1 \qquad\qquad (9.64b)$$

$$B_i^{(2)} = 1 + \lambda\frac{L^2}{2D}\left|u_i^n\right|\Delta t \qquad\qquad (9.64c)$$

$$C_1^{(1)} = C_1^{(2)} = -\frac{1+\mathrm{Cr}_1}{2\mathrm{Cr}_1} \tag{9.64d}$$

$$D_i^{(1)} = -\frac{p_{i+1}^n/\mathrm{Cr}_1 + p_{i-1}^n/\mathrm{Cr}_2}{2} + \frac{\rho c}{2}\left(\frac{u_{i-1}^n}{\mathrm{Cr}_2} - \frac{u_{i+1}^n}{\mathrm{Cr}_1}\right) \tag{9.64e}$$

$$D_i^{(2)} = \frac{\mathrm{Cr}_2 u_{i-1}^n - \mathrm{Cr}_1 u_{i+1}^n}{2} + \frac{\mathrm{Cr}_1 p_{i+1}^n + \mathrm{Cr}_2 p_{i-1}^n}{2\rho c} + g\sin\theta\,\Delta t \tag{9.64f}$$

9.3.4.3 Treatment of boundary conditions

Denote by M the number of computational points in the domain. The formulae (9.56) for an explicit discretization and equations (9.62) for an implicit discretization are applicable only for the points $i = 2, \ldots, M-1$. This means that only $2M-4$ equations can be written for the $2M$ unknown variables (p_i^n, u_i^n) in the computational domain. The missing information must be supplied in the form of boundary conditions at $i=1$ and $i=M$. Only the treatment of the left-hand boundary is detailed hereafter. The treatment of a right-hand boundary can be inferred using symmetry considerations.

Consider the point $i = 1$ in the computational domain (Figure 9.8). Assume first that $\mathrm{Cr}_1 > -1$ (Figure 9.8(a)). Applying the characteristic relationship (9.55a) with the interpolation formula (9.57a) at the point $i = 1$ leads to

$$\rho c u_1^{n+1} - p_1^{n+1} = \rho c\left[(1+\mathrm{Cr}_1)u_1^n - \mathrm{Cr}_1 u_2^n\right] - (1+\mathrm{Cr}_1)p_1^n + \mathrm{Cr}_1 p_2^n \tag{9.65}$$

The boundary condition may be supplied in the form of a known discharge (i.e. velocity), a known pressure, or a known relationship between the pressure and discharge at the left-hand boundary:

$$f_b(p_1^{n+1}, u_1^{n+1}, t_{n+1}) = 0 \tag{9.66}$$

Equations (9.65) and (9.66) can be solved uniquely for p_1^{n+1} and u_1^{n+1}.

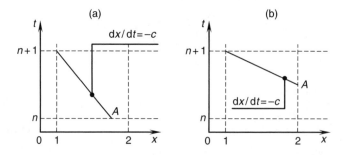

Figure 9.8 Definition sketch for the treatment of the left-hand boundary

If $Cr_1 < -1$ (Figure 9.8(b)), the interpolation formula (9.57a) becomes

$$\rho c \left(u_1^{n+1} - \frac{Cr_1 + 1}{Cr_1} u_2^{n+1} \right) - p_1^{n+1} + \frac{Cr_1 + 1}{Cr_1} p_2^{n+1} = -\frac{\rho c}{Cr_1} u_2^n + \frac{p_2^n}{Cr_1} \quad (9.67)$$

Equations (9.66) and (9.67) supply the missing two equations at the left-hand boundary. Writing similar equations for the point M yields the missing two equations at the right-hand boundary and the system counts $2M$ equations for $2M$ unknowns, thus ensuring solution existence and uniqueness.

9.3.5 Treatment of singularities and junctions

9.3.5.1 Singularities

Singularities such as diaphragms, valves, and sudden pipe widenings or narrowings can be accounted for using head-loss–discharge relationships. From a numerical point of view, they are handled as internal boundaries (Figure 9.9).

Assume that a singularity is located between the points i and $i+1$. The pressures and discharges must be computed at the points i and $i+1$ at the time level $n+1$. Consequently, there are four unknown variables at the singularity at the time level $n+1$. Equation (9.55b) can be used to calculate the flow variables at the point i, and equation (9.55a) is available for the point $i+1$. However, the characteristic equations cannot be used between the points i and $i+1$. The missing two equations are supplied from mass-conservation considerations

$$Q_i^{n+1} = Q_{i+1}^{n+1} \quad (9.68)$$

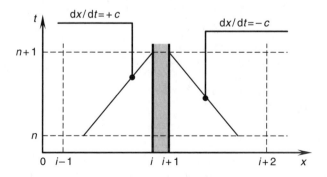

Figure 9.9 A singularity modelled as an internal boundary

and from the head-loss–discharge relationship

$$p_i^{n+1} - p_{i+1}^{n+1} = \rho g \alpha \left| Q_i^{n+1} \right| Q_i^{n+1} \tag{9.69}$$

where α is a known head-loss coefficient. Equation (9.69) is most often linearized to

$$p_i^{n+1} - p_{i+1}^{n+1} = \rho g \alpha \left| Q_i^{n} \right| Q_i^{n+1} \tag{9.70}$$

Note that equation (9.68) is a simplification of the conservation equation $\rho Q =$ constant under Assumption (3) of a nearly constant density.

Equations (9.55), (9.68) and (9.70) allow the system to be closed and a unique solution to be found for the pressures and discharges at the points i and $i+1$.

9.3.5.2 Junctions

Junctions are also classically handled as internal boundaries. The pipes converging to the junction node are treated as separate branches (see Figure 9.10, sketched for three pipes). Without loss of generality, the

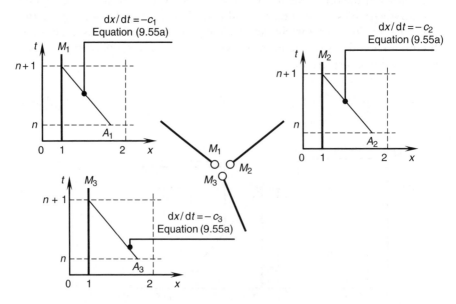

Figure 9.10 A junction handled as an internal boundary. The junction is a left-hand boundary for each of the pipes converging to the junction

x-axis along each pipe is oriented positive away from the node. As many equations (9.55a) may be written as there are pipes:

$$\rho c_k u_{k,1}^{n+1} - p_{k,1}^{n+1} = \rho c_k u_{A_k} - p_{A_k} + S_{k,1}'' \Delta t, \quad k = 1, \ldots, P \tag{9.71}$$

where A_k is the foot of the characteristic $dx/dt = -c_k$ passing at the point $i = 1$ in pipe number k, P is the number of pipes of the junction, and $p_{k,1}^{n+1}$ and $u_{k,1}^{n+1}$ are the unknown pressure and velocity at the first point of the pipe k (point M_k in Figure 9.10).

The missing P equations are provided by the continuity equation (law on nodes) at the junction:

$$\sum_{k=1}^{P} Q_{k,1}^{n+1} = 0 \tag{9.72}$$

and a condition involving the pressures at the nodes M_k, $k = 1, \ldots, p$. For instance, a uniform pressure condition may be used:

$$p_{1,1}^{n+1} = \ldots = p_{k,1}^{n+1} = \ldots = p_{P,1}^{n+1} \tag{9.73}$$

9.3.5.3 Surge tanks and air vessels

The sharp pressure transients induced by waterhammer may be efficiently damped by devices such as surge tanks or air vessels. The role of such devices is to absorb and damp the energy of the transient, thereby reducing the amplitude of the pressure waves. The common feature of a surge tank or an air vessel is to allow for substantial volume variations under small pressure changes. From the point of view of wave propagation, connecting such a device to a pipe is equivalent to inserting a highly deformable, short pipe (i.e. a pipe with a very low sound speed). In surge tanks, the pressure at the free surface of the water is the atmospheric pressure. In air vessels, the air or inert gas above the free surface is pressurized.

Surge tanks and air vessels may be represented using the general sketch in Figure 9.11(a). In the case of a surge tank, the free surface is at the atmospheric pressure

$$p = p_{atm} \tag{9.74}$$

while for an air vessel the pressure p is a function of the volume of the gas above the water. Under the assumption of adiabatic behaviour:

$$p = p_0 \left(\frac{V_T - V_0}{V_T - V} \right)^\gamma \tag{9.75}$$

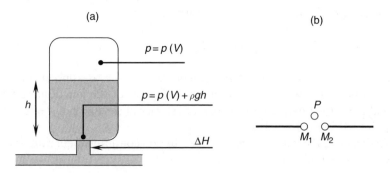

Figure 9.11 Definition sketch for a surge tank/air vessel (a) and discretization of the numerical solution (b)

where V is the volume of water in the vessel, V_T is the total volume of the vessel, V_0 is the volume occupied by the water at the reference pressure p_0, and $\gamma = 1.4$ is the polytropic constant for perfect gases.

Assuming a hydrostatic distribution, the pressure at the bottom of the chamber is $p_T = p + \rho gh$, where h is the water depth in the tank or vessel. Assuming a head loss at the inlet of the chamber of the form

$$\Delta H = \alpha_T |Q_T| Q_T \tag{9.76}$$

where Q_T is the discharge into the tank. Using equations (9.74)–(9.76) and the hydrostatic distribution assumption, one obtains

$$p_i^{n+1} = p_{atm} + \rho gh^{n+1} + \alpha_T |Q_T| Q_T \text{ (tank)} \tag{9.77a}$$

$$p_i^{n+1} = p_0 \left(\frac{V_T - V_0}{V_T - V^{n+1}} \right)^{\gamma} + \rho gh^{n+1} + \alpha_T |Q_T| Q_T \text{ (vessel)} \tag{9.77b}$$

Moreover, continuity and the equality of pressures at the points i and $i+1$ yield the following equations:

$$Q_i^{n+1} - Q_i^{n+1} = Q_T \tag{9.78a}$$

$$p_i^{n+1} = p_{i+1}^{n+1} \tag{9.78b}$$

The discharge Q_T is the time derivative of the volume V of water in the tank:

$$\frac{dV}{dt} = A\frac{dh}{dt} = Q_T \tag{9.79}$$

Equation (9.79) can be discretized into

$$V^{n+1} = V^n + Q_T \Delta t \tag{9.80}$$

The unknowns are the pressures and discharges p_i^{n+1}, p_{i+1}^{n+1}, Q_i^{n+1} and Q_{i+1}^{n+1}, the volume V^{n+1}, the discharge Q_T and the water depth h^{n+1}. The geometry of the tank/vessel being known, V and h are related by a one-to-one relationship, which leaves only six unknowns. Two relationships are provided by equations (9.55) at the points i and $i+1$. The continuity and pressure equations (9.78) provide another two relationships. Closure is ensured by one of the equations (9.77) and equation (9.80). For the determination of the head-loss coefficient at the restricted ('throttled') surge tank/air vessel entry, see the remarks in the following section.

9.4 Physical modelling of closed-conduit flow

9.4.1 Modelling of steady flow and friction head losses

In order to obtain the same value of the friction coefficient λ in *laminar flow*, it follows from equation (4.41) ($\lambda = 64/\text{Re}$) that the *Reynolds number* must be identical in the model and prototype ($M_{\text{Re}} = 1$). This, in turn, means that $M_v = M_\mu/(M_\rho M_l)$ and $M_p = M_\mu^2/(M_\rho M_l^2)$ (see Chapter 5), which satisfies the velocity-distribution equation for laminar flow (Hagen–Poiseuille). However, to obtain similarity of head losses h_f with $M_{h_f} = M_l$ and $M_D = M_l$ we get, from equation (4.36), for the velocity scale

$$M_v = M_{h_f} M_D^2 M_\rho M_l^{-1} M_\mu^{-1} = M_l^2 (M_\rho M_\mu)^{-1} \tag{9.81}$$

This is clearly incompatible with the Reynolds law of similarity, unless

$$M_l = M_v^{2/3} \tag{9.82}$$

For the same fluid in the model as in the prototype, this means that $M_l = 1$ and it is practically impossible to find a suitable fluid that would satisfy all the above conditions with water in the prototype. Therefore, we must conclude that, while on a scale model of laminar closed-conduit flow we can model with the identity of Reynolds number the velocity distribution and friction coefficient, we cannot easily achieve simultaneously similarity of head losses.

For the *smooth turbulent* region it follows from equations (4.46) or (4.48) that the same arguments and conclusions apply as for laminar flow. For flow in the *transition region* ($1/6 < \delta'/k < 4$) the identity of the friction coefficient on the model and in the prototype and similarity of head losses

$(M_{b_f} = M_l)$ can be achieved for a model operated according to the *Froude law* of similarity $(M_v = M_l^{1/2})$ (see equation (4.55)):

$$\frac{k_p}{3.71D_p} + \frac{2.51}{Re_p\lambda_p^{1/2}} = \frac{k_m}{3.71D_m} + \frac{2.51}{Re_m\lambda_m^{1/2}} \tag{9.83}$$

assuming that we can choose the model scale to satisfy the above equation for one discharge only, and only at the expense of adjusting the roughness scale (i.e. M_k is not equal to $M_D = M_l$) (see also below).

If the prototype pipe is hydraulically *rough* $(k > 6\delta')$, which is frequently the case because in the prototype the Reynolds number is often large, the coefficient λ is independent of the Reynolds number and is a function only of the relative roughness (equation (4.49)).

To attain mechanical similarity in the model, i.e. the same velocity distribution as in prototype $(M_\lambda = 1)$, and the similarity of friction losses $(M_{b_f} = M_l)$ with complete geometrical similarity, i.e. with the same relative roughness on the model and in prototype $(M_k = M_D = M_l)$, we must ensure that the model scale is such that, with the preservation of the *Froude law* of similarity, λ on the model is also independent of the Reynolds number. This is only the case if the minimum value of the Reynolds number on the model is greater than (or at least equal to) Re_{sq} (equation (4.52)). For the same liquid on the model as in the prototype $(M_v = 1)$ and for $M_v = M_l^{1/2}$ (Froude law) the following relation will apply for the scale of the model:

$$M_{Re} = M_v M_D M_v^{-1} = M_l^{3/2} < \frac{Re_p k_m \lambda^{1/2}}{400r_m} < \frac{Re_p k_p \lambda^{1/2}}{200D_p} \tag{9.84}$$

Sometimes, however, the condition expressed by the above equation cannot be fulfilled and then the model operates in the transition zone or even becomes hydraulically smooth. For example, in the case of dams one scale model is often used to simulate the spillway, the bottom outlet and the stilling basin, with the outlet being represented by a smooth brass or perspex pipe (see Chapter 13); in such cases the model pipe is usually 'smooth', even if the prototype pipe is 'rough'. The model is operated according to the Froude law of similarity with regard to the decisive part played by inertial forces. By using a smooth pipe on the model the condition of geometric similarity for the roughness is violated; nevertheless, satisfactory results may be obtained by the procedure shown in the following example (Novak and Čábelka (1981)) (Figure 9.12).

Let us consider that using the same liquid as in the prototype we want to simulate on a model friction losses and flow through a prototype pipe with a relative roughness of $k_p/D_p = 0.001$ and $Re_p = 10^7$. From Figure 4.4 (or equation (4.49)), $\lambda_p = 0.0197$ (alternatively, Re_p and λ_p may be given and

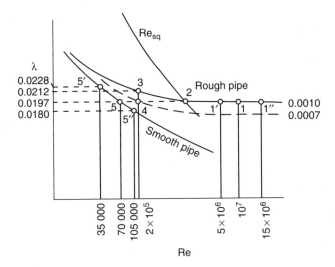

Figure 9.12 Definition sketch for modelling hydraulically rough flow

we determine k_p/D_p). This case is illustrated in Figure 9.12 by point 1. If we satisfy the condition $\mathrm{Re}_m > \mathrm{Re}_p$, the model situation is illustrated in Figure 9.12 by point 2 (or any point between 2 and 1). For the model in the transitional zone for the same relative roughness (0.0010) we get, for example, to point 3 (if $\mathrm{Re}_m = 2 \times 10^5$), with $\lambda_m = 0.0212$ (i.e. $\lambda_m > \lambda_p$). If we want to preserve the condition $\lambda_p = \lambda_m$ we must, for $\mathrm{Re}_m = 2 \times 10^5$, choose $k_m/D_m = 0.0007$ (i.e. point 4 in Figure 9.12). If we want to achieve $M_\lambda = 1$ for a smooth pipe, the model scale is determined by the fact that, for the modelled discharge calculated according to Froude law, Re_m must be on the point of intersection of the horizontal line for the given prototype relative roughness and the curve of the smooth pipe law – point 5 with $\mathrm{Re}_m = 70,000$.

Thus, the model scale (for $M_v = 1$) is given by

$$M_l = M_{\mathrm{Re}}^{2/3} = \frac{10^7}{7 \times 10^4} = 27.5 \tag{9.85}$$

Obviously, we can satisfy the condition $M_\lambda = 1$ for *one discharge only*, as λ changes with discharge (Reynolds number) on the model but not in the prototype. In this case, the model scale is usually chosen so that the condition $M_\lambda = 1$ is valid for the mean or maximum discharge; for the other cases we can compute λ on the model and when scaling the model results onto the prototype apply the appropriate correction numerically. If, in this example, Re_p varies within the limits $5 \times 10^6 - 15 \times 10^6$ (points 1' and 1" in Figure 9.12), with $M_l = 27.5$, Re_m will vary between 35,000 and 105,000 (points 5' and 5") and λ will vary within the limits 0.0228–0.0180.

Apart from the procedure described above, another possibility is to use a distorted model with different length scales in the longitudinal (l) and transverse (D, h) dimensions. For M_v given by the Froude law (i.e. $M_v = \sqrt{M_h} = \sqrt{M_D}$), it follows that

$$M_{h_f} = M_\lambda M_l M_v^2 M_D^{-1} = M_D \qquad (9.86)$$

i.e.

$$M_\lambda = M_D M_l^{-1} \qquad (9.87)$$

For $M_\lambda < 1$ (the model λ is larger than in the prototype) for

$$\text{Re}_{cr} < \text{Re}_m < \text{Re}_{sq} \qquad (9.88a)$$
$$D_p l_p^{-1} < D_m l_m^{-1} \qquad (9.88b)$$

The use of this technique is, however, severely restricted partly because of the distortions it would cause in local losses and partly because, for a given value of M_l, M_D is a function of M_λ, which in turn is a function of the Reynolds number. Thus, M_D varies with the size of the cross-section as well as the discharge, and similarity conditions for fixed values of M_l and computed M_D (or vice versa) can again be satisfied for one discharge only.

Modelling of the prototype closed-conduit water flow in any duct can also be reproduced on any scale model by using a different fluid (e.g. air). This leads to the use of 'air tunnels' for investigating velocity, pressure, flow fields and forces in cases where the model Reynolds number is high enough to avoid viscous effects and the influence of gravity on flow phenomena can be eliminated or is negligible. At the same time, flow velocities should be restricted to values where compressibility effects are absent – generally this means values below 50 m/s (see also Chapter 5).

The special case of using air flow in a duct to simulate flows with a prototype water–air interface is discussed in some detail in Chapter 7.

9.4.2 Modelling of local losses

Referring to Section 4.4.2, it is evident that if similarity of head losses due to friction can be achieved in the model (with $\text{Re}_m > \text{Re}_{sq}$) correct reproduction of all local losses may also be achieved in a model operated according to Froude law as long as complete geometric similarity of the feature causing the local loss is observed and similarity of upstream conditions is maintained.

As in most cases of local losses, independence of viscosity is achieved at lower values of Reynolds numbers than those given by equation (4.53), and this can be taken into account if local losses alone are to be modelled. If only the total head losses in a whole pipeline system (including local losses) are to be illustrated on a model, and if for operational or economic reasons not all changes in direction and cross-sections are reproduced on the model, the effect of local losses may be simulated by a friction loss in an 'equivalent length' of the straight pipe.

9.4.3 Modelling of unsteady flow

9.4.3.1 Waterhammer

By applying to the waterhammer equations (equations (4.76) and (4.77)) the procedure discussed in Chapter 5, we obtain as the criteria of similarity for model and prototype:

(i) $cV/(gH) = $ idem (Allievi characteristic);
(ii) $ct/l = $ idem (Strouhal criterion);
(iii) $V/\sqrt{(gH)} = $ idem; $c/\sqrt{(gH)} = $ idem (Froude number).

The last two equations can also be written as $c/V = $ idem $= Ma$ (Mach number).

The Strouhal criterion, where t is the time of operation T of a gate (or pump or turbine), determines also whether the operation is 'fast' ($T < 2l/c$) or 'slow' ($T > 2l/c$).

It is evident that viscous effects are neglected in the computation of the friction slope and that the assumption is of a sufficiently high Reynolds number (which is correct for most, but not all, times of the unsteady flow process). The assumption that the time scale is unity (i.e. $M_t = 1$ as applied, for example, to the time of closure of the gates) results in $M_c M_v = M_H$, $M_c = M_l$ and $M_S = M_H/M_l = M_v^2/M_D$.

This, in turn, means a distorted model with M_H different from M_l and M_D. As for a Froude model $M_v > 1$, M_c must also be bigger than unity (i.e. the pressure-wave velocity on the model must be only a fraction of the prototype velocity). Such a reduction may be achieved either by suitably increasing the elasticity of the pipe walls (by use of an appropriate material, or by decreasing the thickness), which may be difficult to comply with in practice or, more easily, by inserting a rubber hose filled with compressed air into the model pipeline. The pressure-wave velocity will decrease in proportion to the ratio of the air volume in the hose to the water volume in the pipe, and thus a reduction in 'c' on the model may be achieved over a very wide range.

9.4.3.2 Mass oscillation

Equations (4.79)–(4.81) lead to six conditions of similarity involving ten variables (Novak and Čábelka (1981)):

$$M_L M_v M_t^{-1} = M_z = M_P M_V^2 = M_R M_V^2 \tag{9.89}$$

$$M_V M_D^2 = M_{Vs} M_{Ds}^2 - M_{Q'} \tag{9.90}$$

$$M_{Vs} = M_z M_t^{-1} \tag{9.91}$$

In solving the above equations it is most advantageous and practical to use M_D, M_{Ds}, M_R and M_z as independent variables and compute the rest. Furthermore, to illustrate correctly on the model the loss in the junction of the conduit and the surge tank, which may (and often does) have a complicated shape, the head-loss coefficient ξ, and hence R, should be the same on the model as in the prototype (i.e. $M_R = 1$).

The six remaining variables are then given by

$$M_{Vs} = M_z^{1/2} \tag{9.92}$$

$$M_t = M_z^{1/2} \tag{9.93}$$

$$M_P = M_D^4 M_{Ds}^4 \tag{9.94}$$

$$M_{Q'} = M_z^{1/2} M_{Ds}^2 \tag{9.95}$$

$$M_V = M_z^{1/2} M_{Ds}^2 M_D^{-2} \tag{9.96}$$

$$M_L = M_z M_{Ds}^{-2} M_D^2 \tag{9.97}$$

Obviously, the resulting model is distorted, with the exception of the junction of the conduit (tunnel) and the surge tank. The joining of this undistorted part of the model ($M_D = M_{Ds}$) to the distorted part must be gradual to prevent errors in the oscillations of the water levels in the model tank.

There are two further constraints on the choice of M_{Ds}/M_D (which is also governed by the practical possibilities of commercially available pipelines):

(i) from the above

$$P_m = \frac{1}{2g}\left(1 + \frac{\lambda_m L_m}{D_m}\right) = P_p M_{Ds}^4 M_D^{-4} \tag{9.98}$$

or

$$\frac{1}{2g}\left(1 + \frac{\lambda_m L_p M_{Ds}^2}{D_p M_z M_D}\right) = P_p M_{Ds}^4 M_D^{-4} \tag{9.99}$$

As $\lambda_p(P_p)$, L_p and D_p are all known and λ_m is either known or may be chosen

$$M_z M_{Ds}^{-1} = \phi(M_{Ds} M_D^{-1}) \tag{9.100}$$

or

$$M_z(M_{Ds} \lambda_m) = \psi(M_{Ds} M_D^{-1}) \tag{9.101}$$

This condition practically limits the choice of M_{Ds}/M_D to $0.75 < M_{Ds}/M_D < 1.5$.

(ii) If the tank has an overflow, it is further necessary to fulfil the additional condition $M_Q = M_z^{5/2}$, resulting in $M_z = M_{Ds}$ and one single possibility of M_{Ds}/M_D.

For modelling of cavitation in general and air water flows in closed conduits, see Chapter 13.

9.4.4 Choice of approach

Closed-conduit flow is modelled predominantly by using computational models. This applies both to steady and unsteady flow. The main purpose of *laboratory studies* of closed-conduit flow under pressure is either 'basic' research investigating the physics of multiphase flows (air–water, water–sediment) or the study of cavitation processes. In the case of sediment transport in pipelines, because of the complexities of interaction of the turbulence of the transporting fluid and the particle mechanics, greatly reduced models are seldom used and experiments are often conducted at a 'scale' approaching prototype dimensions.

In applied research, *physical models* are used to investigate head losses and the pressure distribution at transitions with complicated geometry (penstocks, manifolds, conduit–surge-tank junctions, surge-tank throttles, etc.) and of critical cavitation conditions. In addition, sometimes whole pressure systems require physical modelling as a design aid (e.g. in complicated cases of hydropower development).

Physical models are also used to verify the results of numerical calculations in situations where there is uncertainty in the application of the assumption of essentially one-dimensional flow.

Modelling of closed-conduit flow is a good example of hybrid modelling, where, for example, the head-loss coefficients at surge-tank entries are ascertained on a physical model and surge-tank oscillations are found by using computational methods.

9.5 Case study: waterhammer arising from pipe failure

9.5.1 Test case description

The purpose of the present test case is to illustrate the solution properties of the waterhammer equations. The transients arising from the sudden failure of a valve in a pipe with non-uniform, wave-propagation properties are used as an example. The pipe is assumed to be in two parts that have identical cross-sectional areas but different wave-propagation speeds. Denoting the abscissa of the junction between the two parts by x_1, the speed of sound is c_1 for $x < x_1$ and c_2 for $x > x_2$ (Figure 9.13). The water is initially at rest. A closed valve located at the abscissa $x_0 < x_1$ separates two regions with initially different pressures. The pressures on the left- and right-hand sides of the valve are denoted by p_L and p_R, respectively. In the present application, p_L is assumed to be higher than p_R. At $t=0$ the valve fails, and both sides of the pipe are allowed to communicate. For the sake of simplicity, the valve is assumed to fail instantaneously and to induce no local head loss at later times.

After the failure, two waves originating from the abscissa x_0 of the initial discontinuity travel in opposite directions at speeds $+c_1$ and $-c_1$. The pressure in the intermediate region delimited by the two waves is between p_L and p_R, and the flow velocity is positive. When the wave travelling at the speed $+c_1$ reaches the abscissa x_1, part of the energy of the wave is transmitted to the region of the pipe with celerity c_2 and part of it is reflected into the region of the pipe with sound speed c_1. The pressure behind the reflecting wave may be higher or lower than the pressure behind the impinging wave, depending on the ratio c_1/c_2. The parameters retained for the application example are given in Table 9.1. In Option 1, the region of the pipe with the larger celerity is the left-hand region. In Option 2, the left-hand region has the smaller of the two sound speeds.

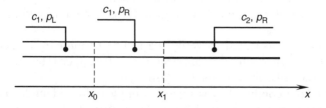

Figure 9.13 Definition sketch for the test case

Table 9.1 Parameters for the application example.

Symbol	Meaning	Value
A	Cross-sectional area of the pipe	$3.14 \times 10^{-2} \, m^2$
c_1	Sound speed in the left-hand region of the pipe	1,000 m/s (Case 1)
		600 m/s (Case 2)
c_2	Sound speed in the right-hand region of the pipe	600 m/s (Case 1)
		1,000 m/s (Case 2)
p_L	Initial pressure for $x < x_0$	2×10^5 Pa
p_R	Initial pressure for $x > x_0$	10^5 Pa
x_0	abscissa of the valve	50 m
x_1	abscissa of the junction	95 m
ρ	water density	1,000 kg/m^3

9.5.2 Simulation results

The waterhammer equations are solved using the MOC presented in Section 9.3.4. The computational time step is $\Delta t = 5 \times 10^{-3}$ s, and the cell width is $\Delta x = c_k \Delta t$ in the region k ($k = 1, 2$). This guarantees that the Courant number is equal to unity everywhere, which eliminates the need for interpolation, and hence leads to a more accurate and faster solution process.

Figure 9.14 shows the pressure profiles computed every fifth time step for Option 1 (whereby the larger of the two sound speeds is on the left-hand side of the junction). At $t = 2.5 \times 10^{-2}$ s, the two waves originating from the abscissa x_0 are clearly visible. At $t = 5 \times 10^{-2}$ s, the rightward wave reaches the abscissa x_1 of the junction and propagates into the pipe at the speed c_2. As $c_2 < c_1$, the pipe on the right-hand side of the junction is more deformable, and this causes the decrease in the pressure observed around $x = x_1$ in the figure. The decrease in the pressure triggers a wave that travels to the left at the speed $-c_1$ and a wave that travels to the right at the speed $+c_2$. Such waves are clearly visible in the graphs at $t = 7.5 \times 10^{-2}$ s and 0.1 s.

Figure 9.15 shows the pressure profiles computed every fifth time step for Option 2 ($c_1 < c_2$). At $t = 2.5 \times 10^{-2}$ s and $t = 5 \times 10^{-2}$ s, the waves travelling in opposite directions from $x = x_0$ are clearly visible. At $t = 7.5 \times 10^{-2}$ s, the rightward wave reaches the junction. The pipe–water system being more rigid on the right-hand side of the junction than on the left-hand side, only part of the energy of the impinging wave can be transmitted to the pipe on the right-hand side. As a consequence, the discharge decreases and the pressure rises at the junction. This yields a return pressure wave that propagates to the left at the speed $-c_1$. This is illustrated by the pressure profile at $t = 0.75$ s, where the return wave is visible, while the wave that propagates

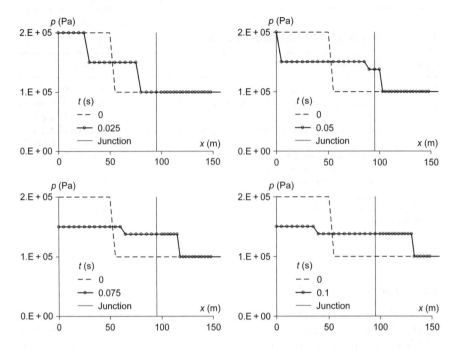

Figure 9.14 Pressure profiles computed using the method of characteristics for Option 1

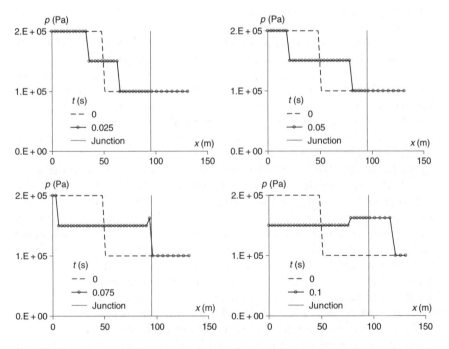

Figure 9.15 Pressure profiles computed using the method of characteristics for Option 2

beyond the junction at the speed c_2 has not yet reached the next computational point. The pressure profile at $t = 0.1\,\mathrm{s}$ clearly shows the propagation of the two waves from the junction.

This computational example shows that the maximum (or minimum) values in the pressure signal are not necessarily reached at the early times of the transient. Depending on the geometric and mechanical characteristics of the pipes, and the combination of the initial and boundary conditions, the maximum and/or minimum values (that condition the protection objectives of the installation) of the pressure and discharge may be reached after one or several reflections of the pressure transient within the network.

References

Allievi, L. (1903), Teoria generale del moto perturbato dell'acqua nei turbi in pressione, *Am. Soc. Ing. Arch. Italiana*. Translated into French by Allievi, 1904. English translation E. E. Halmos (1904). The theory of water hammer, ASME, New York.

Fox, J. A. (1989), *Transient Flow in Pipes, Open Channels and Sewers*, Ellis Horwood, Chichester.

Guinot, V. (2008), *Wave Propagation in Fluids, Models and Numerical Techniques*, Wiley/ISTE, Hoboken, NJ/London.

Jaeger, C. (1933), *Théorie Générale du Coup de Bélier*, Dunod, Paris.

Jaeger, C. (1977), *Fluid Transients in Hydro-Electric Engineering Practice*, Blackie and Sons, Glasgow.

Joukowski, N. (1898), *Water Hammer*. English translation Simin O. (1904), *Proc. Am. Waterworks Assoc.*, 341.

Novak, P. and Čábelka, J. (1981), *Models in Hydraulic Engineering – Physical Principles and Design Applications*, Pitman, London.

Swaffield, J. A. and Boldy, A.P. (1993), *Pressure Surges in Pipe and Duct Systems*, Avebury Technical, Aldershot.

Wylie, E. B. and Streeter, V. L. (1977), *Fluid Transients*, McGraw-Hill, New York.

Chapter 10

Modelling of urban drainage systems

10.1 Introduction

The present chapter is devoted to a presentation of the specific hydraulic features of urban drainage systems and the computational techniques for their hydrodynamical modelling. Prior to reading this chapter, the reader should master a number of notions presented in the previous chapters of this book.

As far as hydraulics is concerned, an indispensable prerequisite to this chapter is Chapter 4. In what follows, the reader is assumed to be familiar with the concepts and equations presented in Section 4.3 on hydraulics and Section 4.4 on flow in conduits. The reader may also refer to Chapter 7 for a presentation of free-surface-flow modelling. More specifically, Section 7.2 on the various models for free-surface flow (the Saint Venant equations, the diffusive and kinematic wave approximations) and Section 7.3 on numerical solution techniques provide useful background reading. For a basic understanding of the features of transient, pressurized pipe flow, the reader is advised to read Section 9.3. For details of hydraulics of various wastewater appurtenances, see Hager (1999).

A specificity of urban drainage systems is the extreme variability of the flow regime. Such variability is expressed in the form of very rapid changes in both time and space (i) from free surface to pressurized flow and (ii) from subcritical to supercritical flow. The variability is mainly due to the large geometric variability of the drainage network (changes in conduit shape and size, singularities, sudden slope transitions, etc.) and the large relative changes in the discharges at conduit junctions (confluences, etc.). Moreover, an urban drainage system may be subjected to very low flow (or no flow) conditions, with dry beds and small depths, which is another source of computational difficulty.

The transition between free-surface and pressurized flow can be modelled by modifying artificially the shape of the pipe (see Section 10.2.3) so that the propagation properties of the solutions of the Saint Venant equations (see Section 7.2) become identical to those of the solutions of the

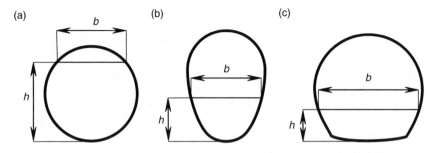

Figure 10.1 Typical conduit shapes: (a) circular, (b) egg-shaped, (c) horseshoe-shaped

waterhammer equations for pipe transients (see Section 9.3). The transition from subcritical to supercritical flow can be modelled provided that the available numerical techniques for free-surface-flow modelling are adapted. The numerical aspects are covered in Section 10.4.

Another specificity of urban-drainage-system modelling compared to classical, open-channel-flow modelling is the artificial character of the channels. The shape of urban drainage conduits is often predefined, as shown in Figure 10.1. Many commercially available software packages incorporate a library of typical conduit shapes.

This chapter does not deal with the movement of sediment in sewers; for an introduction to this subject, see Section 4.6.2.

Although the design of urban drainage systems invariably requires computational procedures, detailed solution of hydraulic intricacies of parts of the system and of many appurtenances can – and often has to – be carried out by means of physical modelling (see Section 13.3.3 and the case study in Section 13.5.4).

10.2 Governing equations of urban drainage systems

10.2.1 Assumptions

The governing assumptions behind modelling of urban drainage systems are similar to those used in open-channel and closed-pipe systems. They can be summarized as follows.

Assumption (1): The longitudinal dimensions of the pipes in the drainage system are much larger than the transverse dimensions. Consequently, the flow can be considered one-dimensional.

Assumption (2): The water is incompressible and the pressure distribution is hydrostatic.

Assumption (3): The velocity distribution within a cross-section can be fully described via the average flow velocity V via the Boussinesq β coefficient (see Chapter 7).

Assumption (4): The angle between the pipes and the horizontal is small.

Assumption (5): The effects of friction can be accounted for via a turbulent head-loss law such as the Manning and Strickler law (under unsaturated conditions) or the Darcy–Weisbach formulation (under saturated conditions); that is, the energy slope S_e is proportional to the square of the flow velocity.

Assumption (6): In contrast with the waterhammer equations, the momentum discharge Q^2/A is not neglected in the momentum equation, even when the system is saturated and the flow is pressurized. This assumption allows mass oscillations to be accounted for.

10.2.2 Governing equations

Following the assumptions in Section 10.2.1, the continuity and momentum equations can be written in conservation form as (see Chapter 7 for the details of the derivation)

$$\frac{\partial \mathbf{U}}{\partial t} + \frac{\partial \mathbf{F}}{\partial x} = \mathbf{S} \qquad (10.1)$$

where the conserved variable $\mathbf{U}$, the flux $\mathbf{F}$ and the source term $\mathbf{S}$ are defined as

$$\mathbf{U} = \begin{bmatrix} A \\ Q \end{bmatrix}, \quad \mathbf{F} = \begin{bmatrix} Q \\ \beta Q^2/A + P/\rho \end{bmatrix}, \quad \mathbf{S} = \begin{bmatrix} q \\ (1-\varepsilon)qu + (S_0 - S_e)gA \end{bmatrix} \qquad (10.2)$$

where A is the wetted cross-sectional area, g is the gravitational acceleration, P is the pressure force exerted onto the cross-sectional area of the pipe, Q is the volume discharge, q is the lateral discharge per unit length, S_0 and S_e are the bottom and energy slopes, respectively, $\varepsilon = \pm 1$ is the sign of q, β is Boussinesq coefficient, and ρ is the water density.

Numerical solution techniques usually solve the non-conservation form

$$\frac{\partial \mathbf{U}}{\partial t} + \mathbf{A}\frac{\partial \mathbf{U}}{\partial x} = \mathbf{S} \qquad (10.3)$$

where the Jacobian matrix $\mathbf{A}$ of $\mathbf{F}$ with respect to $\mathbf{U}$ is given by

$$\mathbf{A} = \begin{bmatrix} 0 & 1 \\ c^2 - \beta u^2 & 2\beta u \end{bmatrix} \qquad (10.4)$$

where the speed c of the waves in still water is given by

$$c \equiv \left[\frac{dP}{d(\rho A)} \right]^{1/2} = \left(\frac{gA}{B} \right)^{1/2} \tag{10.5}$$

where b is the width of the free surface.

10.2.3 Transition between free-surface and pressurized flow – Preissmann's slot

A salient feature of the wave-propagation properties of closed-conduit flow is that c becomes infinite when the pipe is full. Indeed, in a closed conduit, the width B of the free surface tends to zero when the free surface reaches the top of the section (Figure 10.1). For a circular pipe of radius r, A and B are given by

$$A = (\theta - \cos\theta \sin\theta)r^2 \tag{10.6a}$$

$$B = 2r\theta \tag{10.6b}$$

where θ is the angle between the vertical and the straight line connecting the centre of the pipe and the contact between the free surface and the pipe wall (Figure 10.2(a)). It is easy to check that substituting equations (10.6) into equation (10.5) leads to

$$c = \left[\left(1 - \frac{\cos\theta \sin\theta}{\theta} \right) \frac{gr}{2} \right]^{1/2} \tag{10.7}$$

Figure 10.2(b) illustrates the variation in the dimensionless ratio $c/c(r)$ as a function of the dimensionless ratio h/r, where $c(r)$ is the expression of c obtained for $h = r$ (i.e. for $\theta = \pi/2$). Figure 10.2 is drawn for a circular pipe but similar behaviour is observed for the celerity in oval-shaped conduits.

In reality, c does not become infinite because when the flow becomes fully pressurized the (small) elasticity of the pipe-wall material and the (small) compressibility of the water allow the mass ρA per unit length of pipe to vary slightly with the pressure force P. The celerity c of the wave then becomes equal to the (constant) speed of sound c_p for pressurized flow classically used in waterhammer transients (see Section 9.4 on waterhammer). Although this reflects a change in the physical nature of the flow, the transition from the classical expression (10.5) to the speed of sound can be accounted for in the Saint Venant equations provided that the geometry of the pipe section is modified artificially.

(a) 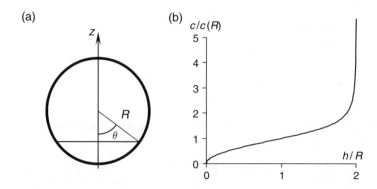 (b)

Figure 10.2 Propagation speed of the waves in still water for a circular pipe: (a) definition sketch; (b) variation in the dimensionless ratio $c/c(R)$ with the dimensionless depth h/R

Figure 10.3 Definition sketch for Preissmann's slot

Preissmann and Cunge (1961) and later Cunge and Wegner (1964) proposed that an artificial slot should be added on top of the pipe (Figure 10.3). This design has two advantages: (i) it allows the water level to rise above the top of the original pipe section, thus allowing pressure surges to be reproduced by the model via the hydrostatic pressure distribution; and (ii) the geometry of the artificial slot can be designed so that equation (10.5) gives $c = c_p$ when the water rises into the slot.

From a theoretical point of view, the mass of water stored in the slot in Figure 10.3 represents the extra amount of water stored in the pipe due to water compressibility and pipe-wall elasticity. Since this amount is very small compared to the volume of water contained in the pipe, the width b_s of Preissmann's slot must be very small. From equation (10.5), one obtains $b_s = gA/c^2$. To give but one example, for a pipe of radius $r = 0.3\,\mathrm{m}$

and a celerity under pressurized conditions of $c_p = 200\,\text{m/s}$, one obtains $b_s = 7 \times 10^{-5}\,\text{m}$.

From a practical point of view, the Preissmann slot does not have a constant width, and a smooth geometrical transition must be ensured between the roof of the pipe and the bottom of the slot when the equations are to be solved numerically (see Section 10.4).

10.2.4 Instability near discharge capacity

10.2.4.1 Conveyance as a non-monotone function of the water depth

Another issue specific to closed-conduit, free-surface flow is the essentially unstable character of the flow when the water rises above a certain threshold level in the pipe. The reasons for flow instability are as follows.

The slope of the energy line in a pipe is classically accounted for by a Manning–Chezy formula in the form (see Section 7.2.2.2)

$$Q = C_{\text{onv}} S_e^{1/2} \tag{10.8a}$$

$$C_{\text{onv}} = \frac{A R^{2/3}}{n_M} = \frac{A^{5/3} P'^{-2/3}}{n_M} \tag{10.8b}$$

where the conveyance C_{onv} expresses the capacity of the pipe to convey the flow under a fixed energy gradient, P' is the wetted perimeter and R is the hydraulic radius.

In contrast with open-channel flow, the conveyance in closed conduits is not a monotone function of the water depth h over the entire range $[0, 2R_p]$. For a circular or ovoid pipe, there exists a depth above which the hydraulic radius increases faster than the cross-sectional area. Indeed, the variation in the conveyance with the depth h is given by

$$\begin{aligned} dC_{\text{onv}} &= \frac{5}{3} A^{2/3} P'^{-2/3} \frac{dA}{dh} dh - \frac{2}{3} A^{5/3} P'^{-5/3} \frac{dP'}{dh} dh \\ &= \frac{1}{3} R^{2/3} \left(5 \frac{dA}{dh} - 2R \frac{dP'}{dh} \right) dh \end{aligned} \tag{10.9}$$

The variation in the cross-sectional area with the depth is given by

$$dA = B dh \tag{10.10}$$

while the variation in the wetted perimeter is given by

$$dP' = \left(\frac{1}{\cos \theta_1} + \frac{1}{\cos \theta_2} \right) dh \tag{10.11}$$

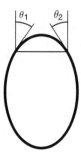

Figure 10.4 Definition of the variations in the conveyance with the water depth

where θ_1 and θ_2 are the angles of the pipe wall with the vertical at the contact line between the pipe and the free surface, respectively (Figure 10.4).

Substituting equations (10.10) and (10.11) into equation (10.9) gives

$$dC_{\text{onv}} = \frac{1}{3}R^{2/3}\left[5B - 2R\left(\frac{1}{\cos\theta_1} + \frac{1}{\cos\theta_2}\right)\right]dh \qquad (10.12)$$

If θ_1 and/or θ_2 tend to $\pi/2$, $1/\cos\theta_1$ and/or $1/\cos\theta_2$ become very large and the derivative of the conveyance with respect to the water depth becomes negative. Besides, in a pipe with a smooth section, there exists a water depth above which B starts decreasing. A straightforward variation analysis shows that when B becomes smaller than $4R/5$, the derivative of the conveyance with respect to h as given in equation (10.12) is necessarily negative, regardless of the value of θ_1 and θ_2. Figure 10.5 illustrates the variations in the conveyance with the water depth for a circular pipe. The conveyance is displayed as a ratio to the conveyance at $h = r$. The graph indicates that the conveyance is an increasing function of the water depth for $0 \leq h/r \leq 1.88$. Above this depth, the conveyance decreases.

The non-monotone behaviour of the conveyance with the flow depth is a source of instability in the flow, as explained in the next section.

10.2.4.2 Flow instability

Consider steady-state flow in a drainage network pipe. Let h_{max} denote the water depth for which the conveyance is maximum. Assume that, for some reason (e.g. a rainy event with a subsequent inflow into the drainage network), the discharge at the upstream end of the pipe increases. Two possibilities arise depending on the initial, steady-state depth.

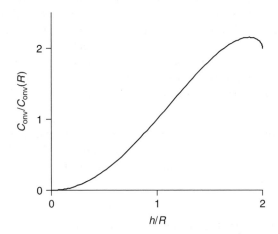

Figure 10.5 The ratio $C_{onv}/C_{onv}(R)$ as a function of the dimensionless water depth h/R for a circular pipe

(1) If the initial water depth is smaller than h_{max}, the increase in the discharge results in an increase in the water depth until the conveyance is large enough to accommodate for the discharge.

(2) If the steady-state water depth is larger than h_{max}, the increase in the discharge results in an increase in the water depth. This triggers a decrease in the conveyance and an increase in the head loss. As a result, the discharge in the section decreases, which in turn yields a decrease in the conveyance. This positive feedback process leads to filling the pipe very quickly. The discharge flowing from upstream can be conveyed only at the expense of an increase in the hydraulic head upstream of the saturated section. As a consequence, the saturation wave travels in the upstream direction until a point is reached where the conveyance needed to accommodate for the discharge falls below the maximum conveyance of the pipe.

Note that, if Preissmann's slot is used in the modelling, the height of the water in the slot must not be taken into account in the calculation of the wetted perimeter P'. It must be kept in mind that Preissmann's slot is a purely virtual construction.

10.3 Solution behaviour – initial and boundary conditions

10.3.1 Initial and boundary conditions

(1) When the drainage system is not saturated, the governing equations are the Saint Venant equations (see Section 7.2.2 for detailed considerations on

the behaviour of the solutions of the Saint Venant equations). As shown in Section 7.2.2, solution existence and uniqueness are ensured provided that the initial flow conditions are known at all points of the solution domain and that a boundary condition is supplied for each inflowing characteristic. Consequently, one boundary condition is needed for a subcritical inflow and two boundary conditions are needed for a supercritical inflow. Subcritical, downstream boundaries require one boundary condition, while supercritical outflow requires no condition.

(2)   Under saturated conditions, the flow velocity becomes negligible compared to the speed c of the waves in still water and the flow is always supercritical. Then only one boundary condition is needed at each end of the solution domain.

10.3.2 Junctions, lateral weirs and overspills

10.3.2.1 Junctions

A junction can be considered as an internal boundary where the law of nodes and the equality of heads (or water levels) are applicable:

$$\sum_{p=1}^{N_p} Q_p = 0 \tag{10.13a}$$

$$\eta_1 = \eta_2 = \cdots = \eta_{N_p} \tag{10.13b}$$

$$\eta_1 + \frac{V_1^2}{2g} = \eta_2 + \frac{V_2^2}{2g} = \cdots = \eta_{N_p} + \frac{V_{N_p}^2}{2g} \tag{10.13c}$$

where N_p is the number of pipes converging to the junction and Q_p, V_p and h_p ($p = 1, \ldots, N_p$) are the discharge, velocity and free-surface elevation, respectively, in the pipe p. Note that in equation (10.13a) the same convention must be used for all the discharges, e.g. the discharge in a pipe is counted positive if the water flows toward the junction.

Equations (10.13a) and (10.13b) (or (10.13c)) yield N_p independent conditions. This matches the number of (internal) boundary conditions needed at the end of each pipe under subcritical conditions. Consequently, solution existence and uniqueness are guaranteed provided that the flow is subcritical in all the pipes at the junction.

10.3.2.2 Lateral weirs

Lateral weirs are often treated as local sink terms along the pipe. If the length of the weir is important, the elevation of the free surface may vary substantially between the upstream and downstream ends of the weir. The outflowing discharge per unit length q is computed as

$$q = C(\eta - z_w)^{3/2} \tag{10.14}$$

where C is the weir discharge coefficient and z_x is the elevation of the crest of the weir. Integrating q with respect to x along the weir leads to the expression of the total outflowing discharge. In modelling packages such as HEC-RAS (USACE (2008)), the discharge coefficient C is computed using Hager's formula (Hager (1987)):

$$C = \frac{3}{5} C_0 \left(g \frac{1-w}{3-2y-w} \right)^{1/2} 1 - (\alpha + S_0) \left(\frac{3-3y}{y-w} \right) \tag{10.15}$$

where the dimensionless numbers w and y are defined as

$$w = \frac{h_w}{H - z_w + h_w} \tag{10.16a}$$

$$h = \frac{\eta - z_w}{H - z_w} \tag{10.16b}$$

where h_w is the difference between the weir crest and the ground elevation and H is the hydraulic head.

10.3.3 Manholes

A manhole can be considered as a junction with a storage capacity. The available equations are the law of nodes, the equality of heads (or water levels) and the continuity equation for the manhole. The equality of the water levels or heads is given by equation (10.13b) or equation (10.13c).

The law of nodes (equation (10.13a)) must be modified so as to account for the variation in storage in the manhole:

$$\sum_{p=1}^{N_p} Q_p = Q_m \tag{10.17}$$

where N_p is the number of pipes connected to the manhole, Q_m is the discharge to the manhole (positive if the water level in the manhole is rising, negative if the water level is falling) and Q_p is the discharge in the pipe p (positive if the water flows to the manhole).

Applying the law of conservation of mass to the water stored in the manhole leads to

$$Q_m = A_m \frac{d\eta_m}{dt} \tag{10.18}$$

where A_m and η_m are the plan-view area and the water level in the manhole, respectively. Moreover, the discharge Q_m is a function of the difference in the heads in the pipes at the junction and the water level in the manhole:

$$H_J - \eta_m = \alpha \frac{|Q_m| Q_m}{2g} \tag{10.19}$$

where $H_J = \eta_p + V_p^2/(2g)$ is the head in the pipes at the junction (identical in all the pipes if equation (10.13c) is used) and α is a head-loss coefficient.

Equations (10.13b) and (10.13c) and equations (10.17)–(10.19) provide $N_p + 2$ independent relationships. The unknowns are the N_p discharges Q_p in the pipes at the junction, plus the water level η_m and the discharge Q_m to the manhole. As the number of equations matches the number of unknowns, the solution is unique.

10.4 Numerical solution techniques

10.4.1 General – finite-difference solution

Although recent advances can be found in the solution of the conservation form of the equations (León et al. (2009)), most existing software packages solve the non-conservation form (equation (10.3)) of the governing equations. The most widely used techniques to do so are the Preissmann scheme described in Section 7.3.3, or Abbott–Ionescu's scheme described in Section 7.3.4. Compared to river systems, however, urban drainage systems exhibit rather specific features that are sources of computational difficulties. As mentioned in the introduction to this chapter, the main three difficulties are the subcritical/supercritical transition, the free surface/pressurized flow transition and the modelling of dry beds and small depths. These issues are addressed in the following sections.

10.4.2 Numerical techniques for transcritical flow

Preissmann's and Abbott–Ionescu's schemes, the most widely implemented schemes in urban drainage-network modelling systems, are not equipped to handle transcritical flow (Johnson et al. (2002), Meselhe and Holly (1997)). A possible solution to this problem consists of modifying the momentum equations so as to minimize the influence of the inertial terms (that are responsible for the possible appearance of supercritical flow). A proper minimization, or alteration, of the inertial terms allows either the flow pattern to remain artificially subcritical for all possible values of the Froude number, or the governing equations to switch from wave-propagation to diffusion-like equations, the solution of which remains unique provided that one boundary condition is supplied at each boundary of the computational domain.

This allows the same algorithms to be used for the computation of the solutions, irrespective of the subcritical or supercritical nature of the flow.

Three approaches are used in practice: the local partial inertia (LPI) approach, the reduced momentum equation (RME) technique, and the partial discretization of the momentum flux. These approaches, as well as the consequences on the wave-propagation properties, are detailed in the following sections.

10.4.2.1 The local partial inertia (LPI) approach

In the LPI technique (Jin and Fread (2000)), the momentum equation is modified to become

$$\left[\frac{\partial Q}{\partial t} + \frac{\partial}{\partial x}\left(\beta\frac{Q^2}{A}\right)\right]\varepsilon + \frac{\partial}{\partial x}\left(\frac{P}{\rho}\right) = (S_0 - S_e)gA \qquad (10.20)$$

where ε is a function of the Froude number

$$\varepsilon = \max\left(1 - |\mathrm{Fr}|^M, 0\right) \qquad (10.21)$$

where Fr is the Froude number and M is a user-defined exponent. Equation (10.21) ensures a continuous transition from $\varepsilon = 1$ for $\mathrm{Fr} = 0$ to $\varepsilon = 0$ for $\mathrm{Fr} = 1$. When the absolute value of the Froude number is equal to or larger than unity, the inertial terms disappear and the diffusive wave approximation is obtained (see Section 7.2). The system of equations is no longer hyperbolic and the notion of wave celerity is meaningless.

As the switch in equation (10.21) is continuous over the whole range $0 \le |\mathrm{Fr}| \le 1$, it acts on the propagation properties of the solution even under subcritical conditions. The LPI technique has consequences on the propagation properties of the solutions of the flow equations. Equation (10.20) can be rewritten in conservation form as

$$\frac{\partial Q}{\partial t} + \frac{\partial}{\partial x}\left(\beta\frac{Q^2}{A} + \frac{1}{\varepsilon}\frac{P}{\rho}\right) = \frac{1}{\varepsilon}(S_0 - S_e)gA \qquad (10.22)$$

The Jacobian matrix **A** of the flux vector with respect to the conserved variable is modified to

$$\mathbf{A} = \begin{bmatrix} 0 & 1 \\ \frac{c^2}{\varepsilon} - \beta u^2 & 2\beta u \end{bmatrix} \qquad (10.23)$$

The wave celerities are given by the eigenvalues of the matrix (see Chapter 2)

$$\lambda_1 = \beta u - \left[(\beta - 1)\beta u^2 + \frac{c^2}{\varepsilon} \right]^{1/2} \tag{10.24a}$$

$$\lambda_2 = \beta u + \left[(\beta - 1)\beta u^2 + \frac{c^2}{\varepsilon} \right]^{1/2} \tag{10.24b}$$

Figure 10.6 shows the variations in the dimensionless ratio λ_p/c $(p = 1, 2)$ with the Froude number for $\beta = 1$ and various values of the power M. The ratios λ_p/c are obtained from equations (10.24) as

$$\frac{\lambda_1}{c} = \beta\mathrm{Fr} - \left[(\beta - 1)\beta\mathrm{Fr}^2 + \frac{1}{\varepsilon} \right]^{1/2} \tag{10.25a}$$

$$\frac{\lambda_2}{c} = \beta\mathrm{Fr} + \left[(\beta - 1)\beta\mathrm{Fr}^2 + \frac{1}{\varepsilon} \right]^{1/2} \tag{10.25b}$$

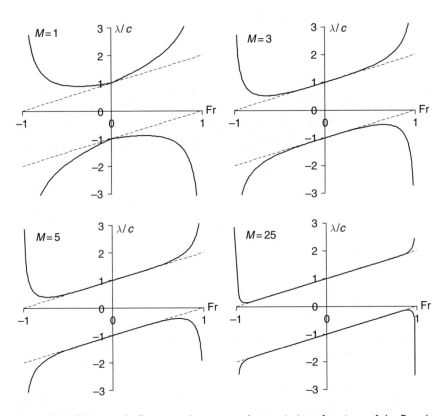

Figure 10.6 LPI approach. Dimensionless wave celerities λ_p/c as functions of the Froude number for $\beta = 1$. Dashed lines: wave celerities of the Saint Venant equations. Solid lines: wave celerities computed from equations (10.21) and (10.25)

As shown in Figure 10.6, λ_1 and λ_2 tend to infinity as the absolute value of Fr approaches unity, regardless of the value of M. Increasing the power M in the switch (equation (10.21)) makes λ_1 and λ_2 closer to the wave-propagation speeds of the original Saint Venant equations over a wider range of the Froude number. Although strongly inaccurate when the absolute value of the Froude number approaches unity, this approach has the advantage that λ_1 is always negative and λ_2 is always positive, therefore allowing classical schemes like Preissmann's or Abbott–Ionescu's scheme to be used throughout all the computational domain without any specific treatment of the boundaries or critical point. Shifting from the subcritical Saint Venant equations to the diffusive wave approximation does not induce any change in the number and location of boundary conditions, because the existence and uniqueness of the solution of the diffusive wave approximation are guaranteed provided that one boundary condition is specified at each end of the computational domain.

10.4.2.2 The reduced momentum equation (RME) approach

In the RME approach, the momentum advection term $\beta Q^2/A$ in the impulse is multiplied by a weighting coefficient ε between 0 and 1. The momentum equation then becomes

$$\frac{\partial Q}{\partial t} + \frac{\partial}{\partial x}\left(\varepsilon\beta\frac{Q^2}{A} + \frac{P}{\rho}\right) = (S_0 - S_e)gA \tag{10.26}$$

which yields the following expression for the matrix $\mathbf{A}$:

$$\mathbf{A} = \begin{bmatrix} 0 & 1 \\ c^2 - \varepsilon\beta u^2 & 2\varepsilon\beta u \end{bmatrix} \tag{10.27}$$

The ratio of the wave celerities to the speed c is given by

$$\frac{\lambda_1}{c} = \varepsilon\beta Fr - \left[(\varepsilon\beta - 1)\varepsilon\beta Fr^2 + 1\right] \tag{10.28a}$$

$$\frac{\lambda_2}{c} = \varepsilon\beta Fr + \left[(\varepsilon\beta - 1)\varepsilon\beta Fr^2 + 1\right] \tag{10.28b}$$

In the original approach, ε was set to zero. It is easy to check that in such a case equations (10.28) yield

$$\lambda_1 = -c \tag{10.29a}$$

$$\lambda_2 = +c \tag{10.29b}$$

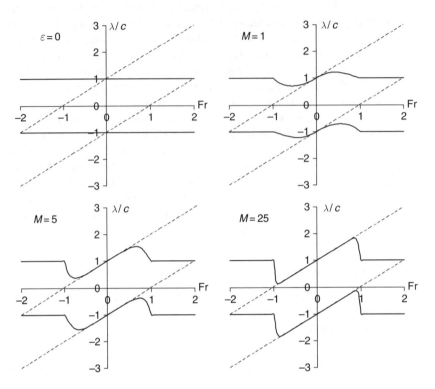

Figure 10.7 RME approach. Dimensionless wave celerities λ_p/c as functions of the Froude number for $\beta = 1$. Dashed lines: wave celerities of the Saint Venant equations. Solid lines: wave celerities computed from equations (10.21) and (10.28)

and the flow is always subcritical. Figure 10.7 shows the variations in the dimensionless wave celerities in the less trivial case where ε is given by the switch (10.21).

Note that other formulations are proposed in (DHI (2005)), such as:

$$\varepsilon = \max\left(1 - Fr^2, 0\right) \tag{10.30a}$$

$$\varepsilon = \min\left[(1 + |Fr| - a)^{-b}, 1\right] \tag{10.30b}$$

where a and b are positive constants to be specified by the user.

10.4.2.3 Partial discretization of the momentum flux

A third possibility for the discretization of the momentum equation is to drop one of the derivatives in the non-conservation form of the equations, as in the Mike11 implementation of Abbott–Ionescu's scheme. As shown in Section 7.3.4, in this implementation, the derivative of the momentum flux is approximated as

$$\frac{\partial}{\partial x}\left(\beta\frac{Q^2}{A}\right) \approx -\beta u^2\frac{\partial A}{\partial t} \tag{10.31}$$

which leads to the following expression for the Jacobian matrix

$$\mathbf{A} = \begin{bmatrix} 0 & 1 \\ c^2 - \beta u^2 & 0 \end{bmatrix} \tag{10.32}$$

The wave celerities of the system are given by

$$\frac{\lambda_1}{c} = -\left(1 - \beta Fr^2\right)^{1/2} \tag{10.33a}$$

$$\frac{\lambda_2}{c} = \left(1 - \beta Fr^2\right)^{1/2} \tag{10.33b}$$

The celerities as given by equations (10.33) are meaningful only if the absolute value of the Froude number is smaller than unity. Under supercritical conditions, the term βu^2 is cancelled in the discretization and the celerities are given by equations (10.29). The resulting wave-propagation-speed diagram is plotted in Figure 10.8.

In industrial implementations of the scheme (DHI (2005)), the partial discretization of the momentum flux is combined with the RME approach.

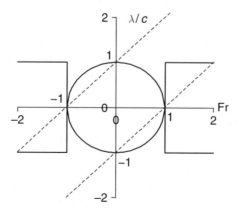

Figure 10.8 Partial discretization of the momentum flux. Dimensionless wave celerities λ_p/c as functions of the Froude number for $\beta = 1$. Dashed lines: wave celerities of the Saint Venant equations. Solid lines: wave celerities computed from equations (10.29) and (10.32)

10.4.2.4   Advantages and drawbacks of the approaches

The LPI and RME approaches yield wrongly computed steady-state flow near critical conditions. Indeed, for steady state, where $\partial Q/\partial t = \partial Q/\partial x = 0$, equations (10.20) and (10.26) become

$$(c^2 - \varepsilon u^2)\frac{\partial A}{\partial x} = (S_0 - S_e)gA \tag{10.34}$$

Introducing the definition $c^2 = gA/b$ and $\mathrm{d}A = b\ \mathrm{d}h$ leads to

$$\frac{\mathrm{d}h}{\mathrm{d}x} = \frac{S_0 - S_e}{1 - \varepsilon \mathrm{Fr}^2}. \tag{10.35}$$

In both the LPI and RME approaches, the coefficient ε tends to zero when the Froude number tends to unity. In practice, this may result in wrongly located hydraulic jumps, critical points, etc.

In contrast, the partial discretization of the momentum discharge presented in Section 10.4.2.3 does not exhibit this problem because the only term dropped in the equations is the term in $\partial Q/\partial x$, that is zero near steady-state simulations. Of course, when the partial discretization approach is combined with the RME approach, the issue of near-critical or transcritical flow modelling arises again.

10.4.3   Transition from free-surface to pressurized flow

10.4.3.1   Non-monotone conveyance curve

The non-monotone character of the conveyance with respect to the flow depth (see Section 10.2.4.1) is a source of numerical difficulty in the solution of the Saint Venant equations. This specific feature of closed-conduit flow may be a source of instability in the numerical solution process, for the following two reasons:

(1) The instability of the exact solution of the equations is naturally transposed to the numerical solution, with strong differences in the flow variables from one time step to the next.
(2) As mentioned in Chapter 7 (see e.g. Section 7.3.3.3 dealing with Preissmann's scheme), solving the flow equations using implicit schemes implies the iterative calculation of a number of parameters, among which the energy slope is a function of the conveyance. The instability described in (1) as occurring from one time step to the next is then transformed into an instability from one iteration to the next within the same time step.

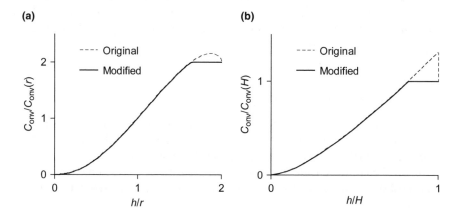

Figure 10.9 Artificial modification of the conveyance curve for solution stabilization: (a) circular conduit, (b) rectangular conduit. $B = 2H$

A standard solution to this problem (USACE (2008)) is to modify the conveyance curve artificially so as to make it monotone with the water depth. Figure 10.9 shows an example of such a modification for a circular conduit (Figure 10.9(a)) and a rectangular conduit, the width of which is twice as much as the height.

10.4.4 Dry beds and small depths

Simulations may start from low flow (or no flow) conditions. Such conditions may cause numerical instability of the solution of the equations. As mentioned in Section 7.3.6, three types of problems arise: (i) problems arising from the discretization, (ii) the non-uniqueness of the relationship between the discharge and the water depth in the discretized equations, and (iii) unphysical computational results induced by too coarse a discretization.

10.4.4.1 Conduits with zero bottom width

In conduits where the channel width $B(Y)$ tends to zero when the depth Y tends to zero, the solution of the equations in non-conservation form fails owing to division by zero when the conduit becomes dry. Similar problems may occur even when the initial depth is not zero, because very small depth usually results in a dramatic overestimation of the variations in the water depth. Repeating this problem over several time steps generally leads to severe mass-conservation problems, if not to numerical instability.

A standard solution, presented in Section 7.3.6, is the so-called 'Abbott slot', an artificial slot introduced in the conduit bottom that prevents the conduit from drying out, even under very low flow. Besides, a small, artificial discharge can be introduced at the upstream end of each pipe in order to preclude the pipe from becoming dry.

10.4.4.2 Wetting and drying fronts

Oscillations may appear near wetting and drying fronts due to the non-uniqueness of the relationship between the discharge and the depth in the discretized equations. As shown in Section 7.3.6, solving the flow equations using Preissmann's scheme under low flow conditions leads to the following relationship between the discharge and the water depth at two consecutive points:

$$Q = \frac{\varepsilon B}{n}[(1 - \psi)Y_i + \psi Y_{i+1}]^{5/3}\left(\frac{Y_{i+1} - Y_i}{\Delta x_{i+1/2}}\right)^{1/2} \tag{10.36}$$

When Q is positive, i is the upstream point and $i+1$ is the downstream point. Q, as given by equation (10.36), is not a monotone function of Y_{i+1} for a given Y_i, unless ψ is set to zero. Conversely, for a negative discharge, Q is a monotone function of the downstream point Y_i only if ψ is set to 1. The non-uniqueness of the relationship between the discharge and the downstream depth may lead to artificial oscillations in the computed flow variables during the (iterative) numerical-solution process.

Cunge et al. (1980) suggest that ψ should be set to zero for positive values of Q and to unity for negative values of Q. In other words, the conveyance $B/nY^{5/3}$ should be estimated using only the upstream point. This numerical stabilization procedure is known as 'conveyance upwinding' and is to be used only for small depths.

10.4.4.3 Computational grid

As shown in Section 7.3.6, a wave travelling into a region with a constant depth A_0 and constant discharge Q_0 yields the following relationship between the variation in the cross-sectional area and the discharge:

$$A_{i+1}^{n+1} = A_0 + 2\theta \Delta t \frac{\Delta Q}{\Delta x} - \Delta A \tag{10.37}$$

where i is the point that has just been passed by the wave, and ΔA and ΔQ are the variations in the cross-sectional area and discharge from the time step n to the time step $n+1$. If A_0 is small and the conveyance of the channel is small (owing to high roughness, narrow free-surface width or a combination of both), even a small value of ΔQ may lead to a large variation in ΔA, and A_{i+1}^{n+1} may drop below the initial value A_0 without any physical reason, thus creating artificial oscillations in the computed water depth and discharge. A_{i+1}^{n+1} may even become negative for some combinations of the initial water depth, geometry, and hydraulic and numerical parameters. This undesirable behaviour can be prevented by keeping Δx sufficiently small to ensure the positiveness of the quantity $2\theta\Delta t\Delta Q/\Delta x - \Delta A$. Alternatively, increasing the value of θ also has a stabilizing effect on the solution.

10.5 Case study

10.5.1 Introduction

The purpose of the present case study is to illustrate the typical behaviour of an urban drainage system subjected to a short rainfall event, as well as the main steps of the construction of an urban drainage model using a commercially available software package. The package used is Mike 11 (DHI (2005)).

Figure 10.10 shows the layout of the network. The geometric characteristics of the conduits are given in Table 10.1.

The boundary conditions are the following: at nodes A, G and I, the inflowing discharge is a known function of time. Node F is prescribed level boundary. The hydrographs are shown in Figure 10.11.

Such hydrographs correspond to short and intense rainfall events occurring over impervious areas (such as urban catchments) with rather steep slopes. In such situations, the concentration times of the catchments are small and high peak discharges can be obtained within short times. In the

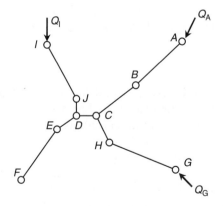

Figure 10.10 Plan view of the network

Table 10.1 Geometric characteristics of the network.

Pipe	Upstream node	Downstream node	Length (m)	Diameter (m)	Upstream node elevation (m)	Downstream node elevation (m)
1	A	B	108.0	0.30	51.00	49.10
2	B	C	5.0	0.35	49.10	49.05
3	C	D	5.0	0.45	49.05	49.00
4	D	E	5.0	0.50	49.00	48.70
5	E	F	72.0	0.50	48.70	48.30
6	G	H	61.0	0.30	49.20	49.10
7	H	C	40.0	0.30	49.10	49.05
8	I	J	60.0	0.30	50.00	49.10
9	J	D	10.0	0.30	49.10	49.00

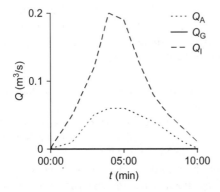

Figure 10.11 Input hydrographs for the example simulation. Note that the solid line for Q_G corresponds to a zero discharge

present case, the numerical difficulty is willingly increased by considering a scenario where the inflowing discharge Q_G remains very small, thus yielding initial conditions with low flows and very small depths in pipes 6 and 7.

10.5.2 Model construction

In defining the computational model, it is necessary to define as many links in the model as there are pipes with different slopes and/or diameter. This is because the geometric tables at the computational points between the nodes are usually precomputed by the model from an interpolation between the conduit geometry at the closest neighbouring nodes. Therefore, it is essential to capture the changes in conduit slope and shape as accurately as possible, otherwise leading to wrongly estimated conduit conveyances, storage volumes or slopes, which in turn may lower the accuracy of the numerical model.

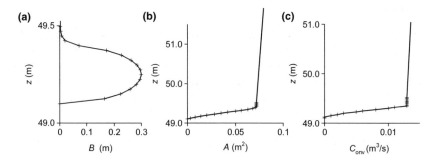

Figure 10.12 Characteristics of the cross-section at the node A as functions of the elevation computed by the software: (a) free-surface width, (b) cross-sectional area, (c) conveyance curve

Commercially available packages incorporate predefined conduit shapes, such as circular or rectangular conduits. This precludes the user from defining each point of the cross-section manually, which would be very time-consuming. The only necessary data to define a cross-section are the shape of the pipe (circular/rectangular), its dimensions (diameter for a circular conduit, height and width for a rectangular conduit) and the bottom elevation of the conduit. After processing the data, the software produces precomputed tables that give the correspondence between the water level in the cross-section and the cross-sectional area, free-surface width, hydraulic radius, conveyance, etc. Figure 10.12 illustrates the precomputed tables for B, A and C_{onv} as computed by the software for the node A. The point values stored in the table are indicated by the crosses. The solid line connecting the crosses indicates how B, A and C_{onv} are interpolated linearly by the software between the point values.

(1)   The bottom of the Preissmann slot is clearly visible in Figure 10.12(a). In order to avoid sharp transitions in the function $B(z)$, the transition between the lid of the conduit and the slot is progressive between $z = 49.4\,\text{m}$ (the lid of the conduit) and $z = 49.5\,\text{m}$. Besides, the width of the slot is automatically set by the software to 1% of the diameter of the conduit. For a conduit diameter $D = 0.3\,\text{m}$, the width of the slot is therefore 3 mm, which is much more than the theoretical 10^{-5} to $10^{-4}\,\text{m}$ that would actually be needed to reproduce the propagation speed of the pressure waves correctly. This overestimated value for the width of the Preissmann slot is reflected by the curve $A(z)$ in Figure 10.12(b), whereby it can be clearly seen that the variations of A with z are not negligible even when z rises above the lid. Note, however, that such an overestimation of the width of Preissmann's slot makes the numerical solution more stable by

reducing the propagation speed of the waves $c = (gA/B)^{1/2}$ compared to the theoretical value.

(2)   As indicated in Section 10.2.4, the conveyance curve is made artificially monotone in order to avoid instabilities (see Section 10.4.3). The classical numerical procedure consists of clipping the conveyance curve at the value of the conveyance for a full conduit (note that, in a circular conduit, the hydraulic radius for a full conduit is half the radius of the conduit):

$$C_{onv} = \min \left[KAR^{2/3}, K\pi r^2 \left(\frac{r}{2} \right)^{2/3} \right] \tag{10.38}$$

If equation (10.38) is applied, the conveyance becomes constant after the full conduit value is reached. However, Figure 10.12(c) shows that the conveyance is an increasing function of the water depth h for $h > 2r$. A close inspection of the numerical values shows that for $h = 10\,\text{m}$, the conveyance is 150% of its value for a full conduit. Although it is very unlikely for the specific hydraulic head to reach 10 m given the overestimated width of the Preissmann slot, this shows that the variations in the conveyance with the head in a saturated conduit are far from negligible.

The governing equations are discretized using a user-specified cell width $\Delta x = 1\,\text{m}$ and a time step $\Delta t = 0.01\,\text{s}$. Although rather small, these numerical parameters allow the propagation of the hydraulic transients throughout the network to be captured accurately. The time-centring parameter θ in Abbott–Ionescu's scheme is set to 0.7, which is a good compromise between solution accuracy and stability, given the strong oscillations that usually occur near wetting or drying fronts in free-surface-flow simulations.

10.5.3   Simulation results

The simulation results are shown in Figures 10.13–10.15.

Figures 10.13(a) and (b) show the water-level profiles along the segments [ABCDEF] and [ABCHG], respectively, at $t = 5\,\text{min}$. Figures 10.13(c) and (d) illustrate the propagation of the hydrograph in the segments [AC] and [DF].

An interesting feature of the computed solution is the return flow in the conduit [GHC]. This return flow is due to the discharges coming from the nodes A and I. The large discharges in the segments [ABC] and [GHC] cause the water level to rise at the node C. As a consequence, the sign of the gradient in the free surface changes between nodes C and H (see Figure 10.13(b)) and a return flow appears. The appearance of a return flow in the simulation results is confirmed by the inspection of the hydrograph at the node H (see Figure 10.14).

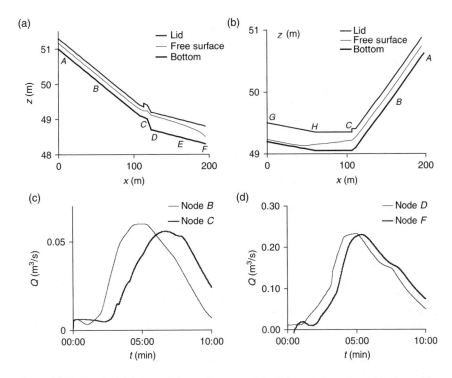

Figure 10.13 Simulated free-surface profiles at $t = 5$ minutes and computed hydrographs along the segment [ABCDEF]

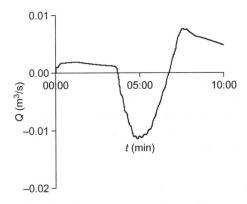

Figure 10.14 Simulated hydrograph at the node H

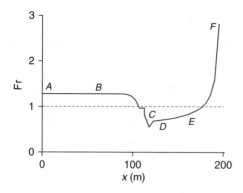

Figure 10.15 Computed Froude number along the segment [ABCDEF] at $t = 5$ min

Note that such features could not have been identified if the flow equations had been solved using, for example, the kinematic wave approximation, which allows neither for backwater effects, nor for changes in the direction of the flow. Although some commercial packages propose the kinematic wave approximation as one of their standard solution schemes, it is strongly advised never to use this approximation in the simulation of flows in complex urban drainage systems with mild slopes or potentially strong backwater effects such as that shown in the present example.

Another aspect of the simulation is illustrated in Figure 10.15, where the simulated Froude number is plotted as a function of the longitudinal coordinate along the segment [ABCDEF]. As in Figure 10.13(a), the profile is drawn at $t = 5$ min. This graph clearly illustrates two direct consequences of the partial discretization of the momentum flux technique combined with the RME approach (see Sections 10.4.2.2–10.4.2.4) on the computed profiles.

(1) The flow is supercritical upstream of the junction node C and subcritical downstream of it. If the full Saint Venant equations were solved, a hydraulic jump would be visible in the free-surface profile. However, the RME approach being applied in the Mike 11 software for the computation of supercritical flow, the diffusive wave approximation is obtained for the supercritical part of the profile, thus eliminating the possibility of a hydraulic jump appearing in the computed profile.

(2) The Froude number becomes larger than unity in the downstream part of the segment [EF]. Once again, if the full Saint Venant equations were solved using the software, it should not be possible to prescribe the water level at the node F because point F is a supercritical, outflowing

boundary. However, because the computational code uses the RME approach for supercritical flow, the diffusive wave approximation is applied, thus allowing a downstream boundary condition to be prescribed (see Section 7.2.3.2).

References

Cunge, J. A., Holly, F. M., Jr and Veswey, A. (1980), Practical Aspects of Computational River Hydraulics, Pitman, London.

Cunge, J. A. and Wegner, M. (1964), Intégration numérique des équations d'écoulement de Barré de Saint-Venant par un schéma implicite de différences finies, *La Houille Blanche*, (1), 33–38.

DHI (2005), Mike11 Reference manual. DHI water and Environment, Hørsholm.

Hager, W. H. (1987), Lateral outflow over side weirs, *J. Hydraulic Eng.*, 113(4), 491–504.

Hager, W. H. (1999), *Wastewater Hydraulics*, Springer-Verlag, Berlin.

Jin, M. and Fread, D. L. (2000), Discussion on the application of relaxation scheme to wave-propagation simulation in open channel networks, *J. Hydraulic Eng.*, 126(1), 89–91.

Johnson, T. C., Baines, M. J. and Sweby, P. K. (2002), A box scheme for transcritical flow, *Int. J. Numer. Methods Eng.*, 55, 895–912.

León, A. S., Ghidaoui, M. S., Schmidt, A. R. and García, M. H. (2009), Application of Godunov-type schemes to transient mixed flows, *J. Hydraulic Res.*, 47(2), 147–156.

Meselhe, E. A. and Holly, F. M. (Jr) (1997), Invalidity of the Preissmann scheme for transcritical flow, *J. Hydraulic Eng.*, 123(7), 605–614.

Preissmann, A. and Cunge, J. A. (1961), Calcul des intumescences sur machines électroniques. In *Proceedings AIHR 9th General Assembly*, Belgrade, pp. 5.1–5.9.

USACE (2008), *HEC-RAS. River Analysis System. Hydraulic Reference Manual*, Report, US Army Corps of Engineers.

Chapter 11

Modelling of estuaries

11.1 Introduction

Practical applications of hydraulics require numerical and graphical results, and if analytical expressions that describe phenomena of interest cannot be derived from a mathematical model, it becomes necessary to obtain the results by physical modelling or by purely numerical methods. This can involve the solution of various types of mathematical problems, some of the most important of which involve the solution of ordinary and partial differential equations (see Chapters 2 and 3). In this chapter, the application of hydraulic theories to problems arising in estuary management are discussed. Before delving into detailed mathematical models, this introductory section describes some of the key features and characteristics of estuaries that are later used to develop appropriate mathematical models. The discussion focuses on a particular estuary but the issues raised are fairly generic and serve to illustrate key points.

Figure 11.1 shows a photograph of the Exe estuary on the south coast of the UK. Numerous features evident in the photograph are common to many estuaries:

- the entrance is partially enclosed by spits and bounded by large offshore sandbanks;
- it has a narrowing funnel shape;
- it covers a large surface area, (in the case of the Exe estuary ~1,810 ha);
- there are extensive intertidal flats;
- there are large areas of salt marsh habitat;
- it has large ebb (Pole Sand) and flood (Bull Hill Bank) deltas;
- the estuary is shallow and characterized by a long meandering tidal channel (in the case of the Exe estuary, 16.7 km in length);
- there is a sizeable freshwater source feeding the estuary;
- there is significant development along and around the estuary.

The mean spring tidal range at the mouth of the Exe estuary is ~3.8 m while the mean neap tidal range is ~1.48 m, and the estuary is therefore

Figure 11.1 Aerial photograph of the Exe estuary in Devon, UK, looking approximately north–north-west. The village of Exmouth is at the mid-right-hand edge, the village of Dawlish just out of view to the front left, the nature reserve Dawlish Warren is at front centre and the city of Exeter towards the back and left. The mainline railway track from London to Penzance runs through Exeter and stations at Star Cross, Dawlish Warren and Dawlish (courtesy of Halcrow)

considered macrotidal (i.e. a tidal range in excess of 4 m). Tidal currents as strong as 3.0 m/s have been reported immediately seaward of the mouth of the estuary. The estuary is ebb dominant and the tidal currents are concentrated at its narrow inlet between Dawlish Warren and Exmouth beach (the terms 'tidal range' and 'ebb dominant' are explained in the next section).

The Exe estuarine system is dominated by Atlantic swell waves from the south-west, although the shallow water offshore attenuates incoming waves. In addition, wave refraction takes place due to headlands, thereby altering the wave approach angle at the mouth. Wave diffraction takes place around the outer estuary banks and ebb shoal while the ebb tidal delta and the Dawlish Warren spit limit wave penetration into the estuary. The terms 'refraction' and 'diffraction' are explained further in Chapter 12.

Marine-sediment input dominates the sediment influx into the Exe estuary system, with both waves and tidal currents moving sediment in suspension and as bedload towards the entrance of the estuary. In a historical context, the Exe estuary is an extremely dynamic system continuously responding to changes in natural processes and human intervention. During the past 150 years, Dawlish Warren spit has been recorded as undergoing sporadic cycles of erosion and accretion. The intertidal flats and sandbanks have long been accreting due to wave-induced sediment transport along the coast and sediment supply from the river Exe.

The evolution of the Exe estuary has been significantly influenced by anthropogenic changes. Human interventions on the estuary include: reclamation of intertidal areas, dredging to maintain channel navigability, construction of the south-west railway line, construction of coastal defences along the Dawlish Warren and Exmouth frontages, light industrial development, fishing and recreational activities. The majority of land reclamation took place between the 18th and 19th centuries, largely through impounding salt marshes on the west bank of the estuary. This has reduced the tidal prism (the volume of water entering the estuary from low tide to high tide), with a resultant effect on the hydrodynamics, sediment-transport and pollutant-dispersion characteristics (SCOPAC (2003)). The approach channel to Exmouth docks accretes, and regular dredging is undertaken to maintain the navigability of the channel. The navigation channel is currently dredged to a depth of about −12 m CD. The width of the channel is maintained so as to be between 100 m and 200 m (SCOPAC (2003)).

Construction of the embankment for the south-west railway line between Exeter and Penzance took place in 1849. The embankment and associated groynes constrict the shoreline and interrupt littoral transport along Dawlish Warren. A significant portion of the beach frontage at Dawlish Warren and Exmouth is protected by coastal defence structures. This includes the flood and erosion defence scheme at Dawlish Warren and Exmouth beach, which comprises a combination of sea walls, revetments, groynes and dune creation works (see Figure 11.2).

Figure 11.2 (Left) View from Dawlish Warren towards Dawlish showing timber groynes and concrete seawall; (Right) Erosion of sand dunes exposing a rock gabion underlayer at Dawlish Warren

The example of the Exe estuary illustrates some of the complexity that needs to be considered when modelling such a situation. The hydrodynamics alone are complicated due to the combined influence of tides, waves and river flow, as well as the potential for salinity gradients, temperature gradients and stratified flow phenomena. The hydrodynamics driven by these factors will determine the nature of the dispersion and/or accumulation of pollutants entering the estuary from run-off, light industrial activity and waste-water disposal. Introducing considerations of sediment transport and morphological change compounds this complexity because of the feedback between the morphology of the estuary, the tidal hydrodynamics and the waves. The tides and waves may cause the spits and bars to migrate or change, but in turn a new configuration of banks and bars may alter the tidal flows and wave propagation. For those attempting to manage the estuarial area in terms of its sustainable development there are even further considerations to take into account, such as fishing, trade, travel, tourism and the terrestrial and marine environments.

It is easy to feel overcome by the complexity of trying to model the workings of an estuary, but in the next few sections some ways of simplifying the problem are described. Many of these simpler models provide a means of understanding important elements of the natural dynamics of estuaries and yield invaluable knowledge to inform effective estuary management practices.

11.2 Hydrodynamic equations

11.2.1 Basic equations

In this section the general equations of motion (see Chapter 4) are reviewed together with the equations for the temperature and salinity fields. The scale of medium-to-large estuaries is such that effects arising from the rotation of the earth become important. Other important factors in estuary circulation are tidal motion, and wind-driven and bathymetrically driven circulations. Tidal motions are regular and predictable, while wind-driven circulations are ephemeral and episodic. Bathymetrically driven flows arise from the physical constraints of the estuary bottom, and change only as the morphology of the estuary changes.

From Chapter 4 (equation (4.7a)) we recall that the equation describing the balance of momentum in the x-direction may be written as

$$\frac{du}{dt} = X - \frac{1}{\rho}\frac{\partial p}{\partial x} + \nu \nabla^2 u$$

where u is the component of velocity in the x-direction, X the body force, p the pressure, ρ the fluid density and ν the kinematic viscosity. Similar equations apply for the y- and z-directions. By denoting x by x_1, y by x_2, and z by x_3 (and similarly for the components of velocity), the three equations of motion may be written compactly as

$$\frac{du_i}{dt} = X_i - \frac{1}{\rho}\frac{\partial p}{\partial x_i} + v\nabla^2 u_i \tag{11.1}$$

for $i = 1, 2, 3$, corresponding to the x-, y- and z-directions, respectively. Similarly, the equation of continuity can be written as (see also equation (4.1a)):

$$\frac{1}{\rho}\frac{d\rho}{dt} + \frac{\partial u_i}{\partial x_i} = 0 \tag{11.2}$$

with the repeated index i denoting summation. Or, for an incompressible fluid,

$$\frac{\partial u_i}{\partial x_i} = 0 \tag{11.3}$$

The velocity components are understood to be measured relative to a rotating earth, so that the body forces include centrifugal and Coriolis forces associated with the earth's rotation. The former is absorbed into the gravity force, as is customary, so that the body force is given by

$$\begin{pmatrix} X_1 \\ X_2 \\ X_3 \end{pmatrix} = \begin{pmatrix} 2\Omega_3 u_2 - 2\Omega_2 u_3 \\ 2\Omega_1 u_3 - 2\Omega_3 u_1 \\ 2\Omega_2 u_1 - 2\Omega_1 u_2 - g \end{pmatrix} \approx \begin{pmatrix} 2\Omega_3 u_2 \\ -2\Omega_3 u_1 \\ -g \end{pmatrix} \tag{11.4}$$

where Ω_i are the components of the earth's angular velocity in a local-coordinate system (see Figure 11.3). The magnitude of the earth's angular velocity is 2π radians in 24 h or $\Omega = 0.7292 \times 10^{-4}$/s. The vertical component of the angular velocity at a latitude θ is $\Omega \sin(\theta)$. From a consideration of the scale of motions we are concerned with in estuaries it is clear that $|u_3| \ll |u_1|, |u_2|$; hence the approximation of the first two components of the body force in equation (11.4).

For many classes of motion the momentum balance can be simplified in the vertical, as it is dominated by the gravitational force. The gravitational acceleration is close to 10 m/s^2. Typical horizontal velocities in nearshore and estuarine waters are of the order of 1 m/s, so that the Coriolis accelerations are of the order of 10^{-4} m/s^2. The viscous stress term is of the order of 10^{-5} m/s^2. Vertical accelerations can be comparable to g in surface-wave orbital motions, but in estuarine-scale flows the vertical velocities are more typically of the order of 10^{-2} m/s and time scales of the order of 10^3 s, giving

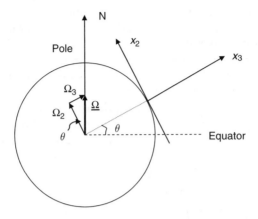

Figure 11.3 Local horizontal (x_1, x_2) and vertical (x_3) coordinate system at latitude θ, and components of the earth's angular velocity. x_1 is directed into the page at the point of intersection of the x_2- and x_3-axes

vertical accelerations of the order of $10^{-5}\,\mathrm{m/s^2}$. From these considerations, the estuarine-scale flows are quasihorizontal and the gravitational force is in approximate balance with the vertical pressure gradient:

$$\frac{\partial p}{\partial x_3} = -\rho g \tag{11.5}$$

This balance is identical to that in a fluid at rest, and its application to moving fluids is termed the 'hydrostatic approximation'. Whether the hydrostatic approximation is made or not, the gravitational force is many orders of magnitude greater than the vertical component of the Coriolis forces; hence the approximation in the vertical component of equation (11.4).

The hydrostatic equilibrium position of the free surface is taken as $x_3 = 0$. In a moving fluid, the free surface will be displaced slightly upward or downward to a position $\eta(x_1, x_2, t)$. Integrating equation (11.5) from the flat seabed (taken as $x_3 = -h$) to the free surface yields

$$p = p_a + \int_{-h}^{\eta} \rho g\, dx_3 \tag{11.6}$$

where p_a is atmospheric pressure. Substituting this expression into the equation for the horizontal pressure gradients gives (for $i = 1, 2$)

$$\frac{\partial p}{\partial x_i} = \frac{\partial p_a}{\partial x_i} + \rho_s g \frac{\partial \eta}{\partial x_i} + \int_{-h}^{\eta} g \frac{\partial \rho}{\partial x_i}\, dx_3 \tag{11.7}$$

where ρ_s is the density at the sea surface. Equation (11.7) shows that the horizontal pressure gradients in the water column arise from gradients in the atmospheric pressure, gradients in the sea-surface elevation and interior horizontal density gradients. For a fluid with constant density, the last term is, of course, zero. Furthermore, for many applications the atmospheric-pressure gradients are a second-order term, so that the first term is also often neglected in analytical treatments.

In shallow coastal seas the density is primarily a function of temperature T and salinity S. Equations governing the evolution of both these quantities can be derived from concepts of continuity, and take the following form:

$$\frac{\partial T}{\partial t} + \frac{\partial}{\partial x_i}(u_i T) = -\frac{\partial}{\partial x_i}\left(\frac{q_i}{\rho c_p}\right) \tag{11.8}$$

$$\frac{\partial S}{\partial t} + \frac{\partial}{\partial x_i}(u_i S) = -\frac{\partial}{\partial x_i}\left(\frac{s_i}{\rho}\right) \tag{11.9}$$

where c_p is specific heat, q_i are heat-flux components ($\text{Js}^{-1}\text{m}^{-2}$), and s_i are salt-flux components ($\text{kgs}^{-1}\text{m}^{-2}$). In many cases the terms on the right-hand sides of equations (11.8) and (11.9) are negligible, and the equations reduce to a statement that temperature and salinity are conserved following the motion; hence also density is conserved following the motion. Clearly, such an approximation is not valid when considering large seasonal heating or freshwater surface run-off.

An important driver of fluid motion in the sea is the boundary stress. At the surface, this may be stress due to the wind blowing over the sea surface. The wind stress (at the surface) is exerted along the direction of the surface wind and is usually described by a quadratic drag law of the form

$$\tau_{i3} = \rho_a C_{10} W_i W \tag{11.10}$$

where W is wind speed, W_i is the component of wind velocity along x_i, ρ_a and ρ_w are the densities of air and water, and C_{10} is a drag coefficient related to the wind speed at a height of $10\,\text{m}$ above the undisturbed sea level. The drag coefficient is determined empirically, and Amorocho and DeVries (1980) suggest

$$C_{10} = \begin{cases} 1.6 \times 10^{-3} & W \leq 7\,\text{ms}^{-1} \\ 2.5 \times 10^{-3} & W \geq 10\,\text{ms}^{-1} \end{cases} \tag{11.11}$$

Similarly, bottom stresses are also represented by a quadratic law that applies at the seabed, with

$$\tau_{i3} = \rho_w C_D u_i u \tag{11.12}$$

where C_D is a drag coefficient, often taken to be 2×10^{-3}. As $\rho_a/\rho_w \approx 10^{-3}$, wind stress in hurricane-force winds is approximately equivalent to a bottom stress induced by currents of the order of 1 m/s.

Stresses applied at the boundaries of a fluid are rarely distributed uniformly over the entire water body. The physical mechanisms for preventing this are stratification and the earth's rotation. The process whereby the distribution occurs is through turbulence and, in particular, the vertical transfer of horizontal momentum through the 'Reynolds stresses' that act within the interior of the fluid (see Section 4.3.3). A common parameterization of the turbulent stresses is to relate the stresses to the velocities:

$$\tau_{i3} = \rho K_V \frac{\partial u_i}{\partial x_3} \tag{11.13}$$

where K_V is an eddy viscosity (see also equation (4.35)). Similar relations are used for the stresses in the other directions with horizontal eddy diffusivity K_H.

At this point it is convenient to drop the suffix notation and use (x, y, z) coordinates, with the z-axis vertically up, and velocity components (u, v, w). Combining equations (11.10), (11.3)–(11.5) and (11.7) (neglecting atmospheric pressure and density gradients), and neglecting viscosity but including an eddy viscosity to account for turbulent stresses as in equation (11.13), we have

$$\frac{\partial u}{\partial t} + u\frac{\partial u}{\partial x} + v\frac{\partial u}{\partial y} - fv = -g\frac{\partial \eta}{\partial x} + \frac{\partial}{\partial z}\left(K_H\frac{\partial u}{\partial z}\right) \tag{11.14a}$$

$$\frac{\partial v}{\partial t} + u\frac{\partial v}{\partial x} + v\frac{\partial v}{\partial y} + fu = -g\frac{\partial \eta}{\partial y} + \frac{\partial}{\partial z}\left(K_H\frac{\partial v}{\partial z}\right) \tag{11.14b}$$

$$\frac{\partial p}{\partial z} = -\rho g \tag{11.15}$$

$$\frac{\partial u}{\partial x} + \frac{\partial v}{\partial y} + \frac{\partial w}{\partial z} = 0 \tag{11.16}$$

where $f = 2\Omega \sin(\theta)$ and is often termed the 'Coriolis parameter' in meteorological and oceanographic literature. These equations are supplemented by boundary conditions at the surface and seabed:

$$w_s = \frac{\partial \eta}{\partial t} + u\frac{\partial \eta}{\partial x} + v\frac{\partial \eta}{\partial y} \tag{11.17}$$

$$w_b = -u\frac{\partial H}{\partial x} - v\frac{\partial H}{\partial y} \tag{11.18}$$

where H is the total water depth. The first is a statement that water particles do not leave the surface, and the second that water particles do not pass through the seabed. Additional boundary conditions at the free surface and at the bottom can be used through the inclusion of the wind-stress and bottom-stress terms in the momentum equations (11.14a) and (11.14b). Equations (11.14)–(11.18) adequately represent the dynamics of fluid flow in a well-mixed shallow sea, and are termed the 'shallow-water equations' (see Sections 2.12 and 4.6.2).

For many applications it is convenient to integrate vertically between the seabed and the free surface. The depth-averaged equations retain the same form, except that:

(1) the non-linear advective terms are multiplied by factors that depend on the assumed vertical structure of u and v, but are often taken as unity for simplicity;
(2) the eddy-viscosity terms are replaced by terms involving the surface and bed stress (see equations (11.10), (11.12) and (11.13)). Thus, for the x-component of momentum the eddy-viscosity term is replaced by a term $(\tau_{sx} - \tau_{bx})/\rho H$, where the subscripts s and b refer to the stresses at the surface due to wind and at the bottom, respectively;
(3) the boundary conditions (11.17) and (11.18) may be used to yield the depth-averaged continuity equation:

$$H\frac{\partial \eta}{\partial t} + \frac{\partial}{\partial x}(HU) + \frac{\partial}{\partial y}(HV) = 0 \tag{11.19}$$

where capital letters denote depth-averaged velocities; thus, U and V are the depth-averaged components of velocity in the x- and y-directions, respectively. The depth-averaged momentum equations may be combined into a single equation for the vertical component of vorticity ζ (see Section 4.2.4):

$$\frac{\partial \zeta}{\partial t} + U\frac{\partial \zeta}{\partial x} + V\frac{\partial \zeta}{\partial y} + v\beta + (\zeta + f)\left(\frac{\partial U}{\partial x} + \frac{\partial V}{\partial y}\right) =$$
$$\frac{1}{\rho}\left(\frac{\partial}{\partial x}\left(\frac{\tau_{sy} - \tau_{by}}{H}\right) - \frac{\partial}{\partial y}\left(\frac{\tau_{sx} - \tau_{bx}}{H}\right)\right) \tag{11.20}$$

where $\beta = df/dy$.

In situations where the flow velocities and gradients are small, the non-linear (advective) terms in equations (11.14)–(11.20) may be neglected, leading to the linearized form of the equations. These will be discussed further in the following sections.

11.2.2 Some elementary conceptual models

11.2.2.1 Geostrophic balance

Consider the very simplified situation when wind stress and bottom stress are negligible, the water is homogeneous (constant density), the motion is steady (all time derivatives vanish) and non-linear terms are neglected. The momentum and continuity equations reduce to

$$-fV = -gH\frac{\partial \eta}{\partial x}$$

$$fU = -gH\frac{\partial \eta}{\partial y}$$

$$\frac{\partial U}{\partial x} + \frac{\partial V}{\partial y} = 0 \tag{11.21}$$

This is known as 'geostrophic balance'. The Coriolis force is balanced by the pressure gradient, the flow is along contours of constant depth (termed 'isobaths') and the sea level varies in the cross-isobath direction only. The 'topography' of the seabed is normally termed the 'bathymetry'. As geostrophic balance predicts zero flow across isobaths it satisfies the coastal constraint (i.e. at the coast there is zero flow across the boundary) and hence, this can be considered an elementary coastal flow model.

11.2.2.2 Inertial oscillations

Now, far away from any boundaries we may postulate that $\eta \approx 0$. Suppose also that there are zero stresses and that any motion is a remnant of earlier forcing. The equations then become

$$\frac{\partial U}{\partial t} - fV = 0$$

$$\frac{\partial V}{\partial t} + fU = 0 \tag{11.22}$$

$$\frac{\partial U}{\partial x} + \frac{\partial V}{\partial y} = 0$$

These have solution $U = U_0 \cos(ft)$, $V = -U_0 \sin(ft)$, where $U_0 = \text{constant}$ and the coordinates have been chosen so that $V = 0$ at $t = 0$. This motion is periodic with period $2\pi/f$, which is the half-pendulum day or ~ 17 h at midlatitudes. The entire water mass oscillates in phase, the particle paths being circles of radius U_0/f. This type of motion is known as an 'inertial oscillation'. It is not consistent with the presence of a coast however; the balance expressed in equation (11.21) is between the local acceleration and the Coriolis force. The first unambiguous observation of this type of motion in the ocean was presented by Gustafson and Kullenburg (1936) from observations recorded in the Baltic Sea.

11.2.2.3 Ekman drift

Suppose $\eta \approx 0$ and that there are no nearby boundaries. Let the motion be forced by a wind stress τ_{sy} applied along the y-axis, and further suppose the motion is steady. Bottom stress, density variations and non-linear terms are neglected. Thus,

$$-fV = 0$$

$$fU = \frac{\tau_{sy}}{\rho} \qquad\qquad (11.23)$$

$$\frac{\partial U}{\partial x} + \frac{\partial V}{\partial y} = 0$$

The solution of these is $U = $ constant and $V = 0$. This corresponds to flow in a direction to the right of the wind (in the northern hemisphere, where f is positive). This unexpected result is known as 'Ekman transport', after the scientist who discovered it (Ekman (1905)). In fact, Ekman considered the vertical structure of the wind-driven flow. The full beauty of the solution is appreciated using the non-depth-averaged equations:

$$-fv = K_H \frac{\partial^2 u}{\partial z^2}$$

$$fu = K_H \frac{\partial^2 v}{\partial z^2}$$

Appropriate boundary conditions are:

$$K_H \frac{du}{dz} = 0 \qquad z = 0$$

$$K_H \frac{dv}{dz} = u_*^2 \qquad z = 0$$

$$\frac{du}{dz} = \frac{dv}{dz} = 0 \quad z \rightarrow \infty$$

corresponding to: zero wind stress in the x-direction; a wind stress in the y-direction that obeys a quadratic drag law, where u_* is the scaled surface current $(= C_{10}^{0.5} W)$; and vertical velocity gradients that vanish in the limit of large depth.

These equations have the solution:

$$u = \pm \frac{u_*^2}{fD} e^{z/D} \left(\cos\left(\frac{z}{D}\right) - \sin\left(\frac{z}{D}\right) \right) = \pm \frac{u_*^2 \sqrt{2}}{fD} e^{z/D} \cos\left(\frac{\pi}{4} + \frac{z}{D}\right)$$

$$v = \frac{u_*^2}{fD} e^{z/D} \left(\cos\left(\frac{z}{D}\right) + \sin\left(\frac{z}{D}\right) \right) = \frac{u_*^2 \sqrt{2}}{fD} e^{z/D} \sin\left(\frac{\pi}{4} + \frac{z}{D}\right)$$

$$(11.24)$$

where $D = \sqrt{(2K_H/f)}$ is known as the 'Ekman depth'. The solution is illustrated in Figure 11.4. At the surface the solution reduces to flow at 45° to the right (left) of the wind direction in the northern (southern) hemisphere. Below the surface the current speed decreases while the direction changes clockwise (anticlockwise) in the northern (southern) hemisphere. The negative solution for u applies in the southern hemisphere.

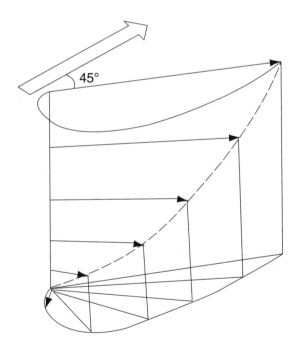

45°

Figure 11.4 Diagram of the horizontal velocity vector for the Ekman spiral. The block arrow shows the direction of the wind and the thin arrows show the water flow at different depths through the water column

11.2.2.4 Wind set-up

Wind stress is a significant driver of coastal ocean circulations. One well-known result of this is known as 'wind set-up'. This is essentially a balance between the wind stress and the pressure gradient force. At the coast, wind set-up is a well-known phenomenon for which there are empirical prediction formulae. In fact, on the open coast wind set-up is a minor contributor, as a process known as 'wave set-up' is usually dominant (see Chapter 12). It is a similar process in that waves create a 'radiation stress' that is balanced by a gradient in surface elevation. Here we use a simplified form of the equations to illustrate the physics of wind set-up.

Consider a closed sea of constant depth H and let a constant wind stress $\tau_{sy} = \rho_a C_{10} V^2$ act along the y-direction. Assuming that the bed stress and non-linear terms are negligible and that the velocity at the coast is zero, the depth-averaged equations become

$$0 = -gH\frac{\partial \eta}{\partial x}$$

$$0 = -gH\frac{\partial \eta}{\partial y} + \frac{\tau_{sy}}{\rho} \qquad (11.25)$$

$$\frac{\partial \eta}{\partial x} = 0$$

These equations have the solution

$$\eta = \frac{\tau_{sy}}{\rho g H}y$$

This describes a steady-state balance between wind stress and the horizontal pressure gradient. For open coasts, wind set-up is usually of the order of centimetres. However, in confined basins, where a storm-force wind blows over a long, narrow, deep (say 30 m) fetch for a reasonable duration, there is limited scope for water to recirculate by flowing along the coast, and wind set-up can be significant (a few metres).

Various extensions of this argument are possible. For example, by including the time derivatives, Coriolis and viscosity terms, it is possible to derive solutions for the time evolution of the set-up and its vertical structure. Also, by defining the particular shape of the basin, specific solutions are possible (see Csanady (1973), (1974), Csanady and Scott (1974)).

11.2.2.5 Western boundary current

Models of the currents on an ocean-basin scale (thousands of kilometres) were developed in the 1940s. Perhaps the best known of these was the simple model due to Stommel (1948). It had been thought that the ocean currents were primarily driven by convection. However, Stommel demonstrated that wind-driven circulations had a number of characteristics that matched the observations, in particular, the tendency for the circulation to be concentrated on the western boundary. Stommel's model was a linear steady-state model driven by wind stress, in which the horizontal flow was incompressible (i.e. the divergence is zero) and the ocean was of a fixed depth H. In this case, the governing equations can be reduced to a single equation for the vorticity (from equation (11.20)):

$$v\beta = \frac{1}{\rho}\left(\frac{\partial}{\partial x}\left(\frac{\tau_{sy} - \tau_{by}}{H}\right) - \frac{\partial}{\partial y}\left(\frac{\tau_{sx} - \tau_{bx}}{H}\right)\right) \tag{11.26}$$

The ocean is confined to the region $0 \le x \le L$, $0 \le y \le L$, and $\tau = \tau_0 \cos(\pi y/L)\mathbf{I}$, which corresponds to westerly wind in the top half of the domain and a return easterly wind in the bottom half. With these simplifications we can write the vorticity in terms of a stream function (see Chapter 4), and the vorticity equation (11.26) becomes

$$\beta\frac{\partial \psi}{\partial x} + K\nabla^2\psi = -\frac{\tau_0\pi}{\rho HL}\sin\left(\frac{\pi y}{L}\right) \tag{11.27}$$

where $K = $ constant, depending on the formulation of the bottom stress. With the boundary conditions on the stream function being

$$\psi = 0 \begin{cases} x = 0, L; & 0 \le y \le L \\ y = 0, L; & 0 \le x \le L \end{cases}$$

equation (11.27) can be solved to yield

$$\psi = \frac{\tau_0 L}{\rho_0 \pi KH}\left\{1 - \frac{e^{-ax}\left\{(e^{-2rL} - e^{(a-r)L})e^{rx} + (e^{(a-r)L} - 1)e^{-rx}\right\}}{e^{-2rL} - 1}\right\}\sin\left(\frac{\pi y}{L}\right)$$

where $a = \beta/(2K)$ and $r = (a^2 + (\pi/L)^2)^{0.5}$. This circulation is highly asymmetric, with a strong jet from south to north along the western boundary and a weak flow southwards in the rest of the domain. If the change in the Coriolis parameter with latitude is omitted ($\beta = 0$), then the resulting solution is symmetrical about the east–west and north–south axes, without any jet on the western boundary. Figure 11.5 illustrates the shape of the streamlines in both cases.

This simple model demonstrates both the importance of the change in the Coriolis force with latitude on the ocean-scale dynamics and also the wind in driving the ocean circulation. Further refinements of this model by Munk (1950), who included an additional diffusion term but neglected

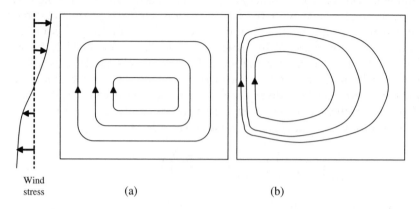

Figure 11.5 Illustrative diagram of the stream-function solution in Stommel's model for (a) $\beta = 0$, (b) $\beta = 10^{-11}$ m/s

bottom friction, and Bryan (1963) and Veronis (1966), who retained the time-derivative terms, led to a better understanding of ocean circulation and the Gulf Stream's fluctuating behaviour. This has since been further improved by detailed numerical modelling of the ocean.

11.2.2.6 Kelvin wave

Kelvin waves are gravity waves that are distorted by the earth's rotation. They are an important form of wave for coasts and estuaries. Both the tides and surges that affect the coast are Kelvin-type waves, although they are caused by different forces. For the moment we consider here the simplest 'free' Kelvin wave (i.e. the means of wave generation is not included). We consider the linearized equations of depth-averaged flow, including the Coriolis terms but excluding external and internal stress terms. Furthermore, we consider a sea of constant undisturbed depth. The governing equations become:

$$\frac{\partial U}{\partial t} - fV = -gH\frac{\partial \eta}{\partial x}$$

$$\frac{\partial V}{\partial t} + fU = -gH\frac{\partial \eta}{\partial y}$$

$$\frac{\partial \eta}{\partial t} + H\frac{\partial U}{\partial x} + H\frac{\partial V}{\partial y} = 0$$

(11.28)

Equations (11.28) are known as the Laplace tidal equations. Solutions are sought that have a harmonic time dependence, that is, $U = e^{i\omega t}U'(x, y)$ and similarly for V and η, where ω is the wave frequency. Substituting these expressions into equation (11.28) yields solutions of U' and V' in terms of η', and an equation for η':

$$\left(\nabla^2 + \frac{\omega^2 - f^2}{gH}\right)\eta' = 0 \tag{11.29}$$

This is known as the Helmholtz wave equation, the solutions of which depend on the boundary conditions. For a Kelvin wave (named after Lord Kelvin as it was he who first cast the Laplace tidal equations in the form of equation (11.28) and found the following solution), we seek solutions for which there is no flow perpendicular to the coast. Taking the coast to run north–south, then $U = 0$. With this condition, equations (11.28) and (11.29) lead to the Kelvin wave solution:

$$\eta_1 = A_1 e^{-fx/\sqrt{gH}} \cos\left(\frac{\omega y}{\sqrt{gH}} + \omega t\right) \tag{11.30}$$

where the coast is at $x = 0$ and A_1 is the amplitude of the wave at the coast. If there is another coast at $x = L$, then the solution is

$$\eta_2 = A_2 e^{-f(x-L)/\sqrt{gH}} \cos\left(\omega t - \frac{\omega y}{\sqrt{gH}}\right) \tag{11.31}$$

The superposition of these two solutions describes a Kelvin wave in an estuary or canal. Figure 11.6 shows the shape and structure of a Kelvin wave. In an estuary or shallow basin, Kelvin waves propagate up and down in a manner similar to waves in a bathtub, although the wavelength of coastal Kelvin waves is very large, typically hundreds or thousands of kilometres. In the northern (southern) hemisphere they propagate around basins with the coast on the right (left), with the highest amplitudes at the coast.

When marked on a chart, the lines of constant amplitude and phase of the Kelvin wave are termed 'corange' and 'cophase' lines. Figure 11.7 shows such a chart for a Kelvin-type wave associated with the tides in the North Sea. This is the chart for the M_2 tidal constituent.

The corange and cophase lines show that the largest amplitudes of the tide wave appear close to the coast and that the wave progresses in an anticlockwise manner around the southern North Sea basin. Also evident are spoke- or wheel-type structures in the cophase lines. These indicate amphidromes – locations at which the amplitude is zero and about which the wave propagates in an almost circular path. In fact, Figure 11.7 shows the result of a complex wave-interference pattern induced by the

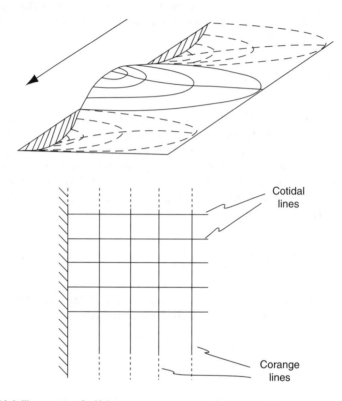

Cotidal
lines

Corange
lines

Figure 11.6 Illustration of a Kelvin wave propagating along a straight coast in the northern hemisphere

shape of the land masses and the varying water depth due to the complicated bathymetry of the North Sea. If the pattern were animated it would show a wave crest starting to the north of Scotland and propagating southward, with its largest amplitude along the east coast of the UK. When it reaches the Norfolk coast, part of the wave crest would refract southward toward the Channel and back up along the French, Belgian and Dutch coasts – setting up a standing wave with a node at the amphidromic point in the southern North Sea. A small part of the energy of this branch of the wave would penetrate through the Channel and interact with tidal waves propagating northward along the French coast and then westward along the UK south coast. Meanwhile, the other portion of the wave that was not refracted southward at Norfolk makes its way across the Dutch coast and then up to Denmark, where a second amphidromic point forms. A tertiary amphidromic point is evident off the Norwegian coast. Cophase and corange charts for various tidal constituents

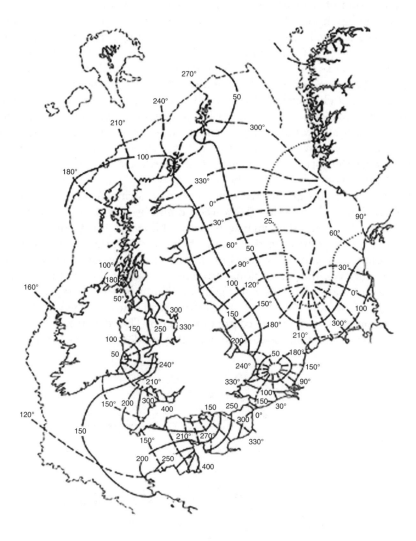

Figure 11.7 Tidal chart of the M_2 tidal constituent in the North Sea. Solid lines: corange lines(centimetres). Dashed lines: cotide lines (degrees relative to zero)

are published by specialist national hydrographic and oceanographic offices (e.g. Howarth (1990)).

Tides and surges are similar wave-like phenomena. However, tides are caused by gravitational forces due to the earth, moon and sun and are predictable to a high degree of accuracy, whereas surges are driven by the weather and are much less predictable. Storm surges are due to a combination of sea-surface pressure changes, wind stress and wave set-up close to the shore where the waves break. Tidal constituents are now explained briefly, but for details the interested reader is referred to specialist texts on tides (e.g. Cartwright (1999), Defant (1961), Godin (1972)).

The tides arise primarily as a result of the gravitational attraction between the earth, moon and sun. The rotation of the earth about its axis leads to additional characteristics of the tides we measure. The tide-generating force may be defined as the attractive force that does not affect the motion of the earth as a whole. Consider first the earth–moon interactions. The tide-generating forces arise because the resultant attractive force is not uniform over the surface of the earth. If the earth and moon are considered as point masses, then it can be shown that they will rotate about their common centre of gravity with the outward centrifugal force balanced by the gravitational attraction. However, on an object with finite dimensions (such as the earth) this balance is only achieved at the earth's centre of mass, E. At a point P on the zenith of the earth's surface, (i.e. where the line joining the centres of mass of the moon and the earth intersects the earth's surface), the gravitational attraction is greater than at the centre of the earth because P is closer to the moon than is E (see Figure 11.8). Conversely, at the point N on the nadir of the earth's surface, the gravitational attraction is less than at the centre of the earth. The centrifugal force is the same at each point on the earth because they all describe circles of identical

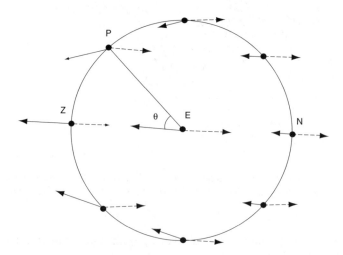

Figure 11.8 Illustration of the tide-generating force arising as the result of gravitational attraction and centrifugal forces. The moon is to the left of the earth (shown). The centre of the earth is E, Z and N are the zenith and nadir, and P is a point on the surface at latitude θ. The full lines represent the gravitational attraction due to the moon and the dashed arrows are the centrifugal force

radius as the earth rotates about the common centre of gravity of the moon and earth.

The resultant tide-generating force is such that at P the gravitational force exceeds the centrifugal force, so there is a net force towards the moon, while at point N the centrifugal force exceeds the gravitational force and the net force is away from the moon. Thus, the tide-generating force gives rise to 'bulges' (or high tides) in the water surface in regions around the zenith and nadir, and depressions (low water) at the poles. Imagining Figure 11.8 in three dimensions, there are also depressions in the surface above and below the page. That is, the fluid on the earth will tend to adopt the shape of a rugby ball, rather than a (spherical) football, a shape that is described as a prolate spheroid.

The rotation of the earth about its axis means that the tide-generating force applied to a fixed point on the earth's surface varies throughout the day. So, following point P we start off with a maximum (high tide), rotate out of the page to a minimum (low tide), continue to where point N started (high tide again), continue into the page to a minimum and eventually return to our starting point at P (high tide). We would thus expect to experience two high and two low tides per day. This is, by and large, what is observed. There are, however, important influences on the progression of the tide 'wave', such as continents, varying depth of bathymetry, and frictional and inertial effects. These act to modify the shape, phase and period of the tidal variation at any point on the earth's surface.

Another complication is that the axis about which the earth spins is not normal to the plane in which the moon orbits the earth. This angle is known as the declination. An observer on the earth's surface at latitude θ will be moved relative to the prolate spheroid, and will observe the height of the free surface to be given by

$$\eta = \frac{r_e}{2} \left(\frac{M}{m} \right) \left(\frac{r_e}{r} \right)^3 \left[(3 \sin^2 \theta \sin^2 \delta - 1) + \frac{3}{2} \sin 2\theta \sin 2\delta \cos \lambda \right.$$
$$\left. + 3 \cos^2 \theta \cos^2 \delta \cos^2 \lambda + \cdots \right] \tag{11.32}$$

where δ is the declination, λ is the earth's angular displacement ($0 \to 360°$ in 24 h), θ is the latitude, r_e is the radius of the earth, M is the mass of the moon, m is the mass of the earth and r is the distance between the centres of mass of the earth and moon. The expression for $\eta(\theta)$ (equation (11.32)) may be rewritten in terms of $\cos(n\lambda)$ using standard trigonometrical relationships:

$$\eta_m = K_0 + K_1 \cos \lambda + K_2 \cos 2\lambda + \cdots \tag{11.33}$$

In this form it is clear that the tidal water-level changes may be considered as the superposition of tidal harmonics that have distinct groups of periods related to the day length. In equation (11.33) K_0 corresponds to long-period tides, which are generated by the monthly variations in the lunar declination δ, the K_1 term corresponds to diurnal tides, with frequencies close to one cycle per day, and the K_2 term corresponds to semi-diurnal tides with frequencies close to two cycles per day. The frequencies in each tide group are termed 'tidal constituents' or 'tidal harmonics'.

A similar analysis applies to the earth–sun system. However, although the mass of the sun is many times that of the moon, the moon is much closer to the earth than the sun. By virtue of the force of gravitational attraction being proportional to the inverse square of the distance between the two masses, the tide-generating force of the moon is approximately twice that of the sun. The equilibrium tide due to the sun may be represented in an analogous form to equation (11.33), but having somewhat different periods. Table 11.1 summarizes the four main tidal constituents.

The diurnal constituents arise from the declination in the moon's orbit about the earth O_1 and the corresponding solar declination K_1. Returning to equations (11.28), we see these are linear, and so the combined tide can be found by superposition of the solutions for the different tidal constituents. In reality, there are continents, the undisturbed water depth is not constant, and there are frictional and non-linear effects to be considered. It turns out that these are most important in shallow waters, where there may be considerable interaction between the tide waves of different frequencies. These interactions give rise to higher harmonics of the main tidal constituents, as well as to frequencies arising from the sum of and differences between couples of frequencies. For example, the M_4 tidal constituent arises from the non-linear advective term, and has a period of 6.21 h.

This simplified account of tidal theory is sometimes termed the 'equilibrium theory' of tides. In this theory it is assumed that:

(1) water covers the whole earth, initially at a constant depth;
(2) water has no inertia (i.e. responds instantaneously);
(3) water is in equilibrium so that the water surface is normal to the imposed force.

Table 11.1 The main tidal harmonics.

	Symbol	Period (hours)	Description
Semidiurnal tides	M_2	12.42	Main lunar constituent
	S_2	12.00	Main solar constituent
Diurnal tides	K_1	23.93	Soli-lunar constituent
	O_1	25.82	Main lunar constituent

The dynamic theory of tides requires the numerical solution of the non-linear shallow-water equations (the non-linear version of equation (11.28)). Such computations have been performed using fairly realistic continental shapes and bathymetry (e.g. Accad and Pekeris (1978)) by using the tide-generating force to drive the equations of motion.

The same equations can be used to forecast surge waves, only in this case the generating mechanism is surface winds and pressure gradients. These have to be specified through wind-stress and pressure-gradient terms. This is normally done by using forecast wind and pressure fields from a meteorological model (e.g. Flather (1984), (1987)). Further discussion of tidal analysis and surge modelling from an engineering perspective can be found in Reeve *et al.* (2004).

Tides and surges provide the major forcing mechanisms for water-level oscillations in estuaries. The bathymetry of an estuary (its depth and cross-sectional shape) have a profound effect on the propagation of Kelvin-type waves in estuaries, governing the tidal range (and navigability for port operations) and the magnitude of surge waves (and flood risk).

11.3 One-dimensional modelling of estuaries

The approach and methods used in mathematical modelling of open-channel systems have been dealt with in some detail in Chapter 7. In the following section an introduction to the issues relevant to tidally forced flows in estuaries is described.

The simplest treatment of estuary hydrodynamics considers an estuary to be of fixed cross-sectional geometry. The starting point is the non-linear shallow-water equations:

$$\frac{\partial U}{\partial t} + U\frac{\partial U}{\partial x} + V\frac{\partial U}{\partial y} - fV = -g\frac{\partial \eta}{\partial x} + \frac{\partial}{\partial z}\left(K_H \frac{\partial U}{\partial z}\right)$$

$$\frac{\partial V}{\partial t} + U\frac{\partial V}{\partial x} + V\frac{\partial V}{\partial y} + fU = -g\frac{\partial \eta}{\partial y} + \frac{\partial}{\partial z}\left(K_H \frac{\partial V}{\partial z}\right) \qquad (11.34)$$

$$H\frac{\partial \eta}{\partial t} + \frac{\partial}{\partial x}(UH) + \frac{\partial}{\partial y}(VH) = 0$$

The first two of these are the depth-integrated versions of equations (11.14a) and (11.14b) and the last is simply the same as equation (11.19). Let the x-axis be aligned along the length of the estuary and assume that transverse velocities may be neglected (i.e. $V = 0$). Then, integrating across the breadth of the estuary yields

$$\frac{\partial U}{\partial t} + U\frac{\partial U}{\partial x} = -g\frac{\partial \eta}{\partial x} + \frac{\tau_{sx} - \tau_{bx}}{H} \qquad (11.35)$$

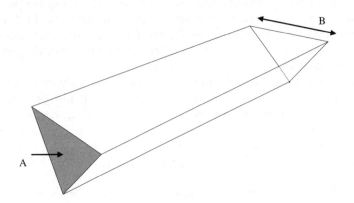

Figure 11.9 Idealized estuary and definitions

$$BH\frac{\partial \eta}{\partial t} + \frac{\partial}{\partial x}(BUA) = 0 \qquad (11.36)$$

where B and A are the breadth and cross-sectional area of the estuary, respectively, see Figure 11.9.

In analytical models it is customary to neglect the non-linear advective term in equation (11.35) and to replace the stress term with a simple linear friction term UF, where F is a constant coefficient. Equations (11.35) and (11.36) then support wave-like solutions with wave speed $c = \sqrt{(gH)}$ (see also equation (4.90)). In confined areas such as estuaries or bays, waves may be reflected from boundaries and interfere with the incoming waves. In this case a phenomenon known as a 'standing waves' occurs. This is a wave pattern in which the superposition of the incoming and reflected wave leads to a wave surface that has twice the amplitude of the incoming wave. The reason for these standing waves is that progressive waves propagating along an estuary are reflected at the upper end and interfere constructively with the unreflected waves if the wave speed and the length of the estuary are such that the time taken for a wave to travel from one end to the other and back is a whole number of wave periods. If the length of the estuary is L, then this time is $2L/c = nT$, where n is a positive integer and T is the period of oscillation. For the simplest case $n = 1$, L is one-half of the wavelength and the oscillation is termed a 'half-wave oscillation'. This type of wave can occur in closed basins and lakes, where it is known as a 'seiche'. In estuaries, there is one open boundary (the sea) and one closed boundary (upstream). In this situation it is possible to get resonant behaviour with an estuary length equal to one-quarter of a wavelength – i.e. a 'quarter-wave oscillator', with a natural period of oscillation equal

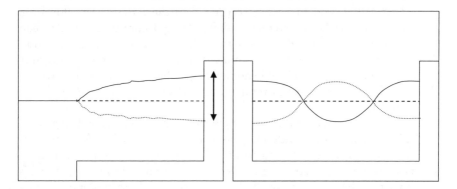

Figure 11.10 Quarter-wave (left) and half-wave (right) oscillations in an open- and closed-
water body, respectively

to $4L/c$. Figure 11.10 illustrates these types of oscillation. It is important
to note that the largest range in water levels occurs at the upper end of
the estuary (or at the shoreline in a closed basin). In other words, there
is an amplification of the tidal range. It is worth noting that the equilibrium theory of tides suggests a tidal range of about 0.8 m. This figure is
increased when the dynamic theory is used, and on open coasts can be as
much as 5 m.

Seiches are not always caused by tidal oscillations. Long-period swell
waves (see Chapter 12) and some meteorologically induced oscillations can
also be close enough to the resonant frequency to trigger seiches.

However, 0.8 m is small in comparison with the tidal ranges in some estuaries, which are close to satisfying the resonance condition. Perhaps the
best known example is the Bay of Fundy in eastern Canada. This bay is
relatively shallow ($\sim 100\,\text{m}$) and its length ($\sim 300\,\text{km}$) is quite close to the
length for resonance for semi-diurnal tidal constituents. The tidal range is
about 5 m at the mouth of this estuary, but a staggering 15 m near the head.
This is clearly not the factor-of-two amplification that might be expected
from the simple argument above. Indeed, friction, narrowing and shoaling
of the estuary and non-linear effects all combine to modify the wave as it
propagates in the estuary. A number of these complications are discussed
in detail by Prandle (2009) and in the references therein. Nevertheless,
it remains quite surprising how well the simple linear analysis works in
reality.

This section concludes with a short discussion of another important characteristic of most estuaries, i.e. the asymmetry between the flood- and
ebb-tide flows. The 'flood' flow corresponds to the time when the tide

is rising and floods the estuary. The ebb flow corresponds to the time when the tide is falling and the water recedes from the estuary. The flows are termed 'asymmetric' if the average peak flood or ebb current is stronger than its opposing current (flood stronger than ebb or ebb stronger than flood). The asymmetry is called 'flood dominant' if the flood current is stronger than the ebb current, and 'ebb dominant' if the ebb current is stronger than the flood current. An example of an ebb-dominant current is shown in Figure 11.12.

Convention defines flood current as positive and ebb current as negative. Peak ebb current is -1.15 m/s and peak flood current is 0.97 m/s. In an estuary there is usually a contribution to the flow from a river. In the absence of this flow the net discharge is zero through the estuary, even though there is asymmetry in the tidal current. At first sight this might seem contradictory. However, there are two properties that control the relationship between discharge and tidal current. One is the shape of the velocity curve, and the other is the difference in phase between the water level and the current. Phase relationships between tidal constituents determine the flood–ebb asymmetry (Aubrey and Speer (1985), DiLorenzo (1988)), in the absence of significant non-tidal forcing. In a case where the river flow is a significant component, ebb dominance will clearly be expected.

As an example, consider an M_2 tidal current and its first harmonic M_4, having amplitudes of 1.0 and 0.1 m/s, respectively. The phase relationship between the two constituents is described as

$$\alpha = 2\phi_{M_2} - \phi_{M_4}$$

where α is the phase difference between the two constituents, ϕ_{M_2} is the phase of the M_2 constituent, and ϕ_{M_4} is the phase of the M_4 constituent. This phase relationship holds because the frequency of the M_4 constituent is twice that of the M_2 constituent. If the M_4 tide lags the M_2 tide by $30°$, the combined tidal current has the ebb-dominated shape shown in Figure 11.11. The duration of the flood tidal current (6.6 h) exceeds that of the ebb tidal current (5.9 h). The greater peak ebb speed balances the longer flood duration such that there is a zero mean velocity through the inlet.

It is possible for the mean discharge to be zero but the mean velocity not to be zero, if the tidal curve for water levels and currents are not in phase. In this case, as discharge is a function of water level and current, the phase difference between the two variables creates a time-varying discharge that has a net value of zero over a tidal cycle.

Ebb and flood dominance is important because it can affect the preferred sand-transport direction in estuaries and tidal inlets. The differential in peak velocity creates a difference in the amount of material being transported on flood and ebb tides. For example, a flood-dominated estuary may deposit sand onto flats and bars and can also infill channels, whereas an

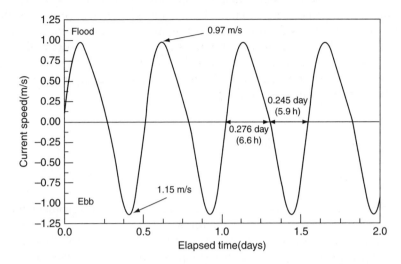

Figure 11.11 Asymmetric tide due to the superposition of two tidal constituents

ebb-dominated inlet will tend to transport sediment seaward, eroding bars and scouring channels.

11.4 Two- and three-dimensional modelling of estuaries

As discussed in Section 11.2, it is not always possible to consider the sea to be a homogeneous fluid with constant density. Both temperature and salinity affect the density of water. There are essentially three approaches to modelling this:

(1) assume the flow is sufficiently turbulent so that the water column is well-mixed, and thus the density is constant to a good approximation;

(2) assume that the changes in density are sharp so that the water column comprises 'layers' of fluid that are essentially immiscible but momentum can be transferred between layers through turbulent fluxes (see e.g. Osment (2004));

(3) solve the dynamic temperature and salinity transport equations that allow a continuously varying density field to evolve.

In the previous section, the 'one-dimensional' depth and cross-sectionally averaged approach was described, in which the density was assumed constant and a 'well-mixed' assumption was implicit. A similar condition applies when the two-dimensional shallow-water equations are used to model coastal and estuarine flow. In conditions where the assumption that the water column is well mixed is not appropriate, the water column is termed 'stratified'. This tends to allude to vertical density gradients, but horizontal density gradients can also have important effects.

This raises the question: 'How can I tell if an estuary is well-mixed?'. Dyer and New (1986) proposed a classification based on the 'layer Richardson number' (see Sections 4.6.2 and 5.8.1):

$$R_i = \frac{\dfrac{g}{\rho}\dfrac{\partial \rho}{\partial z}}{\left(\dfrac{\partial u}{\partial z}\right)^2}$$

which represents the ratio of buoyancy forces to vertical turbulent force. Vertical mixing becomes important for $R_i < 0.25$, and for such cases an assumption of well-mixedness *may* be sufficient. However, the Richardson number is likely to vary significantly along the length of an estuary, with the stage of the tide and with river discharge; so a global Richardson number should be used with care (Simpson *et al.* (1990)).

We conclude this section with a discussion of the wave-like motions that are possible in a two-layer fluid, and a qualitative extrapolation from this to the case of continuously stratified fluids. The situation is shown in Figure 11.12. Subscripts 1 and 2 are used to denote quantities relating to the upper and lower layer, respectively. There is an upper layer of density ρ_1 and depth H_1 overlaying a more dense layer of density ρ_2 and depth H_2. The total depth of fluid is $H = H_1 + H_2$. The undisturbed level of the upper fluid is taken to be at $z = 0$ and that of the lower fluid to be at $z = -H_1$. The free surface of the upper fluid is $z = \eta_1(x, y, t)$ and the interface displacement (upward) is $\eta_2(x, y, t)$. Assuming that the fluid remains in hydrostatic

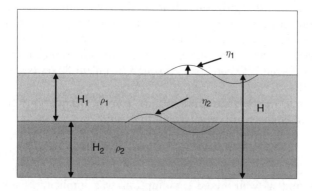

Figure 11.12 Illustration of a two-layer stratified fluid

equilibrium and that the surface pressure is zero ($p = 0$), the pressure in the two layers is given by

$$p_1 = \rho_1 g(\eta_1 - z) \qquad\qquad\qquad -H_1 + \eta_2 < z < \eta_1$$
$$p_2 = \rho_1 g(\eta_1 + H_1 - \eta_2) + \rho_2 g(-H_1 + \eta_2 - z) \quad z < -H_1 + \eta_2 \qquad (11.37)$$

Neglecting non-linear terms, Coriolis effects and friction, the momentum equations become:

$$\frac{\partial u_1}{\partial t} = -g\frac{\partial \eta_1}{\partial x} \qquad\qquad\qquad \frac{\partial v_1}{\partial t} = -g\frac{\partial \eta_1}{\partial y}$$

$$\frac{\partial u_2}{\partial t} = -\frac{\rho_1}{\rho_2}g\frac{\partial \eta_1}{\partial x} - g'\frac{\partial \eta_2}{\partial x} \qquad \frac{\partial v_2}{\partial t} = -\frac{\rho_1}{\rho_2}g\frac{\partial \eta_1}{\partial y} - g'\frac{\partial \eta_2}{\partial y} \qquad (11.38)$$

where

$$g' = g\frac{(\rho_2 - \rho_1)}{\rho_2}$$

For the upper layer, equation (11.38) is just the same as for a single homogeneous fluid. The equations for the lower layer are obtained by integrating the hydrostatic equation over depth and using continuity of pressure at the interface. The continuity equations for each layer are

$$\frac{\partial(\eta_1 + H_1 - \eta_2)}{\partial t} + H_1\left(\frac{\partial u_1}{\partial x} + \frac{\partial v_1}{\partial y}\right) = 0$$

$$\frac{\partial \eta_2}{\partial t} + H_2\left(\frac{\partial u_2}{\partial x} + \frac{\partial v_2}{\partial y}\right) = 0 \qquad (11.39)$$

Using equations (11.38) and (11.39), the velocities can be eliminated to derive an equation for the displacements η_1 and η_2:

$$\frac{\partial^2(\eta_1 - \eta_2)}{\partial t^2} = H_1 g \nabla^2 \eta_1$$

$$\frac{\partial^2 \eta_2}{\partial t^2} = H_2 \nabla^2(g\eta_1 - g'\eta_1 + g'\eta_2) \qquad (11.40)$$

These equations can then be used to eliminate one or other of the displacements to obtain a fourth-order equation governing the other. However, rather than doing this, it is customary to look for solutions for which the displacements are proportional to each other. This leads to two solutions. In the first, the displacements in both fluids are in phase with each other and waves propagate with a phase speed $c_0 \approx \sqrt{(gH)}$. In the second, the free

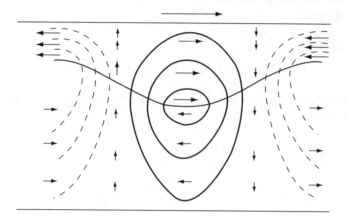

Figure 11.13 Internal modes wave at the interface of two fluids of different density

surface of the upper layer remains virtually undisturbed while the interface has a strong wave-like appearance. The velocity fields in the two layers are in anti-phase. The phase speed of the wave at the interface is $c_1 \approx \sqrt{(g'H_1)}$ if $H_2 >> H_1$. The first wave is termed the 'barotropic' or 'external' mode, as its characteristics are similar to those of a wave on a homogenous fluid. The second wave is termed 'baroclinic' or 'internal', as it depends upon a density gradient for its existence and the wave propagates within the fluid. Figure 11.13 illustrates the second solution. There is negligible disturbance of the upper surface but a large wave at the interface. The flows in the upper and lower layers oppose each other.

In a continuously stratified fluid there will be a continuum of modes as well as a barotropic mode. Further details can be found in Csanady (1982) and Gill (1982). Whether such modes are actually present in a given situation depends upon how the fluid column is forced. It is worth noting that similar behaviour is found if the Coriolis and friction terms are retained, so that one can have barotropic and baroclinic Kelvin waves. Thus, a full description of the tidal dynamics requires stratification to be included. However, the nature of tidal forcing (gravitational body force) is such that it acts throughout the water column so that the barotropic mode is strongly excited and reasonable predictions of tides can be obtained with depth-averaged models. In some estuaries, where stratification is strong, baroclinic modes may be excited and therefore affect the tidal characteristics.

In general, the effects of stratification are not easy to treat in analytical models except for the very simplest cases. There are numerous numerical models available that are capable of simulating forced stratified fluid flow. As an example, a recent study by Marques *et al.* (2009) investigated the freshwater plume that develops as two rivers discharge to the Atlantic through the Patos Lagoon in the southernmost part of Brazil, between 30–32°S and 50–52°W, being connected to the Atlantic Ocean by a single channel less than 1 km wide (see Figure 11.14). The principal rivers contributing at the north of the lagoon have a mean annual discharge of 2,400 m³/s. A (finite-element) model was set up to investigate the movement of the freshwater plume as it exited the channel and spread out into the ocean. Tides, Coriolis force and wind stress were included in the set of calculations. Figure 11.15 shows some of the results obtained by Marques

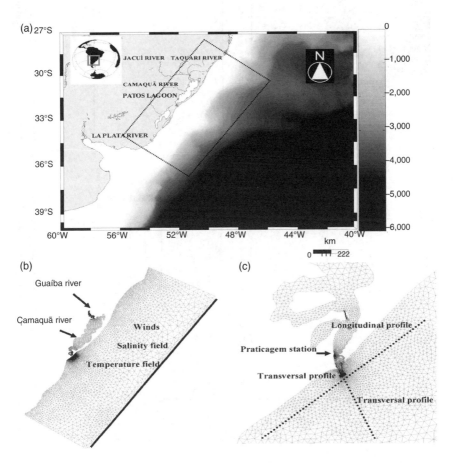

Figure 11.14 (a) The southern Brazilian shelf (dotted rectangle), the Patos Lagoon and its principal tributaries. (b) The finite-element mesh highlighting the fluid and surface boundaries and (c) the lower Patos Lagoon estuary and adjacent coastal area. The positions of the transversal and longitudinal profiles used in the study are also shown (from Marques *et al.* (2009))

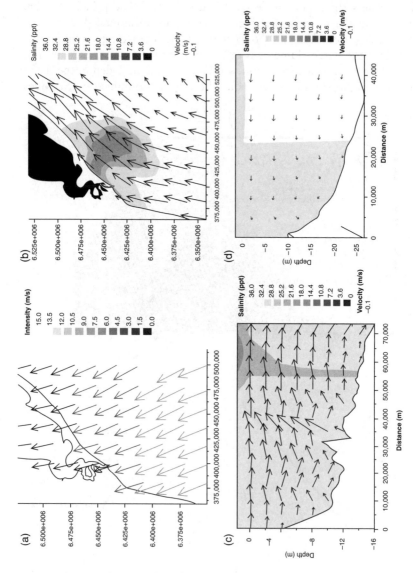

Figure 11.15 (a) Wind field and (b) calculated surface salinity and velocity field. Along-shore (c) and cross-shore (d) salinity and velocity profiles after 135 days of simulation considering the Coriolis force, the tides and the winds (from Marques *et al.* (2009))

et al. (2009). The depth variation in flow and salinity are evident. It is also clear from Figure 11.15(c) that, although the plume remains fairly concentrated in the horizontal, a significant amount of mixing has occurred in the vertical, with a 'pillar' of relatively low-salinity water penetrating all the way to the seabed.

11.5 Environmental modelling of estuaries and lakes

11.5.1 Water quality

Estuaries are very often the focus of human activity, whether for trade, tourism, fishing or other activity. Most such activity will involve either the abstraction of water from the estuary (i.e. for cooling in a power station, or for processing prior to use in a chemical plant), or the deposit of waste water into the estuary (i.e. sewage and stormwater outfalls, discharge from power stations – in this case the 'pollutant' is hot water). As a result, the estuary 'environment' is largely defined by the quality of the water available in the estuary and the movements of sediment within the estuary. In this section we cover some of the issues around modelling the water and bottom layers in terms of their 'chemical quality'. Pollutants are first classified into those that are dissolved and are transported and diffused in solution, and those that are solids and which may undergo some settling as well as interaction with dissolved chemical compounds. The simplest pollutants are those whose behaviour is independent of all other model water-quality variables or determinants. Examples of such quantities might include:

- salinity
- dissolved metals
- pesticides
- radioactivity
- faecal and total coliforms (bacteria)

These can be modelled by using an expression of the form (see also Section 4.6.4)

$$\frac{\mathrm{d}P}{\mathrm{d}t} = -(K_{1P} + K_{2P})P + \frac{K_3}{H} \tag{11.41}$$

where P is the pollutant concentration, K_{1P} is a decay rate for the pollutant P, K_2 is a settling rate for P, K_3 is a release rate for P and H is the water depth. The first term on the right-hand side represents solute loss due to the combined processes of decay and settling, while the second term represents solute gain due to the addition from benthic sources. The rate constant due to decay is often considered to be temperature dependent, following a law such as the van't Hoff equation:

$$K_{1P}(T) = K_{1P}(20)\theta_P^{(T-20)} \tag{11.42}$$

where θ_P is a temperature coefficient. If the speed of a chemical reaction is roughly doubled by a rise in temperature of $10°C$, then $\theta_P = 1.07177$ (the 10th root of 2).

In order to model the complex network of chemical and biochemical reactions it is customary to consider the main reactions that particular element groups undergo. These are often referred to as 'cycles'. Thus, for example, the nitrogen in the system will be defined by a set of chemical-reaction equations that define the cycle of nitrogen exchanges within the system.

11.5.1.1 Nitrogen and sulphur cycles

The main chemical processes involved in the nitrogen cycle are fairly well understood and the models are well developed (see e.g. Orlob (1983)). The model described here consists of the following processes: the oxidation of ammonia to nitrate; the oxidation of nitrate to nitrite; and the reduction of nitrite to nitrogen. The reduction of nitrate, which occurs when the dissolved oxygen concentration falls below 5% saturation, results in nitrogen loss from the model system in accordance with its reduction to molecular nitrogen and evolution as a gas:

$$2NO_3^{2-} \ldots \rightarrow N_2(\text{evolved}) + 3O_2 \text{ (used in oxidation reactions)}$$

At dissolved oxygen concentrations below 5% saturation, when the oxygen demand cannot be met by the reduction of nitrate, further oxygen is made available by the reduction of sulphate to sulphide:

$$H_2SO_4 \ldots \rightarrow H_2S + 2O_2$$

The reaction rate is governed by the oxygen demand. It is usually assumed that the availability of sulphate is not a limiting factor, as sea-water has a high concentration of sulphates. Sulphate is thus not explicitly included in the model as a water-quality determinant. As long as dissolved oxygen concentrations remain above 5% of saturation it is assumed that denitrification will not occur and nitrification will dominate. This is consistent with many field observations which show that the rate of denitrification in water bodies is insignificant when compared to the rate of nitrification if the dissolved oxygen concentration is not below 5% of saturation.

The loss of ammonia by oxidation to nitrate proceeds at a rate that is proportional to the ammonia concentration, so that

$$\frac{dNH_3}{dt} = -K_{1NH_3}NH_3 \tag{11.43}$$

Nitrite is formed by the oxidation of ammonia, and lost by oxidation to nitrate. The rate of ammonia oxidation is given above, this being the rate at which nitrite is formed in the process. The rate of nitrite oxidation can be assumed to be proportional to the nitrite concentration, so that

$$\frac{dNO_2}{dt} = K_{1NH_3}NH_3 - K_{1NO_2}NO_2 \tag{11.44}$$

Nitrate is formed by the oxidation of nitrite, at a rate proportional to the nitrite concentration. However, in the absence of denitrification, nitrate is not lost by any modelled mechanism, so that only the nitrite oxidation term appears in the equations:

$$\frac{dNO_3}{dt} = K_{1NO_3}NO_2 \tag{11.45}$$

Both K_{1NH_3} and K_{1NO_2} are temperature dependent and can be modelled using equation (11.42) with $K_{1NH_3}(20) = 0.16$ to 0.35/day, $K_{1NO_2} = 0.43$/day, $\theta_{NH_3} = 1.106$ and $\theta_{NO_2} = 1.072$.

11.5.1.2 Oxygen balance

The large number of factors determining the dissolved oxygen concentration requires a degree of simplification in order to have a manageable model. A simple representation of the oxygen balance is adopted in which the dissolved oxygen level is governed by the relative rates of oxygen consumption by carbonaceous oxidation, ammonia oxidation, nitrate oxidation and oxygen supply through surface reaeration:

$$\frac{dO_2}{dt} = R_{O_2}(O_2(sat) - O_2) - K_{1BOD}BOD - \alpha K_{1NH_3}NH_3 + \beta K_{1NO_2}NO_2 \tag{11.46}$$

This latter process is described using an oxygen-exchange-coefficient formulation (see e.g. Klein (1962)), which relates the reaeration to the ambient temperature and current speed. (R_{O_2} is the reaeration coefficient.) The saturation concentration (in milligrams per litre) of dissolved oxygen at a temperature $T°C$ can be expressed using the empirical function proposed by (Dysart (1970))

$$O_2(sat) = 14.652 - 0.41022T + 0.007991T^2 - 0.000077774T^3 \quad (11.47)$$

11.5.1.3 Carbonaceous biochemical oxygen demand

The carbonaceous biochemical oxygen demand (BOD), a measure of the amount of oxygen that is used during biochemical oxidation of organic compounds, is also often modelled using an equation of the form (11.41). However, in order adequately to resolve the behaviour of such a wide range of compounds, the total BOD is considered as the sum of two independent 'fast' and 'slow' components. BOD is not 'non-interacting' because it is linked to the sulphur, nitrogen and oxygen cycles. Most of the fast BOD exerted by crude sewage is removed by biological treatment, so the BOD of well-treated sewage effluent is predominantly slow; likewise, the residual BOD of riverine, estuarine and marine waters is almost all slow. The loss of carbonaceous BOD (or, what amounts to the same thing, the uptake of dissolved oxygen by carbonaceous material) may be modelled by means of a first-order reaction (i.e. the more of it there is, the faster it disappears, and the rate at which it disappears is directly proportional to the concentration) (see also equation (4.118)):

$$\frac{dBOD}{dt} = -(K_{1BOD} + K_{2BOD})BOD \quad (11.48)$$

for both fast and slow forms of BOD. The rate coefficient K_1 is usually taken to depend on temperature, as in equation (11.42).

The set of equations above describes an admittedly simplified biochemical system, but one that is relatively standard within the water engineering industry. Some key points to note are: variables that are not limiting, such as sulphate, are not modelled explicitly; phosphate and ecological modelling of plants and animals is not included; and the temperature dependence of reaction rates can be significant and therefore must be modelled. These equations can be solved simultaneously, with the hydrodynamic equations and equations governing the transport of determinants (usually an advection–diffusion equation) to predict the evolution of water quality in an estuary (see e.g. Cheng et al. (1984)).

11.5.2 Ecological modelling

The previous section described how the chemical cycles associated with
waste-water treatment could be included within a hydrodynamic model
to predict how pollutants would disperse and diffuse within a complex estuarine environment. In some environments, particularly those
where swimming, bathing and water sports are encouraged, additional
considerations are required. The primary criterion relates to the concentration of phytoplankton. These are microscopic marine plants that
rely on nutrients (dissolved nitrogen and phosphorus compounds) as
well as sunlight for photosynthesis. An abundance of nutrients together
with favourable water temperature and sunlight can lead to explosive
growth of phytoplankton to create 'blooms'. These in themselves may
not be harmful, but the bloom will eventually be followed by a col-
lapse. This leads to a huge rise in the BOD as the dead plankton rot
and there is a sharp drop in dissolved oxygen levels. When the dissolved oxygen concentration falls below 5% saturation, further oxygen
is released through the reduction of sulphate to sulphide. This releases
hydrogen sulphide, which is both foul-smelling and noxious. Local author-
ities will seek to avoid blooms and consequent collapses, as they have
adverse effects on bathing-water quality and subsequently on health and
tourism.

To model this type of system requires additional cycles to be included
in a water-quality model. In fact, ecological-system models have been in
existence for many decades (e.g. Evans and Parslow (1985)), some concen-
trating on deep-water ecology and some on shallow-water ecology, some on
freshwater environments and some on marine environments. In this section,
equations for modelling a shallow marine environment are described, which
include phytoplankton, macroalgae (seaweed) and a sediment layer that can
act as a store of nutrients.

The model encompasses phytoplankton, macroalgae, zooplankton
(microscopic animals that feed on phytoplankton), suspended particulate
matter (detritus), dissolved nutrients (nitrogen, phosphorus and oxygen)
and sediment layers. Equations governing the biochemical interactions
essentially express the conservation of carbon, phosphorus and nitrogen
within the system. For brevity, the carbon, nitrogen and phosphorus concentrations in the plankton population are written in vector notation as
$\underline{P} = (PC, PN, PP)$. Similarly, for macroalgae, zooplankton, solutes, detritus
and sediment $\underline{M} = (MC, MN, MP)$, $\underline{Z} = (ZC, ZN, ZP)$, $\underline{A} = (AC, AN, AP)$,
$\underline{D} = (DC, DN, DP)$ and $\underline{S} = (SC, SN, SP)$, respectively. The changes in the
above state variables are expressed by a set of coupled ordinary differen-
tial equations. The details of the source and sink terms in any equation
are often very complicated and based on empirical relationships determined

from laboratory experiments. The general structure of the equations is given below:

$$
\begin{aligned}
\frac{dM}{dt} &= M_{gro} - M_{de} \\
\frac{dP}{dt} &= P_{gro} - (P_{grz} + P_{de} + P_{sed}) \\
\frac{dZ}{dt} &= Z_{gro} - Z_{de} \\
\frac{dD}{dt} &= (1 - V_m)(P_{de} + M_{de}) + Z_{ex} + Z_{de} - D_{min} - D_{sed} \\
\frac{dA}{dt} &= D_{min} + Z_{resp} + S_{rel} + \frac{S_{rels}}{depth}) - P_{gro} - M_{gro} + V_m(P_{de} + M_{de}) \\
\frac{dDO}{dt} &= V_0 \left[PC_{gro} + MC_{gro} - DC_{min} - Z_{resp} - SC_{rel} \right. \\
&\quad \left. - V_m(PC_{de} + MC_{de}) \right] + DO_{re} \\
\frac{dS}{dt} &= (P_{sed} + D_{sed} - S_{rel})\, depth - S_{rels} - S_{rem}
\end{aligned}
$$

(11.49)

where

$$
Z_{ex} = P_{grz} - Z_{gro} - Z_{resp} \quad \text{and} \quad Z_{resp} = K_R P_{grz}
\tag{11.50}
$$

The subscripts gro, de, grz, resp, rel, rels, rem, sed, ex, re and min refer to the rates of change due to growth, death, grazing, respiration, instantaneous release, release from the sediment pool, removal from the sediment, excretion, reaeration and mineralization due to bacterial decay, respectively. The units of all variables are grams per cubic metre (g/m^3), except for S which is in grams per square metre (g/m^2). K_R is a respiration parameter for zooplankton, V_m is the proportion of nutrients that is released immediately for use from the dead algal biomass and V_0 is a reaction rate constant. Carbon is normally available in abundance, and so equations for its concentration in solution and in the sediment are not solved explicitly.

Many of the rates for phosphorus and nitrogen are directly related to the associated rates for carbon, for example:

$$
P_{grz} = PC_{grz} \left(1, \frac{PN}{PC}, \frac{PP}{PC} \right)
\tag{11.51}
$$

and this is the case for M_{de}, P_{grz}, P_{de}, P_{sed}, Z_{gro}, Z_{de}, Z_{resp}, Z_{ex}, D_{min} and D_{sed}. Figure 11.16 is a schematic representation of the processes described by the model equations.

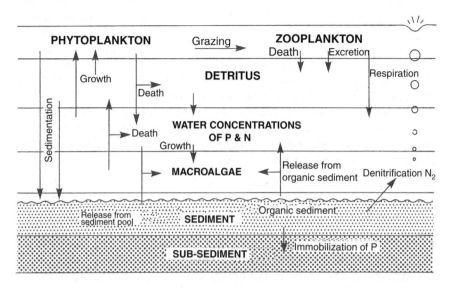

Figure 11.16 Illustrative diagram showing the processes included in the ecological model
system

Equations (11.49)–(11.51) describe the main ecological processes of accumulation and release of nutrients from the sediment and the competition between macroalgae and plankton.

A variety of equations for modelling the processes described above are available (a useful summary is given by Swartzman and Bentley (1979), although this review is now somewhat dated). Detailed equations proposed for modelling the ecological processes can be found in the literature, but the rationale of some of the modelling approaches is outlined below.

The dependence of phytoplankton growth on external nutrient availability and on the intracellular concentration of nitrogen and phosphorus has been well documented (see e.g. Nyholm (1977), (1978)). A similar dependence for macroalgae has been found in experiments by Hanisak (1979) and McPherson and Miller (1987), and has been included explicitly in numerical models (e.g. Auer and Canale (1982)). Equations describing this process may be summarized as follows: if the internal concentration is less than a specified 'critical' concentration, the uptake is proportional to the external concentration. If the external concentration is low, the uptake is set equal to nutrient mineralization rates within the water column and from the sediment. If the internal concentration is above the critical level, the uptake is dependent on the external concentration with an upper bound equal to the uptake with maximal content of nitrogen or phosphorus. This approach has four main advantages. First, the process kinetics are in agreement with laboratory experiments. Second, 'luxury consumption' (i.e. organisms absorbing more nutrients than they require at a particular time in order to store it for use in times of stress) of nutrients is included. Third, for a fixed nutrient loading a higher biomass production is obtained than under the assumption of fixed nitrogen/carbon and phosphorus/carbon ratios in the algae. Finally, resource competition between plankton and macroalgae is simulated. Experimental evidence of the importance of this mechanism in local coastal regions has been presented by Smith and Home (1988). Nutrient uptake is modelled using the Michaelis–Menten relationship, which has the general form for a determinand X:

$$\frac{dX}{dt} = \frac{\alpha X}{k + N} \tag{11.52}$$

where α is the upper limit on the uptake rate and k is a rate constant that corresponds to the nutrient concentration when the uptake reaches half of the maximum. The shape of the uptake curve is approximately linear for small values of nutrient concentration, then levelling off asymptotically to α as the concentration increases. The reader is referred to Dyke (2007) and the references therein for further discussion.

All growth and reaction rates are temperature dependent, and are usually modelled using the van't Hoff equations. However, Lassiter and Kearns (1974) proposed the following form for macroalgae:

$$X(T) = \frac{(T_{max} - T)e^{a(T - T_{opt})}}{(T_{max} - T_{opt})^{a(T_{max} - T_{opt})}} X \tag{11.53}$$

which reflects the relative insensitivity of macroalgae to subtropical temperatures and their marked sensitivity to heat stress.

The growth rates of algae depend strongly on light intensity, which can vary for the following reasons:

- seasonal variations in average daily light intensity;
- seasonal changes in the day length;
- attenuation from dissolved and suspended matter;
- self-shading.

Seasonal variations in light intensity and day length are discussed by Strahler (1971). The depth dependence in the model is limited but can be represented by constraints on the possible interactions, as shown in Figure 11.16 (e.g. phytoplankton cannot draw nutrients directly from the sediment), and also by calculating a depth-average light intensity using

Beer–Lambert's law. The light intensity I at a depth d may be determined from an extended form of the model proposed by Lorenzen (1972):

$$I = I_0 e^{\{-(a_P PCh + a_D DC + a_B + \max(0, a_M(MC - MIMC)))d\}} \tag{11.54}$$

where PCh is the chlorophyll content of plankton (often taken to be in a fixed ratio to the carbon concentration), DC and MC are the carbon content of detritus and macroalgae, and a_P, a_D, a_B and a_M are the light-attenuation coefficients for chlorophyll, detritus, water and macroalgae, respectively. $MIMC$ is the maximum concentration of macroalgae that can exist without any self-shading. In addition, the growth rate of macroalgae is significantly dependent on light intensity. As a first approximation, this can be modelled using a simple piecewise linear function for a multiplicative growth factor lying between 0 and 1:

$$IB = \begin{cases} ID/IKB & ID < IKB \\ 1 & IKB < ID < IHB \\ \max(0, 1 - (ID - IHB)/IKB) & ID > IHB \end{cases} \tag{11.55}$$

where ID is the light intensity at a depth d, IKB is a light saturation intensity and IHB is a photoinhibition constant. Physically, this models the fact that for maximum growth the light intensity needs to be above a certain threshold (IKB), while if the light intensity becomes too great (above IHB), the growth rate is inhibited.

Zooplankton are included in the model as they make an important contribution to phytoplankton depletion and the production of detritus. The only slightly non-standard element in the zooplankton model is the expression for zooplankton death. This is taken to be proportional not only to the zooplankton concentration (death proportional to ZC) but also density (death proportional to the square of ZC). This allows 'closure' of the ecological model, as higher forms of animal life such as zooplankton predators are excluded.

Accumulation of pools of nutrients in the sediment is achieved through the process of settling under gravity. Both detritus and phytoplankton settle, but zooplankton are considered able to avoid this. The settling rates are often taken to depend on concentration alone, but Farr (1983) included a dependence on depth too. Golterman (1980) discussed the importance of sediments acting as a nutrient-storage mechanism in eutrophic (i.e. well-nourished) lakes. In ecology, it is used to describe water bodies that have significant nutrient concentrations to support significant plankton and algal growth. The interchange of nutrients between sediment and the water column is an extremely complex process. A relatively straightforward means of modelling this is to have a sediment submodel, sufficient to model the accumulation and subsequent release of nutrients. The governing equation for the pool of exchangeable phosphorus in the sediment PEX is

$$\frac{\mathrm{d}PEX}{\mathrm{d}t} = (SP_{\mathrm{add}} - SP_{\mathrm{rel}})\mathrm{depth} - SP_{\mathrm{rels}} - PEX_{\mathrm{rem}} \tag{11.56}$$

where $SP_{\mathrm{add}} = (PP/PC) \cdot PC_{\mathrm{sed}} + (DP/DC) \cdot DC_{\mathrm{sed}}$ is the addition of nutrients due to settling, SP_{rel} is the release rate of newly sedimented phosphorus, which is dependent on temperature and the dissolved oxygen level, SP_{rels} is the release rate of phosphorus from the pool of exchangeable phosphorus, and PEX_{rem} is the rate of removal from the pool of exchangeable phosphorus through immobilization and is a fixed percentage (40%) of the net addition to the pool. A completely analogous equation for nitrogen applies, with denitrification replacing immobilization as a process depleting the pool.

In a pond or lake the equations governing this ecosystem might be solved for a single 'box'. That is, the water body would be considered to be well mixed, with any stratification having negligible effect on the ecology. In lakes with larger horizontal extent or in estuaries and coastal regions the effects of advection become more important. The region may be represented by a collection of linked, well-stirred boxes. However, there will be an exchange of dissolved nutrients, detritus and plankton between boxes that is determined by the water motion. It is also possible that there will be additions of nutrients at fixed locations (e.g. waste-water outfalls) as well as abstraction (e.g. at treatment plants). Considering a linear arrangement of boxes as shown in Figure 11.17, the equation governing the change in concentrations due to water exchange between boxes is an expression of the conservation of mass:

$$\frac{\mathrm{d}C_n V_n}{\mathrm{d}t} = Q_n C_{n-1} + R_{n+1} C_{n+1} + \sum_k Q_{nk} C_{nk} - \sum_l Q_{nl} C_{nl} - Q_{n+1} C_n - R_n C_n$$

$$\tag{11.57}$$

where Q_n is the flow rate into box n from box $n-1$; R_n is the flow rate from box n into box $n-1$; Q_{nk} is the inflow from the kth source into box n of concentration C_{nk}; Q_{nl} is the outflow from the lth abstraction in box n, and V_n is the volume of box n. For cases where the tidal variation is small, V_n will be approximately constant for each box and may be removed from the derivative on the left-hand side by dividing the whole equation by it.

Equations (11.49) and (11.57) may be integrated forward in time using a standard ordinary differential equation integrator to predict the evolution of a marine ecosystem. Two illustrations of ecosystem- and water-quality modelling are given in Section 11.6.

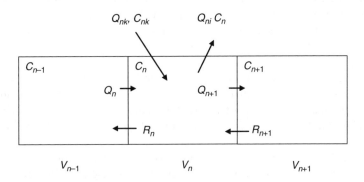

Figure 11.17 Linked-box model, showing the inputs and outflows to box *n*

11.5.3 Morphological modelling

Sediment transport affects many situations of practical importance for engineers. Movement of sediment can lead to erosion or accretion, usually occurring over relatively short periods of time and in localized areas. Over longer periods of time the cumulative effect of local sediment movement can have larger scale impacts, such as the silting up of ports and harbours, transgression of sandbanks and sandbars, etc. (see Section 4.6.3).

In coastal waters most sediment transport is due to tidal currents, and transport occurs via 'bedload', i.e. grains of sediment being rolled or bounced along the seabed surface as the drag force due to the bottom current exceeds friction and gravity forces. On the open shoreline, and to a lesser extent in estuaries, transport is also caused by the enhanced flow associated with waves. Where the flow is particularly turbulent or violent, sediment grains can be picked up and entrained into the water column, where they can be transported in the body of the flow before settling out when the drag forces no longer overcome the gravitational force.

There are several physical properties of sediment that are important in the study of coastal sediment transport. The first is the nature of the sediment itself. Sediments that contain muds are termed 'cohesive'. This type of sediment can aggregate in suspension because of its cohesive nature, forming flocs of sediment and thus changing the effective grain size. Sediments that contain no muds are termed 'non-cohesive' and are, in general, easier to model. The second property is the sediment density ρ_s, typically 2,650 kg/m^3 for quartz. The remaining properties are required in recognition of the fact that the shoreline contains a mixture of materials interspersed with voids, which may be filled with air or water. Thus, the bulk density ρ_b is defined as the *in situ* mass of the mixture/volume of the mixture, the porosity p_s as the volume of air or water/volume of the mixture (typically ~ 0.4 for a sand beach), the voids ratio e as the volume of air or water/volume of the grains, and the angle of repose (φ), which is the limiting slope angle at which the grains begin to roll. This angle is typically $35°$ in air, and in water reduces to about $30°$.

Bedload transport is the dominant mode for low-velocity flows and/or large grain sizes. It is controlled by the bed shear stresses. Conversely, suspended-load transport is the dominant mode for high-velocity flows and/or small grain sizes, and is controlled by the level of fluid turbulence.

A detailed discussion of sediment-transport equations for non-cohesive material can be found in Reeve *et al.* (2004), Svendsen (2006) and in Section 4.6.3. Here, only the general principles for bedload transport of non-cohesive sandy material are outlined, but the reader should also refer to Chapter 8, and particularly to Section 8.3 dealing with computational models for morphological processes in open-channel systems.

The general equation relates the transport rate (a vector quantity) to a representative flow velocity. This has the form

$$\underline{q} = \alpha \left| \underline{u}^{n-1} \right| \underline{u} \tag{11.58}$$

where the coefficient α is dependent on seabed and sediment properties and has to be determined experimentally for practical applications. The exponent n and the coefficient α are generally found to lie in the ranges

$$n = 3 - 6$$
$$\alpha = (0.5 - 5) \times 10^{-4} \, \text{m(m/s)}^{1-n}$$

If bedload is the dominating transport mode, it is customary to take $n = 3$. To acknowledge the fact that there is zero transport for small, non-zero flow a threshold for movement is usually incorporated with equation (11.58) so that there is zero transport if $|\underline{u}| < u_{cr}$. The bed slope may also play an important role, as downhill transport is enhanced relative to uphill transport because of gravity. This is expressed by a modification to equation (11.58):

$$\underline{q} = \begin{cases} 0 & |\underline{u}| \leq u_{cr} \\ \alpha |\underline{u}|^{n-1} \left\{ \dfrac{\underline{u}}{|\underline{u}|} - \beta \nabla h \right\} & |\underline{u}| > u_{cr} \end{cases} \tag{11.59}$$

where β is a proportionality constant and $h(x, y, t)$ is the seabed elevation relative to a fixed datum. Direct measurements of this are not available but comparisons of predicted and measured transport rates suggest it is of the order of 1. The transport equation governing conservation of sediment is then

$$(1-p)\frac{\partial b}{\partial t} + \nabla \cdot \underline{q} = 0 \tag{11.60}$$

In a numerical model, equation (11.60) is solved along with the hydrodynamic equations. This allows the interaction between the hydrodynamics and the seabed morphology to be captured. However, the time scale of changes in morphology is usually much longer than the hydrodynamic time scale, so a multiple scale integration of the equations is common, with the seabed being updated less frequently than the flow equations.

In practice, morphodynamic models usually fall into one of two categories. The first is the dynamical approach described above, where the equations governing fluid flow and conservation of sediment are solved in a deterministic manner. An example of this is shown in Chapter 12 on coastal processes. This approach is computationally expensive and the results can be quite sensitive to both the choice of sediment-transport equation and initial conditions. The alternative approach is termed 'behaviour-oriented' or 'hybrid' modelling, which seeks to simulate changes in the morphology rather than compute the sediment transport. This approach usually focuses on one or two physical processes and thereby simplifies the governing equations. Some successes have been achieved with this approach but it is often difficult to get exact rates and quantities of sediment. The two approaches are described below.

- **Deterministic process models.** These solve the equations of motion expressing conservation of mass and momentum for water and mass conservation for sediment. They include detailed descriptions of the sediment-transport process, including suspension, transport and settling. The models are iterative, requiring sequential solution of the hydrodynamics, sediment transport q and bathymetric updating. The time steps for the hydrodynamics Δt are usually much shorter than for the bathymetric updating Δt_{morph}. Hence, the seabed is held fixed for the hydrodynamic step until a 'sufficient' change occurs. At this stage the bathymetry is updated and the hydrodynamics run with the new bathymetry. The many uncertainties in the sediment-transport formulae, as well as cumulative errors in the iterative scheme, make the predictions highly uncertain. These models can also be prone to instability due to feedback between the hydrodynamics and bathymetric changes. This is usually solved by controlling the bed steepness through an 'avalanching' step to prevent unrealistically steep slopes developing. An example of the structure of a dynamic process model is shown in Figure 11.18.

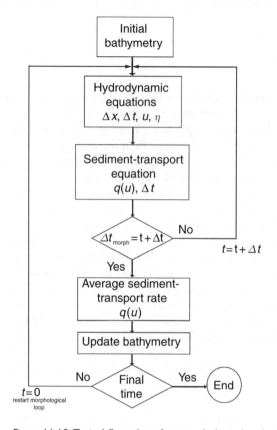

Figure 11.18 Typical flow chart for morphological updating

- **Hybrid models.** This type of model employs a simplification of the physical processes to derive a few evolutionary equations (e.g. Karunarathna and Reeve (2008), Larson *et al.* (1997), Stive and De Vriend (1995), Van Goor *et al.* (2001)). Predictions are made on the basis of parameterizing all but a few processes as a source function. These models have had reasonable success in predicting changes in morphology but there is no established method for defining the parameterization.

11.6 Physical modelling of estuaries

11.6.1 General

As stated in Chapter 1, hydraulic models have been in use since the latter part of the 19th century. It is interesting to note here that the first estuarine model soundly based on physical principles was a tidal model of the Upper Mersey constructed by Osborne Reynolds at Manchester University in 1885.

The introduction and rapid development of computational models in the second half of the 20th century led to some suggestions that physical models, particularly of rivers and estuaries, would no longer be needed. Nevertheless, with time it became clear that, in spite of the obvious advantages of computational models (see Chapter 1), there are problems that can introduce uncertainty into their use. These include the effects of grid resolution on computed results, the treatment of turbulence, sensitivity of the solutions to boundary conditions, potentially unpredictable and chaotic behaviour occurring in non-linear models and the huge computational resource required for detailed simulation of large areas over medium- to long-term time periods. Thus, at present, a combination of physical and mathematical approaches is often used, i.e. hybrid modelling.

The background to physical modelling has been covered fairly comprehensively in Chapter 5. The reader is referred in particular to Sections 5.4–5.7, which deal with the development of laws of similarity, the main similarity laws, the limits of similarity and the scale effects. The application of these laws to hydraulic models of open-channel flow is developed in Section 7.5 and to the modelling of movable beds in Section 8.5; these are, of course, relevant to physical modelling of estuaries. Section 6.1.2 then gives some details of special equipment used in estuarine models (e.g. tide and wave generators). Section 6.2 deals with the materials and construction of river and estuary models; Section 6.3 describes the appropriate measuring methods and instrumentation.

For physical modelling of estuaries, the following points must be noted (for further details, see Novak and Čábelka (1981)):

- Estuary models usually represent substantial areas and thus vertical distortion is almost the norm; sometimes this is quite large, and a ratio $M_l/M_b = 10$ is not unusual.
- In spite of this, distortion models may cover large laboratory areas and, to save space, the upstream parts of the tidal area are sometimes schematized and folded into a labyrinth, while preserving the tidal volume.
- In some situations it is possible to concentrate on the ebb and flood flow; in this case it is possible to dispense with the tide generator and to use steady-state sea conditions. It is also possible to simulate the whole tidal cycle by a series of steady-state conditions (see e.g. the Dargle river and estuary model in Section 7.6).
- If density differences and stratification are important, they can be controlled on the model by brine injection into the 'seawater' circulation, and in some instances the salinity distribution may be maintained by extracting at the model periphery the surface layer of 'fresh water' flowing into the estuary.

11.6.2 Scaling procedure

The following is a brief summary of the additional scaling procedures applicable to estuary hydraulic models; for a more detailed discussion, see, for example, Kobus (1980) or Novak and Čábelka (1981). In a distorted model operated according to the Froude law the horizontal velocity scale is given by

$$M_v = M_h^{1/2} \tag{5.28}$$

From the Strouhal criterion it follows that the time scale that must apply in all directions throughout the model is

$$M_t = M_l \, M_h^{-1/2} \tag{5.32}$$

Thus, in the vertical direction the velocity scale is

$$M_w = M_h M_t^{-1} = M_h^{3/2} M_l^{-1} \tag{(5.33)(11.61)}$$

The difference in the vertical and horizontal velocity scales is acceptable in the case of relatively slow vertical motion (i.e. for the rise and fall of water levels in tidal models and/or the settling of suspended sediment). The above equations have to be observed when programming a tide generator. A horizontal acceleration will be reproduced on the model to a scale

$$M_a = M_v M_t^{-1} = M_h^{1/2} M_l^{-1} M_h^{1/2} = M_h M_l^{-1} \tag{11.62}$$

However, the geostrophic acceleration is reproduced as

$$M_{ag} = M_v = M_h^{1/2} \tag{11.63}$$

Thus, this acceleration, when reproduced on the model, is too small by a factor of $M_l M_h^{-1/2}$. This discrepancy, which is significant only in large bodies of water, can be rectified by using a number of rotating cylinders (Coriolis tops) on the model (see Section 6.1.2).

In models where the density differences have to be considered, the densimetric Froude number (equation (4.96)) or the gradient Richardson number (equation (4.97)) have to be the same in the model as in the prototype. From the definition of these numbers it follows that

$$M_v = (M_h M_{\Delta \rho / \rho})^{1/2}$$

As $M_v = M_h^{1/2}$ (equation (5.28)) and M_ρ is usually 1, it follows that

$$M_{\Delta \rho} = 1 \tag{11.64}$$

The correct modelling of dispersion of an effluent in a three-dimensional flow in a model operated according to Froude law requires an undistorted model; in a distorted model, scale effects inevitably arise; these can be mitigated to a certain extent in special cases and small areas by manipulating the model bed roughness or by the use of two models – an undistorted one for the near field and a distorted one for modelling the convective spread of the effluent over the surface of the recipient, the modelling of surface cooling and the mass transport of the effluent by ambient currents (which, of course, can equally be modelled using a geometrically correct model).

11.7 Case studies

11.7.1 Seine estuary model

An extension of Le Havre harbour (known as the Port 2000 scheme) was constructed in the Seine estuary. To study the impact of the scheme on the morphodynamics of the estuary a distorted physical hydrosedimentological model was built in 1997 at the Sogreah laboratory in Grenoble (Cazaillet, personal communication, (2008)) at scales:

* horizontal: 1/1,000;
* vertical: 1/100.

The model represented the Seine estuary over 35 km from the sea to the Tancarville bridge, and also the coast between Deauvelle and St Adresse (see Figure 11.19). The purpose of the model was to study:

* the sedimentological impacts of the Port 2000 works (sedimentation, erosion, evolution of navigation channels and other tidal channels);
* the impact of the dredging of the harbour access channel.

Existing harbour

Extent of port in 1910

Figure 11.19 A panoramic view of the physical model with the existing and the future (now built) Le Havre harbour (courtesy of Sogreah, Grenoble)

Various port layouts were studied as part of the physical modelling study, leading to the choice of the final scheme.

In 1997, the sedimentation in the Seine estuary was about 3 million m³/year, mainly in the mouth, progressing seawards by 50 m/year. The calibration of the model was carried out by reproducing the evolution of the estuary between 1975 and 1994. One year of morphological evolution in the prototype was run in 5.6 h in the model, giving a sedimentological scale of around 1/1,600 (see also Section 8.5). The model was equipped with:

- a tide generator (controlled by computer, to generate the complex tidal curve in the Seine estuary);
- a Seine discharge generator;
- a wave generator to reproduce agitation at the mouth;
- a sediment-supply system at the downstream boundary of the model (to provide sediment according to the tide at flood tide).

The sediment consisted mainly of treated sawdust of specific characteristics to reproduce the very fine sand of the estuary (0.1–0.2 mm in the prototype). Another artificial sediment was used to simulate the global behaviour of the large north mudflat of the Seine estuary. The choice of the final Port 2000 scheme, as well as other specific goals such as the creation of a meandering north tidal channel as a compensatory measure, maintained by groynes and breaches in the low north dyke of the navigation channel, was reached as a result of the modelling study. The whole scheme is now completed and works well.

It is interesting to note that training works and port development have been ongoing in the Seine for many years. Hunter (1913) describes some of the history of the construction of the high and lower training walls that shape the entrance to the Seine, which were built to improve the navigability of the estuary, but which also led to accelerated siltation along the banks of the river, particularly upriver of the Havre harbour, as it then was. It is not possible to follow here the further details of the continuing development of the Seine, except to mention that the training walls were eventually extended to the meridian of St Sauveur on the north side and to Honfleur on the south, with the result that the navigation to Rouen and the depth in the outer estuary were improved, which had a significant effect on the prosperity of Rouen. The solution for Havre was to create a port and entrance to the open sea to the east, thereby circumventing the need for access from the estuary.

Note: Not all attempts at estuarial training are as successful, particularly those undertaken before the relatively modern understanding of hydrodynamics and morphodynamics. The construction of a breakwater in the Gulf of Ephesus, about 3,000 years ago, is a case in point. It was sheltered from all winds except those from the west. At the head of the Gulf, emigrant

Greeks founded the city of Ephesus, which over time became one of the wealthiest and populous cities of the ancient world. Today, the site of the Gulf of Ephesus is silted up and consists of a mixture of fertile land and salt marshes – and the name of the once great city of Ephesus has all but disappeared from the map.

11.7.2 Tunis north lake

The north lake of Tunis is a shallow seawater lagoon to the east of Tunis connected to the Mediterranean Sea via a narrow inlet known as the Kherredine Canal. The lake was heavily polluted by centuries of sewage discharge from the city. It is approximately $26\,km^2$ (~$10\,km$ long, $2.6\,km$ wide) with a small island (Chekli Island) approximately $2\,km$ from its western end (Figure 11.20).

Prior to 1980, the lake had become eutrophic, with problems of excessive growth of algae and weed, shallowing of the lake due to the deposition of nutrient-rich muds, and release of hydrogen sulphide gas during extended periods of calm winds and high temperatures. A programme of restoration was launched in 1981. This involved:

- diversion of sewage effluents away from the lake;
- dredging of the lake to deepen it and to remove nutrient-rich sediment;

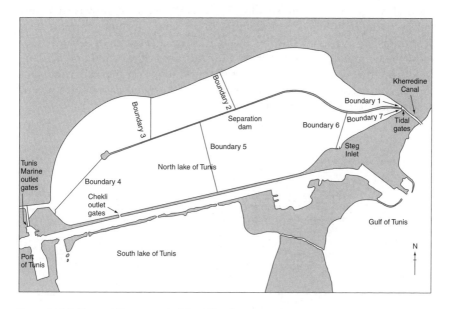

Figure 11.20 Plan of Tunis north lake. The boxes in the model were numbered anti-clockwise, starting at the Kherredine Canal (adapted from Reeve *et al.* (1991a))

- harvesting of macroalgae;
- installation of a 'separation dam' running from Chekli Island to the Kherredine Canal;
- installation of tide-driven gates to promote an anticlockwise residual flow around the lake;
- initiation of a water-quality monitoring programme;
- mathematical modelling studies of the water quality and ecological evolution of the lake.

The mathematical model was intended to be used for verification of the efficacy of the restoration programme, for prediction of future conditions and as a design tool for further water-quality improvement measures, while being sufficiently economical in terms of computing power to run on a modest microcomputer.

The modelling system consisted of two parts: a depth-averaged hydro-dynamic flow model, together with a water-quality module to predict the dispersion of waste-water inputs over the order of a few days; and an eco-logical model driven by the net tidal flow, wind, temperature and light variations. In addition to the flap gates at Kherredine, there was a cooling-water discharge into the lake from the STEG electricity-generation plant, and smaller outlet flap gates at Tunis Marine and Chekli gates. All gates function independently, based on the local lake water levels. The tidal vari-ation of water levels is very small (~15–20 cm), so the net flow is small and can be strongly influenced by the prevailing wind. Residual currents were determined by averaging the flow over a tidal cycle. Figures 11.21 and 11.22 show the computed residual currents for the cases of no wind and a 10 m/s wind from the south-west. The effect of the wind is very marked, producing complex circulation patterns in the lake. The no-wind case shows several dead zones in the southern half of the lake.

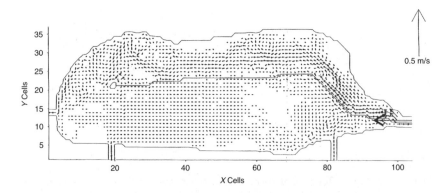

Figure 11.21 Residual flow with no wind (adapted from Reeve et al. (1991b))

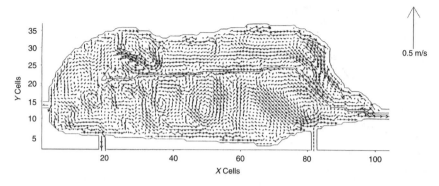

Figure 11.22 Residual flow with a constant 10 m/s wind from the south-west (adapted from Reeve *et al.* (1991b))

Residual flows for typical monthly wind conditions were computed using the hydrodynamic model. In addition, the net positive and negative flows across each of the boundaries shown in Figure 11.20 were computed for use in the ecological model, which was set up as a sequence of six boxes. The flows through the tide gates, across each box boundary and the inflows (STEG) and outflows (Tunis Marine and Chekli outlet) were set at typical monthly values. Figure 11.23 shows examples of field measurements of dissolved oxygen at the locations marked 1, 2, …, 5 in Figure 11.20. Of particular note are the drop in the dissolved oxygen concentration below target levels in September 1988 and again in July 1989 at location 4 and the quite low levels at location 5. The ecological model was set up with typical initial concentrations of the state variables based on the values obtained in the field monitoring and simulations computed for 5–10 years. Figure 11.24 shows the computed time series of dissolved oxygen and nitrogen in the sediment in the six boxes.

The cyclical nature of the population density of phytoplankton and macroalgae are accurately reproduced. In the spring the phytoplankton population rises rapidly. This is followed several weeks later by a slower growth in the macroalgae population. If phytoplankton predominate in the ensuing competition there is a large-scale bloom throughout the summer months, which diminishes if heat stress is excessive.

The bloom often lasts into the autumn, when the lower water temperatures and reduced light intensity reduce the algae growth rate and the population returns to its relatively low winter levels. If macroalgae predominate they will flourish throughout the summer, limited only by self-shading and nutrient supply. In times of high water temperature (>26°C) the heat stress can cause a high mortality rate in macroalgae, resulting

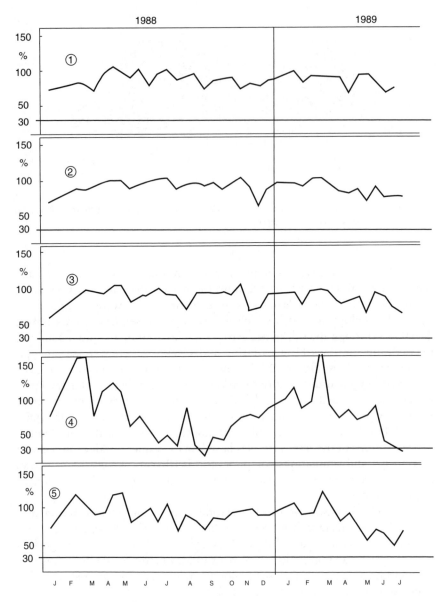

Figure 11.23 Measured dissolved oxygen levels at locations in boxes 1–5 from January 1988 to July 1989 (also marked is the 30% dissolved oxygen concentration, corresponding approximately to the 5% saturation level) (adapted from Osment *et al.* (1991))

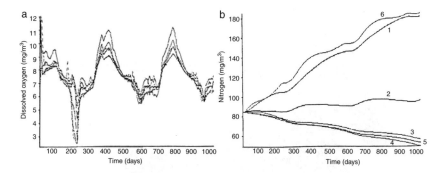

Figure 11.24 (a) Computed time series of dissolved oxygen (mg/m³) in the six boxes. (b) Computed nitrogen in the sediment in each box. (adapted from Reeve *et al.* (1991a))

in very depressed dissolved oxygen levels. In years when a catastrophic event (widespread release of hydrogen sulphide) occurs, the release of nutrients from decaying macroalgae can cause a brief secondary bloom in phytoplankton, while the high temperatures prevent the macroalgae from re-establishing itself. As observed from measurements, the model pre-dicted macroalgae growth in those parts of the lake that are shallow and that have a steady supply of nutrients. In this particular ecosystem, in regions where macroalgae dominate the phytoplankton, substantial reduc-tions in the nutrient pools in the sediments occurred over the course of three years. This is due to nutrient uptake by the macroalgae and then transport out of the system by tidal pumping when the macroalgae die and return nutrients into solution. The converse is true in areas where phyto-plankton dominates. The modelling suggested that encouraging macroalgae growth coupled with harvesting could be used to counter eutrophication. However, very close controls are needed on nutrient inputs to ensure an improvement.

11.7.3 Rivers Don and Dee dispersion and ecological modelling

This example, courtesy of Halcrow Group Ltd, combines coastal tidal modelling (see Chapter 12) and water-quality and ecological modelling. Aberdeen, one of the main cities on the east coast of Scotland, has an extensive sandy beach bounded by the Rivers Dee and Don, which is used for water sports and bathing (Figure 11.25). The most popular stretch of the beach was designated as Bathing Water in 1987 under EC Direc-tive 76/160/EEC, which provides the quality standards that the beach

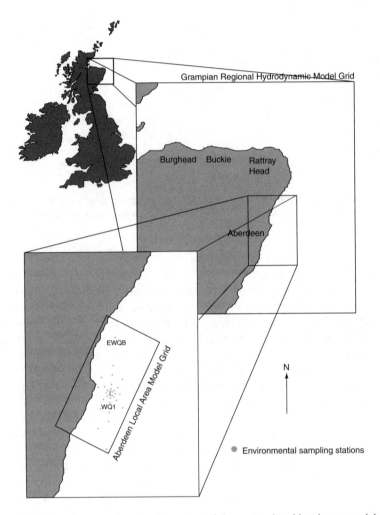

Figure 11.25 Location map showing the extent of the regional and local area models, and the location of the water-quality sampling stations

must meet. There are two sets of standards: Mandatory standards have to be achieved, and Guideline standards that are more stringent. EU Directive 91/271/WWC (CEC 1991) concerns urban waste-water treatment. It stipulates that secondary (biological) treatment of waste water should be the normal practice. However, the Directive states that for population equivalents below 150,000 a reduced level of treatment is acceptable for exceptional circumstances, including regions that are designated as Less Sensitive Zones. The Urban Waste Water Treatment (Scotland) Regulations

1994 (Scottish Office 1994) expand on and clarify the EU Directive as it applies to Scotland. In particular, it defines Areas of High Natural Dispersion, which are considered to meet the criteria for formal identification as a Less Sensitive Zone. Thus, certain coastal waters of the Grampian region, including the waters north and south of Aberdeen, have been designated as Areas of High Natural Dispersion. Any application under Article 8.5 of UWWTD requires detailed investigations to discover whether or not any adverse environmental impact will be experienced by adopting primary rather than secondary treatment standards (MPMMG, CSTT 1996).

In the 1990s, the North of Scotland Water Authority commissioned studies to determine whether the coastal waters around Aberdeen might be classified as a Less Sensitive Zone. Preliminary treatment of waste water from Aberdeen and its environs was provided by a headworks and a long sea outfall. This served a (then) current population of about 180,000 (an estimated population equivalent of 400,000 based on standard BOD loads) and had been in operation since 1988. The outfall discharged the raw sewage effluent after preliminary treatment of screenings maceration and grit removal.

Tidal-flow information included tidal ranges and Tidal Diamonds from Admiralty Charts and Tide Tables, and tidal constituents from Proudman Oceanographic Laboratories' coastal-shelf model. Additional flow data were provided by drogue tracks obtained during two separate field surveys carried out in 1972 and 1973 for the Aberdeen and Balmedie long sea outfalls.

Field-sampling exercises provided measurements of physical and chemical parameters, including currents, temperature, salinity, chlorophyll and other water-quality determinants. Measurements were taken under spring- and neap-tide conditions over the periods July–September 1994 and March–May 1995. Seabed-ecology monitoring was undertaken in the area surrounding the outfall in 1988, 1991 and 1995, and covered the period both prior to the construction and commissioning of the outfall (1988) and following its entry into operation (1991, 1995).

Monthly average values of water-quality determinants (including phosphates, nitrogenous compounds and chlorophyll) were derived from data gathered over the course of many years, and are reported by Turrell and Slesser (1992). Good agreement was achieved between the long-term monthly average values of dissolved available inorganic nitrogen (DAIN) and the sample data collected during the two surveys.

Inputs from the nearby Rivers Don, Dee and Ythan, together with outfall flows and loads are summarized in Table 11.2.

The assessment of the initial dilution achieved at the outfall was based on the design drawings of the Aberdeen Long Sea Outfall diffusers section and the methodology of Metcalf & Eddy Inc. (1991). This relates initial dilution to diffuser design, discharge rate, water depth, current speed, the

Table 11.2 Summary of inputs and loads.

Source	Flow (m³/s)	DO (mg/l)	BOD (mg/l)	SS (mg/l)	DAIN (mg/l)	Escherichia coli (coliforms/100 ml)	DAIP (mg/l)
Aberdeen long sea outfall (dry-weather flows)							
Untreated (1991)	0.84	0.5	270	434	30.0	10^9	6.00
Primary (2001)	1.11	3.5	152	130	27.0	10^7	5.00
Secondary (2001)	1.11	7.0	12	17	25.0	10^4	3.00
River Ythan	2.70	10.6	–	–	7.5	–	0.05
River Don	9.50	10.6	–	–	2.5	–	0.19
River Dee	23.60	10.6	–	–	0.5	–	0.02
Fish factory	0.06	–	–	–	–	–	–

BOD, biochemical oxygen demand; DAIN, dissolved available inorganic nitrogen; DAIP, dissolved available inorganic phosphorus; DO, dissolved oxygen; SS, suspended sediment.

density difference between effluent and ambient, and the vertical density structure of the water column.

In order to assess the hydrographic characteristics of the receiving waters a three-tier system of numerical models was set up. The primary model was a depth-averaged tidal flow covering the whole Grampian Region coastline and having a grid mesh size of 1 km. This regional model provides boundary data to drive a local tidal flow and water-quality models constructed within its boundaries. The regional model had three open-sea boundaries along which values of surface elevation were applied to each model cell along the boundaries at each time step. These boundary data were generated from harmonic data for 61 tidal constituents defined at 12 locations spaced at approximately equal distances around the open boundary.

The secondary hydrodynamic model was used for short/medium-term investigation of dissolved oxygen, BOD and suspended solids. This was driven along its boundaries by the primary hydrodynamic model. The secondary model was approximately 50% larger than a single tidal excursion either side of the discharge and covered a 36 km length of coastline (see Figure 11.24) using a grid mesh size of 200 m. This model was also used to compute the residual flow information to drive the third model in the system: the ecological model which predicts interannual variations in nutrient and chlorophyll concentrations.

It is necessary to calibrate and validate the models against data independent of those used to drive the model. For the regional tidal model,

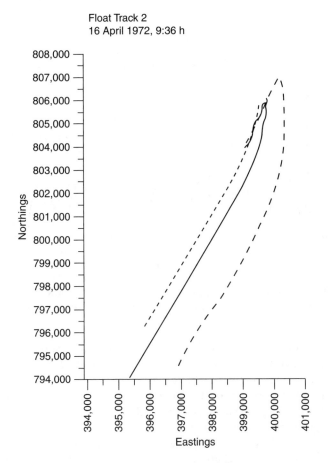

Figure 11.26 Validation of the local hydrodynamic model. Dashed line: observed float track. Solid line: track predicted using the local model. Dotted line: track predicted by the regional model

surface-elevation data computed from Admiralty Tide Tables and current velocities presented on Admiralty Charts for specific locations shown by Tidal Diamonds were used. For the local model, drogue measurements were used for validation (Figure 11.26). The track predicted by the local model is up to 1 km away from that observed in some places. This is not uncommon because the actual drogue can be affected by surface winds (not included in the hydrodynamic model) and will drift with the surface current, not the depth-averaged current. The agreement is considered good, and the results provide a conservative estimate of the dispersion.

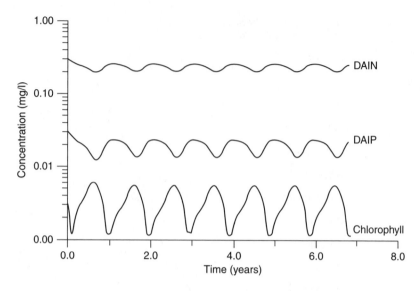

Figure 11.27 Example of the output from the ecological model. DAIN, dissolved available
inorganic nitrogen; DAIP, dissolved available inorganic phosphorus

The equations for the water-quality variables were solved in conjunction with a two-dimensional version of equation (11.57) to describe the
exchange of water between the boxes. The model was run to create
predictions for several years. Figure 11.27 shows an example of the type
of prediction that can be achieved. Note that, in this case, there is a strong
annual cycle, with gradual changes between successive years.

Figure 11.28 shows the good level of comparison achieved between
the ecological model predictions and the sampling results at one of the
most frequently monitored sampling stations. Dissolved available inorganic
phosphorus (DAIP) and dissolved oxygen are slightly overestimated, while
chlorophyll and particulate organic carbon are slightly underestimated. In
comparison with the variability in measured values, these differences are
modest. Dissolved oxygen levels remain very satisfactory, suggesting that
the waters are not eutrophic.

Also of interest was the rate at which enterobacteria present in the waste
water would be dispersed to safe concentrations in the event of a leakage
from the outfall arising from, for example, a fault in the diffuser or failure
of a pipe joint. To investigate this, the local model was used to calculate
the bacterial concentrations over a few tidal cycles, on the assumption of a
large continuous release of waste water from the long sea outfall. Worst-case
conditions (for the beaches) would be an onshore wind, which was included

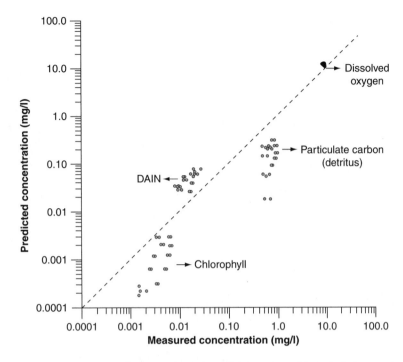

Figure 11.28 Comparison of the ecological model output and sampled determinants. [DAIN-dissolved available inorganic nitrogen]

in the simulation shown in Figure 11.29. Even with an onshore wind the tidal action moves the plume predominantly north–south, which has a rapid diluting effect.

As a result of these studies, secondary treatment at the waste-water plant was postponed while further data gathering continued and the local authority considered further options. It was known that bacteria had also been present in the Rivers Don and Dee due to point-source discharges of sewage or to rainfall run-off from urban or agricultural land. Quantification of these loads was felt to be quite uncertain.

Nevertheless, Aberdeen beach has always met at least the Mandatory standards, and achieved Guideline standards in 2006 for the first time since 1999. In March 2006 the revised Bathing Water Directive (2006/7/EC) came into force and was enacted in the UK by Regulations in March 2008. As a result of the technical and legal considerations, the main waste-water treatment plant serving Aberdeen was upgraded to provide secondary treatment. Furthermore, part of the Dee and Don catchments was designated as a

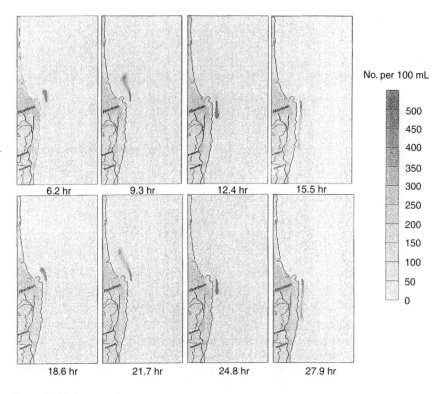

Figure 11.29 Predicted bacterial concentrations over 28 h under adverse (onshore) wind
conditions in the event of a spill from the outfall

Nitrate Vulnerable Zone. Farmers with land within the zone are required
to develop Farm Action Plans to reduce the polluting impact of their activities. The plans are specifically aimed at controlling nutrients, but they may
lead to a reduction in bacteria levels as well.

References

Accad, Y. and Pekeris, C. L. (1978), Solution of the Tidal Equations for the M2 and
    S2 Tides in the World Oceans from a Knowledge of the Tidal Potential Alone,
    Phil. Trans. Roy. Soc., London, A, 290, p. 235.
Amorocho, J. and DeVries, J. J. (1980), A new evaluation of the wind stress
    coefficient over water surfaces, *J. Geophys. Res.*, 85, 433–442.
Aubrey, D. G. and Speer, P. E. (1985), A study of non-linear tidal propagation in
    shallow inlet/estuarine systems. *Part I: Observations. Estuarine, Coastal and Shelf
    Science*, 21, pp. 185–205.

Auer, M. T. and Canale, R. P. (1982), Ecological studies and mathematical modelling of Cladophora in Lake Huron: 3. The dependence of growth rates on internal phosphorus pool size, *J. Great Lakes Res.*, 8(1), 93–99.

Bryan, K. (1963), A numerical investigation of a nonlinear model of a wind-driven ocean, *J. Atmos Sci.*, 20, 594–606.

Cartwright, D. E. (1999), *Tides: A Scientific History*, Cambridge University Press, Cambridge.

Cheng, R. T., Casulli, V. and Milford, S. N. (1984), Eulerian–Lagrangian coordinates, *Water Resource Res.*, 20(7), 944–952.

Council of the European Communities (CEC) (1991), *Council Directive Concerning Urban Wastewater Treatment (91/271/EEC)*, Brussels, 21 May 1991.

Csanady, G. T. (1973), Wind-induced barotropic motions in long lakes, *J. Phys. Oceanogr.*, 3, 429–438.

Csanady, G. T. (1974), Barotropic currents over the continental shelf, *J. Geophys. Res.*, 4, 357–371.

Csanady, G. T. (1982), *Circulation in the Coastal Ocean*, p. 280, Reidel, Dordrecht.

Csanady, G. T. and Scott, J. T. (1974), Baroclinic coastal jets in lake Ontario during IFYGL, *J. Phys. Oceanogr.*, 4, 524–541.

Defant, A. (1961), *Physical Oceanography*, Pergamon Press, London.

DiLorenzo, J. L. (1988), The Overtide and Filtering Response of Small Inlet-Bay Systems, *Hydrodynamics and Sediment Dynamics of Tidal Inlets,* (ed. D. G. Aubrey and L. Weishar), 24–53 Springer-Verlag Publishing, New York, N.Y.

Dyer, K. R. and New, A. L. (1986), Intermittency in estuarine mixing. In *Estuarine Variability. Proceedings of the 8th Biennial International Estuarine Research Conference*, pp. 321–339, University of New Hampshire, Durham, Academic Press, Orlando, FL.

Dyke, P. (2007), *Modelling Coastal and Offshore Processes*, p. 400, Imperial College Press, London.

Dysart, B. C. (1970), Water quality planning in the presence of interacting pollutants, *Water Pollut. Control Fed. J.*, 42(8), 1515–1529.

Ekman, V. W. (1905), On the influence of the Earth's rotation on ocean-currents, *Arkiv. Mateematik, Astronomi och Fysik*, 2(11), 52.

Evans, G. T. and Parslow J. S. (1985), A model of annual plankton cycles. *Biol. Oceanogr.,* 3 pp. 327–347.

Farr, J. A. (1983), *General Description of the Mathematical Model for Eutrophied Coastal Areas*, Danish Water Quality Institute (VKI), Denmark.

Flather, R. A. (1984), A numerical model investigation of the storm surge of 31 January and 1 February 1953 in the North Sea, *Q. J. R. Meteorol. Soc.*, 110, 591–612.

Flather, R. A. (1987), Estimates of extreme conditions of tide and surge using a numerical model of the northwest European continental shelf, *Estuar. Coast. Shelf Sci.*, 24, 69–93.

Gill, A. E. (1982), Atmosphere-ocean dynamics, Academic Press, New York, p. 662.

Godin, G. (1972), *The Analysis of Tides*, Liverpool University Press, Liverpool.

Golterman, H. (1980), Quantifying the eutrophication process: difficulties caused, for example, by sediments, *Prog. Water. Technol.*, 12, 63–80.

Gustafson, T. and Kullenburg, B. (1936), Untersuchungen von Trägheitsstromungen in der Ostsee. *Sven. Hydrogr.-Biol. Komm. Skr., Hydrogr.*, 13, 28.

Hanisak, M. D. (1979), Nitrogen limitation of *Codium fragile* ssp. *Tomentosoides* as determined by tissue analysis, *Mar. Biol.*, 50, 333–337.

Howarth, M. J. (1990), *Atlas of Tidal Elevations and Currents Around the British Isles*, Report OTH 89 293, HMSO, London.

Hunter, W. H. (1913), *Rivers and Estuaries or Streams and Tides*, Longmans, Green & Co., London.

Karunarathna, H. and Reeve, D. E. (2008), A Boolean approach to prediction of long-term evolution of estuary morphology, *J. Coast. Res.*, 24(2B), 51–61.

Klein, L. (1962), *River Pollution II. Causes and Effects*, Butterworths, London.

Kobus, H. (ed.) (1980), *Hydraulic Modelling*, Bulletin No. 7, German Association for Water Resources and Land Improvement, Bonn.

Larson, M., Hanson, H. and Kraus, N. C. (1997), Analytical solutions of one-line model for shoreline change near coastal structures, *J. Waterways, Port, Coast. Ocean Eng.*, 123(4), 180–191.

Lassiter, R. R. and Kearns, D. K. (1974), Phytoplankton population changes and nutrient fluctuations in a sample aquatic ecosystem model. In *Modelling the Eutrophication Process*, pp. 131–138 (ed. E. J. Middlebrook), Ann Arbour Science, Ann Arbour, MI.

Lorenzen, C. J. (1972), Extinction of light in the ocean by phytoplankton, *Journal du Conseil International d' Exploration de la Mer*, 34(1), 210–246.

Marques, W. C., Fernandes, E. H., Monteiro, I. O.,Moller, O. O.(2009), Numerical modeling of the Patos Lagoon coastal plume, Brazil, Continental Shelf Research, 29, pp. 556–571.

McPherson, B. F. and Miller, R. L. (1987), The vertical attenuation of light in Charlotte harbour: a shallow subtropical estuary, south western Florida, *Estuar. Coast. Shelf Sci.*, 25, 721–737.

Metcalf & Eddy Inc. (1991), *Wastewater Engineering*, McGraw-Hill, New York.

MPMMG, CSTT (1996), Comprehensive Studies for the Purpose of Article 6 of Dir 91/271/EEC, The Urban Wastewater Treatment Directive.

Munk, W. H. (1950), On the wind driven ocean circulation, *J. Meteorol.*, 7, 79–93.

Novak, P. and Čábelka, J. (1981), *Models in Hydraulic Engineering – Physical Principles and Design Applications*, Pitman, London.

Nyholm, N. (1977), Kinetics of nitrogen-limited algal growth, *Prog. Water. Technol.*, 8(4–5), 347–358.

Nyholm, N. (1978), Dynamics of phosphate-limited algal growth: simulation of phosphate shocks, *J. Theor. Biol.* 70, 415–425.

Orlob, G. T. (ed.) (1983), *Mathematical Modeling of Water Quality: Streams, Lakes, and Reservoirs*, Wiley, New York.

Osment, J. (2004), A simple 3D model for coastal and estuarine applications, *J. Hydroinformat.*, 06.2, 123–132.

Osment, J., Reeve, D. E., Maiz, N. B. and Moussa, M. (1991), A PC-based water quality prediction tool for Tunis North Lake, *Techniques for Environmentally Sound Water Resources Development*, pp. 229–239, HR Wallingford/Pentech Press, London.

Prandle, D. (2009), *Estuaries: Dynamics, Mixing, Sedimentation and Morphology*, Cambridge University Press, Cambridge.

Reeve, D. E., Chadwick, A. J. and Fleming, C. A. (2004), *Coastal Engineering: Processes, Theory and Design Practice*, Spon, London.

Reeve, D. E., Hoggart, C. R. and Brown, S. R. (1991a), Simulation of bio-ecological and water quality processes in enclosed coastal seas, *Mar. Pollut. Bull.*, 23, 259–263.

Reeve, D. E., Phillips, S. J., Hoggart, C. R. and Osment, J. (1991b), Computer modelling of water quality in coastal and inland ecosystems, In *Computer Modelling in the Environmental Sciences*, Vol. 28 (eds D. G. Farmer and M. J. Rycroft), IMA Conference Series; Pergamon Press, Oxford.

SCOPAC (2003), *Lyme Bay and South Devon Sediment Transport Study*, http://www.scopac.org.uk/scopac%20sediment%20db/default.html (accessed 2/11/09).

Scottish Office (1994), *The Urban Waste Water Treatment (Scotland) Regulations*, November 1994, The Stationary Office Limited.

Simpson, J. H., Brown, J., Matthews, J. and Allen, G. (1990), Tidal straining, density currents and stirring in the control of estuarine stratification, *Estuaries*, 13(2), 125–132.

Smith, D. W. and Home, A. J. (1988), Experimental measurement of resource competition between planktonic microalgae and macroalgae, *Hydrobiologica*, 159, 259–268.

Stive, M. J. F. and De Vriend, H. J. (1995), Modelling shoreface profile evolution, *Mar. Geol.*, 126, 235–248.

Stommel, H. (1948), The westward intensification of wind-driven ocean currents. *Trans. Am. Geophys. Un.*, 29, 202–206.

Strahler, A. N. (1971), *The Earth Sciences*, 2nd Edn, Harper & Row, New York.

Svendsen, Ib. A. (2006), *Introduction to Nearshore Dynamics*, World Scientific, Singapore.

Swartzman, G. and Bentley, R. (1979), A review and comparison of plankton simulation models, *ISEM J.*, 1(1–2), 30–81.

Turrell, W. R. and Slesser, G. (1992), *Annual Cycles of Physical, Chemical and Biological Parameters in Scottish Waters*, Scottish Fisheries Working Paper 5/92, Scottish Office Agriculture and Fisheries Department.

Van Goor, M. A., Stive, M. J. F., Wang, Z. B. and Zitman, T. J. (2001), Influence of relative sea level rise on coastal inlets and tidal basins. In *Proceedings of the 4th Coastal Dynamics*, pp. 242–251, ASCE, Lund.

Veronis, G., (1966), Wind-driven ocean circulation: Part 1. Linear theory and perturbation analysis, *Deep Sea Res.*, 13, 17–29.

Chapter 12

Modelling of coastal and nearshore structures and processes

12.1 Introduction

Accurate predictions of storms are required to design sea defences to protect against flooding and erosion. These predictions are also used by governments to support cost-effective schemes and by operational agencies for emergency flood planning. Novel approaches to coastal defence that harness, rather than combat, the energy of the sea are being developed but are in relative infancy. A better understanding of their long-term environmental impacts is necessary.

The threat of rising sea levels and the associated coastal erosion give an increased impetus to the modelling – both mathematical and physical – of coastal processes. In this chapter the focus is on the associated physics and processes forming the necessary background to any modelling procedure, and on the practical modelling aspects, with a few examples. The complex interaction between waves, currents, coastal, nearshore and offshore structures and their foundations is the subject of ongoing research. Key principles of the underlying theory and approaches to this topic are given in this text. The reader is referred to the papers in the references for further detail – and, indeed, to some other topics in this chapter – can be given in this text.

12.2 Physics and processes

12.2.1 Water-level variations

Water-level variations at the coast very often take the form of long-period waves, and can be classified as:

- Astronomical tide – periodic variations due to the tide-generating forces. These are well understood and can be predicted with good accuracy many years in advance;
- Storm surge – variations in water level due to the passage of atmospheric weather systems across the surface of the sea. Storm systems

are significant because of their frequency and their potential for causing large water-level variations in conjunction with large wind waves;

- Basin oscillations or seiches – resonant responses of partially enclosed water bodies to external forcing.

Tsunamis are also surface waves; they are associated primarily with subsea seismic disturbances. Apart from a few examples in Japan, defences against tsunamis are rarely constructed. There are several reasons for this. It is extremely difficult to predict when and where an earthquake will occur, and if it has occurred, it is still very difficult to determine whether it will generate a tsunami. Although there are models that can predict the general propagation characteristics of tsunami waves, they are not so good at predicting the details of the wave height and period on a local scale. For small tsunamis, existing defences can provide a partial defence; for large tsunamis, the cost of constructing a suitably robust defence is exorbitant. The design of tsunami defences is in its infancy, and there is little practical or tested experience of successful tsunami defence design. A detailed discussion of tsunamis is beyond the scope of this book, and the rest of this section focuses on tides and surges.

12.2.1.1 Tides

Sea charts (known as Admiralty Charts in the UK) provide depth, tide and landmark information for seafarers and mariners. There are thousands of charts available from the UK Hydrographic Office which cover most of the world in various levels of detail. Approaches and entrances to harbours are usually covered in the greatest detail, with medium-scale charts covering coastal areas and small-scale charts covering the open seas. Primarily meant for navigational assistance, the charts are also a useful first source of information for coastal engineers. They provide depth contours and spot heights of the bathymetry as well as information on the tidal currents (in the form of 'tidal diamonds'), tide levels and other information useful for navigation. Depths are usually given relative to 'Chart Datum' (CD). By convention, on UK and many other charts, the zero of Chart Datum is the level of Lowest Astronomical Tide (LAT). A notable exception is the United States' National Oceanic and Atmospheric Administration (NOAA), which uses as a datum the mean lower low-water level, defined as the average of the lowest tide recorded at a tide station each day during the recording period.

The tide levels commonly quoted on tidal charts and tables are:

- HAT – highest astronomical tide, the maximum tide level possible given the harmonic constituents for that particular location;
- MHWS – mean high water of spring tides;
- MHWN – mean high water of neap tides;
- MTL – mean tide level;

- MLWN – mean low water of neap tides;
- MLWS – mean low water of spring tides;
- LAT – lowest astronomical tide, the minimum tide level possible given the harmonic constituents for the location.

These levels are illustrated in Figure 12.1(a). The Mean Sea Level (MSL) is calculated as the average level of the sea at a given site over a long period of time (usually years). If atmospheric effects do not cancel out to zero, the

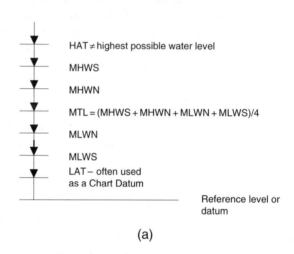

(a)

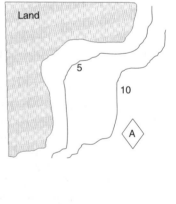

(b)

Figure 12.1 (a) Commonly used levels related to tides. (b) Illustration of the type of information available on sea charts – tidal diamonds and depth contours

MSL will be different from the MTL as it contains sea-level fluctuations due to atmospheric and wave effects as well as the tidal forces. Tidal diamonds are symbols on UK Admiralty Charts that indicate the direction and speed of the tidal flow at different stages of the tide. The symbols consist of a capital letter in a 'diamond'. There may be many or no tidal diamonds on a particular chart, depending on the information available to the chart compositor. Each tidal diamond has a unique letter, starting from A and continuing alphabetically. Somewhere on the chart will be a table similar to the one shown in Figure 12.1(b), which contains thirteen rows and three columns for each diamond. The rows are the hours of the tidal cycle showing the 6 h from low water to high water, high water itself and then the 6 h from high water to low water. The columns show the bearing relative to north of the tidal flow and its speed (in knots) at both spring tide (Sp) and neap tide (Np).

Tide tables are normally produced by national oceanographic or port organizations. For example, in the USA tide tables are published by NOAA, and in the UK they are produced by The Admiralty. These tables will often contain predictions for high and low water for the main ports in a defined region of the world. Also included in the tables is information on the main tidal constituents from which a tidal curve can be constructed.

Water levels are measured by a variety of means. In estuarine and port locations it is quite common for recording to be only semi-automated. A popular method used to be to take manual readings from a tide board (a timber or metal rule that has levels relative to a known datum marked on it). Nowadays, many tide gauges are automated and the data recorded digitally in a form suitable for analysis. ('Tide gauge' is actually a misnomer because the gauge measures total water level, not the water level due to tides alone.)

Tides display an inherent regularity, due to the regularity of astronomical processes. As a result, certain harmonics can be identified easily from observations of tide levels. Harmonic analysis describes the variation in water level as the sum of a constant mean level, contributions from specific harmonics and a 'residual':

$$\eta = Z_0 + \sum_{i=1}^{n} a_i \cos(\Omega_i t - \phi_i) + R(t) \tag{12.1}$$

where η is the water level, Z_0 is the mean level above (or below) local datum, Ω_i is the frequency of the ith harmonic (obtained from astronomical theory – see Chapter 11), a_i is the amplitude of the ith harmonic (obtained from astronomical theory), ϕ_i is the phase of ith harmonic, n is the number of harmonics used to generate the tide, t is the time, and $R(t)$ is the residual water level variation or 'surge'.

Given a sequence of water-level measurements, equation (12.1) may be used to determine a_i, ϕ_i and $R(t)$ for a selected group of i tidal harmonics.

The numerical procedure involves fitting a sum of cosine curves to the measurements. Values of Ω_i are taken to be known from equilibrium theory, and a_i and ϕ_i are determined by choosing the values that give the best fit to the measurements. The difference, or residual, is $R(t)$. This represents a combination of numerical errors arising from the fitting calculations, measurement errors and water-level fluctuations not attributable to the selected tidal harmonics.

The general form of the tidal variations at a given location can be categorized by the tidal ratio F, which is defined as

$$F = \frac{K_1 + O_1}{M_2 + S_2} \tag{12.2}$$

where the tidal symbols denote the amplitudes of the corresponding tidal constituent. This ratio is a measure of the relative importance of the diurnal constituents to the semi-diurnal constituents. The forms of tide may be classified as follows and are also shown in Figure 12.2:

(i) $F = 0.0$–0.25 (semi-diurnal form). Two high and low waters of approximately the same height. Mean spring tide range is $2(M_2 + S_2)$.

(ii) $F = 0.25$–1.50 (mixed, predominantly semi-diurnal). Two high and low waters daily. Mean spring tide range is $2(M_2 + S_2)$.

(iii) $F = 1.50$–3.00 (mixed, predominantly diurnal). One or two high waters per day. Mean spring tide range is $2(K_1 + O_1)$.

(iv) $F > 3.00$ (diurnal form). One high water per day. Mean spring tide range is $2(K_1 + O_1)$.

The tidal curves in Figure 12.2 all exhibit modulation of the diurnal or semi-diurnal tide. This modulation arises from the superposition of two (or more) harmonics. Times when the maximum amplitudes occur are known as 'spring tides' and, conversely, times when the smallest amplitudes occur are known as 'neap tides'.

12.2.1.2 Surges

Figure 12.3 shows a typical set of measurements taken on the Norfolk (UK) coast, the reconstructed tidal trace determined using the computed harmonics, and the residual or surge. The time period is 1–16 April 2001 and the vertical axis shows the level in metres relative to a local datum.

The residual, which is considered to be the contribution of all non-tidal effects on the total water level, can be both positive and negative. For example, a storm will be associated with low surface pressure, and consequently a positive residual. Conversely, periods that are dominated by high surface pressure are likely to coincide with negative surge.

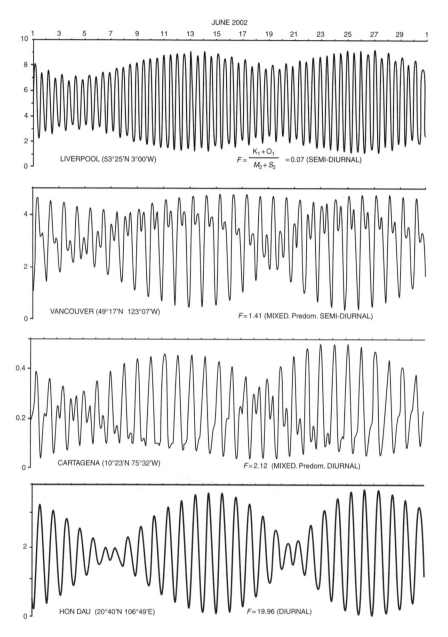

Figure 12.2 Tidal traces constructed from tidal harmonic amplitudes and phases quoted in the Admiralty Tide Tables 2002 for the month of June 2002 at Liverpool, Vancouver, Cartagena and Hon Dau. The tides at these ports are semi-diurnal, mixed predominantly semi-diurnal, mixed predominantly diurnal, and diurnal, respectively (Reeve *et al.* (2004))

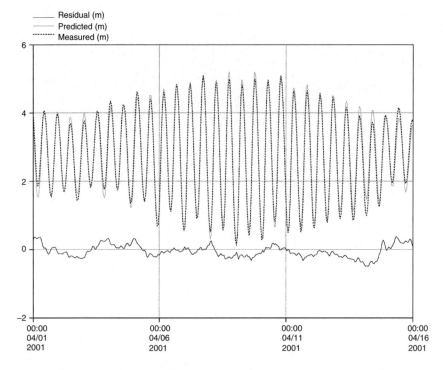

Figure 12.3 Water-level time series, reconstructed tidal curves and the residual (Reeve *et al.* (2004))

As noted above, the residual computed from a harmonic analysis will be the aggregation of measurement and numerical errors, errors arising from any truncation of the harmonic series, as well as the 'surge'. Surge is a generic term that covers all non-tidal water-level variations. In a coastal context, it can include static surge, dynamic surge, wind set-up, wave set-up and seiches. *Static* or *barometric surge* is simply the barometric effect of low surface air pressure leading to a slight upward 'bulging' of the sea surface. In a hydrostatic fluid at rest, equation (11.7) gives

$$\frac{\partial p}{\partial x} = -\rho g \frac{\partial \eta}{\partial x} \tag{12.3}$$

for a single space dimension. Integrating with respect to x and making the appropriate dimensional adjustments for units we obtain

$$\eta_B = 0.01(1013 - p_a) \tag{12.4}$$

where η_B is the barometric surge in metres and p_a is the atmospheric pressure in millibars.

Dynamic surge refers to propagating Kelvin wave disturbances that are triggered by moving storm systems. To forecast this component of surge it is necessary to use numerical prediction with, for example, the shallow-water equations as described in Chapter 11. Another component of surge is *wind set-up*. Wind blowing over the surface of the sea induces a surface stress. This force is balanced by a gradient in the sea surface, as discussed in Chapter 11, equation (11.27):

$$\frac{d\eta_w}{dx} = \frac{\tau_s}{\rho_w g d} = C_{10} \frac{W^2}{\rho_w g d} \tag{12.5}$$

where η_w is the wind set-up, τ_s is the surface stress, W is the surface wind speed, ρ_w is the density of water, d is the undisturbed water depth and C_{10} is a drag coefficient of the order of 10^{-3}. In engineering manuals, the wind set-up is usually estimated by first computing the gradient in equation (12.5) and then noting that the set-up at the downwind coast will be the gradient multiplied by half the fetch length. (The reason for this is that, by continuity, there will be a corresponding wind set-down at the upwind boundary.) That is,

$$\eta_w = C_{10} \frac{W^2}{\rho_w g d} \frac{F}{2} \tag{12.6}$$

where F is the fetch length. Under severe gale conditions with $W \approx 50 \, \text{m/s}$, a fetch of 100 km, a water depth of 10 m and $C_{10} = 0.0025$, the wind set-up is ~3.2 m. In practice, wind set-up is rarely as large on the coast because storms there are relatively small in relation to the scale of the oceans and the assumptions leading to equation (12.6) are not fully met; storms are relatively localized, so the water rarely has time to take up the static equilibrium shape suggested by equation (12.6). Nevertheless, under hurricane conditions ($W > 73 \, \text{m/s}$) wind set-up on the coast can be several metres.

Seiches have been discussed in Chapter 11, and before wave set-up can be described some background on ocean waves is required.

12.2.2 Waves

Ocean waves are mainly generated by the action of wind on water. The waves are formed initially by a complex process of resonance and shearing action, in which waves of different height, length and period are produced, travelling in various directions. Once formed, ocean waves are a very effective mechanism for transferring energy over large distances. As they propagate away from the area of generation, waves spread and reduce in height, but maintain their frequency and wavelength. The wave speed depends on the frequency, and therefore outside the storm-generation area the sea state evolves as waves of different frequencies and directions separate. In the generation area waves of different frequencies are not separated. This is termed a *wind sea* condition. Thus, wind waves may be characterized as irregular, short crested and steep containing a large range of frequencies and directions. The low-frequency waves travel faster than the high-frequency waves, resulting in a *swell sea* condition. Consequently, away from the generation area, long-period (low-frequency) waves will predominate. Swell waves tend to be fairly regular, long crested, directional and not very steep, and contain a narrow range of low frequencies.

As waves approach a shoreline, their height and length are altered by the processes of refraction and shoaling before breaking on the shore. The region where waves break is termed the *surf zone*. It is in this region that some of the most complex wave-transformation processes occur. The flow is highly turbulent and energetic, which can create large amounts of sediment transport. The energy in the waves is also converted into cross-shore and longshore currents, as well as a set-up of the mean water level.

Where coastal structures are present, waves may also be diffracted and reflected, resulting in additional complexities in the wave motion. Figure 12.4 shows some of the main wave-transformation and wave-attenuation processes that must be considered by coastal engineers in designing coastal defence schemes.

Figure 12.4 Port at Ehoala, Madagascar, showing wave diffraction, reflection and breaking near the breakwater head, and shoaling and breaking at the shore (courtesy of RTZ)

A brief introduction to the main points underlying much of the wave theory used in coastal engineering is given here (see also Sections 4.2 and 4.5). A more detailed discussion can be found in texts devoted to wave theory. Consider first the idealized situation of a small-amplitude wave propagating in the absence of any forcing. This simple case is known as Airy wave theory. As the restrictions of this simple case are removed, so the complexity of the solutions increases.

Under the following assumptions:

(1) The water is of constant depth and the wave is of constant period;
(2) The wave motion is two-dimensional;
(3) The waves are of constant form;
(4) The fluid is incompressible;
(5) The effects of viscosity, turbulence and surface tension are neglected;
(6) The wave height H is small in comparison to the wavelength L and water depth h;

the conservation of mass can be written in terms of the Laplace equation

$$\nabla^2 \phi = 0$$

The velocity potential ϕ is defined in terms of the horizontal and vertical components of the velocity (u and w, respectively), as

$$u = \frac{\partial \phi}{\partial x} \quad w = -\frac{\partial \phi}{\partial z} \tag{12.7}$$

where x and z are the horizontal and vertical coordinates, respectively, and

$$\frac{\partial u}{\partial x} + \frac{\partial w}{\partial z} = 0$$

The conservation of momentum is governed by the unsteady Bernoulli equation. These equations, together with linearized dynamic and kinematic boundary conditions for the bed and free surface, constitute the basic set of equations for linear wave theory. Their solution can be found using separation of variables (Airy (1845), Stokes (1847)):

$$\phi(x, z, t) = \frac{Hg}{2\omega} \frac{\cos h\{k(h+z)\}}{\cos h\{kh\}} \cos(kx - \omega t) \tag{12.8}$$

where H is the wave height, ω is the wave frequency ($= 2\pi/\text{wave period}$), k is the wave number ($= 2\pi/\text{wavelength}$), g is the acceleration due to the earth's gravity and h is the undisturbed depth of water in which the wave is propagating. Note that corresponding expressions for the components of velocity can be obtained by substituting equation (12.8) into equation (12.7). We denote the wave period by T (and write $f = 1/T$) and the wavelength by L. The speed of propagation of the wave (or phase speed) is given by

$$c = \frac{L}{T} = \frac{\omega}{k} = \frac{g}{\omega} \tan h(kh) \quad \text{or} \quad \omega^2 = gk \tan h(kh) \tag{12.9}$$

The latter expression is known as the 'dispersion relation'. It demonstrates that the wave speed depends on water depth, wave period and wavelength. It may be solved to find the wavenumber (and hence wavelength and phase speed) given the wave period and water depth. The wave energy is transmitted at the group wave velocity c_g. For dispersive waves such as linear water waves, the group velocity is not identical to the phase velocity. The group wave velocity is given by

$$c_g = \frac{c}{2} \left(1 + \frac{2kh}{\sin h(2kh)} \right)$$

and the wave power, or rate of transmission of wave energy P, is given by

$$P = c_g E$$

where E is the energy of the wave given by $\rho g H^2/8$, and ρ is the density of seawater. Two important wave-transformation processes can be described using linear theory. These are *shoaling* and *refraction*. Shoaling occurs when waves travel over a seabed of varying depth. Consider a wavefront travelling parallel to the seabed contours. Under the assumption that the wave propagates without energy loss, the wave power transmission in deep water and in shallow water is identical. Thus

$$\frac{H}{H_0} = \left(\frac{c_{go}}{c_g} \right)^{1/2} \equiv K_S$$

where K_S is termed the 'shoaling coefficient'. Physically, what happens is that wave heights tend to increase as waves propagate into shallower water. When a wavefront travels obliquely to the seabed contours, Snell's law may be used in conjunction with the assumption of no energy loss, to derive the relationship

$$\frac{H}{H_0} = \left(\frac{\cos(\alpha_0)}{\cos(\alpha)} \right)^{1/2} \left(\frac{c_{go}}{c_g} \right)^{1/2} \equiv K_R K_S \tag{12.10}$$

where K_R is termed the refraction coefficient. Physically, what happens is that a wavefront crossing a contour line obliquely will start to realign itself so that it is more closely parallel to the seabed contours. The process of diffraction occurs when a wave passes an obstruction such as the end of a harbour breakwater. The wavefront is curved in the lee of the obstruction, as is evident in Figure 12.4. Diffraction cannot be treated in quite such a straightforward manner as shoaling or refraction. However, nomograms are available for evaluating combined refraction and shoaling effects, as well as accounting for diffraction (e.g. *Coastal Engineering Manual* (2009)), and provide a quick and reasonably accurate and practical method.

Wave *reflection* can also be explained using linear theory by considering the superposition of two waves of identical amplitude and frequency but propagating in opposite directions, and a boundary condition of zero horizontal flow at the location of the reflecting structure. The resulting expression for the sea-surface elevation describes a standing wave that has a maximum amplitude at the reflecting surface and at locations (antinodes) spaced regularly every half-wavelength away from the reflector.

There are two criteria that determine wave *breaking*. The first is a limit to wave steepness and the second is a limit on the wave-height/water-depth ratio. The two criteria are given by:

(1) Steepness $H/L < 1/7$. This normally limits the height of deep-water waves;
(2) Ratio of wave height to water depth. This ratio is known as the 'breaker index' and, as a rule of thumb, can be taken as 0.78. In practice, the breaker index can vary from about 0.4 to 1.2, depending on beach slope and breaker type.

Goda (2000) provides a design diagram for the limiting breaker height of regular waves, which is based on a compilation of a number of laboratory results. Wave breaking is generally a difficult process to simulate numerically because the overturning of the wave is difficult to capture using methods that are based on a fixed grid of points. Various techniques have been developed to simulate breaking, ranging from a simple restriction of the wave energy to a maximum given by a breaker index, through more sophisticated roller models that assume wave breaking of a particular form (see Svendsen (2006)), to direct numerical simulation. The manner in which waves break depends on the steepness of the slope they are propagating over, as well as the wave period. Breaking waves may be classified as one of three types, as shown in Figure 12.5. A reasonable guide to the type of

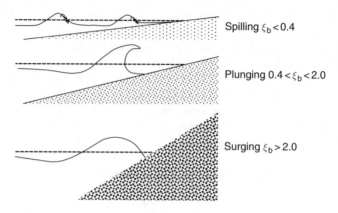

Figure 12.5 Wave-breaking classification (Reeve (2009))

breaking to expect can be determined by the value of the surf similarity parameter (or Iribarren Number)

$$\xi_b = \tan\beta / \sqrt{\frac{H_b}{L_b}} \qquad\qquad (12.11)$$

where $\tan\beta$ is the beach slope and the subscript b refers to values at the point of wave breaking. A common form of classification of breaking-wave types is shown in Figure 12.5, although there are variations in the literature.

12.2.3 Wave set-up

A phenomenon known as *wave set-up* arising from the wave radiation stress is also important for coastal locations. This is defined as the excess flow of momentum due to the presence of waves (with units of force/unit length), and is due to the orbital motion of individual water particles in the waves. These particle motions produce a net force in both the direction of propagation and at right angles to the direction of propagation. The original theory was developed by Longuet-Higgins and Stewart (1964). Its application to longshore currents was subsequently developed by Longuet-Higgins (1970). Further derivations, as well as the application of this idea to a simple model in which the radiation stresses are balanced by the sea surface gradient, can be found in Reeve *et al.* (2004). In this case it is shown that the wave set-up may be approximated by ~25% of the breaking-wave height.

12.2.4 Random waves

The distribution of wave energy across frequency and direction is described by the directional wave spectrum $E(f, \theta)$, where f is the wave frequency (in hertz) and θ is the wave direction (in radians). The directional wave spectrum, the units of which are usually metres squared second $(m^2 s)$, is often written in the form

$$E(f, \theta) = S(f)G(\theta, f) \tag{12.12}$$

where $S(f)$ is the frequency spectrum and G is a directional spreading function that satisfies

$$\int_0^{2\pi} G(\theta, f)d\theta = 1 \tag{12.13}$$

and the frequency spectrum therefore satisfies

$$S(f) = \int_0^{2\pi} E(f, \theta)d\theta \tag{12.14}$$

Various analytical forms have been proposed for both $S(f)$ and $G(\theta, f)$ on the basis of theoretical and observational considerations. Some of the more widely used forms are the Pierson–Moskowitz, JONSWAP and TMA spectra, which are applicable to open seas, fetch-limited conditions and depth-limited conditions, respectively. For reference, the JONSWAP spectrum is given here:

$$S(f) = \frac{\alpha g^2}{(2\pi)^4 f^5} \exp\left[-1.25\left(\frac{f}{f_m}\right)^{-4}\right]\gamma^q \tag{12.15}$$

where $\alpha = 0.076(gF/U^2)^{-0.22}$, with U the wind speed at $10\,m$ above the sea surface and F the fetch length, and

$$q = \exp\left(-\frac{(f - f_p)^2}{2\sigma^2 f_p^2}\right)$$

with

$$\sigma = \begin{cases} 0.07 & f \le f_p \\ 0.09 & f > f_p \end{cases}$$

and $\gamma = 3.3$. The frequency at which the spectrum attains its maximum value is denoted by f_p. The value for the peak enhancement parameter γ is an average figure derived by Hasselmann *et al.* (1973). Gaussian or cosine-squared functions are often employed for the directional spreading function. Here, we give the modified cosine power law proposed by Mitsuyasu *et al.* (1980):

$$G(\theta, f) = \frac{1}{2\sqrt{\pi}} \frac{\Gamma(s+1)}{\Gamma(s+1/2)} \cos^{2s}\left(\frac{\theta - \theta_m}{2}\right) \tag{12.16}$$

where

$$|\theta - \theta_m| < \pi \quad s = s_m \left(\frac{f}{f_p}\right)^{\mu} \quad s_m = 9.77 \tag{12.17}$$

and

$$\mu = \left\{ \begin{array}{ll} -2.33 & \text{for} \quad f \geq f_p \\ 4.06 & \text{for} \quad f < f_p \end{array} \right\}$$

$\Gamma(x)$ is the gamma function (see e.g. Gradshteyn and Ryzhik (1980)) and θ_m is the mean wave direction. The use of a Gaussian spread in wave directions has since been supported by the theoretical work of Reeve (1992), who considered the directional scattering effect of a plane wave propagating over a randomly varying seabed.

Figure 12.6 illustrates the directional spectrum and an idealized directional spreading function. Random waves are often described by the significant wave height H_s and mean period T_m. The significant wave height is the mean height of the largest third of the waves and corresponds (approximately) to the wave height by which an experienced observer would characterize the conditions. The mean wave period is the average wave period taken over a sequence of individual waves.

12.2.5 Wave overtopping

When waves meet a dune or wall, they will run up the slope. If the waves are large enough, the wave will run up to the crest level and over the top. This is known as wave 'overtopping' and can lead to damage of the structure, erosion of the dune and flooding. When waves meet a structure, they may be unbroken, already broken or actually break on the structure. This last case provides the most spectacular displays of overtopping, like the one shown in Figure 12.7. For the majority of coastal structures it is the amount of overtopping that determines the elevation of the crest of

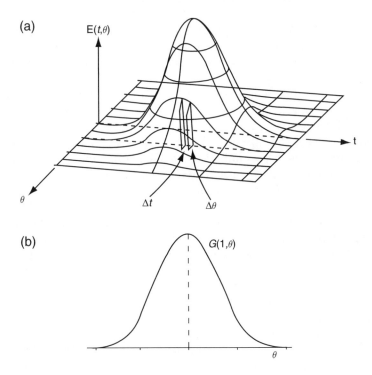

Figure 12.6 Directional spectra: (a) idealized directional spectral density, (b) idealized directional spreading function (Reeve *et al.* (2004))

Figure 12.7 Waves breaking over Alderney breakwater, October 2002 (courtesy of Dr C Obhrai)

the structure. Modern design practice uses the mean overtopping rate as a criterion rather than wave run-up, whereas early designs, in the absence of reliable experimental measurements, used wave run-up as a surrogate for overtopping.

Formulae for estimating overtopping rates are normally couched in terms of a mean discharge (l/sec per metre run) and can appear to be relatively small values. However, the actual discharge occurs as a sequence of individual events. The formulae are empirical, being based on laboratory experiments, so there are limitations to their accuracy, and it should be borne in mind that the formulae represent the best fit to points that exhibit considerable scatter. Some of the earliest work on wave overtopping was carried out by a number of investigators, most notably Owen (1980), who established the formulation framework that continues to be used today. It is generally accepted that even the most reliable methods cannot provide absolute discharges, and they can only be assumed to produce overtopping rates that are accurate to within one order of magnitude. Indeed, there are many ways to fit a curve to data, and Hedges and Reis (1998) provide an alternative model to Owen's that is based on the same set of data but incorporates additional physics-based constraints.

The most recent definitive and comprehensive work, which addresses overtopping for different structural forms, has been carried out and published as the on-line European Overtopping Manual (EurOtop), which retains Owen's framework as an option suitable for UK designs. For reference, the formula due to Owen is given below.

The mean overtopping discharge (m³/s/m) for a plain, rough, armoured slope may be calculated using the equation

$$Q_m = T_m g H_s A e^{-B \frac{R_c}{r T_m \sqrt{g H_s}}}$$ (12.18)

where R_c is the freeboard (defined as the height of the crest above the still water level), H_s is the significant wave height, g is the acceleration due to gravity, T_m is the mean period of the wave at the toe of the structure, A and B are empirical coefficients dependent on the slope of the structure, and r is the roughness coefficient (lying between 0 and 1). Values of A, B and r can be found in Owen (1980) or EurOtop for a range of different structure types and configurations.

12.2.6 Wave forces

Wave forces on coastal structures depend on both the wave conditions and the type of structure being considered. Wave forces are highly variable, and three different cases of wave impact need to be considered: unbroken; breaking; and broken waves. Coastal structures may also be considered as belonging to one of three types: vertical walls (e.g. quay walls and caisson breakwaters), sloping structures (e.g. rubble-mound breakwaters) and individual piles (e.g. offshore wind-turbine towers, oil rigs and jetties). Here, only a brief discussion of some of the main concepts is given, together with references for further details. For a discussion of forces acting on cylindrical structures (cylindrical members of offshore structures and submarine pipelines), see, for example, Novak *et al.* (2007).

The forces exerted on a vertical wall by waves comprise three parts: static pressure, dynamic pressure forces and impact forces. When the incident waves are unbroken, a standing wave will exist seaward of the wall and only the static and dynamic forces will exist. These can be readily determined from linear wave theory (see Dean and Dalrymple (1991)).

However, more commonly, the structure will need to resist the impact of breaking or broken waves. One of the most widely used formulae for estimating the forces in such situations is that due to Goda. The paper by Burcharth in Abbott and Price (1994) provides a fairly recent review of methods. Waves that break on the structure can give rise to extremely high impulsive shocks due to breaking waves trapping pockets of air. This is an ongoing area of research (see Allsop *et al.* (1996), Bullock *et al.* (2000)), and there are no really well-established design formulae at present.

For sloping structures, waves will generally break on the structure itself, and their energy is partly dissipated by turbulence and friction, with the remaining energy being reflected and/or transmitted. Many breakwaters are constructed using large blocks of rock or concrete units. Methods for determining the suitable weight of rock to ensure stability under a given design wave condition are well established (see e.g. *Coastal Engineering Manual* (2009), Reeve *et al.* (2004)). In the case of concrete units, the design criteria are often provided by the manufacturers on the basis of extensive laboratory experiments.

For the case of unbroken wave forces on piles, an equation proposed by Morison *et al.* (1950) may be used. This equation is empirical and describes the contribution of two separate components. These are a drag force induced by flow separation around the pile and an inertial force due to the accelerating flow associated with the passage of a wave. For a vertical pile, Morison's equation applies, as only the horizontal flow creates a force. From linear wave theory it is noted that these forces are 90° out of phase, so the maximum total force does not occur at the peak (or trough) of the wave, as illustrated in Figure 12.8. As the velocity varies with depth, so too does the force. The total force acting on the pile is found by integrating over the length of the pile. Accurate values of C_D and C_I are difficult to establish from field measurements, but recommended values are quoted in the Shore Protection Manual (1984) and BS6349 (1984). (C_D is a function of the Reynolds number, varying between about 1.2 and 0.7 as the Reynolds number increases. C_I has a relatively constant value of about 2.0.)

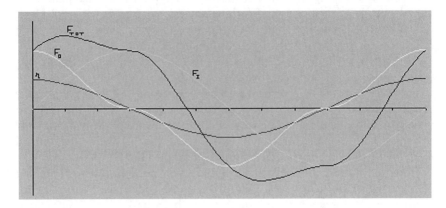

Figure 12.8 Time plot showing the variation in the sea surface (η), drag force (F_D), inertial force (F_I) and total force (F_{TOT}) over a wave period for the case when C_D and C_I are equal

12.2.7 Sediment transport

The equation for sediment transport given in Chapter 11 (equation (11.58)) is for bedload transport of non-cohesive material. In highly turbulent flow a significant proportion of the sediment transport is in suspension. In practice it is often impossible to make a clear distinction between suspended load and bedload transport. As a result, some success can be found by using one of the so-called 'total load formulae' that account for both types of transport (see also Section 4.6.3). Engelund and Hansen (1972) proposed such a formula for q, the sediment volume per metre per second, m²/s:

$$q = \frac{0.04 C_D^{1.5}}{g^2 \left(\dfrac{\rho_w - \rho_s}{\rho_w} \right)^2 d} U^5 \tag{12.19}$$

where C_D is a drag coefficient, d is the grain size and U is the depth-averaged velocity. For $U > 1\,\text{m/s}$ the transport rate rises rapidly. Various sediment-transport formulae for currents and waves and for oscillating currents have been proposed. The details of this transport process are quite complex, and the interested reader is referred to the monograph by Van Rijn (1993).

12.3 Computational modelling

Until fairly recently, simulating wave overtopping using the Navier–Stokes equations was beyond both the computational power of available comput-ers and the numerical methods required to capture the intricacies of wave

breaking. Not only was it necessary to be able to describe wave overturning, but it was also necessary to describe the rapid energy dissipation caused
by the highly turbulent flow in broken waves. The development of the 'volume of fluid' method, described by Hirt and Nichols (1981), was a major
breakthrough in modelling highly distorted flows. The simulation of turbulence required the solution of 'turbulence equations' simultaneously with
an averaged form of the Navier–Stokes equations, the 'Reynolds averaged
Navier–Stokes equations', (RANS), (see also Section 4.3.3).

Briefly, for a turbulent flow, both the velocity field and the pressure field
can be split into mean component and turbulent fluctuations as follows:

$$u = \langle u_i \rangle + u_i' \tag{12.20}$$

$$p = \langle p \rangle + p_i' \tag{12.21}$$

The mean flow is governed by the RANS equations as follows:

$$\frac{\partial \langle u_i \rangle}{\partial x_i} = 0, \tag{12.22}$$

$$\frac{\partial \langle u_i \rangle}{\partial t} + \langle u_j \rangle \frac{\partial \langle u_i \rangle}{\partial x_j} = -\frac{1}{\langle \rho \rangle} \frac{\partial \langle p \rangle}{\partial x_i} + g_i + \frac{1}{\langle \rho \rangle} \frac{\partial \langle \tau_{ij} \rangle}{\partial x_j} - \frac{\partial \langle u_i' u_j' \rangle}{\partial x_j} \tag{12.23}$$

in which $\langle \ \rangle$ denotes the mean quantities, the prime represents the turbulent fluctuations, u_i denotes the ith component of the velocity vector, p is
the pressure, ρ is the density, g_i is the ith component of the gravitational
acceleration, and τ_{ij}^m is the molecular viscous stress tensor. The product of
the density and the correlation of velocity fluctuation, $\rho \langle u_i' u_j' \rangle$, is called the
Reynolds stress. The correlation is modelled by a non-linear eddy viscos-
ity model (modified k–ε equations, where k is the turbulent kinetic energy
and e is the turbulent dissipation rate). More details of the mathematical
formulation can be found in Lin (1998), Lin and Liu (1998) and Liu *et al.*
(1999).

Figure 12.9 shows an example of the type of output that can be obtained
with this type of simulation. This approach has been used to simulate wave
overtopping of seawalls and good agreement with experimental results has
been obtained (see e.g. Soliman and Reeve (2003)).

Simulation using the RANS equations is still very computationally expensive, typically requiring several hours of computer time (on a desktop PC)
to simulate a few minutes of real time. For this reason, simpler forms of
equation to predict wave propagation are valuable.

The type and sophistication of numerical wave-transformation model
employed in the design of coastal structures has been, and continues to be, a
function of many factors. These include: the available models; their ease and
practicality of use; the models' computational requirements; and the nature

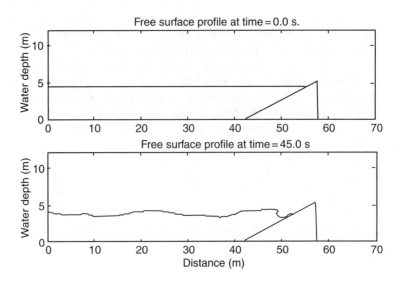

Figure 12.9 Output from the RANS model showing the initial condition and waves running up a sloping seawall after 45 s

of the engineering study (e.g. conceptual, preliminary or detailed design). Broadly speaking, ray models are often used in situations where diffraction is not significant (but consideration of spectral behaviour may be). For investigations of wave penetration around breakwaters and into harbours, models based on a wave-function description are required. For situations where wave–structure interactions are important, models based on non-linear Boussinesq or shallow-water equations have been employed. Only in the last few years have fully three-dimensional numerical wave models been developed for application to coastal engineering problems (Li and Fleming (2000)). Some of these models that are in widespread use are discussed briefly below before describing some test cases.

12.3.1   Ray tracing

During the 1970s, numerical ray models were developed (e.g. Abernethy and Gilbert (1975)) that allowed the transformation of deep-water wave spectra to inshore locations, accounting for refraction and shoaling. These models relied on the principle of linear superposition and worked as follows. The offshore wave spectrum was discretized in both direction and frequency. A refraction and shoaling analysis was performed for each direction–frequency combination, and the resulting inshore energies were summed to assemble an inshore directional spectrum. From this, the standard wave parameters such as H_s and T_m could be computed. In short, the inshore spectrum was computed from

$$S(f, \theta_i) = E(f, \theta_0)K_R^2(f, \theta_0)K_S^2(f) \qquad (12.24)$$

where the subscripts 0 and i refer to offshore and inshore, respectively. The refraction coefficient was determined from a numerical ray tracing over a digital representation of the seabed.

With the development of the TMA shallow-water spectrum (Bouws *et al.* (1985)), which provided an upper bound on the energy content of the frequency spectrum, ray-tracing models could be extended to incorporate wave braking and other surf-zone processes in an empirical manner. This was done by reducing the energy content of the computed inshore frequency spectrum to the value predicted by the TMA spectrum.

Such models could provide a spectral description of the nearshore wave climate at a point and account for refraction and shoaling, together with an empirical treatment of wave breaking. However, they could not account for diffraction, and were limited to describing conditions at a selected position. Nevertheless, ray models remain in current use because of their very modest computational requirements and because they can provide a spectral description of nearshore wave conditions. Figure 12.10 shows an example of this type of model. Figure 12.10(a) shows the location of Marsaxlokk Bay (southeast end of Malta) together with a contour plot of the bathymetry around the bay. Note the submerged spur protruding seaward from the western side of the bay.

Figure 12.10(b) shows the results of some ray-tracing calculations. In the top diagram, forward tracing is shown. This involves computing the ray path of a wavefront from offshore towards the coast. Rays start at a selected number of points along the wavefront. As they pass over (slightly) different seabed elevations, the rays gradually diverge or converge. Indeed, in this case, the submerged spur acts to focus wave energy in the entrance to the bay.

12.3.2 Mild slope equation

As computer hardware has gained in power and our understanding of wave processes has increased, so the sophistication of numerical wave-transformation models has developed. A significant step in this development was the introduction of the mild-slope equation by Berkoff (1972). The mild-slope equation is derived from the linearized governing equations of irrotational flow in three dimensions under the assumption that the bottom varies slowly over the scale of a wavelength. The mild-slope equation has been used widely to date to predict wave properties in coastal regions. The equation, which can deal with generally complex wave fields

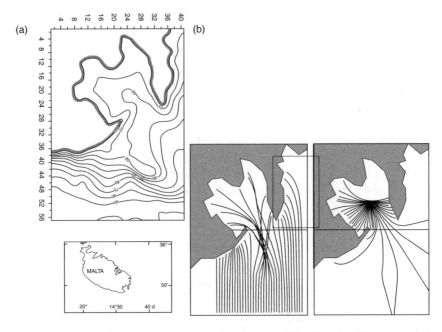

Figure 12.10 (a) Location map of Marsaxlokk Bay, Malta and bathymetry of the bay. (b) Ray-tracing calculations illustrating forward tracking (parallel rays propagated forward from offshore) and backward tracking (rays traced from a single inshore point back out to sea) (Al-Mashouk *et al.* (1992))

with satisfactory accuracy, accounts for refraction, shoaling and diffraction (and, in some forms, reflection as well). The mild-slope equation may be written as

$$\nabla.(cc_g\nabla\Phi) + \frac{\omega^2 c_g \Phi}{c} = 0 \qquad (12.25)$$

for the complex two-dimensional potential function Φ. In a three-dimensional Cartesian coordinate system, Φ is related to the water wave velocity potential of linear periodic waves $\Xi(x, y, z, t)$ by

$$\Phi(x, y) = \Xi(x, y, z, t)\frac{\cos h(\kappa h)}{\cos h\left(\kappa(h + z)\right)}e^{-i\omega t} \qquad (12.26)$$

where the frequency ω is a function of the wavenumber $\underline{k} = (k, l)$ with $\kappa = |\underline{k}|$ by virtue of the dispersion relationship

$$\omega^2 = g\kappa \tan h(\kappa h)$$

which is the three-dimensional version of equation (12.9).

The local water depth is $h(x, y)$, the local phase speed $c = \omega/\kappa$ and the local group velocity $c_g = (\partial\omega/\partial k, \partial\omega/\partial l)$. Writing $\psi = \Phi\sqrt{(cc_g)}$ allows the mild-slope equation to be cast into the form of a Helmholtz equation. Under the assumptions of slowly varying depth and small bottom slope, Radder (1979) showed that the equation for ψ may be approximated as the following elliptical equation:

$$\nabla^2\psi + \kappa^2\psi = 0 \tag{12.27}$$

Several numerical models are available that solve the elliptical form of the mild-slope equation by means of finite elements (e.g. Liu and Tsay (1984)). However, a finite-differences discretization is generally easier to implement. This approach produces reasonably good results, provided that a minimum of 8–10 grid nodes is used per wavelength. This requirement precluded the application of this equation from the modelling of large coastal areas (i.e. with dimensions greater than a few wavelengths) due to the high computational cost. As a result, a number of authors have proposed models based on different forms of the original equation.

Copeland (1985) has transformed the equation into a hyperbolic form. This class of model is based on the solution to a time-dependent form of the mild-slope equation and involves the simultaneous solution of a set of first-order partial differential equations. In practical applications, numerical convergence can be difficult to achieve with this approach. An alternative simplification was proposed by Radder (1979). This involved a parabolic approximation that relied on there being only small variations in wave direction. The advantage of such an approach is that a very computationally efficient time-stepping approach can be adopted, and this allows solutions to be obtained over much larger areas. The disadvantages include an inability to deal with reflections and neglect of diffraction effects in the direction of wave propagation.

More recently, computationally efficient and stable solution procedures for the elliptical form of the mild-slope equation have been developed (e.g. Li and Anastasiou (1992), Li (1994)). These later developments have obviated the need to make approximations regarding wave angles, and, as a result, models based on the parabolic and hyperbolic forms of the equation have fallen from favour. These elliptical models have been extended to account for irregular waves (i.e. a wave spectrum) by Al-Mashouk et al. (1992) and Li et al. (1993) using the model to compute solutions for individual direction–frequency pairs. The results are then combined, following Goda (2000), as a weighted integral to provide a combined refraction–diffraction–shoaling coefficient. This can then be used to derive the wave spectrum at any grid node, given the offshore wave spectrum.

Figure 12.11 shows the output from a solution of the elliptical mild-slope equation run for a spectrum of waves with a relatively narrow directional

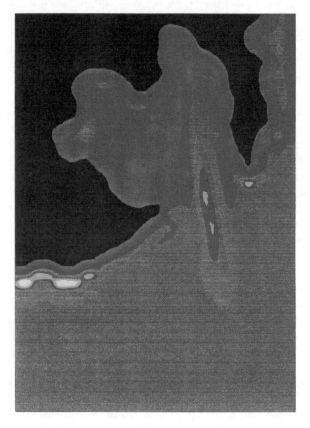

Figure 12.11 Mild-slope computation of spectral wave conditions in Marsaxlokk Bay (Al-Mashouk *et al.* (1992))

spread corresponding to a swell-type condition. The case is for the same example shown in Figure 12.10, Marsaxlokk Bay in the Mediterranean Sea. The focusing effect of the submerged spur is perhaps more clearly evident. More importantly, by including diffraction effects, the formation of cusps that can occur in ray models no longer occurs, with the attendant problem of dealing with infinite wave amplitudes.

Another example of the power of the mild-slope equation for port and harbour modelling is shown in Figure 12.12. In the 1990s, the author-ities at the Port of Sohar on the east coast of Oman were considering various options for extending the number and size of berths in the port. Figure 12.12(a) shows the bathymetry around the main port, one option for deepening and extending the berths within the harbour, and the exist-ing fishing harbour to the northwest of the main port. In this elliptical

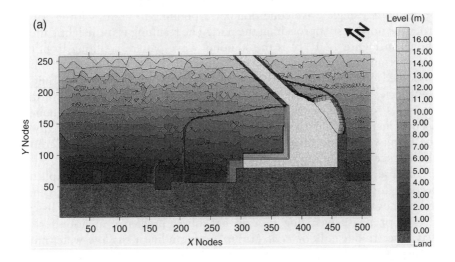

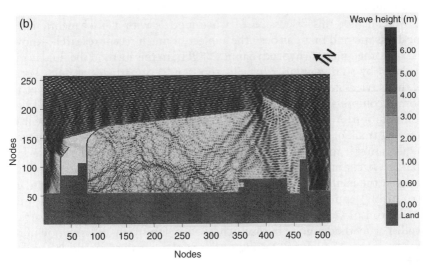

Figure 12.12 (a) Bathymetry for one of the options for an extension of the Port of Sohar. (b) Wave heights, computed from the mild-slope model, driven by storm waves (courtesy of Halcrow)

model wave reflection is included so that reflections from the port break-waters and quays can be incorporated. Figure 12.12(b) shows the results of a computation for a slightly different port layout, in which the fishing har-bour is extended and protected by an extension of the existing port break-water. Furthermore, there is relatively little overlap between the two arms of the main port breakwaters. The breakwaters have a reflection coefficient of 0.6 and the port quays have a reflection coefficient of 0.8. The model is set up with a single wave condition corresponding to waves with $H_s \approx 6\,\mathrm{m}$ from the northeast approaching the port. The resulting wave-height pattern shown in Figure 12.12(b) is the standing-wave pattern resulting from the complex interaction of incoming and reflected waves; this would be much smoothed out because, in practice, the waves would have a spectrum of frequencies and directions. However, if the design were to proceed, further measures to reduce the wave penetration into the port would be required, such as increasing the breakwater overlap and decreasing the reflectivity of the breakwaters and quays. Note also that the extended fishing harbour shows excellent conditions in this case.

The disadvantage of linear mild-slope models is that they do not explicitly account for non-linear processes such as wave breaking, harmonic generation or wave–wave interaction. Boussinesq models (e.g. Beji and Battjes (1994)) or 'phase-averaging' models that solve a predictive equation for the wave spectrum (e.g. Booij et al. (1996)) and non-linear shallow-water models (e.g. Dodd (1998), Hu et al. (2000)) have been developed to simulate wave run-up and overtopping of beaches and seawalls.

We conclude this discussion of wave models with a brief mention of a meshless method that has been the subject of much recent research. Known as 'smoothed particle hydrodynamics' (SPH), in this method the fluid is represented by a large number (usually thousands) of particles of fluid. Forms of the Navier–Stokes equations that govern these particles can be derived for both incompressible and compressible flow. Each particle is tracked in the computation. A significant amount of the computational effort is spent on keeping track of the particles. However, one distinct advantage of the technique is that it appears to be able to capture the distortion of the free surface that can occur in wave breaking and impacts with structures. Some applications of the method to coastal engineering are presented in Monaghan and Kos (1999) and Shao et al. (2006). Figure 12.13 shows an example SPH simulation of waves running up a beach and hitting the base of a vertical wall. The method is good in cases where there is a strong deformation of the free surface, although full testing of the method against careful laboratory experiments is lacking.

12.3.3 Sediment-transport modelling

There are two main schools of research in respect of sediment-transport modelling. The first, known as the 'bottom-up' approach, tries to solve the hydrodynamic equations and the sediment-transport equation simultaneously, updating the seabed depth as the computation proceeds. This approach works reasonably well when applied to sediment transport due to tides. When surface waves are considered, the time scales become much shorter and the whole computational strategy becomes extremely

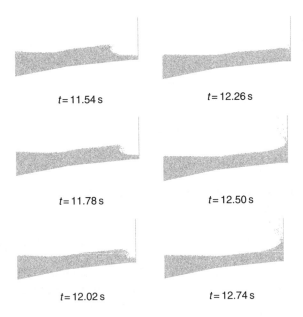

$t = 11.54\,s$ $t = 12.26\,s$

$t = 11.78\,s$ $t = 12.50\,s$

$t = 12.02\,s$ $t = 12.74\,s$

Figure 12.13 SPH calculation of a wave breaking against a vertical wall

computationally expensive. Also, due to the uncertainties in many of the sediment-transport formulae, as well as the numerical inaccuracies that can accumulate because many small time steps are taken, the results obtained using this approach can be extremely sensitive to the initial conditions and assumptions made. The advantage of this approach is that of determinism; if the equations that are being solved include all the necessary physical processes, then the correct solution will be obtained as long as the computational procedure is sufficiently accurate and the model parameters are specified appropriately.

The second approach, known as 'top-down', dispenses with the detailed hydrodynamic and sediment-transport equations and instead uses equations that describe a small subset of what is assumed to be the physical processes important for describing the evolution of the coastal morphology. Typically, such models have a small number of governing equations that are amenable to efficient numerical solution, and so can be run to simulate long periods of time (e.g. decades), which are of most interest to designers and coastal managers. The advantages of this approach are: solutions can be obtained very quickly; repeated calculations with slightly varying conditions can be performed to investigate sensitivity to particular parameters; and the simplified physics means that the predicted morphological responses can be understood in a relatively straightforward manner. The disadvantage is that,

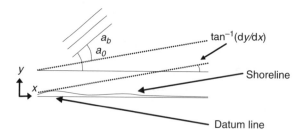

Figure 12.14 Schematic of the one-line model

without a long period of observations against which to validate the model, it is not possible to be sure that all the necessary physical processes have been included.

We conclude this section with a brief description of a top-down model that has been used widely for predicting the response of beaches to waves. It is known as the 'one-line' model and stems from work undertaken by the French engineer Pelnard-Considère (1956). The primary assumption is that the beach profile (cross-section) remains unaltered but can move seaward or landward depending on the net alongshore sediment transport (Figure 12.14). The alongshore transport of sediment is considered to be driven by the action of waves alone.

The continuity of sediment is expressed as

$$\frac{\partial y}{\partial t} = -\frac{1}{D_c}\frac{\partial Q}{\partial x} + q \tag{12.28}$$

where y is the position of the shoreline from a fixed datum line, usually taken to be the x-axis, which runs parallel to the shoreline trend so that x is the alongshore distance, t is the time, Q is the longshore particulate sediment-transport rate, D_c is the depth of closure (i.e. the depth below which no appreciable profile change takes place), and q denotes line sources/sinks of sediment along the shoreline. To solve equation (12.28) an expression for Q is needed. One such expression is the CERC formula, the general form of which may be written as

$$Q = Q_0 \sin\left\{2\left[\alpha_0 - \arctan\left(\frac{\partial y}{\partial x}\right)\right]\right\} \tag{12.29}$$

with

$$Q_0 = K\frac{\rho}{16(\rho_s - \rho)\sigma}H_b^2 c_{gb} \tag{12.30}$$

where Q_0 (m^3/s) is the amplitude of the longshore sand-transport rate, α_0 is the angle between the breaking wave crests and the x-axis, $\partial y / \partial x$ is the local shoreline orientation, K is a proportionality coefficient, H is the wave height, c_g is the wave-group velocity, ρ is the seawater density, ρ_s is the sediment density, and σ is the sediment porosity. The subscript b denotes quantities at breaking. Equations (12.29) and (12.30) are the elementary equations of a one-line model.

Analytical solutions can be derived on the basis of assuming that wave crests approach the shore at small angles from shore parallel. This reduces the continuity of sediment equation to a diffusion-type equation; analytical solutions to the latter are common, and have been derived in a number of studies for different cases of shoreline change using simple wave-driven, sediment-transport models (e.g. Grijm (1961), Larson *et al.* (1997), Le Méhauté and Soldate (1977), Pelnard-Considère (1956), Wind (1990)). Apart from the assumptions of a small local shoreline orientation (i.e. of a smooth shoreline) and of a small angle of wave approach, analytical solutions are also limited by the assumption that waves are constant in time and in space (i.e. the diffusion coefficient is constant). This constraint was addressed to some extent by Larson *et al.* (1997), who allowed for a sinusoidally time-varying breaking wave angle at a single groyne and at a groyne compartment. Dean and Dalrymple (2002) discuss time-varying wave conditions in the context of the longevity of beach nourishment on an initially straight shoreline. They describe a solution technique based on a Fourier decomposition in terms of the longshore dependence, and present a solution for an individual Fourier component. Reeve (2006) used a formal Fourier cosine transform to develop a new closed-form solution for the case of a single groyne, in which the wave conditions could be specified as a time series. This approach has since been extended by Zacharioudaki and Reeve (2008) to solutions for a free shoreline and a groyne compartment.

Numerical solutions for the one-line model are, by their nature, more flexible, and can include nearshore wave transformation, diffraction effects, and so on. Gravens *et al.* (1991) and Hanson and Kraus (1989) describe one such modelling system.

12.4 Physical modelling

12.4.1 General

Physical models of coastal engineering structures and processes have been in operation at many major hydraulic laboratories since about 1930. However, only fairly recently has modern technology in wave generators enabled a more correct modelling of the sea state and the resultant impact on coastal structures. The increasing threat of coastal erosion and the development of ever larger offshore structures have given a new impetus for research in this area.

The necessary background to physical modelling has been covered in preceding sections of this chapter and in previous chapters. The reader should, in particular, refer to Sections 4.5 and 12.2 for the hydraulics background and to Sections 5.4–5.7 for the development of the main laws of similarity, their limits and the scale effects. Section 6.1.2 gives some detail of relevant equipment used in coastal engineering models (wave flumes and basins, tide generators, and wave generators reproducing a range of wave spectra and directions) and Section 6.3 deals with the associated measuring methods and instrumentation. Section 7.5 gives details of modelling friction losses in fixed-bed models and Section 8.5 deals with the physical modelling of movable beds.

Coastal models, including models of harbour engineering, are invariably three-dimensional. Models of parts of offshore structures or of forces acting on coastal-protection elements, including wave breaking and overtopping, may be placed in wave flumes, and can also be combined with wind simulation. Because most coastal and offshore models are undistorted or, at best, only slightly distorted (see below), their space requirement is large and the establishment of their acceptable boundaries is important.

12.4.2 Scaling laws and scale effects

Although the predominance of gravity and inertial forces indicates that models have to be operated according to the Froude law, Reynolds, Weber and Mach (Cauchy) numbers and associated similarity laws and scale effects must also be considered. Only the basic approach to scaling procedures can be outlined here; for further discussion, see, for example, Kolhase and Dette (1980), Novak and Čábelka (1981), Oumeraci (1984) and the associated references therein.

Inspection of equations (4.89), (4.90) and (12.9) indicates that the scale of the wave celerity M_c is given by

$$M_c = M_h^{1/2} \tag{12.31}$$

for a shallow-water wave (consistent with the Froude law) and

$$M_c = M_L^{1/2} \tag{12.32}$$

for a deep-water wave.

For intermediate waves it follows from equation (4.84) that

$$M_c = M_L^{1/2} = M_h^{1/2} \qquad (12.33)$$

Also, from equation (4.82),

$$M_c = M_L \, M_T^{-1} \qquad (12.34)$$

Thus, equations (12.33) and (12.34) do not, in principle, exclude distorted models, as the horizontal scale M_l can be chosen independently (of the wavelength scale); however, in any case, such distortion should be small (in contrast to estuarine models – see Section 11.6). For reproduction of wave refraction, equations (12.31) and (12.34) are sufficient and a distorted model is permissible; this applies also to movable-bed wave models. However, for reproduction of wave diffraction, the wave height at any point along the obstacle must be reproduced correctly and the scale of the wavelength must be equal to the horizontal scale of the model ($M_L = M_l$); thus, an undistorted model is required if scale effects are to be avoided. For reproduction of wave breaking, overtopping and wave-impact forces acting on obstacles and coastal protection works, undistorted models are also generally required.

In Section 4.5 it was shown (equations (4.87) and (4.88)) that for the effects of surface tension to become negligible a minimum wave celerity $c = 0.23\,\text{m/s}$ and minimum (gravity) wavelength $L = 0.017\,\text{m}$ are required; usually substantially larger values are considered when choosing model scales.

Surface tension can also affect models of breaking waves and the simulation of air entrainment, which plays an important part in the turbulence-induced mixing by breaking waves; sufficiently large scales are thus required to avoid surface tension and viscosity effects. A critical Reynolds number of 3×10^4 is often quoted as necessary to maintain turbulent flow in a protective armour layer. Special materials (epoxy resin, plastics, porcelain) may also be required for model units to reduce their mutual friction, as traditional materials (mortar) may overestimate their stability.

As in very shallow water the effect of friction becomes important, similarity dictates large enough Reynolds numbers ($\text{Re} < \text{Re}_{sq}$), or additional roughness may have to be used.

In general, in three-dimensional models the depth should not be less than $0.05\,\text{m}$ and the wave height not less than $0.02\,\text{m}$ in order to avoid wave attenuation by surface tension. Also, a minimum depth of $0.02\,\text{m}$ is usually specified to avoid viscous effects.

For the reproduction of the elasticity of a structure subjected to wave impact, the Mach (Cauchy) numbers and similarity laws have to be considered (see Sections 13.2 and 13.3). It is generally accepted that in order not to distort unduly the pressure loading on a structure a wave height of 0.3 m is required. The dependence of the C_D coefficient in the Morison equation (equation (12.19)) on the Reynolds number has also to be taken into account.

Based on the above considerations, a model scale in the region of $M_l = 20$ is usual for studies of the stability of breakwaters, $M_l = 50$ for models of structures and wave reflection, and $M_l = 100$ for harbour models; all these values are approximate averages and can be exceeded in either direction depending on the specific conditions.

Figure 12.15 shows an example of a harbour model with a breakwater $M_l = 100$ and Figure 12.16 shows an example of a movable-bed coastal model with groynes.

Figure 12.15 Model of a harbour with a breakwater and wave diffraction (courtesy of HR Wallingford)

Figure 12.16 Movable-bed model with groynes (courtesy of HR Wallingford)

12.5 Practical modelling aspects and case studies

12.5.1 Feasibility study for a port at Vadhavan, India

In the late 1990s, proposals were put forward for the development of a port at Vadhavan in Maharashtra State, India. The primary purpose of the port was to provide berthing, servicing and maintenance for cargo ships carrying minerals for import, processing and export. Figure 12.17 shows the main ports in India and the location of Vadhavan.

In its 1996 Infrastructure Report, the Indian Government Ministry of Surface Transport recommended the adoption of privatization schemes based on the Build–operate–transfer (BOT) approach. BOT project proposals for the development of new berths and/or terminals envisaged a maximum contract duration of 30 years, including the construction period, the assets being transferred back to the port without costs at the end of the lease or license period. In the State of Maharashtra, P&O was awarded a BOT contract to develop the port of Vadhavan. At its final stage the port was envisaged to consist of 29 berths able to handle up to 250 million tonnes/year of cargo.

Prior to any construction, extensive studies were required, not just to determine the port layout, size, orientation and construction quantities, but also for financial forecasting and environmental-impact assessment. The area around Vadhavan is known within India as the 'lungs of Bombay'. It also provides breeding grounds for both fish and endangered Olive Ridley

Figure 12.17 Map of India showing the main ports of India and the proposed port at
Vadhavan

turtles. In the following the focus is on the work undertaken as part of the
studies investigating the potential port layout.

Figure 12.18 shows some of the port layout options that were considered.
The port was conceived as being positioned some distance offshore from
the mainland, connected by a breakwater/suspended link. This reduced the
amount of dredging and blasting that would be required to create sufficient depth for the cargo ships. A preliminary part of the wave-modelling
work was an analysis of offshore wave conditions to determine extreme
wave heights for 30° direction sectors. From Figure 12.17 it is clear that
the largest fetch is to the west and southwest, and hence the options all
provide protection from these directions. It might be thought that waves
from the northwest would be relatively small due to the small fetch, and
indeed they are. However, initial assumptions that wave action from this

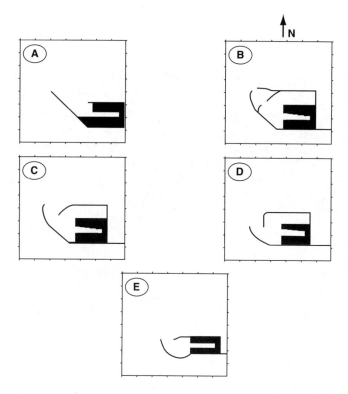

Figure 12.18 Some options for the port layout (courtesy of Halcrow)

direction could be disregarded (as in Option A in Figure 12.18) were soon corrected by numerical wave simulation using a mild-slope model. Storm waves from the northwest were of the order of 2 m. The port developers sought to have a maximum wave height of 0.6 m during such conditions so that port operations were interrupted as little as possible.

Option A was suggested as an easy-to-construct configuration, with straight-line breakwaters and quays. Options B, C and D provided more protection from waves from the northwest, with overlapping breakwater arms to trigger diffraction and spreading and dispersion of wave energy, but the length of the breakwater to the port area would make these options expensive. They also included an additional area for future expansion of the port, as well as a berthing area with non-parallel sides to avoid constructive interference from reflections. Option E has hooked, curved breakwaters and a large turning area in the lee of the northern breakwater.

An example of the wave modelling is shown in Figure 12.19. This shows waves approaching Option A from the northwest. The breakwater arms

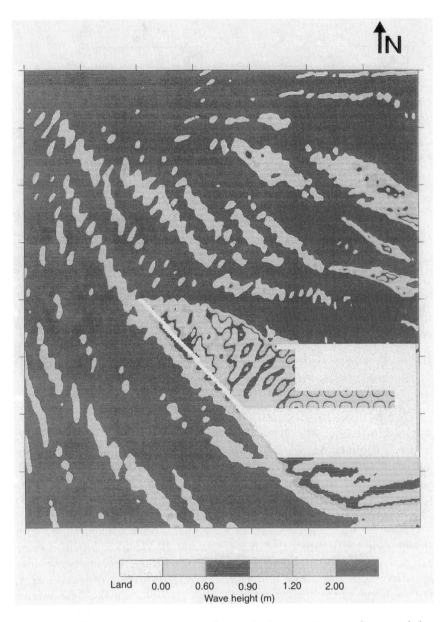

Figure 12.19 Computed wave heights in Option A, demonstrating standing-wave behaviour in the berthing areas (courtesy of Halcrow)

have a reflection coefficient of 0.5 and the quays a coefficient of 0.8. The long, straight breakwater acts to reflect waves in towards the berthing area. The rectangular berthing area acts as a resonant chamber, and a large standing-wave pattern is evident. This type of wave behaviour would be very damaging to moored ships, and this option was discarded.

In contrast, Option E performed much more satisfactorily, as shown in Figure 12.20. Wave diffraction and some interference with waves reflected off the inside of the southern breakwater is evident, but conditions in the berthing area remain quiescent.

Option E provided a good solution for reducing the wave penetration. However, consideration also needed to be given to its flushing by tidal flows and scour or siltation effects. For this a depth-averaged tidal model of the

Figure 12.20 Computed wave heights in Option E demonstrating much better wave-energy absorption for the berthing area. The orientation and shading are the same as in Figure 12.19 (courtesy of Halcrow)

area was set up to investigate the performance of the different options. The tidal flow is reasonably strong, and flushing of the port occurred in an acceptable time. The tidal model was also used to compute potential sediment movements over a tidal cycle. This was done by using a combined current and wave sediment-transport formula with results from the wave model and the tide model to solve a sediment-transport equation over time. Some local erosion and accretion were predicted but nothing so severe that it would undermine the breakwater foundations or silt up the port entrance.

The proposal to construct the port was strongly opposed by the people of Dahanu, an area which is situated near Vadhavan, and which is one of the last green belts along India's rapidly industrializing western coast. Environmental groups argued that the port project would destroy the 'ecologically fragile' region and seriously affect the livelihood of Dahanu's 300,000 inhabitants, who are mainly tribal and fisherfolk. In their argument they pointed out that Dahanu was 'notified', or classified, under the Indian Coastal Regulation Zone (CRZ) by the Federal Ministry of Environment and Forests in 1991. The Notification restricts industrial development and prohibits a change of land use in environmentally sensitive areas. Following the letting of the BOT contract, in 1996 the people of Dahanu took the Maharashtra government to court for failing to implement the Notification. After an extended legal battle, the Supreme Court of India upheld the Dahanu Notification, prohibiting any change of land use in the region. The Supreme Court also appointed the Dahanu Taluka Environment Protection Authority (DTEPA) to ensure that the Notification was implemented, and Dahanu remains a protected region. Along with the Notification, Dahanu's coasts were classified under the most stringent clause of the Coastal Regulation Zone (CRZ) Notification [CRZ I (i)], 1991, which prohibited any development within 500 m of the high-tide line. The DTEPA held a series of hearings with activists and the local communities, passing a landmark order in 1998 that the port could not be permitted in Dahanu.

The environmental regime, along with civil action, prevented the construction of the port. The proposed port development would have altered the coastal environment and, arguably, damaged its communities. A decade after the court proceedings, the Dahanu coastal environment remains protected but the residents of many fishing villages are struggling to live off the natural resources.

12.5.2 Happisburgh to Winterton sea defence scheme, UK

Shore-parallel breakwaters can provide an effective solution for coastline protection, together with substantial recreational development and low environmental impact. They have been successfully used to control shoreline evolution on many coastlines of the world, particularly in areas with a small tidal range. The village of Sea Palling, located on the north Norfolk

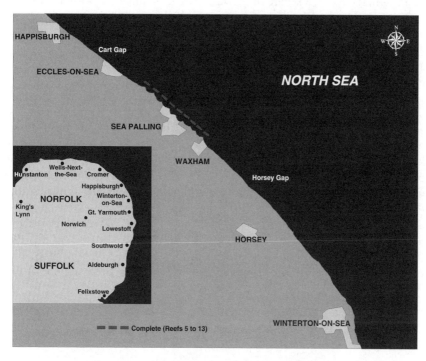

Figure 12.21 Location map of the Happisburgh to Winterton scheme (Reeve (2009))

coast, is a site where a shore-parallel breakwater scheme was implemented between 1993 and 1997. The shore-parallel breakwater scheme provides protection against sea flooding, not only to the village but also to a large part of the hinterland, including the Norfolk Broads (Figure 12.21). The tides at the site have a typical mean spring tide range of 3.2 m and mean neap tide range of 1.58 m. The tidal regime is semi-diurnal, with two high tides each day. The tidal range is from +1.85 m Ordnance Datum New-lyn (ODN) at MHWS to −1.35 ODN at MLWS. The MSL at the site is 0.24 m.

The shoreline at Sea Palling faces toward the northeast. As a result, the coastline is exposed to a wide range of wave directions, ranging from north-northwest to southeast. Historic records show that this area of coastline has suffered a series of major flood events over the course of several centuries. For example, records show that in 1287 nearly 200 people were drowned in extensive flooding. In 1604 a flood inundated some 800 ha of land, destroying over 60 houses and badly damaging the town church. In 1938 the whole village was cut off, and an area of 3,000 ha was flooded, with consequent disruption to the community and damage to agricultural land. On the night of 31 January 1953, a severe northwesterly gale, coincident with a period of spring tides, produced a large tidal surge, which raised the sea level to 2.4 m above normal high tide levels. The sand dunes at Sea Palling were washed away, causing extensive damage to houses and seven deaths. The defences at Horsey and Eccles were also damaged, and an area of almost 500 ha was inundated.

The existing seawall and dunes protect a large area of low-lying hinterland from flooding during storm events. A significant proportion of the coastline is of national importance for its landscape and has been designated an Area of Outstanding Natural Beauty. Inland, the freshwater lakes known as the Norfolk Broads (that developed as a result of the extraction of peat from the area dating back to the 12th century) are recognized for their important landscape, historical interest and wildlife, receiving similar status to a National Park. A breach in the defences would cause extensive damage to properties, agricultural land and these sites of nature conservation importance.

Figure 12.22 is an aerial photograph of the coastline taken in 2008. The breakwater scheme is in the distance. Also evident is the decaying line of old

Figure 12.22 Aerial photograph of the north Norfolk coastline, looking southeastwards. The village of Happisburgh and its church are in the foreground, the Happisburgh lighthouse is to the right mid-ground and the breakwaters are top mid-picture (courtesy of Mike Page)

defences and groynes. At locations where these have been destroyed, rapid erosion of the cliffs can be seen. By the late 1980s the beach in front of the seawalls at Sea Palling had lowered to a critical level. The construction of nine offshore breakwaters (Reefs 5 to 13) was approved by the government. The breakwater scheme was constructed in two stages: Reefs 5 to 8 in Stage 1 (1993–1995) and the remaining reefs in Stage 2 (1997). In addition, beach recharge was undertaken during Stage 2. The reefs constructed in Stage 2 used a modified version of the Stage 1 reef design, which had lower crest levels, were shorter in length and had more closely spaced reefs. A subsequent Stage 3 (2002–2004) was undertaken, which consisted of beach recharge, the construction of a rock revetment south of Reef 13 and improvement of several rock groynes. Reefs 1–4 were reserved for a subsequent stage, and have yet to be constructed; thus, the northernmost reef is Reef 5. The beach started to evolve in response to the initial stages of construction. Once Stage 1 was completed, tidal tombolos (features that are salients at high tides and tombolos at low tide) developed. Landward of the five reefs built during Stage 2, a sinuous shoreline has developed, which allows flow around the breakwater system at all states of the tides. In association with periodic nourishment programmes, the scheme has been recognized as a successful one (Fleming and Hamer (2000)), although a full understanding of the sediment movements, the driving forces and their inherent variability has yet to be gained.

Modelling a site such as this is extremely challenging. For the purpose of this particular case the focus was on the beach shape in and around the structures. A one-line model was chosen for this purpose. To drive the model, offshore wave records were obtained from the UK Meteorological Office, covering the period from 31 December 1994 to 1 January 2008. The records consisted of time series of significant wave height, mean wave period and mean wave direction. Over the period, significant wave heights varied from 0.1 m to 3.9 m, wave periods lay between 1 s and 10 s and wave directions showed a broad spread between north and southeast. At this site the effect of the breakwaters is crucial in determining the beach response. The breakwaters alter the distribution of wave energy and direction in the surf zone, so the impact of energy redistribution has to be captured accurately in order to simulate the morphological evolution. Figure 12.23(a) illustrates the wave diffraction around two adjacent breakwater tips, and Figure 12.23(b) illustrates the type of wave pattern that can occur across the whole scheme.

To achieve this, an elliptical mild-slope wave model (described by Li (1994)) was linked with the one-line model to simulate the complex wave transformations around the offshore breakwaters. In addition to wave-induced currents, there are strong tidal currents. For long-term simulations it is not practical to include a detailed tidal flow process in a morphological model. Thus, prior to the shoreline simulation, tidal flows were simulated using a tidal model. The effect of these tidal currents was then included in the longshore sediment transport using the formula described by Hanson *et al.* (2006). The model was calibrated against historical observations of

(a)

(b)

Figure 12.23 (a) Wave diffraction around the tips of two adjacent breakwaters. (b) After completion (Stages 1 and 2), showing wave diffraction and interference patterns as well as the sinuous evolution of the beach (courtesy of Mike Page)

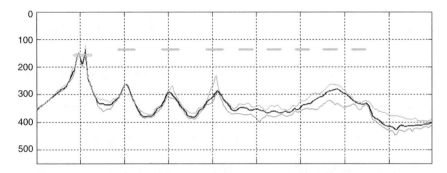

Figure 12.24 Average shoreline (heavy line) with extreme positive and negative excursions (light lines) over a 13-year period. Y-axis is position (in metres) from an offshore reference line. X-axis shows distance along the shore with gridlines every 500m

transport rates (i.e. the constant K in equation (12.30) was adjusted to get the best fit). Repeated simulations for the period 1994 to 2008 were performed with the calibrated model to build up a picture of the variability that might exist in the beach position within the scheme. The outcome is shown in Figure 12.24.

Figure 12.24 shows that the shoreline behind the Stage 1 breakwaters is relatively fixed, whereas in Stage 2 the shoreline shows a greater variability in position. The simulations included the effects of tide-level variation but not long-term sea-level changes due to climate change. Nevertheless, plots such as the one in Figure 12.24 can provide a useful tool for local engineers and planners when formulating coastal-management strategy.

12.6 Concluding remarks

The challenge of providing predictions of how the coastal and nearshore morphology will evolve in response to waves and tides is enormous. Not only are the flows highly turbulent but the process of sediment transport is still very poorly understood, particularly with respect to its aggregated effect, such as in changes in morphology, and when there are multiple grain sizes present. Computing power and numerical techniques are beginning to reach a stage at which relatively sophisticated numerical simulation can be performed over the short to medium term (i.e. days to months). Longer- term simulations are still very onerous in terms of computer time and also suffer from sensitivity to the specification of initial and boundary conditions. To perform simulations over periods of interest to coastal planners, such as a 70-year planning window, is not realistic, and a different approach is necessary. Simplified models, such as the one-line beach model, provide a robust if somewhat simplified means of predicting future changes. Recent research in this area has begun to see the development of 'systems models' that seek to capture the interactions between the elements of a coast–estuary–river system, rather than define equations that describe the detail of the physical processes (e.g. Karunarathna and Reeve (2008), Reeve and Karunarathna (2009), Van Goor *et al.* (2001)).

Our understanding of wave–structure interaction is also at a fairly early stage. Despite the successes of quantifying overtopping rates using empirical formulae, our ability to simulate the details of this is quite limited. New techniques, such as SPH, may provide the means to extend our modelling abilities. However, the role of air bubbles trapped in the water is now being recognized as an important factor in determining the magnitude of the impulsive wave forces that can be exerted on coastal structures. These are many times larger than previously thought, due to the compression that can take place in highly aerated water (Bullock *et al.* (2004)). This has important ramifications for the performance of jointed sea and harbour walls.

References

Abernethy, C. L. and Gilbert, G. (1975), *Refraction of Wave Spectra*, Report INT117, Hydraulics Research Station, Wallingford.

Airy, G. B. (1845), Tides and waves, Encyc. Metrop. 192, 241–396.

Abbott, M. B. and Price, W. A. (1994), Coastal, estuarial and harbour engineers' reference book, E & FN SPON, London, p. 736.

Allsop, N. W. H., McKenna, J. E., Vicinanza, D., Whittaker, T. T. J., (1996) New design methods for wave impact loadings on vertical breakwaters and seawalls. *25th International Conference on Coastal Engineering*, Sept 96, Florida, (Ed. Billy L. Edge, American Society of Civil Engineers), pp. 2508–2521.

Al-Mashouk, M., Reeve, D. E., Li, B. and Fleming, C. A. (1992), ARMADA: an efficient spectral wave model. In *Proceedings of 2nd International Conference on Hydraulic and Environmental Modelling of Coastal, Estuarine and River Waters*, Vol. 1, pp. 433–444, Ashgate, Bradford.

Beji, S. and Battjes, J. A. (1994), Numerical simulation of non-linear wave propagation over a bar, *Coast. Eng.*, 23, 1–16.

Berkoff, J. C. W. (1972), Computation of combined refraction and diffraction. In *Proceedings of the 13th JCCE*, pp. 941–954, American Society of Civil Engineers, Reston, Vancouver, VA.

Booij, N., Holthuijsen, L. H. and Ris, R. C. (1996), The SWAN wave model for shallow water. In *International Conference on Coastal Engineering*, pp. 668–676, American Society of Civil Engineers, Orlando, FL.

Bouws, E., Günther, H., Rosenthal, W. and Vincent, C. L., (1985), Similarity of the wind wave spectrum in finite depth water, *J. Geophys. Res.*, 90, pp. 975–986.

Bullock, G., Obhrai, C., Müller, G., Wolters, G., Peregrine, H. and Bredmose, H. (2004), Characteristic and design implications of breaking wave impacts. In *Proceedings of ICCE*, pp. 3966–3978, American Society of Civil Engineers, Reston, Lisbon, VA.

Coastal Engineering Manual (2009), US Army Corps of Engineers: http://chl.erdc.usace.army.mil/cem (accessed 24 April 2009).

Copeland, G. J. M. (1985), A practical alternative to the mild-slope wave equation, *Coast. Eng.*, 9, 125–149.

Dean, R. and Dalrymple, J. (1991), Water Wave Mechanics for Engineers and Scientists, *World Scientific*, Singapore.

Dean, R. G. and Dalrymple, R. A. (2002), *Coastal Processes: With Engineering Applications*, Cambridge University Press, Cambridge.

Dodd, N. (1998), A numerical model of wave run-up, overtopping and regeneration, *ASCE J. Waterways, Port, Coast. Ocean Eng.*, 124(2), 73–81.

Engelund, F. and Hansen, E. (1972), *A Monograph on Sediment Transport in Alluvial Streams*, 3rd Edn, Technical Press, Copenhagen.

Fleming, C. A. and Hamer, B. (2000), Successful implementation of an offshore reef scheme. In *Proceedings 27th ICCE*, pp. 1813–1820, ASCE, Sydney.

Goda, Y., (2000), Random Seas and Design of Maritime Structures, *Advanced Series on Ocean Engineering*, Vol. 15, *World Scientific*, Singapore.

Goda, Y. (2000), *Random Seas and Design of Maritime Structures*, University of Tokyo Press, Tokyo.

Gradshteyn, I. S. and Ryzhik, I. M. (1980), *Table of Integrals, Series and Products*, Academic Press, San Diego, CA.

Gravens, M. B., Kraus, N. C. and Hanson, H. (1991), *GENESIS: Generalized Model for Simulating Shoreline Change, Report 2, Workbook and System User's Manual*, US Army Corps of Engineers, Vicksburg, MS.

Grijm, W. (1961), Theoretical forms of shoreline. In *Proceedings of the 7th Conference on Coastal Engineering Conference*, pp. 219–235, American Society of Civil Engineers, New York.

Hanson, H. and Kraus, N. C. (1989), *GENESIS-Generalized Model for Simulating Shoreline Change*, Technical Report No. CERC-89-19, USAE-WES, Coastal Engineering Research Center, Vicksburg, MS.

Hanson, H., Larson, M., Kraus, N. C. and Gravens, M. B. (2006), Shoreline response to detached breakwaters and tidal current: comparison of numerical and physical models. In *Proceedings of the 30th International Conference of Coastal Engineering*, American Society of Civil Engineers, Vicksburg, San diego, MS.

Hasselmann, K., Barnett, T. P., Bouws, E., Carlson, H., Cartwright, D. E., Enke, K., Ewing, J. A., Gienapp, H., Hasselmann, D. E., Kruseman, P., Meerburg, A.,

Mller, P., Olbers, D. J., Richter, K., Sell, W. and H. Walden, (1973). Measurements of wind-wave growth and swell decay during the Joint North Sea Wave Project (JONSWAP)' Ergnzungsheft zur Deutschen Hydrographischen Zeitschrift Reihe, A(8)(Nr 12), p. 95.

Hedges, T. S. and Reis, M. T. (1998), Random wave overtopping of simple sea walls: a new regression model. In *Proceedings Institution of Civil Engineers: Water, Maritime and Energy*, Vol. 130, pp. 1–10.

Hirt, C. W. and Nichols, B. D. (1981), Volume of fluid (VoF) methods for dynamics of free boundaries, *J. Comput. Phys.*, 39, 201–225.

Hu, K., Mingham, C. G. and Causon, D. M. (2000), Numerical simulation of wave overtopping of coastal structure using the non-linear shallow water equation, *Coast. Eng.*, 41, 433–465.

Karunarathna, H. and Reeve, D. E. (2008), A Boolean approach to prediction of long-term evolution of estuary morphology, *J. Coast. Res.*, 24(2B), 51–61.

Kolhase, S. and Dette, H. H. (1980), Models of wave-induced phenomena. In *Hydraulic Modelling*, pp. 165–181 (ed. H. Kobus), Verlag Paul Parey/Pitman, Hamburg/London.

Larson, M., Hanson, H. and Kraus, N. C. (1997), Analytical solutions of one-line model for shoreline change near coastal structures, *ASCE J. Waterways, Port, Coast. Ocean Eng.*, 123(4), 180–191.

Le Méhauté, B. and Soldate, M. (1977), *Mathematical Modelling of Shoreline Evolution*, CERC Misc. Report 77-10, USAE-WES, Coastal Engineering Research Center, Vicksburg, MS.

Li, B. (1994), An evolution equation for water waves, *Coast. Eng.*, 23, 227–242.

Li, B. and Anastasiou, K. (1992), Efficient elliptic solvers for the mild-slope equation using the multi-grid technique, *Coast. Eng.*, 16, 245–266.

Li, B. and Fleming, C. A. (2000), Three dimensional model of Navier–Stokes equations for water waves, *J. Waterways, Port, Coast. Ocean Eng.*, 127(1), January/February 2001, 16–25.

Li, B., Reeve, D. E. and Fleming, C. A. (1993), Numerical solution of the elliptic mild-slope equation for irregular wave propagation, *Coast. Eng.*, 20, 85–100.

Lin, P. (1998), *Numerical Modeling of Breaking Waves*, Ph.D. thesis, Cornell University, New York.

Lin, P. and Liu, P. L.-F. (1998), A numerical study of breaking waves in the surf zone, *Fluid Mech.*, 359, 239–264.

Liu, P. L.-F., Lin, P. A. and Chang, K. A. (1999), Numerical modeling of wave interaction with porous structures, *ASCE Waterways, Port, Coast. Ocean Eng.*, 125(6), pp. 322–330.

Liu, P. L.-F. and Tsay, T.-K. (1984), Refraction–diffraction model for weakly nonlinear water waves, *J. Fluid Mech.*, 141, 265–274.

Longuet-Higgins, M. S. and Stewart, R. W. (1964), Radiation stresses in water waves: a physical discussion with application, *Deep sea Research*, 11, pp. 529–562.

Longuet-Higgins, M. S. (1970) Longshore currents generated by obliquely incident sea waves, 1. *J. Geophys. Res.*, 75, 6783–6789.

Mitsuyasu et al. (1980), Observation of the power spectrum of ocean waves using a cloverleaf buoy, *Journal of Physical Oceanography*, (10), 286–296.

Monaghan, J. J. and Kos, A. (1999), Solitary waves on a Cretan beach, *ASCE J. Waterways, Port, Coast. Ocean Eng.*, 125(3), 145–154.

Morison, J. R., Johnson, J. W., O'Brien, M. P. and Schaaf, S. A. (1950), The forces exerted by surface waves on piles, *Petroleum Trans. Am. Inst. Min. Eng.*, 189, 145–154.

Novak, P. and Čábelka, J. (1981), *Models in Hydraulic Engineering – Physical Principles and Design Applications*, Pitman, London.

Novak, P., Moffat, A. I. B., Nalluri, C. and Narayanan, R. (2007), *Hydraulic Structures*, 4th Edn, Taylor & Francis, London.

Oumeraci, H. (1984), Scale effects in coastal hydraulic models. In *Symposium on Scale Effects in Modelling Hydraulic Structures*, paper 7.10, pp. 1–7 (ed. H. Kobus), Technische Akademie, Esslingen.

Owen, M. (1980), *Design of Seawalls Allowing for Overtopping*, Report Ex. 924, Hydraulics Research Station, Wallingford.

Pelnard-Considère, R. (1956), Essai de theorie de l'evolution des forms de rivages en plage de sable et de galets, *Societe Hydrotechnique de France, Proceedings of the 4th Journees de l'Hydraulique, les Energies de la Mer, Question III*, Paris, pp. 289–298, Rapport No. 1.

Radder, A. C. (1979), On the parabolic equation method for water wave propagation, *J. Fluid Mech.*, 95(1), 159–176.

Reeve, D. E. (1992), Bathymetric generation of an angular spectrum, *Wave Motion*, 16, 217–228.

Reeve, D. E. (2006), Explicit expression for beach response to non-stationary forcing near a groyne, *ASCE J. Waterways, Port, Coast. Ocean Eng.*, 132, 125–132.

Reeve, D. E. (2009), *Risk and Reliability: Coastal and Hydraulic Engineering*, SPON Press, London, p. 320, ISBN: 978-0-415-46755-1.

Reeve, D. E., Chadwick, A. J. and Fleming, C. A. (2004), *Coastal Engineering: Processes, Theory and Design Practice*, SPON Press, London.

Reeve, D. E. and Karunarathna, H. (2009), On the prediction of long-term morphodynamic response of estuarine systems to sea level rise and human interference, *Continental Shelf Res.*, 29, 938–950.

Shao, S., Ji, C., Graham, D. I., Reeve, D. E., James, P. W. and Chadwick, A. J. (2006), Simulation of wave overtopping by an incompressible SPH model, *Coast. Eng.*, 53(9), 723–735.

Soliman, A. and Reeve, D. E. (2003), Numerical study for small freeboard wave overtopping and overflow of sloping sea walls. In *Proceedings of Coastal Structures 2003*, pp. 643–655, American Society of Civil Engineers, Portland, OR.

Stokes, G. G., (1847) On the theory of oscillatory waves, *Trans. Camb. Phil. Soc. 8*, 441–445.

Svendsen, Ib. A. (2006), *Introduction to Nearshore Hydrodynamics, Advanced Series on Ocean Engineering*, Vol. 24, p. 722, World Scientific, Singapore.

US Army Corps of Engineers (1984), *Shore Protection Manual*, Vols. 1 and 2 Vicksbury, MS, USA.

Van Goor, M. A., Stive, M. J. F., Wang, Z. B. and Zitman, T. J. (2001), Influence of relative sea level rise on coastal inlets and tidal basins. In *Proceedings of the 4th Conference on Coastal Dynamics*, pp. 242–251, American Society of Civil Engineers, Lund.

Van Rijn, L. C. (1993), *Handbook Sediment Transport in Rivers, Estuaries and Coastal Seas*, Aqua Publications, Amsterdam.

Wind, H. G. (1990), Influence functions. In *Proceedings of the 21st International Conference on Coastal Engineering*, pp. 3281–3294, American Society of Civil Engineers, New York.

Zacharioudaki, A. and Reeve, D. E. (2008), Semi-analytical solutions of shoreline response to time varying wave conditions, *ASCE J. Waterways, Port, Coast. Ocean Eng.*, 134(5), 265–274.

Chapter 13

Modelling of hydraulic structures

13.1 Introduction

The term 'hydraulic structures' covers a variety of works, ranging from dams and weirs, through hydroelectric development, navigation, irrigation, drainage, water supply, river training and coastal structures, to public-health engineering works. Our concern in this chapter is the aspects of hydraulic engineering design of these structures, which in turn determine the methods of their model investigations aimed at a well-functioning, economical and environmentally friendly design. Investigations of hydraulic structures are a perfect example of the interplay between theoretical analysis, numerical and physical models, and field observations, with emphasis on the last two.

This chapter touches on the various types of hydraulic structure (with the exception of river training, drainage, coastal and offshore structures dealt with in previous chapters) with emphasis on their modelling. It is not the intention to deal with details of the structures or their design but rather to concentrate on the physical basis and processes involved.

In spite of the great diversity in both size and type of hydraulic structures, most of the hydraulic principles associated with their design can be classified under the broad headings used in the following paragraphs. The discussion of these principles is necessarily brief, as the main aim of this chapter is to establish the basis for their modelling as a design tool.

13.2 Physics and processes

13.2.1 General layout and the flow field

The general layout of structures is determined primarily by their function and their relationship to and interaction with the surrounding body of water. This applies particularly to low-head structures (i.e. barrages, weirs, low-head hydroelectric development, intakes, inland navigation structures, etc.).

The flow upstream and downstream of structures, and also within the structure itself, is mostly three-dimensional. The boundary conditions are often too complicated to express the flow field mathematically (except in special cases), or this can usually be done only by means of strict assumptions and/or restrictions. The solution is thus based mainly on engineering judgement and experience or on the results of physical-model studies, sometimes combined with numerical modelling.

With few exceptions, we are dealing with three-dimensional, free-surface flow governed by gravity that is barely influenced by viscosity or surface tension. The design usually aims at an even distribution of flow without vortex formation, and separation of flow from its boundaries.

Any mathematical solution, if at all feasible, will usually be based on the application of the Navier–Stokes or Euler equations (equations (4.3)–(4.7) and (4.33)) and a numerical treatment of finite-difference, finite-element or boundary-element methods. In mathematical modelling of the flow field, hydraulic structures often form internal boundaries to the model, with the relevant equations (continuity, compatibility, stage–discharge relationship, etc.) characterizing the influence of the structure.

Modelling of the *far flow field* on physical models usually concentrates on the flow and velocity distribution and the effect of structures; numerical modelling may use one-dimensional unsteady-flow models (based on the Saint Venant equations) to study the effect of structures on the propagation of waves in systems of channels, steady-flow computations (e.g. of backwater curves) or two-dimensional modelling of steady flow.

Near flow field modelling of structures and their effect on flow – apart from the usual physical modelling – includes the generation of vortices and their driving forces (turbulent exchanges of momentum, viscous stresses, non-uniform velocity distribution in the vertical) and/or turbulence models. For further details, see Section 13.4.

The general layout, as well as the detailed design, often substantially influences the flow-induced forces. For example, in control structures we are interested not only in the hydrostatic forces acting on gates in their closed position, but also in the dynamic loading induced by the flow through the control structure and, more importantly, during the operation of the gates (see also Section 13.2.10).

13.2.2 Discharge capacity

One of the most frequent design problems is the provision of adequate discharge capacity at free or gated dam spillways, barrages, outlets, culverts, etc. The solutions of the discharge equations are invariably based on the continuity and energy or momentum equations (equations (4.14)–(4.16)), which can be found in many hydraulics textbooks (see e.g. Chadwick *et al.* (2004)).

The hydraulics of *spillways* (and gates and energy dissipators) is well documented in the literature (see e.g. Novak *et al.* (2007) and associated references). This applies to overfall, shaft, siphon, side-channel and chute spillways (and to various types of notches); the need for modelling arises mainly in non-standard shapes of these spillways, or the design of other less frequent types (e.g. labyrinth spillways) or sometimes of gated spillways.

In essence, models used for the determination of discharge capacity of *overfall* spillways concentrate mainly on the evaluation and optimization of the discharge coefficient c in the equation for the discharge Q:

$$Q = c\sqrt{2g}bH^{3/2} \tag{13.1}$$

where H is the head on the spillway crest and b the size (length) of the spillway. The coefficient c is not only a function of the spillway shape and the ratio of the actual and design heads but also of the mutual interaction between the spillway, gates, piers, approach flow conditions, turbulence (see Section 13.2.10) and friction (Section 13.2.3), and (for small discharges) can be influenced by surface tension and viscosity.

The discharge capacity of *gated* spillways depends, among other things, primarily on the shape and position of the gate(s) relative to the spillway. For a partially opened gate, equations of the type (13.1) or (13.2) may be used, i.e.

$$Q = c_1 b\sqrt{2g}(H^{3/2} - H_1^{3/2}) \tag{13.2a}$$

or

$$Q = c_2 ba\sqrt{2gH_e} \tag{13.2b}$$

where a is the height of the gate opening, H_1 is the head on the lip of the gate and H_e is the effective head on the spillway ($= H$) (Figure 13.1). For further details, see Hager (1994), Kolkman (1994), Lewin (2001) and Naudascher (1987). For flow over the gate, equation (13.1) applies.

Model investigations of gated spillways, apart from cavitation and vibration problems (see Sections 13.2.6 and 13.2.10), concentrate in special cases on the shape of the water passage, the effect of piers and three-dimensional effects of neighbouring gates on the discharge capacity, and hydrodynamic forces acting on the gate(s) (see also Section 13.2.10).

Equation (13.1) assumes that the crest of the spillway controls the flow, i.e. free-flow conditions. The discharge over *low weirs* may be substantially influenced by the downstream water level if the crest is submerged and the flow is non-modular. In this case, the shape of the weir and its lateral walls and/or of the downstream part of the piers influence the discharge capacity, which may require model studies. Submerged flow conditions can

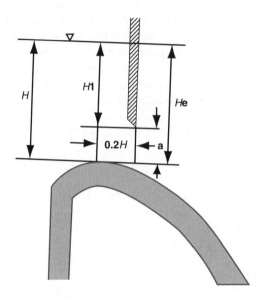

Figure 13.1 Gated spillway (Novak et al. (2007))

also be described by equation (13.1), with Q multiplied by a parameter $\psi (0 < \psi < 1)$, which is a function of the submergence ratio $H/H - H_1$ (the ratio of the upstream and downstream head above the spillway or weir crest) and the channel and structure geometry. For further details, see, for example, Chadwick *et al.* (2004), Chow (1983) and Henderson (1966).

Equation (13.1) applies also to *rectangular sharp-crested weirs*. For a fully aerated lower surface of the overfall jet with head h, the coefficient c can be expressed by equations involving the effect of the velocity of approach and the effect of surface tension. Thus, for example (Hager (1994)),

$$c = 0.3988 \left[1 + \frac{0.001}{h} \right]^{3/2} \left[1 + 0.150 \left(\frac{h}{w} \right) \right] \tag{13.3a}$$

where h is in metres, with the limits $h > 0.02$ m (see also Section 5.8.2f), the weir height $w > 0.15$ m and $h/w < 2.2$. A similar equation using a Weber number We in the form $\mathrm{We} = \rho g h^2 / \sigma$ is

$$c = 0.409 \left(1 + \frac{2.33}{\mathrm{We}} + 0.122 \frac{h}{w} \right) \tag{13.3b}$$

The discharge per unit width (specific discharge) $q = Q/b$ is one of the most important parameters in spillway design as it affects flood routing through the reservoir, the depth of flow on the spillway, self-aeration (see Section 13.2.5), cavitation protection (see Section 13.2.6), energy dissipation (see Section 13.2.7) and downstream erosion (see Section 13.2.8). For the application of numerical techniques dealing with overfall spillway design and capacity, see Section 13.4.

Investigations of discharge at *shaft* spillways are usually associated not only with problems of capacity, flow regime and control (free flow with crest control, or drowned with orifice control) but also with issues of vortex formation, cavitation, aeration, vibration and flow control in the tunnel downstream of the shaft proper. In a free-flowing shaft spillway (with $H/D_c < 0.225$), the circumference of the crest (in a circular shaft πD_c, with D_c the crest diameter) takes the place of the spillway length b in equation (13.1). For a drowned-flow regime, the normal orifice equation applies:

$$Q = c_d A \sqrt{2gH_*} \qquad (13.4)$$

where A is the shaft cross-sectional area and H_* is the difference between the upstream water level and the level of the control orifice or the downstream water level, if the outflow (from the tunnel) is submerged; Figure 13.2(a) shows a cross-section of the spillway and Figure 13.2(b) a stage–discharge curve.

The performance of the *side-channel* spillway is closely related to the shape and size of the channel receiving the discharge from a (usually) standard free-overfall spillway and to the determination of the control section where the flow in the channel changes from subcritical to supercritical. The flow in a side-channel spillway is a typical case of spatially varied,

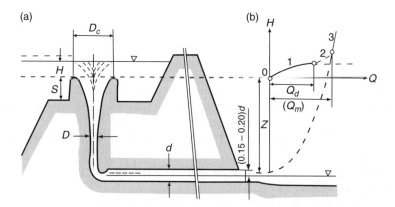

Figure 13.2 Shaft spillway (Novak et al. (2007)). (a) Cross-section of the spillway; (b) stage–discharge curve

non-uniform flow governed by the equation (all the terms have their usual meaning)

$$\frac{dy}{dx} = \frac{S_0 - S_f - 2Q/(gA^2)dQ/dx}{1 - \text{Fr}^2} \qquad (13.5)$$

The numerical integration of the finite-difference form of equation (13.3) (Chow (1983)) yields the water surface in the side channel (which must not be so high as to influence the water level in the reservoir). An important part of the procedure is to determine the critical section (usually at the outflow from the side-channel spillway), which is the starting point of the computation.

The discharge capacity of *siphon* spillways is often investigated using physical models for more complicated shapes of the water passage and for the air regulation of the flow. The basic form of the discharge equation is given by equation (13.2), where H_* is the difference between the upstream water level and the siphon outlet (or the downstream water level if the outlet is submerged) and the coefficient of discharge includes all head-loss coefficients from entry to exit from the siphon.

The discharge over spillways is, of course, closely connected with the pressure and velocity distribution in the overflow stream, which in special cases (siphon, shaft and non-standard overfall spillways) may require model investigations.

For a brief discussion of *chute* and *stepped (cascade)* spillways, see Sections 13.2.4 and 13.2.5.

The complex hydraulic conditions at *side weirs*, which are another example of spatially varied, non-uniform flow, are essentially governed by the differential equation

$$\frac{dy}{dx} = \frac{S_0 - S_f - Q/(gA^2)dQ/dx}{1 - \text{Fr}^2} \qquad (13.6)$$

and equation (13.1).

The solution of this system of equations is complicated because of the uncertainties in the coefficient in equation (13.1), which is strongly influenced by the boundary conditions and regime of flow at the side weir (see e.g. Chow (1983), Henderson (1966), Jain (2001)). This is, therefore, a fairly frequent case of physical modelling, particularly when the side weir and/or the whole conduit are of an unusual shape as, for example, in some storm-sewer separators.

The discharge capacity of (dam) *outlets, culverts*, etc., is, in principle, governed by equation (13.3), and the physical-model studies concentrate mainly on the entrance and exit conditions, and possibly on the effect of the plan layout on the value of the coefficient of discharge. Other possible considerations (e.g. cavitation, aeration, vortex formation) are discussed briefly in the following paragraphs.

The flow through culverts and outlets flowing full is governed by the equations discussed in Section 4.4 (see, in particular, equations (4.37) and (4.65)).

13.2.3 Friction

The head losses due to frictional resistance and surface roughness have been adequately discussed in Section 4.4.1 and expressed in equations (4.36)–(4.64) covering laminar and turbulent flow and hydraulically smooth- and rough-surface conditions of flow in conduits flowing full or with a free water surface. One aspect, however, which has a bearing on the modelling of structures, has to be elaborated further. In Section 4.3.2 has been demonstrated that the growth in the turbulent boundary layer δ and the boundary resistance are a function of the Reynolds number, the shape of the conveyance and the surface roughness. It has also been shown that the turbulent boundary layer consists of a turbulent, transitional and laminar part – see Figure 4.3, with a definition of the thickness of the laminar sublayer δ' (equation (4.31)). As a free flow comes into contact with a structure (surface) a laminar boundary layer develops, which under the influence of surface roughness quickly changes into a turbulent one (see Figure 4.2) – the rougher the surface, the closer to the leading edge this process occurs, and the faster the boundary-layer thickness δ (and the laminar sublayer) will grow. If the height of the surface roughness k is such that the value k/δ is significant, the boundary resistance is mainly due to eddies caused by the flow over the surface roughness. These eddy losses become insignificant as the boundary layer develops and the surface roughness becomes submerged in the laminar sublayer; thus, at high Reynolds numbers, the roughness of the surface quickly becomes unimportant (this process must not be confused with the transition from smooth- to rough-flow conditions as described in equations (4.51)–(4.53)).

13.2.4 Supercritical flow

The flow through many parts of hydraulic structures is supercritical (i.e. Froude number > 1). The flow is often associated with aeration (see Section 13.2.5) and may exhibit waves of translation and interference. A typical example is the flow in *chute* spillways.

Translatory waves (waves of translation, roll waves – see Novak *et al.* (2007)) originate under certain conditions from the turbulent structure of supercritical flow and, as their name implies, move downstream with the flow. Their main implications are the requirement for (higher) freeboard and possible (regular) impulses to the receiving downstream pool, which in extreme cases may result in its failure. However, they occur only at shallow flows with a Vedernikov number (see Section 5.8.1) of Ved > 1 – or usually with the ratio of depth to the wetted perimeter below 0.1 – and long chutes with length L

$$L > -9.2 \left(\frac{V_0^2}{gS_0} \right) \frac{(1 + 2/3(\varphi/\text{Ved}))}{1 - \text{Ved}} \tag{13.7}$$

where φ is the channel shape factor ($\varphi = 1 - RdP/dA$) and $\text{Ved} = k\varphi\text{Fr}$ ($k = 2/3$) (see e.g. Jain (2001)) ($\varphi = 1$ for very wide channels). They thus rarely present a real problem in the hydraulic engineering design of chute spillways designed for the maximum discharge capacity.

On the other hand, a more serious design situation is presented by *interference waves* (cross-waves, standing waves), which are shock waves occurring whenever the supercritical flow is 'interfered' with, such as at inlets, at changes of section, direction or slope, at spillway gates or bridge piers, etc. Interference waves are dependent on the geometry and the flow Froude number, and thus their position and size will change with discharge, but for a given flow situation they are stationary. At points where the waves meet obstacles (e.g. the side walls of a chute), water will 'pile up', requiring substantially increased freeboard. Interference waves may also create difficulties in energy dissipation downstream of chutes should they persist so far (which is rarely the case, because once the flow becomes aerated (see Section 13.2.5) the celerity of the shock wave is greatly reduced and interference waves practically disappear).

The hydraulics of interference waves is reasonably well established, particularly for flow in rectangular channels (Chow (1983), Henderson (1966), Vischer and Hager (1998)). For example, for a channel contraction with a side-wall deflection θ, an angle of inclination of the wave to the flow direction β, upstream and downstream depths y_1 and y_2 and Fr_1 an upstream Froude number Fr_1, the ratio y_2/y_1 is given by

$$\frac{y_2}{y_1} = \frac{\tan \beta}{\tan (\beta - \theta)} = \frac{1}{2 \left((1 + 8\text{Fr}_1^2 \sin^2 \beta)^{1/2} - 1 \right)} \tag{13.8a}$$

For small values of θ (or y_2/y_1 approaching 1) equation (13.8a) reduces to

$$\text{Fr}_1 \sin \beta = 1 \tag{13.8b}$$

For small values of β and $y_2/y_1 > 2$ equation (13.8a) can be simplified to

$$\frac{y_2}{y_1} = 1 + \sqrt{2\text{Fr}_1\theta} \tag{13.8c}$$

(Equation (13.8a) can be used in an iterative way in the design of transitions with known upstream conditions (Q, b_1, y_1) and known width or depth downstream.)

The reduction of interference waves can be achieved by the reduction of the product $Fr_1\theta$ (*the shock number*), wave interference by channel geometry (applicable to one approach flow only), or by bottom and/or side reduction elements (or by a combination of these methods) (see e.g. Vischer and Hager (1998)). The best method of avoiding shock waves in hydraulic design is to remove their cause whenever feasible (e.g. not using bridge piers in supercritical flow) or by using only very gradual transitions should these be unavoidable.

Nowadays, studies on physical models concentrate mainly on waves of interference in non-rectangular channels, shock diffractors, and special cases of large waves and supercritical inlet conditions.

Stepped spillways have recently received increased attention, mainly because of new material techniques (RCC dams and prefabricated blocks) and their enhanced energy dissipation, which contributes to the economy of overall design. The crucial problems encountered in their design are the flow regime (nappe flow or skimming flow, with a transitional zone between the two), air entrainment and energy dissipation. All investigations of cascade spillways (see e.g. Novak *et al.* (2007)) indicate that the flow regime is a function of the critical depth y_c ($= (q^2/g)^{1/3}$, which, of course, denotes a Froude number $= 1$). For a more detailed discussion of cascade (stepped) spillways, see, for example, Boes and Hager (2003a) and Chanson (2001).

Problems of air entrainment, cavitation and energy dissipation are dealt with in the following sections.

13.2.5 Aeration

Aeration and air entrainment form one of the most frequent, but also intractable, problems encountered in the design of hydraulic structures, both large and small. They also form one of the most frequent causes of scale effects in the physical modelling of structures (see Section 13.3.1).

A brief description of the various forms of air–water flows in free-surface and closed-conduit systems is given in Chapter 4 (Section 4.6.5); the physical parameters and dimensional numbers involved have been mentioned in Chapter 5 (Sections 5.8.1 and 5.8.2). It may be useful to state here that, in hydraulic structures, air entrainment may have beneficial as well as detrimental effects. The main beneficial aspects are prevention of excessive negative pressures and cavitation, improved energy dissipation and (in most cases) improved water quality. The principal negative effect is the increased depth of flow ('bulking' of flow) requiring, for example, higher side walls of chute spillways and the carrying of aerated flow into situations where the flow becomes pressurized (e.g. in tunnels and culverts), with the resulting difficulties in flow and pressure fluctuations, 'blowouts', etc. (unless measures are taken to release the air, e.g. in deaeration devices). The most pronounced effect of entrained air is usually in vertical flow configurations (e.g. in drop shafts and shaft spillways).

The main *mechanism of aeration* at hydraulic structures is turbulence. For example, initiation of surface aeration at high-speed flows on spillways is primarily caused by the turbulent boundary layer spreading to the free water surface; local aeration (e.g. by impinging jets or at hydraulic jumps) is caused by turbulent shear layers. Another mechanism is air entrainment through a vortex (e.g. at transitions to pressurized flow) (see also Section 13.2.9).

The general *controlling conditions* for aerated flow (see also Section 4.6.5) are the *inception limit* (i.e. the minimum velocity that has to be exceeded for air entrainment to take place), the *entrainment limit* (a function of the approach Froude number, which must exceed a critical value), the *air supply limit* (a ducted air-supply system may limit the air entrainment, e.g. at spillway aerators) and the *transport limit*, which is governed by downstream conditions (usually a function of velocity, turbulence, wall shear and bubble size, and in closed conduits a function of the conduit length/diameter ratio).

The *bubble-rise velocity* v_b was discussed in Section 4.6.5, mainly as a function of bubble size d_b; for a more detailed discussion of v_b, see, for example, Kobus (1991). The majority of large bubbles in turbulent flow are $1 < d_b < 10$ mm, with the mean bubble diameter usually in the range 2.2–3.5 mm but decreasing with increasing turbulence (the above values are valid for tap-water quality as they are also dependent on the liquid parameter z (equation (13.11) – see also the discussion of oxygen transfer below). High-speed flows ($v >> v_{ibe}$) usually exhibit a smaller range of bubble size. There is also evidence that the mean diameter of entrained air bubbles decreases as the upstream (jet) velocity increases (Ervine (1998)).

Neglecting air properties, the air–water flow will be a function of the geometry, conduit size, velocity of flow, turbulence, air-bubble size, characteristics of the air-supply system and physical properties of the water (density, viscosity, surface tension and, in pressure transients, also compressibility). For A denoting the air flow and Q the water flow their ratio β can be expressed (e.g. by dimensional analysis) as (see also Sections 4.6.5 and 5.8.1)

$$\frac{Q_a}{Q} = \beta = f\left(\text{geometrical ratio, Tu, Eu, Fr, Re, We, } z, \frac{d_b}{y}\right) \qquad (13.9)$$

where Tu denotes turbulence, Eu characterizes the air-supply system and z is the liquid parameter

$$z = \frac{\text{We}^3}{\text{Fr}^2\text{Re}^4} = \frac{g\mu^4}{\rho\sigma^3} \qquad (13.10)$$

The average air concentration in aerated flow $C = Q_a/(Q + Q_a)$ and the parameters β and $\rho = Q/(Q + Q_a)$ are linked through

$$C = 1 - \rho = \frac{\beta}{\beta + 1} \qquad (13.11)$$

The computation of *surface aeration* in various flow situations (particularly on chutes) has been the subject of hydraulic research for many years, and purely empirical as well as semi-empirical methods of computation based on model experiments and/or field observations are available. For a summary of methods of the computation of depth of aerated flow on spillways (and in steep partially filled tunnels), see, for example, Novak *et al.* (2007) and the references therein.

The simplest computations link the parameter β and the Froude number of the non-aerated flow (i.e. they neglect all other parameters in equation (13.9)). For example, for flow in a chute (spillway) the empirical equation

$$\frac{y_a - y_0}{y_0} = \beta = 0.1(0.2\text{Fr}^2 - 1)^{1/2} \qquad (13.12)$$

gives reasonable results (y_a is the depth of the aerated uniform flow depth and y_0 is the depth of the non-aerated flow).

Equally, a very simple and approximate method for estimating the average air concentration on a chute spillway is given by

$$C = 0.75(\sin\theta) \qquad (13.13)$$

One of the difficulties of a more sophisticated computation of the depth of aerated flow and air concentration is the fact that the friction coefficient of the flow with entrained air (in contact with a rough or smooth boundary) is smaller than it would be for non-aerated flow under the same conditions ($\lambda_a < \lambda$); this is why the mean velocity of aerated flow is larger than the velocity of the same water flow without air. The ratio λ_a/λ depends on the air concentration, and a simplified equation (based on the data reported by Anderson and Straub, Ackers and Priestly 1985) is

$$\frac{\lambda_a}{\lambda} = 1 - 1.9C^2 \quad \text{for} \quad C < 0.65 \qquad (13.14)$$

with $\lambda_a/\lambda = 0.2$ for $C > 0.65$.

On the other hand, the difference between the non-aerated depth of uniform flow y_0 and the (imaginary) depth of the water component of aerated flow y_0' (reduced because of the increased velocity) becomes significant only for $C > 0.4(y_0'/y_0 < 0.95)$.

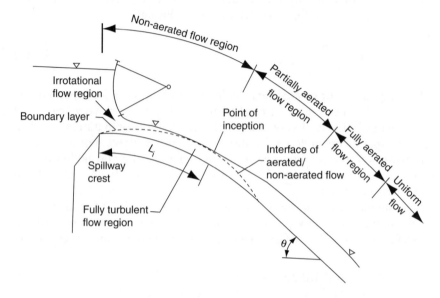

Figure 13.3 Aerated flow over a spillway (Novak *et al.* (2007))

The distance L_i from the crest of a spillway to the point of inception (Figure 13.3) is basically given by the point where the turbulent boundary layer (see Sections 4.3.2 and 13.2.3) penetrates the full depth of flow (see Figure 13.3), and can be determined by combining the equation for the non-uniform non-aerated flow with the equation for the turbulent boundary growth. Ackers and Priestley (1985) quote a simple equation for the boundary-layer growth δ with distance L:

$$\frac{\delta}{L} = 0.0212 \left(\frac{L}{H_s}\right)^{0.11} \left(\frac{L}{k}\right)^{-1.10} \tag{13.15}$$

where H_s is the potential flow velocity head.

$L_i(m)$ can also be estimated as a function of the unit discharge q (m²/s) from

$$L_i = 15q^{1/2}(m) \tag{13.16}$$

(sometimes a critical velocity $v_c = 6$ m/s is also used, see Section 4.6.5).

Local surface aeration processes (e.g. plunge-pool aeration at weirs, drop structures, siphons, shafts) depend on many parameters, and the boundary geometry is extremely important. No generally applicable design criteria can be formulated (Kobus (1991)). In the following text a few examples of local aeration are briefly quoted, together with the relevant relationships.

Experimental evidence suggests that for water jets plunging into a downstream pool the minimum velocity that has to be exceeded for air entrainment – the *inception limit with a critical velocity* v_c – is a function of the liquid parameter z, the relative turbulence intensity ($\sqrt{v'^2}/v$) and the dimensionless parameter $v_c^3/(gv)$. For constant values of z and turbulence intensity $\sqrt{v'^2}/v > 4\%$, experimental evidence suggests that $v_c^3/(gv)$ is constant $= (0.5–1.0)10^5$, resulting in $v_c = 0.8–1.0$ m/s (Kobus and Koschitzky (1991)) (see also Section 4.6.5).

Ervine (1998) discusses in some detail the mechanisms of plunge-pool aeration and quotes a 'broad-brush' equation for aeration rates that is valid for flat-jet velocities U up to 15 m/s, averaged over a range of conduit slopes, for a jet thickness greater than 30 mm with q_a (m²/s) being a function entirely of the velocity at the plunge point v:

$$q_a = 0.00002(v-1)^3 + 0.0003(v-1)^2 + 0.0074(v-1) - 0.0058 \quad (13.17)$$

Another general relationship for β (for air entrainment on one side of the jet) (Kobus and Koschitzky (1991)) is

$$\beta = k\mathrm{Fr}^2\left(1 - \frac{v_c}{v}\right)^3 \qquad\qquad (13.18)$$

At hydraulic jumps the entrainment process usually begins at Fr > 1.7.

The length of fall of a jet required for disintegration can be approximately expressed for flat jets issuing horizontally (e.g. under a gate) as

$$L = 6q^{1/3}\ (\mathrm{m}) \qquad\qquad (13.19)$$

For *hydraulic jump* configurations, the expression

$$\beta = k_l(\mathrm{Fr} - 1)^a \qquad\qquad (13.20a)$$

can be used (k, k_1, and a are constants) (Kobus and Koschitzky (1991)).

Equation (13.20a) can also be used for dimensioning air vents for conduit flow regulating gates that have a hydraulic jump or ring jump downstream.

Rajaratnam and Kwam (1996) studied air entrainment at *drops* both for plunging jets and for the case when a hydraulic jump is formed downstream at a section where the deflected stream below the drop becomes parallel to the bed. If h is the height of the drop (the difference between the upstream and downstream beds) and y_t the downstream water depth, for the case of the hydraulic jump

$$\beta_x = \frac{q_a}{q} = 1.44 \left(\frac{q^2}{gh^3}\right)^{0.226} \tag{13.20b}$$

and $\beta_p = f(y_t/h)$ for the plunging jet with $\beta_{pmax} = 0.4$ and $\beta_x = \beta_p$ when the downstream depth falls to a value y_{tx} required for a free jump just below the drop.

For *spillway aerators* a similar equation has been developed:

$$\beta = k(\mathrm{Fr} - 1)^a \left(\frac{K}{y}\right)^b \tag{13.21}$$

where y is the non-aerated supercritical flow depth and K is a factor that is the function of the aerator-control orifice area and the chute width (Pinto (1991)).

Air entrainment at *various types of structures* (weirs, gates, spillways) is often a welcome means of improving the oxygen content in the water. Apart from the hydraulic processes mentioned above, the actual *transfer of oxygen* from the entrained air bubbles into solution will depend on the water properties – temperature, salinity (i.e. the liquid parameter z) – and on the initial dissolved oxygen content. The resulting *oxygen uptake* is best described by the deficit ratio, i.e. the ratio of the upstream to the downstream oxygen deficit

$$r = \frac{C_s - C_u}{C_s - C_d} \tag{13.22a}$$

or the oxygen-transfer efficiency

$$E = \frac{1 - 1}{r} = \frac{C_d - C_u}{C_s - C_u} \tag{13.22b}$$

where C_u, C_d and C_s are the upstream, downstream and saturation oxygen concentrations, respectively.

When comparing the results of mass transfer for various situations, the data must be reduced to standard conditions, mainly by applying a temperature, and sometimes also a water-quality, correction.

On the basis of extensive laboratory experiments with jets from weirs falling into a downstream pool, Avery and Novak (1978) developed the equation (r at 15°C):

$$r_{15} - 1 = k\mathrm{Fr}_j^{1.78}\mathrm{Re}_j^{0.53} \tag{13.23}$$

where $Fr_j = (gh^3/(2q_j))^{0.25}$ and $Re_j = q_j/v$. In this equation q_j is the unit discharge (discharge per jet perimeter) at impact into the downstream pool and h is the difference between the water levels above and below the weir (i.e. the height of the jet fall). For wide jets of width b with air access on one side only (e.g. flow over a spillway) $q_j = q = Q/b$; for free-falling jets (e.g. flow over a gate or weir) $q_j = q/2$. The boundary conditions for equation (13.21) are a solid (not disintegrated) jet, i.e. the height of fall $h < 6q^{11/3}$ (m) (see equation (13.17)) and the downstream pool depth d must be sufficiently deep to allow full unimpeded penetration of the entrained air bubbles (and therefore maximum contact time). This depth was given by Avery and Novak (1978) as

$$d = 0.00433\,Re_j^{0.39}Fr_j^{1.787}(m) \tag{13.24}$$

These boundary conditions are not extremely rigid, as good results have been obtained by applying equation (13.21) to heights of fall somewhat bigger and downstream depths smaller than those given above.

The coefficient k in equation (13.21) is 0.627×10^{-4} for tap water but varies considerably with the salinity of the water, rising by 100% (1.243×10^{-4}) for a concentration of 0.6% sodium nitrite. The reason for this was shown to be the reduction in the mean bubble size from 2.53 mm for water with zero salinity to 1.57 mm for the saline water, and hence an increased air/water interfacial area resulting in greater oxygen transfer.

Gulliver et al. (1998) extensively tested a number of predictive equations for oxygen uptake at various types of structure by comparing the predicted results with many field measurements. They concluded that, for weirs, equation (13.21) performed best. For ogee spillway crests they recommended the equation

$$E = 1 - \exp\left(\frac{-0.263h}{1+0.215q} - 0.203y\right) \tag{13.25}$$

where y (in metres) is the downstream pool depth, and for gated sills they recommended the equation

$$E = -1 - \exp\left(\frac{-0.0086hq}{s-0.118}\right) \tag{13.26}$$

where s (in metres) is the submergence of the gate lip.

13.2.6 Cavitation

The phenomenon of cavitation, its causes and prevention have been discussed in Section 4.6.5. At hydraulic structures, cavitation may occur mainly (but not exclusively) in high-velocity spillway and conduit flows, at gate slots and at stilling basins with large velocity and pressure fluctuations. The tendency for cavitation is greatest where the boundary configuration gives rise to flow separation; this occurs at any discontinuity of the boundary, such as steps (into and away from the flow), abrupt changes in bed (wall) direction, slots, large roughness elements and uneven joints. For example, for a step of only 3 mm height (into the flow) cavitation begins at velocities of 11 m/s (at ambient atmospheric pressure) (Cassidy and Elder (1984)).

Cavitation occurs if the cavitation number $\sigma = (p - p_v)/(\rho U^2/2)$ (see equation (4.12)) falls below a critical value σ_c (which is strongly dependent on the boundary geometry). A value of around $\sigma_c = 0.25$ is sometimes considered when assessing the critical velocity on 'smooth' concrete surfaces, or for uniformly rough surfaces $\sigma_c = 4\lambda$, where λ is the Darcy–Weisbach friction factor (see equation (4.39)). The onset of cavitation will also depend on water quality, particularly on the presence of dissolved gases and suspended particles.

According to the ICOLD survey (Cassidy and Elder (1984)), of 123 spillways, about 60% of which operated in excess of 100 days, the danger of cavitation damage (unless exceptional care is taken in design and construction) is acute at velocities in excess of about 35 m/s and at unit discharges over 100 m³/sm. This does not mean that cavitation cannot occur at lower velocities or discharges or that it must occur at these or even higher values; its occurrence depends very much on the detail of the spillway design.

The analysis of instantaneous pressure fluctuations underneath the hydraulic jump provides an insight into the possibility of cavitation occurring (e.g. on a stilling basin floor), even if the average pressure is well above atmospheric. In the case of a free jump, experiments have shown that a cavitation number in the form $\sigma = \dfrac{\sqrt{p'^2}}{(\rho V_1^2 \,/2)}$, where V_1 is the supercritical velocity at the toe of the jump with depth y_1, and p' is the deviation of the instantaneous pressure from its time-averaged mean value, attains a maximum value of about 0.05 at a distance 12 y_1 from the toe of the jump. From this an indication of the probability of cavitation may be obtained from the value of k:

$$k = \frac{p'}{\sqrt{p'2}} = \frac{(p_0/(\rho g) + 3y_1)\,2g}{0.05\ V_1^2} = f(\mathrm{Fr}_1) \tag{13.27}$$

$(1 < k < 5$; for $k > 5$ there is practically no danger of cavitation). For further details, see Locher and Hsu (1984) and Novak et al. (2007).

13.2.7 Energy dissipation

Energy dissipation occurs at all flows through and over hydraulic structures. Its most common manifestation is at dam spillways and stilling basins or other types of energy dissipators.

Energy dissipation is a continuous process, but often it is useful to analyse it in distinct stages. For example, at dam spillways five phases can be considered (Novak *et al.* (2007)): energy dissipation on the spillway surface, in free-falling jets, at impact into the downstream water surface, in the stilling basin or plunge pool, and at the outflow from the basin into the river. The energy loss is usually expressed as a function of the velocity head ($\xi v^2/(2g)$) and may best be judged from the value of the velocity coefficient φ, which is the ratio of the actual velocity at any section of the flow to the theoretical velocity.

From the energy equation (see Section 4.3.1)

$$\frac{1}{\varphi^2} = 1 + \xi \tag{13.28}$$

and the ratio of the energy loss e to the total energy E (relative energy loss) is

$$\frac{e}{E} = \frac{\xi}{1+\xi} = 1 - \varphi^2 \tag{13.29}$$

For a ratio of the height S of a smooth spillway crest above its toe to the overfall head H of $S/H < 30$ (or in the case of the spillway ending in a free-falling jet above the jet take-off point which may be S' above the basin floor, $(S - S')/H < 30$) (see Figure 13.4), the velocity coefficient for the first phase of energy dissipation is given by

$$\varphi_1 = 1 - 0.0155 \frac{S}{H} \tag{13.30}$$

For a given spillway crest position φ_1 increases with an increase in H (i.e. for a given discharge with the decrease in the spillway length). Thus, for $S/H = 5$, $\varphi_1 = 0.92$ and the relative head loss e/E is only 15%, whereas for $S/H = 25$, $\varphi_1 = 0.61$ and the head loss is 62%. The head loss could be increased by using a rough spillway; however, the increased energy dissipation could also result in an increase in the cavitation danger unless aeration is provided at the spillway surface, either artificially or through self-aeration (see Section 13.2.5).

For free-falling jets (stage 2) the energy loss will be a function of turbulence, jet geometry, length of fall and degree of jet disintegration.

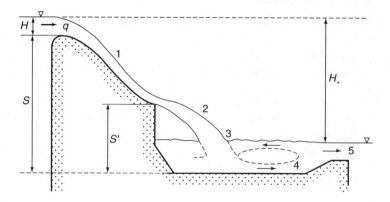

Figure 13.4 Spillway with free-falling jet (Novak *et al.* (2007))

An energy loss of about 12% is achieved in this phase unless the jet collides with another one or an obstacle, when the loss may be considerably higher.

The main benefit in energy dissipation for spillways with free-falling jets is in the third phase at impact into the downstream pool; here the collision of masses of water and the compression of air bubbles (both entrained in the jet and drawn into the pool at impact) contribute significantly to the resulting energy loss.

Generally,

$$\varphi_{1-3} = (S'/S, \ q, \ \text{geometry}) \tag{13.31}$$

The lowest (optimum) value of φ_{1-3} for a given q and geometry will be at about $S'/S = 0.6$.

Stilling basins are the most common form of energy dissipators, converting the supercritical flow from spillways into the subcritical flow compatible with the downstream river regime. The straightforward, and often the best, method of achieving this transition is through a simple submerged hydraulic jump formed in a stilling basin of rectangular cross-section (Figure 13.5). The depth of basin below the downstream river bed is given by the difference of the submerged jump subcritical depth y^+ and the downstream river depth y_0, where y^+ is in turn given by the conjugate depth y_2 of a free jump multiplied by a safety factor (about 1.10). From the energy equation for energy E above the basin floor

$$E = y_1 + q^2/(2gy_1^2\varphi_{1-3}^2) \tag{13.32}$$

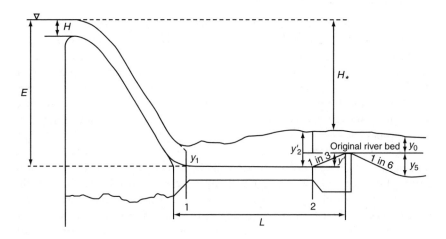

Figure 13.5 Hydraulic jump stilling basin (Novak et al. (2007))

y_2 is then obtained from the free-jump equation (derived from the momentum equation (4.16); see also equation (4.71)):

$$y_2 = \frac{y_1}{2\left(-1 + \sqrt{(1 + 8\mathrm{Fr}_1^2)}\right)} \tag{13.33}$$

The required length of the stilling basin is best expressed as a multiple of the jump height

$$L = K(y_2 - y_1) \tag{13.34}$$

With $4.5 < K < 5.5$; the lower value applies for Fr > 10 and the higher one for Fr < 3. The energy dissipation in a jump basin may be estimated from (see also equation (4.75)):

$$e = \frac{(y_2 - y_1)^3}{(4y_2 y_1)} \tag{13.35}$$

and the energy 'loss' at the outflow from the basin alone from

$$e = 0.25 y_{cr} \tag{13.36}$$

where y_{cr} is the critical flow depth.

For further details of the design of stilling basins, see Novak *et al.* (2007) or Vischer and Hager (1998).

13.2.8 Erosion at structures

Local erosion occurs at all types of hydraulic structures built on erodible stream beds. Breusers and Raudkivi (1991) give a comprehensive survey of the computation of scour at river constrictions, spur dikes, abutments, bridge piers, culvert outlets, weirs and dams. It is clearly not practical in this brief overview to give a full account of the various cases. Only the two problems most frequently investigated by modelling are briefly discussed here: the scour downstream of weirs and dams, and the depth of scour at bridge piers.

The scour downstream of *weirs and dams* is caused primarily by the residue of the energy not dissipated in the structure itself before the flow enters the river (see Section 13.2.7), by excessive turbulence, and by the flow structure of the (in most cases) two-dimensional stream with a different velocity distribution from that in the receiving water. Downstream of weirs and low dams with a hydraulic jump-stilling basin, Novak *et al.* (2007) propose for the depth of scour y_s (in metres) below the river bed the equation

$$y_s = 0.55 \left(6 H_*^{0.25} q^{0.5} \left(\frac{y_0}{d_{90}} \right)^{1/3} - y_0 \right) \tag{13.37}$$

where H_* (in metres) is the difference between the upstream and downstream water levels, y_0 (in metres) is the depth of flow in the river and d_{90} (in millimetres) is the 90% grain size of the river bed.

For higher dams with a plunge pool, Mason (1989) suggested the equation

$$y_s = 3.27 \left(\frac{q^{0.6} H_*^{0.05} y_0^{0.15}}{g^{0.3} d^{0.1}} \right) \tag{13.38}$$

For further details, see Breusers and Raudkivi (1991) and Novak *et al.* (2007).

The depth of scour at *bridge piers*, mainly due to the three-dimensional flow structure with a horseshoe vortex, is a function of the fluid properties, the bed size and bed material, the pier size, and the geometry of the pier and its alignment with the direction of flow. Furthermore, it will depend on whether the river-flow velocity is below the critical velocity for initiation of sediment transport – resulting in 'clear water scour' – or whether scour takes place in a 'live-bed' situation, and on the flow regime (steady or unsteady).

There have been many experimental and some field studies of local scour at bridge piers, some of which are summarized in, for example, Breusers and Raudkivi (1991), Cop *et al.* (1988) and Melville and Coleman (2000).

The equations for the normalized ('maximum, equilibrium') scour depth y_s/b (where b is the pier width – for circular piles of diameter D) – assuming the absence of viscous effects – range from a numerical constant (e.g. 1.6) to the form

$$\frac{y_s}{b} = f\left(\frac{V_0}{V_C}, \frac{y_0}{b}, \frac{V_0}{(gy_0)^{1/2}}, \frac{V_0}{(g\Delta d)^{1/2}}\right) \qquad (13.39)$$

The pier form and angle of attack by the flow are usually expressed by multiplying equation (13.39) by constants K_1, K_2, etc.

Several observations have to be added to the above:

(1) The scour will be influenced not only by the form and size of the structure but also by the general flow configuration.
(2) For scour at *culvert outlets* the effects of the tailwater depth are clearly important.
(3) For scour at *spur dykes* the relative length and inclination of the dyke as well as the downstream recirculating zone are relevant.
(4) In *unsteady flows* (e.g. *floods*) the shape of the hydrograph, particularly the duration of the maximum flow, has to be considered.
(5) For scour in *oscillatory flow* (e.g. at *piles and pipelines* in water waves) the Keulegan–Carpenter number is an additional important parameter. At *sea walls* scour rates depend on the type of wall, the initial water depth, the wave height and period, and the sediment size. (For further details, see Sumer and Fredsoe (2002) and Whitehouse (1998).)

A particularly important point concerns the definition of 'maximum', 'final' and 'equilibrium' scour. All these terms (as well as item (4) above) indicate that scouring is a process with time-dependent results. This has been the subject of many investigations for all types of scour. Generally it has been established that a 'final clear water' scour depth is reached only asymptotically with time. To express this process, a reference scour or a finite rate of increase in scour has to be chosen. Thus, Melville and Chiew (1999) define the time required for equilibrium scour at bridge piers t_e as the time at which the scour hole develops to a depth y_{se} at which the rate of its increase does not exceed 5% of the pier diameter in the succeeding 24 h period, i.e.

$$\frac{d(y_{se})}{dt} < \frac{0.05b}{24\,(\text{h})} \qquad (13.40)$$

The authors suggest the following equation for the temporal development of y_s:

$$\frac{y_s}{y_{se}} = \exp\left(-0,03\left(\frac{V_c}{V_0}\right)\ln\left(\frac{t}{t_e}\right)^{1.6}\right) \tag{13.41}$$

where y_{se} has to be determined from an equation of the type of (13.39), which can also be rewritten as

$$\frac{y_{se}}{b} = f\left(\frac{V_0}{V_c}, \frac{y_0}{b}, \frac{d_{50}}{b}, \frac{t}{t_e}\right) \tag{13.39a}$$

According to the authors, the scour depth y_s at 10% of t_e varies between 50% and 80% of the equilibrium depth y_{se}.

Oliveto and Hager (2002) suggest the use of a reference scour depth $(y_0b^2)^{1/3}$ for both bridge piers and spur dykes (perpendicular to the direction of flow), and the equation

$$\frac{y_s}{(y_0b^2)^{1/3}} = 0.086K\mathrm{Fr}_d^{1.5}\log\frac{t}{t_{ref}} \tag{13.42}$$

where K is a form factor ($K=1$ for cylindrical piers and 1.25 for rectangular spur dykes), $\mathrm{Fr}_d = \left(\frac{d_{84}}{d_{16}}\right)^{1/6} - \frac{V_0}{(\Delta g d_{50})^{1/2}}$ and $t_{ref} = (y_0b^2)^{1/3}(\Delta g d_{50})^{1/2}$.

Analysing model experiments on local scour below weirs, Novak postulated that, after a relatively fast initial development of scour, the depth of scour increases asymptotically to its 'final' value independently of sediment diameter. Thus, taking the reference scour depth y_{s1} as the depth after $t=1$ h (t_1), the equation for $y_s > y_{s1}$ becomes

$$\frac{y_s}{y_{s1}} = a - be^{-c\left(t/t_1+d\right)} \tag{13.43}$$

where a, b, c and d are numerical constants depending on the type of stilling basin used. For example, for a plain basin with a sloping end sill (see Section 13.2.6) $a=1.65$, $b=1$, $c=0.23$ and $d=1$, and the 'final' scour is $y_{se} = 1.65y_s$ (for further details, see Novak and Čábelka (1981)).

An overview of scour in hydraulic engineering, including riprap failure, has been given by Hager (2007), who emphasizes the densimetric Froude number of the approach flow as the dominant parameter.

13.2.9 Swirling flows and vortex formation

Outflows from orifices or under spillway gates and, more importantly, all types of intake structures are prone to swirling-flow problems and vortex formation. The main features are air entrainment and swirl intrusion at the transition from the free surface to the pressure flow. The aim of the design is usually to avoid the first and minimize the swirl, which is most frequently caused by eccentricity of the approach flow and velocity gradients. The difficulties associated with the vorticity of flow are increased head losses, reduced flow rates, reduced efficiency of hydraulic machinery, stimulation of vibrations and cavitation, operational problems caused by suction of debris and, of course, problems caused by entrainment of air into the pressure flow.

The conditions for the prevention of swirling flow are difficult to quantify as they are to a great degree design-dependent and, apart from the geometry of the intake and its orientation to the approach flow φ, will be influenced by gravity (Froude number), viscosity (Reynolds number), surface tension (Weber number) and vorticity (circulation or Kolf number $N_\Gamma = \Gamma d/Q$, where Γ is the circulation $\Gamma = 2\pi c$ (see also Section 4.2.4) ($c = v_t r$ with v_t the tangential velocity at radius r). In very general terms, air entrainment can be avoided and swirl entrainment reduced by increasing the submergence of the intake. This concept leads to a critical submergence value h for different types of intake (e.g. diameter d). Thus

$$\left(\frac{h}{d}\right)_c = f(\varphi, \mathrm{Fr}, \mathrm{Re}, \mathrm{We}, N_\Gamma) \qquad (13.44)$$

Ignoring the Weber number (which is not likely to be important at *critical* submergence conditions) and the Reynolds number (if sufficiently large – see Section 13.2.1), and combining Fr and N_Γ as $\mathrm{Fr}N_\Gamma = v/(gd)^{1/2}2\Pi c/(\Pi dv/4) = 8c/(g^{1/2}d^{3/4})$ leads to

$$\left(\frac{h}{d}\right) = f_1\left(\varphi, \frac{c}{(g^{1/2}d^{3/4})}\right) \qquad (13.45)$$

The above equation illustrates the difficulties of its application, as the result depends on knowledge of the circulation constant c. Furthermore, not taking the Weber and Reynolds numbers into account is not always correct (this applies particularly to the effect of viscosity – see Section 13.3.1).

In simplified versions of the critical submergence (e.g. for pump intakes), the recommendations range from $h = 2.5d$ (rectangular pump sump), to $(h/d)_{cr} = $ constant $\times$ Fr or $(h/d)_{cr} = $ constant $\times$ Fr^x (with moderate vortex formation, the constants and x are geometry-dependent), and to $h/D = a + bFr_D^n Fr_D^n$, where D relates to the suction-bell diameter and a, b and n are constants (usually $a = 1.5$, $b = 2.5$ and $n = 1$).

For a detailed state-of-the-art review of this complicated subject, see Knauss (1987) (which contains contributions from 12 authors).

There are numerous ways in which vortices may be suppressed, all depending on the details of the design and the approach flow. Free-surface vortices – apart from 'sufficient' submergence – may be reduced by guide walls eliminating stagnant regions of water, vertical 'curtain' walls, horizontal grids below the free surface, hoods, vanes, etc. Submerged vortices may also be suppressed by increasing the distance between a boundary and the intake and/or by increasing the boundary roughness.

Swirling flows are not necessarily always undesirable. For example, in dropshafts, as encountered in storm sewer systems, or in shaft spillways, they can result in increased energy dissipation, reduced vibration and controlled air flow. The detailed design of these vortex-flow intakes is a frequent subject of model studies. For further details, see again Knauss (1987).

13.2.10 Turbulence

From the point of view of modelling, the most important aspect of reproducing turbulence and its effects is that turbulence is a system of eddies – large ones generated by the main flow, with a size of the order of the flow field (e.g. depth of flow), where viscous effects are negligible, and small eddies (generated by the large ones) dissipating energy due to viscosity.

In the study of hydraulic structures using modelling as a design tool, attention to turbulence is mainly concerned with fluctuating velocities and the resulting forces (i.e. the outcome of turbulence). This applies, for example, to the study of forces acting on stilling-basin appurtenances, and floors and gates and their components. Section 4.3.3 described some aspects of turbulence relevant to the interaction of flow and structures, and dealt briefly with some physical and statistical aspects of turbulence; Sections 13.2.5, 13.2.6, 13.2.8 and 13.2.11 are also particularly relevant.

13.2.11 Hydrodynamic forces and vibrations

The main *force* acting on hydraulic structures is usually due to hydrodynamic pressures caused by non-uniform turbulent flow, with subsidiary forces caused by waves, ice, impact of floating bodies, etc. The hydrodynamic forces are usually considered in two parts: the time-averaged mean component (or steady flow part), and the fluctuating components induced by various excitation mechanisms, which in turn are closely related to structural vibrations. In the design of structures, often a probabilistic approach based on the statistical distribution of loading (and knowledge of the strength and failure mechanism of parts of the structure) is required.

Taking a weir/dam crest gate as an example, the *mean hydrodynamic forces* acting on a partially opened gate with overflow and/or underflow will depend on the geometry of the gate and approach passage, the geometry and position of the gate accessories (supports, seals, etc.), the surface roughness, the upstream and downstream water levels (if the flow is submerged), the flow Froude, Reynolds, Weber and Mach numbers, the degree of turbulence and the unsteadiness of the incoming flow (Tu), the aeration of the space downstream of the gate (Ae), the gate elasticity (Ca) and the cavitation number (Ca'). Denoting the mean piezometric head by h (i.e. the sum of the mean pressure head and elevation above datum), and including the roughness and water-level positions in the geometry term, we can write a very general statement

$$\frac{(h - h_0)/v_0^2}{2g} = f(\text{geometry},\text{Fr},\text{Re},\text{We},\text{Ma},\text{Ca},\text{Ca}',\text{Tu},\text{Ae}) \qquad (13.46)$$

It is obvious that the proper determination, even of the mean forces acting on a gate, is a difficult task, partly because of the complexity of the hydraulic conditions but also because the design often has to satisfy conflicting demands (the same applies to hydrodynamic forces acting on other parts of hydraulic structures): vibration damping may conflict with keeping the forces required for the gate operation to a minimum; the need to avoid vibrations may conflict with the optimum shape and strength required by the flow conditions and loading; the optimum shape of the gate edges and seals may conflict with the demand for water tightness of the closed gate; etc.

In a preliminary design it is useful to determine the loading by means of potential-flow and/or finite-element analysis. In very simple cases the momentum equation can be used. A combination of theoretical analysis and experience, from field observations and measurements as well as model experiments, is often required to achieve satisfactory determination of the hydrodynamic loading.

Turning now to the *fluctuating components of the hydrodynamic forces*, these can be induced by different excitation mechanisms and are closely related to the structural vibrations, which they in turn may induce. The main categories of these excitation mechanisms are *extraneously-induced excitation* (EIE) (e.g. pressure fluctuations due to the turbulent flow downstream of a control gate and impinging on the gate), *instability-induced excitation* (IIE) (e.g. flow instabilities due to vortex shedding downstream of a cylinder), *movement-induced excitation* (MIE) (e.g. self-excitation due to the fluctuation of a gate seal) and *excitation due to a (resonating) fluid oscillator* (EFO) (e.g. a surging water mass in a shaft downstream of a tunnel gate).

It is clear that, in a general case, equation (13.46) will also be applicable for the fluctuating components of forces, with the inclusion of parameters according to the individual case and excitation mechanism.

For a detailed analysis of fluctuating and mean hydrodynamic forces acting on hydraulic structures, see Naudascher (1991). (Flow-induced) *vibrations* can be *forced* by turbulence or flow structure (see EIE and IIE above); they can also be amplified by body vibrations synchronizing with random turbulence excitation, or the flow excitation can be induced purely by the body vibration itself – *self-excitation* or *negative damping* (see MIE and EFO above). Vibrations can endanger the structure, and have many troublesome environmental effects (e.g. noise).

The basic equation for an oscillator (or resonator) with mass m, displacement y under a force F is

$$\frac{m\delta^2 y}{\delta t^2} + \frac{c\delta y}{\delta t} + ky = F(t) \qquad (13.47)$$

where c denotes damping and k the rigidity (the first term in equation (13.47) is the inertial force and the second is the friction force). The natural frequency f is given by $f = \sqrt{(k/m)}/2\Pi$ and the ratio γ of the damping factor and critical damping is

$$\gamma = \frac{c}{2\sqrt{(km)}}$$

In the design of hydraulic structures it is desirable for excitation frequencies to be remote from resonance frequencies (unless there is high damping), and negative damping (self-excitation) should be avoided.

For a body submerged in a fluid, the mass m in equation (13.47) consists of two components: the mass of the body itself, and the mass of the surrounding fluid, which is accelerated (or decelerated) during vibration (i.e. the *virtual or added mass*). If the external force applied to the system is of a relatively short duration, equation (13.47) with $F = 0$ denotes free oscillation, and it can be used to determine the damping characteristics of the system.

As the vibration of a structure will be dependent on its elastic properties, the Mach (Cauchy) number in equation (13.46) must apply both to the bulk modulus of the fluid K and to Young's modulus E of the structure material (see also equation (3.19e)). Furthermore, the dynamic part of any acting external force (i.e. the part additional to the time-averaged force) will have a dominant (excitation) frequency f associated with the Strouhal number $S = fl/v$. In the same way, we can define the Strouhal number of the natural frequency $S_n = f_n l/v$, where v is the velocity of the oncoming flow and l is a representative length of the (vibrating) body.

In summary, for flow-induced vibrations we have to consider equation (13.46) with two Cauchy numbers $K/\rho v^2$ and $E/\rho v^2$ and the Strouhal numbers S and S_n.

For further detailed treatment of flow-induced vibrations, see, for example, ICOLD (1996), Kolkman (1984) and Naudascher and Rockwell (1994).

13.3 Physical (hydraulic) modelling

13.3.1 Scaling laws and scale effects for models of hydraulic structures

As the flow in and at hydraulic structures is usually three-dimensional, vertical accelerations cannot be neglected and distortion of the velocity distribution is not acceptable; thus, *undistorted* models are the norm. In some cases, where the flow is essentially two-dimensional (e.g. flow over a straight overflow spillway), *sectional* geometrically correct models may be used. In exceptional cases, where the investigated problem is not dependent on the local effect caused by the structure, *distorted* models including the structure may be used (see e.g. Chapter 7).

As can be seen from the previous section, the compliance with *Froude* conditions is also generally required (with the exception of some cavitation experimental studies and the use of aerodynamic modelling). It has to be appreciated that in a model of scale M_l, operated according to Froude law and using the same fluid as in the prototype, the *velocities will be reduced* by $M_l^{1/2}$ and hence the Reynolds (vl/v), Weber $(\rho v^2 l/\sigma)$ and Mach $(v\sqrt{(\rho/K)})$ numbers will all be reduced: the Reynolds number by $M_l^{3/2}$, the Weber number by M_l^2 and the Mach number by $M_l^{1/2}$.

The reduction in the Reynolds number is particularly important as another general requirement is that there should be *fully turbulent flow* on the model, i.e. $\mathrm{Re}_m > \mathrm{Re}_{sq}$ (see equations (7.11) and (4.52)).

The choice of scales is, in most cases, subjected to certain *limits* (if scale effects are to be avoided or minimized) to account for the effects of viscous and surface-tension forces. The choice is further reduced by the presence of erodible beds (e.g. in modelling of scour) or the need to reproduce the elastic behaviour of a structure or its parts and to observe the boundary conditions imposed by critical velocities.

Referring to the discussion of the physics and processes in Section 13.2, we can reach the conclusions laid out in the following sections.

13.3.1.1 General layout

Modelling of both the far- and near-flow field requires undistorted models operated according to the Froude law (see Section 5.7.1) (with the exception of aerodynamic models). The scale of the conventional hydraulic models should be large enough to ensure fully turbulent flow, i.e. the Reynolds number of the flow in the approach channel of the structure should comply with the conditions set out in Section 7.4.3 for open-channel flow ($\mathrm{Re} > \mathrm{Re}_{sq}$ as defined by equation (7.11)).

Turning now to aerodynamic models of the general layout of hydraulic structures, apart from geometrical similarity, these only require fully turbulent flow ($\mathrm{Re} > \mathrm{Re}_{sq}$, see equations (4.52) or (7.10)) and a velocity below

50 m/s (see Sections 5.8.2 and 7.4.4). Aerodynamic modelling is simple in this case, as the water-surface elevation in the approach channel is known and the iterative procedure outlined in Section 7.4.4 is not required.

13.3.1.2 Discharge capacity

From Section 13.2.2 and equations (13.1)–(13.5) it is evident that, apart from geometrical similarity and the Froude law, special attention must be paid to the coefficients of discharge and their dependence on viscosity and surface tension.

For example, for flow over sharp-crested weirs it is easy from equations (13.3a) or (13.3b) to estimate when the terms involving $1/h$ or We become insignificant or what their influence on discharge is likely to be.

In Section 5.8.2.6 some further limits applicable to the choice of scale for modelling discharge and the shape of the outflow jets have been stated (these are important for the determination of the optimum shape of the dam bottom outlet intakes, bottom gate seals, the shape of overflow spillways, etc., which are relevant for the determination of the coefficients of discharge). Briefly, for the extrapolation of the shape of the nappe from a sharp-edged rectangular notch, the minimum head should be about 40 mm, a limit that 'can be ascribed just to surface tension' (Ghetti and D'Alpaos (1977)); at 20 mm the overflow parabola becomes almost a straight line. In the case of flow under a gate the shape of the outflow jet can be extrapolated for a gate opening a bigger than 60 mm and the head $h > 3.3a$; the outflow jet shape from a circular orifice of diameter D is independent of the head for $h > 6D$ and $D > 70$ mm. These conditions translate roughly to a limiting Reynolds number of Re > 10^5. For further details, see, for example, Novak and Čábelka (1981), and for a detailed discussion of the nappe shape from a sharp-edged rectangular weir, see Hager (1995).

As these conditions may lead to very expensive large models, a departure from the above limits may be necessary, as long as it is not too large and the consequences are taken into account in the final analysis.

13.3.1.3 Friction

Equations (4.36)–(4.64) cover the effect of friction in laminar to fully turbulent flow, and particularly the effect of the Reynolds number and the parameter k/δ (i.e. the ratio of roughness size and boundary layer thickness). Sections 7.5 and 9.4 then deal in detail with physical modelling of open- and closed-conduit flow and associated similarity issues.

In modelling of hydraulic structures the conditions for similarity of flow in conduits has become important mainly when modelling flow in long tunnels, bottom outlets and chutes. However, as has been shown in Section 13.2.3, the growth of the turbulent boundary layer and the value of

k/δ can become important in flow over surfaces of structures. For example, it is likely that even a 'smooth' model spillway is going to be rougher than the prototype, thus influencing the coefficient of discharge. Damle (1952) showed that even with a scale 1:4 the error in the coefficient of discharge of a smooth model weir was 5% smaller than that for a full-scale prototype; the smaller the model, the bigger the 'error', i.e. the coefficient decreases as the scale M_l increases, which, of course, acts in the direction of safety. However, this local frictional effect is usually neglected in models of large spillways.

13.3.1.4 Supercritical flow

From equations (13.8) it follows that for flow on *chutes* – the surface aeration issue apart (see next paragraph) – geometrical similarity, the observance of the Froude law and a Reynolds number Re > Re_{sq} (equation (7.11)) are sufficient to ensure similarity of *interference waves*. As far as the *waves of translation* are concerned, the same conditions apply, as geometrical similarity generally ensures also equality of the Vedernikov number; the condition for the necessary length of the chute given by equation (13.6) is then satisfied by the equality of the Froude number as $M_L = M_l = M_v^2$.

As models of concrete (i.e. 'rough') chutes are usually smooth, reproducing 'correctly' the mean values of velocity and depth, more exact modelling of, for example, the velocity distribution (influencing aeration) requires the development of the boundary layer (see Section 13.3.1.3) to be taken into account (see the procedure outlined in Section 9.4). Any uncertainties in this are, however, usually small compared with the scaling problems arising from aeration.

As the criterion for the type of flow over *stepped spillways* is a function of the critical depth, the condition of similarity will automatically be satisfied on a geometrically similar Froude model.

13.3.1.5 Aeration

As evident from Section 13.2.5, aeration at various structures and flow configurations is one of the most intractable problems in design and the cause of scale effects in modelling. Equation (13.9) demonstrates the complexity of the problem for expressing and modelling even the average air concentration in a general case of aerated flow (and even more so when considering the distribution of air). Furthermore, the limiting conditions for aeration inception, entrainment and transport have to be added to the parameters in equation (13.9).

From the discussion of *surface aeration at, for example, supercritical flow* (see Section 13.2.5) it follows that observing the previously stated general conditions (geometric similarity, Froude law, Re > Re_{sq}) and for the same water properties on the model as in the prototype (z = constant) the main problem in reproducing even the average air concentration and air transport will be the effect of surface tension (Weber number) and air-bubble size, which will remain the same on the model as in the prototype (i.e. the parameter d_b/y will not be modelled correctly). On most models the flow velocity will not be large enough to overcome the effect of surface tension (see inception limit $v_c = 6\,\text{m/s}$). Nevertheless, the point of air-entrainment inception can usually be observed (e.g. on models of chutes) by a sudden rough appearance of a previously smooth water surface. The effect of the ratio v_b/v on the model, too large by a factor $M_l^{1/2}$, will be (should aeration take place at all) faster deaeration of the entrained air and smaller transport rates (according to downstream conditions). Also, the equation (13.16), which is not dimensionless, if applied in the model when converted to the prototype is bound to give 'wrong' results by a factor $M_l^{1/4}$. On the other hand, equations (13.12)–(13.14) indicate that in sufficiently large models (usually $\text{Re}_m > 10^5$) some useful results are possible.

In the case of *local surface aeration* (see Section 13.2.5), equations (13.17) and (13.18) would indicate, in general, the impossibility of correct modelling of the aeration rate q_a, as this is dependent on the velocity to a varying power or the ratio of v_c/v. Equally, equation (13.19) for the length of fall of a jet required for disintegration cannot be correctly modelled, as it will include a scale effect $M_l^{1/2}$.

For *hydraulic jump configurations* and *spillway aerators* (see Section 13.2.5) there is the possibility of correct modelling of β (equations (13.20) and (13.21)) as the entrainment-limit effect is predominant with high turbulence levels and sufficiently high Reynolds numbers. Chanson and Gualtieri (2008), on the basis of experiments conducted with two hydraulic jumps scaled 2:1, concluded that the smaller model showed scale effects both in air entrainment and detrainment for $\text{Re}_1 < 2.5E + 4$ (with identical Froude numbers $\text{Fr}_1 = V_1/\sqrt{(gy_1)}$ on both models).

Kobus and Koschitzky (1991) quote an equation for the limiting condition of the boundary-layer length scale at the location of air entrainment:

$$l_e > \left(\frac{\text{Re}_{sq}v}{\text{Fr}\sqrt{g}}\right)^{2/3} \tag{13.48}$$

For a given configuration, equation (13.48) allows an estimate of the minimum model dimensions required for fully turbulent flow. For example, for a hydraulic jump for the lower limit ($\text{Fr} = 1$) the minimum water depth for $\text{Re}_{sq} = 10^5$ is $l_e = 100\,\text{mm}$.

Turning to *oxygen transfer* from entrained air (see Section 13.2.5) it is obvious from equations (13.23)–(13.26) that a physical model operated according to the Froude law cannot predict the prototype oxygen-transfer efficiency, as all equations also contain the discharge q (and equations (13.25) and (13.26) are not dimensionless). However, as long as the appropriate boundary conditions are observed, good results are obtained by computation using equations (13.23)–(13.26).

Boes and Hager (2003b) conclude that the smallest Reynolds and Weber numbers required to minimize scale effects in modelling two-phase air–water flows on *stepped spillways* are Re $= 10^5$ and We $= 10^4$ (We is defined as $v^2\rho L/\sigma$, where L is the distance between step edges). This results typically in a limit $M_l = 15$. Chanson and Gonzales (2005) stress that physical modelling of stepped chutes is more sensitive to scale effects than that of classical smooth invert chutes; one contributing factor is the strong interaction between free-surface aeration and turbulence (which extends from the stepped invert to the pseudo-free surface).

For estimating aeration rates and scale effects on physical models, one commonly used method is to use a series of models of different scales and try to extrapolate the results to the prototype.

We can conclude that, although scale effects in models involving aeration are often inevitable, useful results can be obtained, particularly when modelling alternative design options and/or by supplementing physical models by computational procedures and applying engineering judgement.

13.3.1.6 Cavitation

Cavitation (see Section 13.2.6) is characterized by the cavitation number $\sigma = 2(p - p_v)/(\rho v^2)$, which is a form of the Euler number. For cavitation to occur, σ must be smaller than the critical value σ_c (for incipient cavitation). In models of hydraulic structures operated according to the Froude law, σ would be identical to that in the prototype (assuming the pressure and velocity values refer to mean values), if the pressures p and p_v were to be reduced to scale by M_l. This, however, clearly is not possible in conventional models, as the atmospheric pressure p_0, included in the value of p, and the vapour pressure p_v are not reduced. Hence, cavitation will usually not occur on the model (due to reduced velocities); however, measuring the pressure distribution on the model and converting this to the prototype can give some idea of the areas where cavitation is likely to occur. In some situations, such as in siphon flow, the local pressures in the model may be low enough to actually produce cavitation.

Kenn (1984) demonstrates that even a small-scale model tested in a *water tunnel* at full-scale heads and velocities can usefully indicate likely *patterns* of cavitation and even cavitation erosion for elements of a large structure (e.g. at gate slots).

For cavitation to be actually observed and to develop a design that avoids, or at least minimizes, it, cavitation tunnels (see Chapter 6, Figure 6.3) generally have to be used. However, particularly when testing flow around submerged bodies (e.g. turbines or ship propellers), substantial scale effects in observing the degree of cavitation (incipient, etc.) have been noticed.

The principal problem that makes comparison of exactly the same experiments in different testing facilities difficult is the water-quality effect, caused by the different tensile strengths of the test liquids, which depends strongly on the gas content and suspended dust particles. However, even when this problem is eliminated by filtering and measuring the tensile strength of the liquid in a vortex nozzle chamber, real scale effects remain.

Based on extensive tests in the laboratory at Obernach (Munich University of Technology), Keller (1994) reported scale effects due to velocity, size, turbulence and viscosity, for which he proposed an empirical equation for the cavitation inception number:

$$\sigma = K_0 \left(\frac{L}{L_0}\right)^{1/2} \left(\frac{v_0}{v}\right)^{1/4} \left(1 + \left(\frac{V^*}{V}\right)^2\right) \left(1 + \frac{K_0 S}{S_0}\right) \tag{13.49}$$

where K_0 is a constant characteristic of the body shape determined by a separate experiment, and L_0, v_0 and S_0 are arbitrarily chosen values of the length of the body, the kinematic viscosity and the standard deviation of the free-stream velocity V^*, which is about 12 m/s.

Although the situation may be less complicated when testing elements of structures with free-surface flow (e.g. the stilling-basin baffles in the configuration shown in Figure 6.3(b)), it is clear that the interpretation of cavitation experiments requires a series of tests and experience in handling of cavitation tunnels.

13.3.1.7   Energy dissipation

From Section 13.2.7 it is evident that the main cause of any possible scale effects, when predicting energy dissipation from geometrically similar models operated according to the Froude law, will be the effects of aeration (see Section 13.3.1.5) and, to a certain extent, of losses due to friction (see Section 13.3.1.3); there may also be scale effects due to the degree of turbulence (see Section 13.3.1.10). Some of these effects will be eliminated, or at least minimized, on large-scale models and with a sufficiently large Reynolds number (at least 10^5).

Energy dissipation involving hydraulic jumps will be well represented on models, as is evident from equations (13.33)–(13.36). The correct representation of the incoming supercritical flow depth will, of course, depend on the value of the velocity coefficient ϕ_{1-3} (equations (13.31) and (13.32)). For this the possible role of aeration and friction has to be assessed, but experience shows that most models of, for example, the performance of stilling basins reproduce prototype behaviour well.

A typical example of possible scale effects in energy dissipation is the modelling of a jet from a 'ski-jump' spillway. It is possible that even a very smooth model spillway will be rougher than required for a correct representation of head losses, and thus the velocity of the jet at its origin may be slightly lower than it would be in the prototype. However, due to the usually short upstream length this effect can be quite negligible. On the other hand, the jet is likely to be less aerated on the model than in the prototype causing its impact into the river downstream to be further away from the structure, resulting (for the correct representation of the river bed – see Section 13.3.1.8) in a somewhat deeper scour than in the prototype. Finally, when interpreting model results one may have to take into account the increased air resistance to the high-velocity prototype jet. The overall effect therefore could be a scour somewhat closer to the structure, but possibly a shallower effect than indicated by the model. Thus, the modeller and designer have to apply some degree of engineering judgement to the interpretation of model results and the consideration of the safety of the structure.

All experience shows that the main value of model testing of energy dissipation lies in comparative tests of alternative design features, where any possible scale effects often become irrelevant.

13.3.1.8 Erosion at structures

Section 13.2.8 briefly discussed local erosion downstream of dams and weirs and at bridges, quoting some equations for the depth of scour and its development over time. Although not all equations, particularly the older ones, are dimensionless (e.g. equation (13.37)), they are useful for scour computations. Equations (13.38)–(13.43) are dimensionless, and their application in modelling seems to be relatively straightforward. However, for modelling the scour on geometrically correct models operated according to the Froude law, there are some major problems.

(a) It is essential to reproduce the turbulence and velocity distribution in the flow, as these are closely connected to the scouring process. This requires a sufficiently large Reynolds number (say $Re > 10^5$) and correct simulation of the bed and structure roughness (see also Section 13.2.10 and the discussion in Section 13.3.1.10).

(b) Many of the relevant equations contain the bed sediment diameter (e.g. equations (13.37)–(13.39) and (13.42)). Modelling of scour of coarse granular material requires the use of the correct density and grading of material, which is of sufficient size to be free of viscous effects (see also Section 4.6.3). In general, a model scale resulting in sand with $d > 0.5$ mm and $d_{50} > 1.5$ mm is acceptable; finer material could give results with substantial scale effects. The use of lightweight materials to avoid too small particles of correct density could be a possibility, but it gives rise to some problems (e.g. in reproducing roughness effects and particle shape). Generally, scour investigations require careful selection of bed material, sometimes with use of cohesion-reduction measures (see Section 6.2.3).

(c) The mechanism of scour of a rocky bed is basically due to pressure fluctuations and pressure propagation into the rock fissures. As it is usually not practicable to reproduce this on a model, comparative studies of the erosion of a suitable substitute material (see Section 6.2.3) have to be used to optimize the design.

In general, model scour experiments produce very good and useful results in comparative studies of various designs. However, as experimental studies in laboratory flumes (Ettema *et al.* 1998), field studies and the use of computational fluid dynamics each have limitations, in particular in more complicated structural configurations, the best results are likely to be achieved by a combination of all three approaches (Ettema *et al.* (2006)).

For a more detailed discussion of modelling turbulence, roughness and sediment in scour experiments, see, for example, Prins (1971).

13.3.1.9 Swirling flows and vortex formation

Section 13.2.9 demonstrated the impossibility of a generally valid design procedure for determining the conditions for vortex formation, and the difficulties of modelling swirling flows without any scale effects. Ranga Raju and Garde (1987) discuss in some detail the various methods used in modelling which, apart from the difficulty in establishing a uniformly valid procedure even for one given geometry, also present problems in measuring circulation swirl velocities.

As the Kolf number N_Γ is dependent on the intake approach flow, overall geometry and discharge, the modelling criteria for a given situation will be primarily dependent on the Froude, Weber and Reynolds numbers. From the discussion of experimental results by various authors (notably Knauss 1987) the following conclusions can be reached:

(1) The Weber number (defined as $V\sqrt{(\rho d/\sigma)}$, where V is the flow velocity through an intake of diameter d) does not affect vortex formation for values over 11. As this will almost always be the case in model studies, the influence of surface tension may be ignored.

(2) Practically all studies show the predominant influence of the Froude number for investigations of critical submergence, with the possibility of some distortion necessary to account for the non-constancy of the Reynolds number.

(3) Viscous effects are likely to be absent for a ratio of $Re/Fr = Vd\sqrt{(gd)}/(Vv) = g^{1/2}d^{3/2}/v > 5 \times 10^4$.

(4) Another way of establishing a limiting Weber and Reynolds number for a horizontal intake is to use the critical submergence h, giving a value $We = Q^2 \rho h / (A^2 \sigma) > 10^4$ and a radial $Re = Q/(h\nu) > 3 \times 10^4$.

(5) A series of models may be used to find a Froude number multiplying factor, which can be used to model an air-entrainment vortex.

(6) The technique of testing a geometrically similar model at the prototype velocity, as adopted in some studies, is likely to lead to a conservative design.

In conclusion, it may be stated that for the most frequent modelling case, i.e. where air entrainment and vortex formation can be avoided, the Weber number may be neglected and a geometrically correct Froude model may be used with useful results. But even situations with air-entraining vortices can be reproduced correctly on models, although the procedure will be strongly dependent on the studied geometry, and as far as possible the viscous effects should be checked.

13.3.1.10   Turbulence

It follows from the fact that turbulence is a system of eddies – large ones with negligible viscous effects and small ones dissipating energy due to viscosity (see Sections 4.3.3 and 13.2.10) – that in small-scale models the dissipation term of the turbulence–energy equation may represent a too large proportion of energy consumption, leading to scale effects. It is important, therefore, to reproduce as far as possible the fully developed range of eddies, i.e. to use a geometrically similar (non-distorted) model of sufficiently large size to reproduce correctly the macroscale of the eddies. In this case a Froude model will reproduce well the mean values of fluctuating velocities (and thus pressures), and with sufficiently large Reynolds numbers may also reproduce the MRS values and energy spectra. Thus, Lopardo (1988), from a study of pressure fluctuations induced by a hydraulic jump on stilling-basin appurtenances, concluded that a model scale $M_l = 50$, with $Re > 10^5$, a lowest model depth of 30 mm and power spectra in a frequency band below 20 Hz reproduced well the amplitude and frequency of the prototype pressure fluctuations.

However, it is difficult to apply all these conditions generally, but a Reynolds number above 10^5 is always advisable when modelling turbulence and its effect on hydraulic structures (see also Section 13.3.1.8).

13.3.1.11   Hydrodynamic forces and vibrations

The scaling laws for a preliminary assessment of hydrodynamic forces with rigid models (e.g. of gates) *without taking into account the consequences of vibration* would follow conventional procedures, with the emphasis on

Froude modelling and large enough models to be able to ignore the effects of viscosity and surface tension but with correct representation of geometry, aeration, approach flow and turbulence.

Modelling of forces *including vibration* will be based on equations (13.46) and (13.47), with consideration of the two Strouhal numbers $S = fl/v^2$ and $S_n = f_n l/v^2$ (taking into account the dominant excitation and natural frequencies). In principle, again the effects of viscosity and surface tension can be ignored if the model is large enough. The size of a 'large enough' model scale will obviously depend on the circumstances; for example, for barrage gates a scale in the range $8 < M_l < 25$ is usually used and a Reynolds number $Re > 10^5$ is aimed at. Although both the Mach and Cauchy numbers are included in equation (13.46) (the Mach number for the fluid and the Cauchy number for the structure), it would be a very rare case where the fluid compressibility has to be taken into account, and thus the Mach number can be ignored (see also comment below).

In general, two different cases have to be distinguished: the modelled structure is submerged (e.g. thrash racks) or the modelled structure is in free-surface flow (e.g. crest gates).

In the first case, it is sufficient to satisfy the identity of the Cauchy number $(\rho V^2/E)$, as the Froude law does not apply (and the model is large enough to permit viscous effects to be ignored). Using the same fluid and material in the model as in the prototype ($M_\rho = M_E = 1$) results in $M_V = 1$, i.e. the same velocity in the model as in the prototype.

In the second case, using Froude similarity, full use must be made of the conditions arising from equations (13.46) and (13.47). Equation (13.47) leads to scale conditions with respect to force for mass, damping and rigidity coefficients for the amplitude of oscillations, and the resonance frequency to have correct scales (with $M_t = M_f^{-1}$ and $M_y = M_l$):

$$M_m = \frac{M_F}{M_f^2 M_l} \qquad (13.50a)$$

$$M_c = \frac{M_F}{M_f M_l} \qquad (13.50b)$$

$$M_k = \frac{M_F}{M_l} \qquad (13.50c)$$

As $M_V = M_l^{1/2}$ (Froude law), the scale for the two Strouhal numbers $M_S = M_{Sn} = 1$ automatically. As far as the cavitation number is concerned, theoretically its scale is also unity, but the problems associated with not reducing the vapour pressure p_v and other scale effects (discussed in Section 13.3.1.6) have to be considered. Thus, ignoring the effects of viscosity and surface tension for a geometrically correct model with correctly modelled turbulence and aeration (see Sections 13.3.1.5 and 13.3.1.10), for the identity of the Cauchy number in model and prototype:

$$M_\rho, M_1 M_E^{-1} = 1 \tag{13.51a}$$

where ρ' is the specific mass of the material of the structure. As the structure and part of the water vibrate together, using water on the model leads to $M_\rho = M_{\rho'} = 1$ and

$$M_l = M_E \tag{13.51b}$$

As for steel $E = 20 \times 10^{10}\,\mathrm{N/m^2}$ and $\rho' = 7,850\,\mathrm{kg/m^3}$, and for concrete $E = 20 \times 10^8\,\mathrm{N/m^2}$ and $\rho' = 2,500\,\mathrm{kg/m^3}$, it is not easy to satisfy the above conditions. The materials most frequently used for modelling steel structures are PVC or one- or two-component epoxy resins, which with $E = 6 \times 10^9\,\mathrm{N/m^2}$ leads to a scale $M_l = 20 \times 10^{10}/6 \times 10^9 = 33$. As ρ' for the model material is less than that for the prototype ($\rho' = 1,400\,\mathrm{kg/m^3}$), extra mass (e.g. lead) has to be added locally to the model to achieve $M_{\rho'} = 1$.

If the fluid compressibility has to be taken into account, similarity would require the Mach number $v\sqrt{(\rho/K)}$ of the model and the prototype to be equal, i.e. $M_{Ma} = 1$. If water is used on the model with $M_\rho = M_k = 1$, the above condition would lead to $M_V = 1$, and thus on a Froude model $M_l = 1$. Clearly, this is not practicable and special models with reduced wave celerity have to be used; this can be achieved by inserting an elastic element into the model (e.g. a plastic container filled with air in the water passage or a rubber penstock in a model of a vibrating gate and penstock). Models of vibrating structural elements require special instrumentation that does not unduly restrict the movement of the model (e.g. strain gauges).

It is clear that in modelling of often fairly complicated phenomena, scale effects are unavoidable, particularly when modelling damping, which is likely to be unduly large in the model, or in studying phenomena strongly dependent on the Reynolds number (e.g. flow past cylinders with flow separation and vibrations). Models of vibrating elements may contain irrelevant information, and recorded data have to be interpreted carefully; a good knowledge of hydrodynamics is essential.

For further information, see the references cited in Section 13.2.11 and, for example, Haszpra (1979) and Naudascher (1984).

13.3.2 Inland navigation models

The most frequently used models for structures on inland waterways are models of locks, their gates and filling systems. Other inland navigation models concern the movement of vessels on restricted-width waterways or through construction sites of barrages.

Models of *lock gates* follow the procedures described in the previous sections. A special case is models used as a design tool for *lock filling/emptying systems*, with the aim of reducing the forces acting on vessels both inside and outside the lock and of increasing the speed of filling or emptying the lock in order to increase the efficiency of the waterway. Sometimes models are also used to monitor the forces acting on *ships moored* in harbours and on vessels in *ship lifts*. Although originally these problems were investigated exclusively using physical models, nowadays mathematical or hybrid modelling is often used.

Physical models of lock-filling systems and of the forces acting on vessels require special instrumentation, particularly dynamometers and recording equipment. The emphasis in lock experiments is on the investigation of the longitudinal forces acting on the model vessel. These consist of frictional and form-resistance forces, forces due to the action of the lock-filling jet(s) and the effect of the (sloping) water levels in the lock, which change during the filling (emptying) process. This is a typical case of modelling complex phenomena, as described in Section 5.9, and if the necessary inputs are determined experimentally the result can also be obtained mathematically. The scale of the physical models is usually in the region of $M_l = 20$.

Models of the *movement of barges* on restricted waterways are governed by the Froude law and the condition that the scale of the model barge (sometimes with remote control) must be the same as the scale of the surrounding waterway. The required minimum Reynolds number for the body of the vessel (related to its length) is about 5×10^6 (or 3×10^6 with artificially induced turbulence), for the rudder 1.5×10^6 and for the propeller 7×10^4 to 3×10^5. These values lead to large, and therefore expensive, models, typically $M_l = 15$. On the other hand, useful, albeit qualitative, results can be obtained with much smaller remotely controlled models (scale $50 < M_l < 100$), assuming that the model barges are 'steered' by persons with prototype-vessel experience.

For further details of instrumentation and models, see Delft Hydraulics Laboratory (1985), Novak and Čábelka (1981) and Renner (1984), and the following case studies.

13.3.3 Models of urban hydraulics structures

Although sometimes of a small scale, the design of the manifold structures connected with water supply and waste-water treatment presents very similar problems to those encountered when modelling large structures, as dealt with in the previous sections, and thus the same physics and processes and scaling laws and scale effects described in Sections 13.2 and 13.3.1 apply to their modelling. In addition, in some cases we have to consider water-quality effects.

If the prototype contains clean water or only some suspended matter, as is the case in water supply or strongly diluted waste units, the previously discussed criteria apply in the physical modelling (in most cases using the Froude law). If the prototype contains a substantial concentration of particles influencing its density without affecting its viscosity, we may have to consider also the need for the identity of the *Froude densimetric number* (see equation (4.96)) (the coefficient of viscosity of sewage is usually the same as that of clean water that is 6°C cooler). When modelling density currents, the identity of the *Richardson number* (see equation (4.97)) is required. If surface-renewal and liquid-film coefficients are involved in the process, such as in oxygen transfer (see also Section 13.2.5 and equations (13.22)–(13.28)), the identity of the *Sherwood number* $Sh = K_L d_b/D_m$ (K_L is the liquid film coefficient, d_b the bubble diameter and D_m the coefficient of molecular diffusion) should be observed or scale effects will arise.

If the prototype fluid is non-Newtonian (e.g. flow of sludges and flow in digestors – Bingham fluids with pseudoplastic properties) any modelling has to take into account the laws of rheology and the preservation of the *Hedström number* $He = \tau \rho D^2/\mu$ (τ is the initial stress required to produce a deformation of the fluid).

Problems connected with sewer flow and design of sedimentation tanks are practically always dealt with using the equations given in Chapter 4 and by mathematical modelling, although some cases (e.g. distribution of inflow into tanks) may require physical modelling. The most frequent cases of physical modelling are connected with the design of sewer interceptors, cross-drainage and drop structures, and pumping sumps and intakes.

Section 4.6 contains further information on, and the equations used in, the mathematical modelling and design of urban hydraulics structures (e.g. equations (4.110)–(4.114) for sewers with sediment transport), and Chapters 9–11 deal with some relevant processes and their modelling.

For a comprehensive review of waste-water hydraulics, see Hager (1999), and for some examples of physical models of associated structures, see Novak and Čábelka (1981).

13.4 Mathematical modelling

The background to mathematical, computational and numerical modelling (see Chapter 1) of hydraulic structures is given in Chapters 2–4. Chapter 4 gives the necessary hydrodynamics and hydraulics background and the relevant equations (continuity, momentum, Navier–Stokes equations for turbulent flow, Saint Venant equation, shallow-water equation, etc.), and

Chapter 3 is a guide to the numerical solutions of the differential equations. Section 13.2 then provides the equations for the physics and processes that are the necessary input to modelling. Chapters 7, 9 and 10 contain sections that are closely related to the computational modelling of hydraulic structures.

Many of the *software packages* mentioned in the previous chapters (e.g. DAMBRK, FLUCOMP, Mike 11, Flowmaster) are relevant in computational fluid dynamics applied to hydraulic structures.

Computational models applied to the design of hydraulic structures fall broadly into two categories: those dealing with the *general layout and (far) flow field* of the structures, and models of a *part and detail* of the structure (e.g. a spillway, piers). The following are examples of some of many references dealing with the above.

For a *general introduction* to the subject, see, for example, Verwey (1983), which includes a case study of the Oosterschelde storm-surge barrage in the Netherlands. For a comprehensive treatment of the modelling of *free-surface flows* (including application to gated spillways), see, for example, Bűrgisser (1999), which contains a large list of references, and Schindler (2001), which includes an application to flow over a raised bed sill. Hervouet (2007) provides a detailed guide for the modelling of free-surface flows using the finite-element method (using, for example, Telemac, developed by Electricité de France).

An example of computational modelling applied specifically to a *component* of hydraulic structures is the simulation of flow over *spillways* (see e.g. Song and Zhou (1999), Assy (2001)). Akoz *et al.* (2009) provide an example of the application of computational fluid dynamics to the modelling of sluice-gate flow. For more general applications, using a FLOW-3D model with a volume-of-fluid technique capable of dealing with rapid spatial and temporal variations in water-surface elevation and providing a comparison with physical model data (e.g. for weirs and pumping stations), see, for example, Spaliviero and Seed (1998). Other examples are the application of $k - \varepsilon$ turbulence modelling to *submerged hydraulic jumps* using boundary-fitted coordinates (Gunal and Narayanan (1998)), and the application of computational fluid dynamics to *sewage storage chambers* (Stovin and Saul (2000)).

In the design of *gates*, computational methods are applied mainly to the determination of the strength necessary to accommodate hydrodynamic loads and to compute added mass if the mode of vibrations can be predicted (ICOLD (1996)).

Sumer (2007) provides a review of the mathematical modelling of *local scour* around piers, below pipelines, and at groynes and breakwaters; the review, which gives the main ideas, general features and procedures in the modelling process, also contains an extensive bibliography for both physical and mathematical modelling of scour.

Leschziner (1995) deals with modelling turbulence in *physically complex flow*, with particular emphasis on Reynolds-averaged modelling, and compares the predictive capabilities of eddy-viscosity and second-moment models.

Omid *et al.* (2005) give an example of the application of *artificial neural networks* to a component of hydraulic structures.

It is important to appreciate that all computational models require good-quality *input data*, and have to be *calibrated* and *validated* to give reasonably reliable results.

13.5 Case studies

13.5.1 Kárahnjúkar dam

The Kárahnjúkar dam in Iceland is part of a 690 MW hydroelectric power project completed in 2008. The 198 m-high, concrete-faced rockfill dam has a spillway designed for a capacity of 1,350 m³/s, which consists of a free, conventional, side-channel spillway followed by a transition bend and a 419 m-long chute with a spillway aerator 125 m from its end (as the flow at that point would have otherwise reached a critical cavitation number), terminating with a free-falling jet falling into a narrow 70–90 m-wide canyon with the river bed about 100 m below. The canyon sides are unstable, with cracks and soft rock, and the spillway jet has to avoid direct contact with them as well as causing minimum possible impact pressures on the river bed. These design requirements could only be met by a detailed hydraulic model study, which was carried out at the hydraulic laboratory of the Versuchsanstalt für Wasserbau, Hydrologie und Glaziologie (VAW) at ETH Zurich (Pfister *et al.* (2008)).

The chosen scale of $M_l = 45$, using Froude law, is large enough to study on one model the flow over the spillway, the issuing jet with its dispersion, and the resulting downstream scour and energy dissipation. Although viscosity and Reynolds-number effects could be neglected, inevitably scale effects due to the aeration, surface tension and turbulence structure of the jet had to be accounted for.

As the reservoir inflow from glacial rivers varies greatly between the seasons, the spillway design had to account for a wide variation in discharge. The model tests resulted in a chute of increasing slope and width in the downstream direction, terminating in a special structure consisting of an oblique upper chute end with wedges and baffles, and a minor lower platform draining small discharges into the narrow canyon. As the velocities at the end of the spillway varied according to the unregulated discharge between 8 m/s and 36 m/s, the jet length and the point of its downstream impact varied over a wide range. A tailwater dam provided a 15–20 m water cushion in the plunge pool.

Detailed model studies of the terminating chute structure, the free-falling disintegrated jet and the plunge pool resulted in a design – the suitability of which was proved by prototype observation – where the jet for the whole range of discharges falls clear of the narrow canyon walls (although in the prototype it is surrounded by mist) and the time-averaged dynamic pressure head at the river bed was reduced to a maximum of 11 m for the design discharge (the corresponding maximum pressure head on the opposite canyon flank within the plunge pool was around 7 m).

Figure 13.6 shows a general view of the model, and Figure 13.7 gives a detailed view of the jet and plunge pool, both with $Q = 600 \, \text{m}^3/\text{s}$.

The whole model study is an example of the processes and the application of the principles outlined in Sections 13.2.5–13.2.8 and 13.3.1.5–13.3.1g.

Figure 13.6 Kárahnjúkar dam model (courtesy of VAW, ETH Zűrich)

Figure 13.7 Kárahnjúkar dam model – jet and plunge pool (courtesy of VAW, ETH Zűrich)

13.5.2 Děčín barrage

This case study deals with a proposed barrage on the River Labe (Elbe) at Děčín in the Czech Republic (Gabriel *et al.* (2007)); its main purpose is to raise the water levels at low river discharges to improve the navigation conditions on this important inland waterway, as otherwise navigation has to be severely restricted or abandoned altogether for long periods, even in an average water year (see also the case study in Section 7.6.3). The design of the barrage is complicated by difficult morphological conditions, the restricted site and by its location in an environmentally and ecologically sensitive area. The barrage raises the water level by 5.30 m for low discharges, without affecting the flood levels; it has three sections with hydrostatic sector gates (43 × 5.2 m). The lock is 24 m wide and 200 m long, and the 9.7 MW power station has two direct-flow Kaplan turbines.

A combination of aerodynamic, computational and, above all, hydraulic modelling was chosen to finalize the design of the barrage and adjoining low-head power station, navigation lock and fish passes. To explore various solutions for the barrage layout and the inlet and outlet parts of the power station and fish passes, two- and three-dimensional mathematical $k - \varepsilon$ turbulence models were used. The results of these studies (as well as those obtained from preliminary aerodynamic models) formed an input to a hydraulic model. For the far-field modelling of the upstream area a two-dimensional model was used; the mean velocities in the verticals obtained from this model just upstream of the barrage were used as a boundary condition for the three-dimensional model of the flow at the structure. Figure 13.8 shows an example of the three-dimensional mathematical model for the design of the power-station forebay, with curved piers directing the flow to the turbine inlets, and the same in the hydraulic model wherein flow conditions were illustrated by cork floats (weighted to show the mean velocities in the vertical).

The aerodynamic models and a large hydraulic model were constructed in the laboratory of the T. G. Masaryk Water Research Institute (VÚV-TGM) in Prague. The two distorted aerodynamic models represented the upstream reservoir and the downstream area of the barrage, with scales $M_h = 150$ and $M_l = 300$. Both models showed in detail the river bed (modelled in polystyrene and putty) and the navigation approaches, with the barrage and power station represented only schematically (in wood and plastic). As the water levels were known and were represented by fixed covers, the model operation was straightforward (see Section 7.4.4). The main purpose

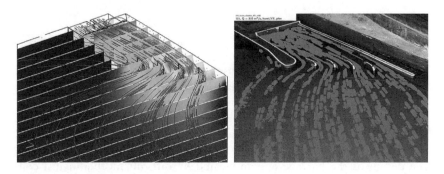

Figure 13.8 Děčín barrage – mathematical model of power station forebay (courtesy of VÚV- TGM, Prague)

Figure 13.9 Děčín barrage – aerodynamic model (courtesy of VÚV-TGM, Prague)

of the models was to determine approximately the optimum layout and orientation of the barrage, with the result forming the input to detailed investigations on the hydraulic model. Flow conditions were tested by flow visualization by introducing burning sawdust into the air stream. Detailed measurements of the velocity field constant were made using temperature anemometers and cylindrical probes (see Chapter 6). Figure 13.9 shows an overall view of the aerodynamic model.

The conventional hydraulic model (Figure 13.10) had a scale $M_l = 70$ and was used to optimize the final layout of the barrage, lock dividing walls and approaches, power station inlets and outlets, and the entry and exit of the fish passes.

An important part of the modelling process were nautical experiments using models of push trains and barges (scale 1:70) to determine the suitability of the design from the navigation point of view, and to test conditions during the barrage construction stages (see Section 13.3.2) (Figure 13.11).

13.5.3 Fusegate stability

The Jindabyne rockfill dam on the Snowy River in New South Wales, Australia (completed in 1967), has a service spillway with two radial gates

Figure 13.10 Děčín barrage – hydraulic model (courtesy of VÚV-TGM, Prague)

and a capacity of 3,000 m³/s. In 1997, the new guidelines for dam safety and spillway capacity showed that substantial new spillway capacity was required, and it was decided to add a new auxiliary spillway to the dam with eight labyrinth-type fusegates, with a total capacity of 5,600 m³/s (with all the gates tipped). The design parameters of the reinforced-concrete gates with stainless-steel well tops are 7.60 m high, 11.24 m long and weight 380 t. Considering the unique size and design of the gates, an extensive model study was undertaken at the Sogreah Laboratory in Grenoble (Jones *et al.* (2006)).

Two models were used:

- A large model, scale $M_l = 45$, to study the flow in the approach channel, the downstream conditions and the spillway performance for various operational possibilities of the fusegate spillway system.
- A small sectional model, scale $M_l = 20$, with two gates to study the fusegate stability. On this model the gates were fully functional and tipped at the design upstream water levels. The fusegates were built

from resin, with a cement and gravel ballast to give the correct weight and position of the centre of gravity. The model was used to measure the water-surface profiles above the fusegates, the pressures, the gate stability (also under impact from boats) and the actual displacement of the gate (see also Section 13.3.1.11).

Figure 13.12 shows the flow in the larger model with a discharge of 2,646 m³/s after the tipping of four gates. The models were operated according to the Froude law and the Reynolds numbers were sufficiently high to ensure a fully rough turbulent flow regime and a good representation of the frictional losses on the models (even on the smaller scale model, the Reynolds number was over 2,000 for the smallest discharge, and rose to 70,000 for the highest discharge before the tipping of the gates and to 200,000 without the gates in position).

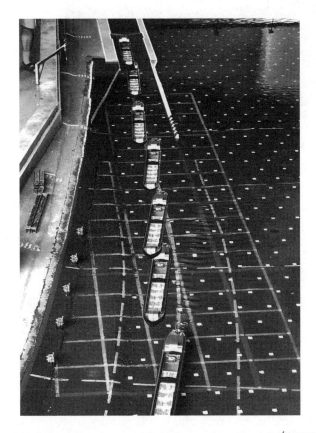

Figure 13.11 Děčín barrage – nautical experiment (courtesy of VÚV-TGM, Prague)

Figure 13.12 Fusegate model (courtesy of Sogreah, Grenoble)

The main results of the model tests were the determination of the spillway discharge coefficient and water-level–discharge relationships for various operational situations, and the confirmation of the design assumptions for the correct tipping of the gates and their stability.

13.5.4 Combined sewer screening chamber

A hydraulic model study of a combined sewer screening chamber for a commercial client was carried out in 2003 in the hydraulic laboratory of the School of Civil Engineering and Geosciences at the University of Newcastle upon Tyne (Valentine (2003)). The main objectives of the study were:

(1) to provide efficient energy-dissipation measures for the flow into the chamber and to ensure a maximum velocity of 0.9 m/s and uniform flow conditions at the rotary screens;

(2) to reduce the length of the chamber in order to minimize the impact on the existing rail embankment;

(3) to achieve self-cleaning conditions, good side-weir performance and reasonable hydrodynamics when all or any screen is isolated.

A model scale of $M_l = 18.4$ was chosen which, when operated according to the Froude law, ensured sufficiently high Reynolds numbers and permitted the use of standard pipe diameters (see also Section 13.3.3). Figure 13.13 shows the layout of the model (including the test results). The rotary screens were represented by flat ones, but with the correct head loss. Downstream control levels were established from tidal levels and sewer line losses.

The objectives of the study were achieved by suitably dimensioning and positioning the dividing piers in front of the screens. An energy dissipater in the form of an equilateral-triangular-sectioned pillar was placed at the chamber inlet in such a position that sufficient energy dissipation occurred to promote uniform flow conditions to the screens with a maximum approach velocity of 0.7 m/s and a possible reduction in the weir length by 8.25 m (from the original length of 21.8 m).

Figure 13.13 Sewer screening chamber model (courtesy of University of Newcastle upon Tyne)

13.5.5 Moislains lock

With a 30 m head the Moislains lock is the highest of the seven locks in the new 106 km-long Seine–Nord canal, which is the central link of the Seine–Scheldt waterway.

The 195 m-long lock, of width 12.5 m, is provided by five thrift basins, each with four culverts leading to a common valve chamber beneath the thrift basins, and a pressure chamber beneath the lock connected to it by an array of nozzles. Each of the 20 culverts from the thrift basins is governed by a fixed-wheel, vertical lift gate; there are also vertical lift gates in the two filling and emptying culverts in the upstream and downstream lock head, governing the filling and emptying of the remaining two-sevenths of the lock-chamber volume. The operation of the 24 gates is computer controlled to achieve the smallest possible safe time for the lock filling/emptying (of the order of 14 min).

To study the design and operation of the lock, a model of scale $M_l = 25$ was constructed at the Sogreah laboratory in Grenoble (O. Cazaillet, personal communication, 2008). The model (Figure 13.14) reproduced the lock chamber and the thrift basins, a portion of the upper and lower canal reach, the 24 culverts and gates, the valve and pressure chambers, and the 1,056 nozzles. The model was used to measure the water levels in the

Figure 13.14 Moislains lock model (courtesy of Sogreah, Grenoble)

lock (six gauges) and one thrift basin (three gauges), the water levels in the upstream and downstream reaches (echo sounding), the forces on hawsers (three boats were studied – one 110 m long, one 85 m long, and a leisure boat – with the boats fixed by two mooring lines on floating bollards), the waves in the downstream reach and thrift basins, and the vortices at the water intakes (see also Section 13.3.2). In-depth studies of the uncertainty in each measured parameter were carried out.

13.6 Concluding remarks

From the above, some generally valid conclusions can be drawn:

(1) Experimental hydraulics still has many roles to perform in elucidating the details of the physics and processes in the flow at hydraulic structures, particularly with regard to the air–water (aeration) and structure–soil–water (erosion) interfaces and to refining the limits of similarity.
(2) Physical modelling, particularly of more complicated situations and non-conventional structures, has, and will continue to have, an important role in the design process.
(3) Hybrid modelling is an economical way to deal with the problems mentioned above, and can considerably speed up the modelling process. This applies to a combination of hydraulic and computational modelling, as well as to a combination of these methods with aerodynamic modelling.
(4) Mathematical modelling plays an important and increasing role in the design of structures, but requires good input data and careful validation and calibration.
(5) The design, operation and, particularly, the interpretation of the results obtained from modelling are both an art and science, and require good engineering judgement and experience, especially where the safety of the structure is of concern.
(6) Field (prototype) measurements are of paramount importance, and are really the only way to confirm the validity of the conclusions drawn from the modelling process.

References

Ackers, P. and Priestley, S. J. (1985), Self-aerated flow down a chute spillway. In *Proceedings of the 2nd International Conference on the Hydraulics of Floods and Flood Control*, Cambridge, Paper A1, British Hydromechanics Research Association, Cranfield.

Akoz, M. S., Kirkoz, M. S. and Oner, A. A. (2009), Experimental and numerical modeling of sluice gate flow, *J. Hydraulic Res.*, 47(2), 167–176.

Assy, T. M. (2001), Solution for spillway flow by finite difference method, *J. Hydraulic Res.*, IAHR, 39(3), 241–247.

Avery, S. and Novak, P. (1978), Oxygen transfer at hydraulic structures, *J. Hydraulics Div. ASCE*, 104(HY11), 1521–1540.

Boes, R. M. and Hager, W. H. (2003a), Hydraulic design of stepped spillways, *J. Hydraulic Eng. ASCE*, 129(9), 671–679.

Boes, R. M. and Hager, W. H. (2003b), Two-phase flow characteristics of stepped spillways, *J. Hydraulic Eng. ASCE*, 129(9), 661–670.

Breusers, H. N. C. and Raudkivi, A. J. (1991), *Scouring: Hydraulic Structures Design Manual*, Vol. 2, Balkema, Rotterdam.

Bűrgisser, M. (1999), *Numerische Simulation der freien Wasseroberflächa bei Ingenierbauten*, No. 162, p. 144, Mitteilungen, VAW Zűrich.

Cassidy, J. J. and Elder, R. A. (1984), Spillways of high dams. In *Developments in Hydraulic Engineering*, Vol. 2 (ed. P. Novak), Elsevier Applied Science, London.

Chadwick, A., Morfett, J. and Borthwick, M. (2004), *Hydraulics in Civil and Environmental Engineering*, 4th Edn, Spon, London.

Chanson, H. (2001), *The Hydraulics of Stepped Chutes and Spillways*, Balkema, Lisse.

Chanson, H. and Gonzales, C. A. (2005), Physical modelling and scale effects of air–water flows on stepped spillways, *Sci. J. Zhejiang Univ.*, 6A(3), 243–250.

Chanson, H. and Gualtieri, C. (2008), Similitude and scale effects of air entrainment in hydraulic jumps, *J. Hydraulic Res. IAHR*, 46(1), 35–44.

Chow, V. T. (1983), *Open Cannel Hydraulics*, McGraw-Hill, New York.

Cop, H. D., Johnson, I. P. and McIntosh, J. (1988), Prediction methods of local scour at intermediate bridge piers. In *Proceedings of the 68th Annual Transportation Research Board Meeting*, Washington, DC.

Damle, P. M. (1952), Role of hydraulic models in determining spillway profiles for low dams. In *Symposium on Role of Models in the Evolution of Hydraulic Structures and Movement of Sediments*, CBIP, New Delhi.

Delft Hydraulics Laboratory (1985), Navigability in restricted waterways. *Hydrodelft*, No. 71, Waterloopkundig Laboratorium, Delft.

Ervine, D. A. (1998), Air entrainment in hydraulic structures – a review, *Proc. ICE, Water, Maritime and Energy*, 130(9), 141–153.

Ettema, R., Kirkil, G. and Muste, M. (2006), Similitude of large scale turbulence in experiments on local scour at cylinders, *J. Hydraulic Eng. ASCE*, 132(1), 33–40.

Ettema, R., Melville, B. W. and Barkdoll, B. (1998), Scale effect in pier-scour experiments, *J. Hydraulic Eng. ASCE*, 124(6), 639–642.

Gabriel, P., Libý, J. and Fošumpaur, P. (2007), *Hydraulic Research of the Děčín Barrage*, p. 67, T. G. Masaryk Water Research Institute, Prague.

Ghetti, A. and D'Alpaos, L. (1977), Effect des forces de capillarité et de viscosité dans les écoulements permanents examinées en modèle physique. In *Proceedings of the 17th IAHR Congress*, Baden-Baden, pp. 389–396.

Gulliver, J. S., Wilhelms, S. C. and Parkhill, K. L. (1998), Predictive capabilities in oxygen transfer at hydraulic structures, *J. Hydraulic Eng. ASCE*, 124(7), 664–771.

Gunal, M. and Narayanan, R. (1998), k–ε turbulence modelling of submerged hydraulic jump using boundary-fitted coordinates, *Proc. ICE, Water, Maritime and Energy*, 130, 104–114.

Hager, W. H. (1994), Discharge measurement structures. In *Discharge Characteristics*, Vol. 8 (ed. D. S. Miller), IAHR Hydraulic Structures Design Manual, A. A. Balkema, Rotterdam.

Hager, W. H. (1995), Überfallstrahlen, *Schweizer Ingenier und Architekt*, 113(19), 442–447.

Hager, W. H. (1999), *Wastewater Hydraulics*, Springer-Verlag, Berlin.

Hager, W. H. (2007), Scour in hydraulic engineering, *Water Manage.*, 160(3), 159–168.

Haszpra, O. (1979), *Modelling Hydroelastic Vibrations*, Pitman, London.

Henderson, F. M. (1966), *Open Channel Flow*, Macmillan, New York.

Hervouet, J.-M. (2007), *Hydrodynamics of Free Surface Flows: Modelling with the Finite Element Method*, Wiley, Chichester.

ICOLD (1996), *Vibrations of Hydraulic Equipment for Dams*, Bulletin 102, International Commission on Large Dams, Paris.

Jain, S. C. (2001), *Open Channel Flow*, Wiley, New York.

Jones, B. A., Cazaillet, O. and Hakin, W. D. (2006), Safety measures adopted to increase spillway potential at Jindabyne Dam in the Snowy Mountains. In *Proceedings of the 22nd Congress, ICOLD*, Barcelona, Q. 86–R. 17, 1–24.

Keller, A. P. (1994), New scaling laws for hydrodynamic cavitation inception. In *Proceedings of the 2nd International Symposium on Cavitation*, Tokyo.

Kenn, M. J. (1984), Flow visualization aids cavitation studies. *Symposium on Scale Effects in Modelling Hydraulic Structures*, Vol. 1.15, pp. 1–6 (ed. H. Kobus), Technische Akademie, Essllingen.

Knauss, J. (ed.) (1987), *Swirling Flow Problems at Intakes*, Vol. 1, IAHR Hydraulic Structures Design Manual, A. A. Balkema, Rotterdam.

Kobus, H. (1991), Introduction to air–water flows. In *Air Entrainment in Free-surface Flows*, Vol. 4 (ed. I. R. Wood), Hydraulic Structures Design Manual, A. A. Balkema, Rotterdam.

Kobus, H. and Koschitzky, P. (1991), Local surface aeration at hydraulic structures. In *Air Entrainment in Free-surface Flows*, Vol. 4 (ed. I. R. Wood), IHAR Hydraulic Structures Design Manual, A. A. Balkema, Rotterdam.

Kolkman, P. (1984), Gate vibrations. In *Developments in Hydraulic Engineering*, Vol. 2 (ed. P. Novak), Elsevier Applied Science, London.

Kolkman, P. (1994), Discharge relationships and component headlosses for hydraulic structures. In *Discharge Characteristics*, Vol. 8 (ed. D. S. Miller), IAHR Hydraulic Structures Design Manual, A. A. Balkema, Rotterdam.

Leschziner, M. A. (1995), Modelling turbulence in physically complex flows. In *Proceedings of the XXVIth IAHR Congress "Hydra 2000"*, IAHR, Madrid Vol. 2, pp. 1–33.

Lewin, J. (2001), *Hydraulic Gates and Valves in Free Surface Flow and Submerged Outlets*, 2nd Edn, Thomas Telford, London.

Locher, F. A. and Hsu, S. T. (1984), Energy dissipation at high dams. In *Developments in Hydraulic Engineering*, Vol. 2 (ed. P. Novak), Elsevier Applied Science, London.

Lopardo, R. (1988), Stilling basin pressure fluctuations. In *Model–Prototype Correlations of Hydraulic Structures, Proceedings of the International Symposium ASCE*, Colorado Springs, pp. 56–73 (ed. P. H. Burgi), American Society of Civil Engineers, New York.

Mason, P. J. (1989), Effect of air entrainment on plunge pool scour, *J. Hydraulic Eng. ASCE*, 115(3), 385–399.

Melville, B. W. and Chiew, Y.-M. (1999), Time scale for local scour at bridge piers, *J. Hydraulic Eng. ASCE*, 125(1), 59–65.

Melville, B. W. and Coleman, S. (2000), *Bridge Scour*, Water Resources, Colorado.

Naudascher, E. (1984), Scale effects in gate model tests. In *Symposium on Scale Effects in Modelling Hydraulic Structures*, Vol. 1.1, pp. 1–14 (ed. H. Kobus), Technische Akademie, Essllingen.

Naudascher, E. (1987), *Hydraulik der Gerinne und der Gerinnebauwerke*, Springer-Verlag, Vienna.

Naudascher, E. (1991), *Hydrodynamic Forces*, Vol. 3, IAHR Hydraulic Structures Design Manual, A. A. Balkema, Rotterdam.

Naudascher, E. and Rockwell, D. (1994), *Flow Induced Vibrations*, Vol. 7, IAHR Hydraulic Structures Design Manual, A. A. Balkema, Rotterdam.

Novak, P. and Čábelka, J. (1981), *Models in Hydraulic Engineering – Physical Principles and Design Applications*, Pitman, London.

Novak, P., Mofffat, A. I. B., Nalluri, C. and Narayanan, R. (2007), *Hydraulic Structures*, 4th Edn, Taylor and Francis, London.

Oliveto, G. and Hager, W. H. (2002), Temporal evolution of clear-water pier and abutment scour, *J. Hydraulic Eng. ASCE*, 128(9), 811–820.

Omid, M. H., Omid, M. and Varaki, M. E. (2005), Modelling hydraulic jumps with artificial neural networks, *Water Manage.*, 152(2), 65–70.

Pfister, M., Berchtold, T. and Lais, A. (2008), Kárahnjúkar dam spillway: optimization by hydraulic model tests. In *Proceedings of the 3rd IAHR International Symposium in Hydraulic Structures*, Vol. VI, pp. 2106–2111, Hohai University, Nanjing.

Pinto de S., N. L. (1991), Prototype aerator measurement. In *Air Entrainment in Free-surface Flows*, Vol. 4 (ed. I. R. Wood), IAHR Hydraulic Structures Design Manual, A. A. Balkema, Rotterdam.

Prins, J. E. (1971), *Phenomena Related to Turbulent Flow in Water Control Structures*, Publication No. 76, p. 84, Delft Hydraulics Laboratory, Delft.

Ranga Raju, K. G. and Garde, R. J. (1987), Modelling of vortices and swirling flows. *Swirling Flow Problems at Intakes*, Vol. 1 (ed. J. Knauss), IAHR Hydraulic Structures Design Manual, A. A. Balkema, Rotterdam.

Rajaratnam, N. and Kwam, A. P. (1996), Air entrainment at drops, *J. Hydraulic Res. IAHR*, 34(5), 579–588.

Renner, D. (1984), *Schiffahrtstechnische Modellversuche für Binnenwasserstrassen – ein neues Messsystem und neue Auswertungsmöglichkeiten*, Bericht No. 48, p. 140, Versuchsanstalt für Wasserbau der T. U. München, Munich.

Schindler, M. (2001), *Mehrdimensionale Simulation von Freispiegelströmungen mit der Finite-Element-Methode*, Mitteilungen No. 68, p. 180, Hydraulik und Gewässerkunde, T. U. München, Munich.

Song, C. C. S. and Zhou, F. (1999), Simulation of free surface flow over spillway, *J. Hydraulic Eng. ASCE*, 125(9), 959–967.

Spaliviero, F. and Seed, D. (1998), *Modelling of Hydraulic Structures*, Report SR545, HR Wallingford.

Stovin, V. R. and Saul, A. J. (2000), Computational fluid dynamics and the design of sewage chambers, *J. CIWEM*, 14, 103–110.

Sumer, B. M. (2007), Mathematical modelling of scour: a review, *J. Hydraulic Res. IAHR*, 45(6), 723–735.

Sumer, B. M. and Fredsoe, J. (2002), *Mechanics of Scour in the Marine Environment*, World Scientific, River Edge, NJ.

Valentine, E. M. (2003), *A Hydraulic Model Study of the Sunderland CSO at Hendon*, Report for Montgomery Watson Harza, Newcastle upon Tyne.

Verwey, A. (1983), The role of computational hydraulics in the hydraulic design of structures. In *Developments in Hydraulic Engineering*, Vol. 1 (ed. P. Novak), Applied Science, London.

Vischer, D. L. and Hager, W. H. (1998), *Dam Hydraulics*, Wiley, Chichester.

Whitehouse, R. (1998), *Scour at Marine Structures – A Manual for Practical Applications*, Thomas Telford, London.

Author index

Subject index

Note: Page numbers in *italic* refer to figures.

As some terms appear throughout the text (Froude number, model etc) only the page numbers where they appear for the first time or where the definitions are given.